CW01390366

# BIOLOGY: The Human Perspective

# BIOLOGY:

**Donald J. Farish**

University of Missouri, Columbia

# The Human Perspective

Harper & Row,     Publishers
New York, Hagerstown, San Francisco, London

Acknowledgment is made to the following for use of illustrative material:

3.19  From "How Animals Run" by Milton Hildebrand. Copyright © 1960 by Scientific American, Inc. All rights reserved.

3.20  a and b  From "The Antiquity of Human Walking" by John Napier. Copyright © 1967 by Scientific American, Inc. All rights reserved.

8.10  From *Biology: A Human Approach* by Irwin V. Sherman and Vilia G. Sherman. Copyright © 1975 by Oxford University Press, Inc. By permission.

Picture in box p. 194, 8.16, and 8.17 From *Human Sexual Response* by Masters and Johnson. Copyright © 1966 by William H. Masters and Virginia E. Johnson. By permission of Little, Brown and Company.

10.20  From *The Cerebral Cortex of Man* by Penfield and Rasmussen. Copyright © 1950 by The Macmillan Company. By permission.

11.1  From *Ethology: The Biology of Behavior,* Second Edition, by Irenaus Eibl-Eibesfeldt. Copyright © 1970, 1975 by Holt, Rinehart and Winston. By permission of Holt, Rinehart and Winston.

13.4  From "The Genetics of Human Populations" by L. L. Cavalli-Sforza. Copyright © 1974 by Scientific American, Inc. All rights reserved.

14.21  From "The Food Sources of the Ocean" by S. J. Holt. Copyright © 1969 by Scientific American, Inc. All rights reserved.

Sponsoring Editor:  Jeffrey K. Smith
Project Editor:  Claudia Kohner
Designer:  Gayle Jaeger
Production Supervisor:  Kewal K. Sharma
Photo Researcher:  Myra Schachne
Compositor:  Progressive Typographers, Inc.
Printer:  The Murray Printing Company
Binder:  Halliday Lithograph Corporation
Art Studio:  Eric G. Hieber Associates, Inc.
Cover Art:  Robert L. Smith

**BIOLOGY: The Human Perspective**
Copyright © 1978 by Donald J. Farish

Library of Congress Cataloging in Publication Data

Farish, Donald J    1942–
  Biology.
  Includes index.
  1. Biology.   2. Human biology.    I. Title.
QH308.2.F28      574      77-21068
ISBN 0-06-041995-4

To my wife, who made it worthwhile,

and to my children, who made it necessary.

# Contents

"Good heavens, not another introductory biology book!" After a number of years of teaching freshman biology, that tends to be my reaction when I receive complimentary copies of textbooks in the mail, and it may have been your reaction, too, upon receiving this book. With so many good texts already available, it is incumbent on any author to explain his reasons for having written one.

I found that the more often I taught freshman biology, the fussier I became in choosing a text. Most of my students are nonmajors, and many of the biology textbooks are simply too detailed to be of use to them. Other books looked too much like a watered-down majors' text, an approach which I felt did not best serve the nonmajors' interests. Another group of texts tried to be "relevant" (I have come to hate that word), and in the process looked like a series of news clippings—there was no hint of biology as an intellectual discipline in them. I rejected still other books simply because I didn't agree with the author's choice of topics for emphasis. My own students were pushing me increasingly into using the human organism as the primary focus for biological principles, and I found the course evolving farther and farther away from available texts. I felt that other instructors might well be having the same problems, so I finally decided to write a text to fill what I interpreted as a void.

These are the criteria I used in writing this text, listed more or less in order of importance:

1  Readability   I have tried to put the students' interests first by attempting to write in such a way as to provoke interest and facilitate understanding. If a student doesn't read the text, either because he or she doesn't understand it or is bored by it, then the text is serving no purpose other than as a security blanket. Moreover, if the instructor is forced to spend all his or her time explaining the text, rather than being able to use the text to supplement lectures, then the tail is wagging the dog.

2  Topic selection   I have focused on the human organism, with a sizable section on population biology as well. Even nonmajors are interested in their own bodies, so I decided to capitalize on this

# Preface to the instructor

interest and introduce them to biological concepts using the human organism as a vehicle. Diseases are not stressed, but neither is a discussion of them avoided.

3  **Terminology**  I have attempted to pare back terminology wherever possible, although it is an impossible task. Biology probably has a larger jargon than any other undergraduate discipline, but surely we can better serve the nonmajors than merely asking them to become transitorially familiar with a set of terms most of which they will never see again.

4  **Chemistry**  I have virtually eliminated chemistry from this text, primarily because, in my experience, nonmajors may be capable of memorizing chemistry, but not of understanding it. As a consequence, most of what we have come to call molecular biology is omitted. As exciting as that material is, I don't think the average nonmajor student is capable of understanding it without a solid chemistry background.

5  **Biological concepts**  By the same token, I don't think nonmajors are any less bright than majors, and they shouldn't be insulted by a watered-down course and text. Therefore, I have tried to present biological concepts in a reasonably complete and sometimes even sophisticated way. Regardless of the motivation of students taking an introductory course, and regardless of the discipline, they should expect to encounter some of the governing principles of the discipline, and I have made no effort to cut back in this area. Homeostasis and evolution, for example, are discussed at length.

6  **Length**  I have structured this book for use in a one-semester (or perhaps two-quarter) course. Obviously, it is not encyclopedic in scope, but I felt this to be a worthwhile tradeoff. There seems to be an inverse relationship between the length of text and the amount read by the student.

Finally, I would welcome any comments you might have about the text, be they favorable or critical.

Donald J. Farish

# BIOLOGY: The Human Perspective

A young woman suffers severe head injuries in an auto
accident. She is rushed into an intensive care unit,
where elaborate machinery keeps her heart beating,
her lungs and kidneys functioning—but her brain shows
no activity. Is she dead? Is the doctor who disconnects
the machinery guilty of murder? The answers to these
questions depend on our definition of "life"—the subject
matter of the science of biology.

chapter **I**

# chapter 1
# Life, death, and history–the development of biological principles

**WHAT IS LIFE?**  Biology is traditionally defined as "the study of life." However, when applied to what biologists are actually doing today, this definition is both too narrow and too broad. It is too narrow to the extent that it suggests that biologists are only interested in the processes underlying "life," that is, those processes that collectively distinguish living from nonliving material. The population biologist may be more mathematician than biologist; the behaviorist may be as much psychologist or anthropologist as biologist; and the molecular biologist may be equally versed in chemistry as in biology. A narrow interpretation of "the study of life" fails to include such diverse activities.

This definition of biology is also too broad. Psychologists, sociologists, and anthropologists, among many other scientists, work on various components of living systems, yet most would bristle at being called a biologist.

The problem, of course, centers on the phenomenon of "life" and its meaning and application within the preceding definition. The simple fact is that although we use this term very frequently, no one is in a position to say just exactly what life is. Intuitively, we accept that living systems possess something more than nonliving systems—but stop a moment and try to explain why a deer that has just been shot through the heart by a hunter is now no longer living. Your immediate response would probably be to say, "Obviously, if the deer has been shot through the heart, it is dead because the heart has ceased to pump blood to the other organs of the body." Yet such a response really begs the question. For years, the standard medical test for determining if life had ceased was to determine whether or not the heart was still beating. Thus, once you are told that the deer's heart has stopped beating, you automatically conclude that the deer must be dead. But let us pursue this reasoning process a bit further. What events have transpired within the cells of the deer, which collectively comprise this formerly living organism, after its heart stopped beating? How do these cells differ from those of a deer that is still alive? If changes have occurred, are they reversible?

These questions are difficult to answer. We know that once the individual cells are no longer provided with a suitable environment, they cease to function and will ultimately die—that is, certain chemical processes ordinarily characteristic of cells fail to occur and the cell ceases to function in any respect. Organismal death, then, results from (or results in, depending on your approach) cellular death. Yet, in a sense, all we have done is to shift our focus from the deer to the individual cells of the deer. We are still begging the question. Certainly the cells undergo change— the whole deer began to change when it was shot, by ceasing to move, for example—so we would expect such changes in organismal behavior to be mirrored by changes in cellular chemistry. Yet we are really no closer to defining, let alone understanding, the difference between life and death. To be sure, we can describe events that, when present, indicate life and when absent, suggest death, but we cannot pin down an understanding of the difference in a single phenomenon, let alone a chemical equation.

Thus, we are left in the embarrassing position of being unable to define the phenomenon that we are supposed to be studying. It is this lack of definability that results in the rather sloppy and generalized use of the term, and which in turn accounts for the problems outlined above in deciding whether our opening definition of biology was too broad or too narrow. To come full circle, to define "biology" as "the study of life" is to do two things. First, it is simply to make a literal translation of the Greek root of the term (bios, meaning "life"); second, it is to use, in a definition, a word which itself cannot be defined.

Although it may strike you as ironic that as supposedly exacting scientists biologists are laboring at understanding the significance and ramifications of a phenomenon that they cannot yet define or explain, in reality it is precisely this difficulty that provides much of the stimulus to study life in the first place. Moreover, even though we know a great deal about the chemical events that occur within living cells, and we have even gone so far as to duplicate certain of these events in the test tube, it does not seem likely that a single chemical process will be identified as the distinction between life and nonlife. The present definition of life is a classic example of the whole being more than the sum of its parts—that is, the collective web of interacting chemical equations within cells is sufficiently complex and complete as to constitute life.

**Characteristics of living systems**  Although life cannot be defined, it can be described. That is, in examining living systems, one finds certain properties which, taken collectively, characterize life. The precise number and description of these characteristics varies among different authorities, but the list includes:

1 The capacity to use and transform materials and to carry out chemical reactions
   (**metabolism**)
2 The ability to be able to respond, in some fashion, to changes in the environment
   (**irritability**)
3 The capability for **maintenance and growth** of the system

According to standard Judeo-Christian ethic, the death of an individual is that instant when the spirit leaves the body. Because spirits cannot be scientifically investigated, by convention, an individual has been regarded as dead when his heart stops beating. However, the selection of the activity of the heart as the sole indicator of death was initially based first on the relative ease by which the activity of the heart could be monitored (i.e., pulse), and second on the exaggerated importance that was placed on the heart in ancient times (e.g., the totally fallacious role of the heart in emotions, most notably love). Gradually, we have come to recognize the heart as nothing more than a pump—a magnificently efficient pump, to be sure, but a pump nonetheless.

This realization has been made especially emphatic by the recent development of medical techniques capable of maintaining the activity of the heart (and the activity of the lungs and kidneys) intact, even in situations in which there is evidence suggesting that other organs of the body (such as the brain) may have permanently ceased to function. The individual in question may never regain consciousness or be able to survive without the machinery responsible for heart, lung, and kidney function.

Faced with two very pragmatic considerations—the need for organ donors and the enormous expense of maintaining individuals on sophisticated machinery—coupled with the recognition of the fact that the activity of the heart (basically a mechanical event) was not the best indicator of life, doctors have recently proposed substituting the absence of brain activity, as measured by an electroencephalogram (*EEG*), for a period of 24 hr, as the primary indicator of death.

**When does death occur?**

It is important to recognize, however, that a flat EEG is only a convention—it does not get to the basic question, "What is life?" As such, it is open to challenge. Is 24 hr long enough? Suppose there are occasional bursts of brain activity, but at a level insufficient to allow the regaining of consciousness, or even the regaining of independence from the heart, lung, and kidney machines? Can such an individual, by prior consent, elect to have the machinery disconnected without violating state laws against suicide? The question of when to "pull the plug" is still very much with us, and will remain so until such time as a more complete understanding of "life" is attained.

4 The ability to synthesize new individuals (**reproduction**)
5 The possession of a quality of **dynamic interaction** with the environment (as opposed to a purely static, and unchanging, system)
6 A formal **organization**, typically with the **cell** as the basic structural unit

To some extent, these terms overlap (e.g., 1, 3, and 5 may be considered merely as different ways of saying the same thing). However these characteristics are listed, it is true that nonliving systems possess, at best, only one or two of them, and as such, they are a useful checklist in describing life.

**PHILOSOPHICAL APPROACHES TO BIOLOGY** Much of the genesis of biology as a science came from early attempts at understanding exactly what distinguished life from nonlife. To some extent, the original inquiries were more philosophical than scientific. Two schools of thought quickly developed. First, there were the **vitalists,** who believed in a *vis vitalis,* or **vital force,** which was incapable of being explained chemically. However, by virtue of insisting on the *vis vitalis* as an unknown and unmeasurable entity, the vitalists effectively removed themselves from the purview of science, as science is concerned *only* with those events that can be measured. Therefore, although it is conceivable that there is such a thing as a *vis vitalis* (perhaps the equivalent of the biblical concept of a spirit), it is no more capable of scientific analysis than are such concepts as truth or beauty.

Associated with the vitalists were the **teleologists,** who accounted for differences in the structure or behavior of species by invoking a sense of purpose. The teleologists were discredited for unnecessarily invoking purposefulness, by implying either individual control over adult form and function on the part of the growing organism or, alternatively, a kind of biological destiny. To most modern biologists, explaining such events as the growth of individual plants by invoking purpose is, at the least, unnecessary and, at the most, wrong. Therefore, rather than saying, "Trees grow roots because they need to obtain water and salts from the soil," most present-day biologists would say, "Trees obtain water and salts from the soil through the roots." At first glance, the distinction may not seem profound, but the underlying philosophical difference between the points of view represented by these two statements is not only enormous, but in a very real sense, has dictated the whole approach to biological investigation.

Remaining on the field after the retreat of the vitalists are the **mechanists,** who see living systems as analogies to machines—that is, that living systems conform to the same physical laws of nature as machines, even though such systems are far more complex than any machine. As such, there is simply no need to hypothesize such an unprovable concept as a *vis vitalis.*

Although the overwhelming majority of modern biologists subscribe basically to the mechanistic point of view, mechanism has been carried even further by the **reductionists,** who believe that life processes will ultimately be explainable completely on the basis of long series of chemical equations. As George Wald[1] has stated, "Living organisms are the greatly magnified expressions of the molecules that compose them." This point of view is useful to the molecular biologist, and, in fact, justifies his work. There is no question that the discovery of how proteins are formed, or how enzymes function, among many other cellular secrets that have been uncovered in the past 25 years, fully supports this approach.

However, the reductionist approach is somewhat less useful at higher levels of biological organization. It will be a long time before the functioning of the human brain, let alone the complexities of ecological interactions, will be capable of being reduced to a series of chemical equations. Therefore, many biologists investigating such levels of organization are not reductionists, but **compositionists,** stressing the

---

[1] 1967 Nobel prize winner in physiology and medicine.

adaptive usefulness of a structure or a process, both to the organism and to its species as a whole, rather than its underlying chemical basis. The two schools of thought might seem complementary, because each is directed at different levels of biological organization, but a certain amount of friction still occurs among those who feel that their particular philosophy is superior to the other person's. Such disagreements are hardly unique to biologists, of course, and reflect not so much on the nature of the subject as on the nature of its investigators.

**THE EVOLUTION OF SCIENTIFIC THOUGHT** To the extent that culture is the transmission of the knowledge, abilities, values, and beliefs of one generation to another, with each generation contributing to the whole, science would seem to exemplify culture at least as well as any other area of endeavor. Was Rodin a better sculptor than Praxiteles? Was Picasso a better artist than Rembrandt? Was Stravinsky a better composer than Bach? Was Faulkner a better novelist than Dickens? Each is debatable and none is capable of satisfactory resolution. Surely each of the latter artists was influenced in some way by his predecessor, but in many respects each of the earlier artists reached a culmination of a certain type of expression which was never surpassed—and, perhaps by being unsurpassable, necessitated new modes of expression within the particular art form. The contribution of the arts to culture consists of a series of dead ends that result in new beginnings.

By comparison, the sciences have been more linear and, in that sense, more progressive. False starts were made, and will continue to be made, but these represent only temporary distractions. Progress in the sciences is demonstrated by a coming together of originally separate disciplines, as the unity of physical laws was discovered and as these laws were recognized as applying to living systems as well. It may well be that the scientific intellects of Newton, Galileo, and Pasteur will never be surpassed, but any of you knows more science now than any of those scientists ever knew. For this reason, an examination of the development of biology is more than a mere historical exercise. At the risk of overstating its importance, the history of biology represents a microcosm of human intellectual development.

**Why the Greeks "blew it"** A seemingly mandatory requirement of all textbooks in science is a brief paean to the great historical figures in the discipline (Figs. 1.1–1.3). Typically, this consists of a number of ancient engravings and a long list of names and dates. It also typically remains unread by the student. This is indeed unfortunate, because of the need for anyone embarking on (and perhaps simultaneously ending) a study of science to develop a sense of perspective. That is, we can best know where we are today by studying where we came from.

It all began with Aristotle, who was the successor to Plato, the tutor of Alexander the Great, and the first of the great Greek philosophers to do extensive work in the natural sciences. Unfortunately, his approach to the study of science was quite different from the methodology common today. That is, he applied logic and deduction as opposed to performing experiments—an unfortunate error, if one's starting point is total ignorance. This approach nonetheless won him many disciples who continued to follow this "Aristotelian" approach, and in the process perpetu-

**Fig. 1.1** Aristotle was a Greek philosopher living in the fourth century B.C. A student of Plato and teacher of Alexander the Great, Aristotle was the first of the great philosophers to study natural science extensively. In many respects, his methods of scientific analysis have been replaced by more rigorous techniques, but in one respect he was entirely modern—his research was financed by grants from the king! (Culver)

**Fig. 1.2** Hippocrates lived in the century before Aristotle and is considered the father of modern medicine because of his concern for facts and his abhorrence of superstitions. Hippocrates began the trend that took medicine out of the hands of priests by showing that disease could have natural causes. His principles formed the basis of the modern medical theories developed in the nineteenth century. (Culver)

ated ignorance and myth for centuries. This may seem a strong statement, to be sure, but the fact remains that the basic Aristotelian approach has been discredited and largely discarded as interjecting too many of the observer's biases. In short, this approach fails to be objective, and that is the ultimate heresy to modern science.

A prime example of the magnitude of error introduced by the Aristotelian approach is provided by an individual named Galen (Figure 1.4), who lived in the second century A.D., some 500 years after Aristotle's death. Galen has traditionally been regarded as second only to Hippocrates in the history of medicine, and was the first to discover that arteries carried blood and not air, as had previously been thought. As an anatomist, he did some important work, but as his fame grew, so did his imagination. Galen promulgated a theory of blood circulation which, to our

**Fig. 1.3**
The Swedish naturalist, Carl Linne, was responsible for initiating the scientific method of naming living things with a Latin binomial (e.g., *Homo sapiens*). So enamored was he of Latin and the classical world that he changed his own name to Latin—Carolus Linnaeus, the name most often used when moderns refer to Linne. He lived and worked in the century before Darwin, and believed in the immutability of species. (Culver)

minds, would seem to defy belief, but which was believed nonetheless for almost 1500 years simply because Galen had said it was so. Lincoln's statement of not being able to fool all of the people all of the time was not applicable to the Middle Ages.

According to Galen, blood was formed in the liver from digested food carried to the liver from the intestine (Fig. 1.5). The blood was then distributed throughout the body via two major veins (which we now know collect blood *from* the body, rather than distributing blood *to* the body). According to Galen, a portion of the blood passing to the upper body passed through the heart to the lungs, which were assumed to require large amounts of blood. Galen believed that at certain times of the day, blood passes from the liver back to the intestines through the same vessel that carries digested food to the liver. In addition, he assumed that air from the lungs was mixed with blood seeping between the right and left side of the heart to provide the "vital spirit" needed to allow the body to operate properly. He also believed that excess blood was "sweated off" by the individual organs. Galen did not believe in circulation of the blood as such, but rather thought that it was continuously being formed by the liver from digested food.

Of course, Galen's view is almost totally inaccurate. In fact, it is so inaccurate that you may have had difficulty reading it with a straight face. Even though you may never have studied blood circulation formally, a basic understanding of circulation is now so much a part of our culture that we give little thought to its discovery. More importantly, the "scientific method" (see next section) is so fully ingrained in our thinking that it must seem preposterous to you to learn that Galen's views were fully accepted for almost 1500 years!

**Fig. 1.4**
Galen was the Roman anatomist and physiologist responsible for giving the world a unique interpretation of blood flow. Galen's version bore little resemblance to reality but was accepted for 1500 years. (Culver)

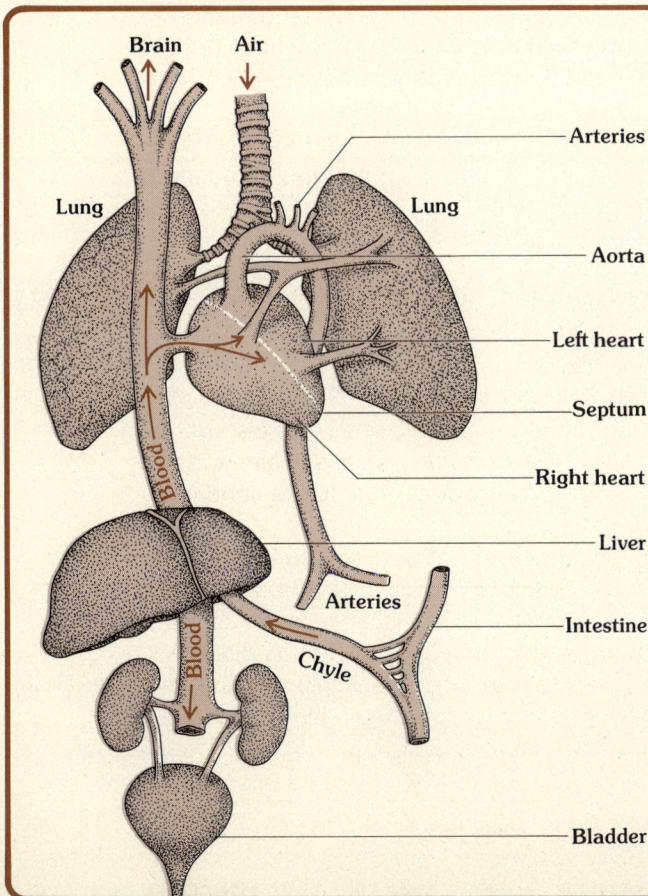

**Fig. 1.5**
The Galenic version of blood circulation.

Brain

Air

Arteries

Lung

Lung

Aorta

Left heart

Septum

Right heart

Blood

Liver

Arteries

Intestine

Chyle

Blood

Bladder

In 1628, one of the great discoveries of science was made when William Harvey, an English physician trained in Italy, published his theories of circulation. His experiments and observations were so simple and logical, at least to our minds, that one can only wonder why it took 1500 years for someone to come along and test Galen's views. Nevertheless, Harvey thought them to be so controversial that he did not expect anyone over the age of 40 to agree with him (this age of credulity has since been lowered to 30, in our own society).

To begin with, Harvey calculated that the heart holds about 60 g (2 oz) of blood and beats about 72 times/min. Consequently, the heart must pump 245 kg (540 lb) of blood per hour. Even assuming that Galen had been a prodigious eater, it is difficult to believe that his liver (or anyone else's) would be capable of synthesizing so much blood in such a short period of time. This observation alone was enough to destroy the major premise in Galen's scheme of circulation, and it led Harvey to the only logical conclusion—blood must be reused, that is, it must circulate in some manner through the body. He went still further and correctly described the route of blood through the heart, and between lungs and heart and body and heart. Harvey also correctly interpreted the significance of the valves in the heart and in the veins, concluding that blood flowed from the heart through arteries and back to the heart through veins. In short, using information either possessed by Galen or readily available to him, Harvey correctly mapped the circulation of the blood (Fig. 1.6).

However, Harvey did not stop here, as "logic" had also been the basis of Galen's view of circulation; instead, he devised an experiment to prove it. Harvey

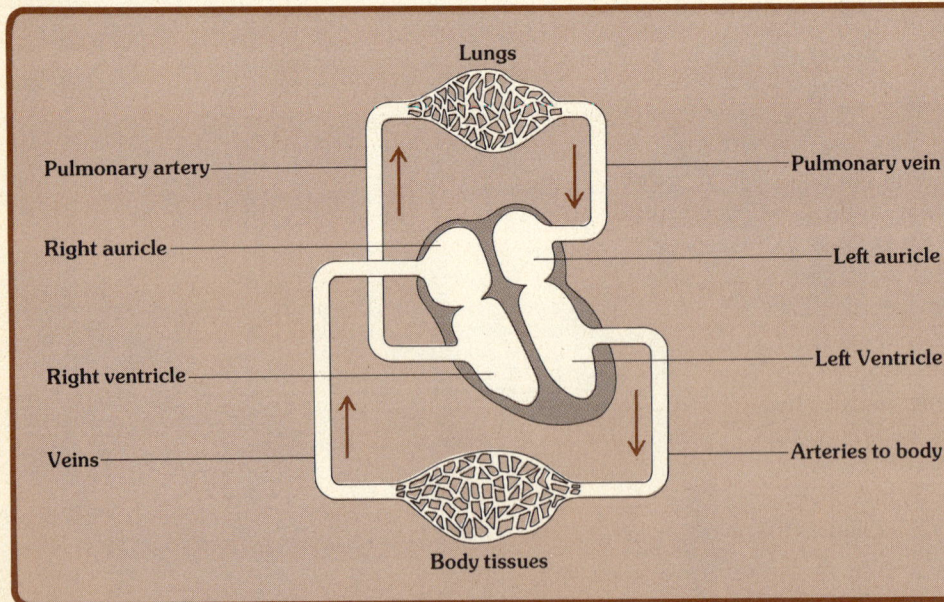

Fig. 1.6
The Harveian (and correct) version of blood circulation.

The development of experimental science

It is odd to think of experiments as being novel, because we now accept them as the hallmark of scientific investigation, but in the time of William Harvey, they were just starting to come into vogue. Aristotle transformed science from a purely philosophical discipline into one that employed direct observation. But the Greeks and Romans never progressed to the point of experimentation, which explains how it was that their astronomy (a nonexperimental science) was so much more advanced than their life sciences.

Francis Bacon, an English philosopher who lived about 75 years before Harvey's famous experiment, was among the first to state that "Nature should be put to the test." This idea was very slow to become accepted, because of the historical importance placed on the abstract idea, and it was not really until the nineteenth century that science became truly experimental.

tied a cord around the arm of a subject, tight enough to cut off venous flow (low blood pressure) but not arterial flow (high blood pressure). He then pointed out how the veins became swollen below the cord (drainage to the heart impeded), but collapsed above the cord (drainage to the heart unimpeded). This proved the existence of circulation, which must be unidirectional within any given vessel. Despite the overwhelming strength of these observations, Harvey's conclusions were not widely accepted until some 30 years later, when the connections between arteries and veins, the vessels we now call **capillaries,** were discovered.

**Louis Pasteur and spontaneous generation**  An equally graphic example of the recent maturity of scientific thought is provided by the long debate over **spontaneous generation.** Modern science had its origins in the myths of primitive societies, and certain of these myths have proved extremely difficult to discard. The adherants of the theory of spontaneous generation believed that given the proper set of environmental circumstances, living organisms could arise from nonliving materials. How else could life have come about? Frogs came from mud; flies developed from spoiled meat; and mice emerged from piles of dirty clothing left in a dark place. We find all of these examples ludicrous today, although each has an inherent (albeit faulty) logic. Nonetheless, it was only a little more than 100 years ago, still within the memory of living people, that the French scientist Louis Pasteur disproved spontaneous generation to general satisfaction.

Attempts at disproving this doctrine had been made by numerous scientists for 200 years, and to be sure many scientists disbelieved it. However, their attempts at proving its impossibility (proving a negative is always difficult) were never completely satisfactory. One early scientist noted that meat protected by gauze did not "develop" flies as did uncovered meat (Fig. 1.7). Perhaps not, said the detractors, but the meat still spoiled and could be seen to have microscopic organisms crawling around inside.

Discounting the spontaneous generation of microorganisms was a larger chal-

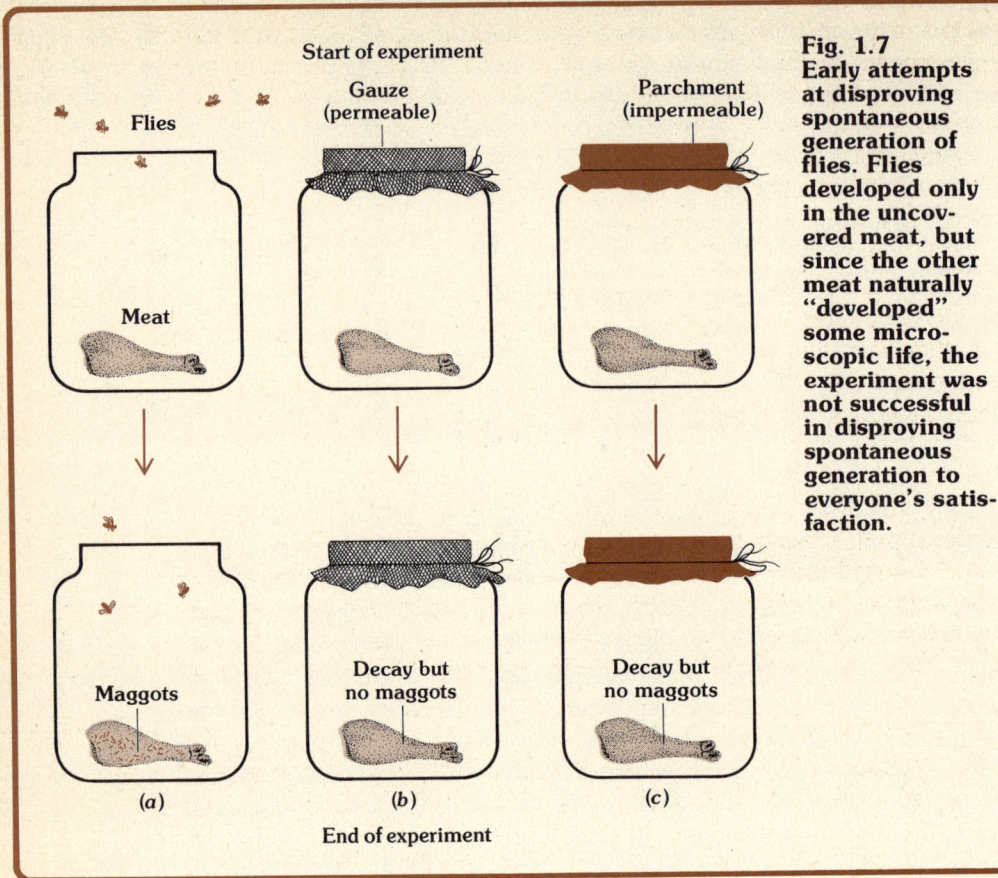

Fig. 1.7 Early attempts at disproving spontaneous generation of flies. Flies developed only in the uncovered meat, but since the other meat naturally "developed" some microscopic life, the experiment was not successful in disproving spontaneous generation to everyone's satisfaction.

lenge, but some time later, an industrious scientist heated a broth in a flask and then sealed the neck of the flask. Unlike the broth in an open flask, the heated broth developed no microorganisms. "Unfair," cried the supporters of spontaneous generation. "Cutting off the supply of air meant that the 'vital force' needed to generate life was prevented from reaching the broth." Undiscouraged, the resourceful fellow gave up science and founded the canning industry.

Louis Pasteur performed a similar experiment, but rather than sealing the flask, he simply drew it out in an S shape (Fig. 1.8). It was therefore open to the air, but of course the spores and bacteria in the air settled in the bottom of the S and the broth remained pure. Again, it seems incredible that such a simple experiment should take two centuries, and one of science's greatest minds, to devise. The only acceptable explanation for our incredulity is the fact that experiments and the scientific method have now so completely become a part of our culture.

**The scientific method**   Two points should now be apparent. First, the development of the scientific method not only has been long and tortuous, but it has been completed only within the last century. Second, its integration into our culture has nonetheless been so complete that we tend to accept it almost intuitively today and wonder what all the fuss is about.

The scientific method typically begins with a series of **observations** which lead to the development of a **hypothesis.** A hypothesis is nothing more than a testable guess based on the observations. Hypotheses are tested by experimentation. It should be noted that experiments do not always take the form of vats of colored liquids percolating through miles of glassware. They are frequently nothing more than a series of observations performed under **controlled** conditions (*vide infra*).

Ideally, the hypothesis is formulated in such a way that a single key experiment can be performed that will affirm or refute the hypothesis. Very often, however, the experiment does not achieve this goal but rather necessitates additional experiments or suggests other possible hypotheses. A series of hypotheses and experiments may then ensue, finally resulting in the formulation of a **theory,** which is nothing more than a hypothesis that has stood up to experimental testing.

Most of the time, the scientist relies on **inductive reasoning** in the formulation of the hypothesis, that is, reasoning from the specific to the general. For example, an investigator may find that lung cancer is found regularly in a population of cigarette smokers, and will therefore construct a hypothesis that cigarette smoke is one cause of lung cancer. Inductive reasoning is at the heart of the scientific method. How-

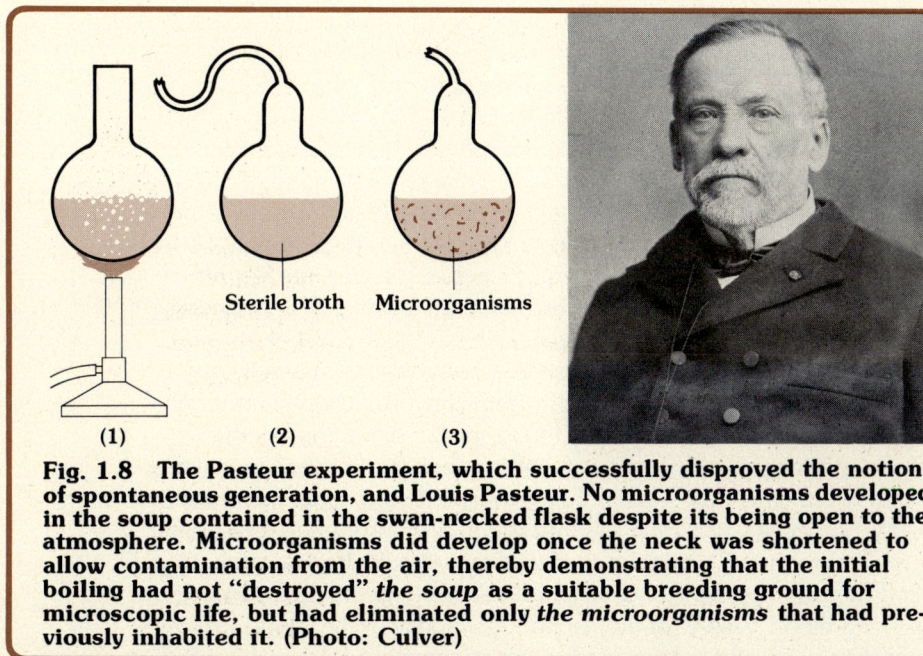

Sterile broth    Microorganisms

(1)        (2)        (3)

**Fig. 1.8  The Pasteur experiment, which successfully disproved the notion of spontaneous generation, and Louis Pasteur. No microorganisms developed in the soup contained in the swan-necked flask despite its being open to the atmosphere. Microorganisms did develop once the neck was shortened to allow contamination from the air, thereby demonstrating that the initial boiling had not "destroyed"** *the soup* **as a suitable breeding ground for microscopic life, but had eliminated only** *the microorganisms* **that had previously inhabited it. (Photo: Culver)**

ever, the fact remains that many scientific breakthroughs have been based on **deductive reasoning** (reasoning from the general to the specific). Deduction, of course, depends on the accuracy of the general principles, and more than one scientist has been sorely embarrassed because of errors in the general principles. Moreover, the Aristotelian method is grounded on deductive reasoning (as is logic generally), and this approach has been largely discredited, in favor of experimentation, as we have just seen. The fact remains, however, that many of the most important discoveries in science have been based on "intuition" or some "gut feeling" on the part of the investigator, which actually translates into deductive reasoning. Moreover, in testing the applicability of a theory to specific situations, deductive reasoning must be employed. Thus, even though the history of the development of the scientific method has been one of a gradual shift from deductive to inductive reasoning, both must remain if the method is to be useful.

Two other elements enter into the application of the scientific method. These are the concepts of **controlled conditions** and **replication,** which are exemplified in the following imaginary situation. Sidney Scientist feeds granulated sugar to a rat and the rat promptly dies. Is Sidney justified in concluding that the sugar killed the rat? Is he justified in assuming that sugar is lethal to rats in general? Suppose you found out Sidney's test rat was 7 years old, and that the average life-span of a rat is less than 3 years—would that affect your answers?

Hopefully, you would reject Sidney's conclusions, and for two reasons. First, if the rat is 7 years old, it may be that it was about to die of old age. In order to validate the findings, we must have another 7-year-old rat that is treated in precisely the same way as our test rat, save only for the feeding of granulated sugar. This second rat is our **control** animal. By subjecting both rats to identical conditions, except for the experimental procedure, we increase the validity of concluding that any change in the experimental rat was attributable to the conditions of the experiment.

Second, there is something unnerving about accepting a general principle (e.g., that granulated sugar is lethal to rats) based on a single experiment. To test the validity of this conclusion, the experiment should be repeated several times—that is, it should be **replicated.** If the same result occurs consistently, the chances of the conclusion being correct are substantially increased. (In actual practice, the test procedure might be done simultaneously on several rats, i.e., concurrently. It is not necessary that the experiment be repeated consecutively.)

Tied in with the idea of the need for replication and control is the necessity of distinguishing between **cause and effect** and **correlation.** This distinction is frequently not made in many scientific investigations even today. In the experiment just described, even with consistent replication, we are not immediately justified in stating that granulated sugar is lethal to rats, that is, that granulated sugar was the **cause** of death (the **effect**). Rather, we can only say that the ingestion of granulated sugar is positively (or even "absolutely") **correlated** with death in rats. For example, if, on autopsy, we find that the mechanical effects of swallowing granulated sugar caused rupturing of the blood vessels of the esophagus, and that the blood loss was fatal, we might then conclude a cause and effect relationship.

This is not a semantic exercise. Consider the evidence that cigarette smoking may cause lung cancer. Even though this conclusion is probably justified, cause and effect in humans has not yet been demonstrated (and may never be, given that we cannot very well grab a bunch of people off the street, stick cigarettes in their mouths, and wait for lung cancer to develop). The relationship between smoking and cancer is still just a correlation. Suppose, for the sake of argument, that everyone possesses one or another of two genes: one gene increases the urge to suck, which frequently leads to smoking in adults; the other gene represses this urge. Let us also suppose that the first gene increases the rate of cell division in the lungs (a necessary part of the growth of a cancer tumor). In such a case, cigarette smoking would have no causative effect on lung cancer whatsoever, but would merely be correlated with the cancer. Although admittedly this example is far-fetched, it is still true that many investigations never really prove cause and effect, despite the fact the results are frequently described as doing so. Increasingly, citizens are going to be called upon to participate in decision-making that involves so-called hazardous (e.g., cyclamates) or beneficial (e.g., megavitamins) materials, and a determination of whether a given relationship is cause and effect or merely correlation is fundamental in making the correct decision.

## SUMMARY

The difficulty of distinguishing life from nonlife, on an abstract or definitional basis, has intrigued people since the earliest times. We are still unable to define life except in a roundabout manner, but in the process of attempting to do so, a completely new and inductive method of analysis was developed, which replaced the earlier deductive philosophical approach first used by the Greeks and which continued until the Renaissance. This new approach was the scientific method, which relies heavily on the testing of hypotheses through carefully controlled experiments. The scientific and technological advances of the past 400 years, which have greatly outstripped advances from all the rest of human history, are owed as much to the acceptance and utilization of the scientific method as to any other single factor. So completely has this approach become ingrained in our culture that it is only with difficulty that we can step aside and contemplate a time when it did not exist.

*"What little interest I have in biology is in how my body works, or in understanding environmental problems. Why should I have to know anything about atoms and molecules?"* If you feel this way, be of good cheer—there is very little chemistry in the following pages. But without some inkling of molecules and how they work, try and explain why diabetes can be fatal if insulin treatment is unavailable? Or why you have the same amount of salt in your blood as your crazy roommate, who inundates every morsel of food with salt? Or why lead poses such a severe threat to ourselves and our environment that the Environmental Protection Agency felt constrained to mandate that all new cars must burn lead-free gasoline? Our world and our bodies are not insulated from atoms and molecules, but in fact are composed of them. Before we know if a bridge will stand or fall, we must first know something about the girders.*

chapter **II**

# chapter II
# From atoms to the biosphere—the scope of biological organization

**BIOLOGICAL ORGANIZATION**  Because living systems consist of complex series of chemical reactions, and at the same time interact with other living systems and with their environment, biological organization runs the full spectrum from the **atom** to the **biosphere.** In order to understand subsequent topics in this book, it is first necessary to have some grasp of biological organization. Therefore, we shall briefly examine each organizational step.

**ATOMS AND ELEMENTS**  All the matter in the universe is composed of various combinations of only 92[1] distinct substances, which we call **elements** (Fig. 2.1). By way of analogy, all the words in the English language are composed of combinations of only 26 letters (Fig. 2.2). A chunk of any given element is composed of many single units of that element, called **atoms,** the smallest unit still possessing all the characteristics of that element. Splitting the atom would change its properties, and it would no longer be the element in question. To continue the analogy, suppose we had a huge ball of Es (Fig. 2.2b). We could divide that ball into successively smaller portions until we were left with a single E. However, to split that one letter any further would change its character, and it would no longer be an E.

**Elements of life**  Table 2.1 lists those elements known to be required for our continued existence. (There are several others found in minute traces in our bodies that may ultimately be proved essential, and two or three others that are required by other species of plants and animals, but the elements required by our-

---

[1] The figure 92 refers to naturally occurring elements. An additional 13 or 14 elements have been created in the laboratory, but, because of their very large atoms, they are unstable and quickly break apart, forming the atoms of other elements in the process.

**Table 2.1**
**The relative composition of the human body, the oceans, and the earth's crust in terms of number of atoms, in percent**

| Atomic no. | Symbol of element | Name of element | Human body | Oceans | Earth's crust |
|---|---|---|---|---|---|
| 1 | H[a] | Hydrogen | 63 | 66 | 0.22 |
| 6 | C[a] | Carbon | 9.5 | 0.0014 | 0.19 |
| 7 | N[a] | Nitrogen | 1.4 | tr[b] | tr |
| 8 | O[a] | Oxygen | 25.5 | 33 | 47 |
| 11 | Na | Sodium | 0.03 | 0.28 | 2.5 |
| 12 | Mg | Magnesium | 0.01 | 0.033 | 2.2 |
| 15 | P[a] | Phosphorus | 0.22 | tr | tr |
| 16 | S[a] | Sulfur | 0.05 | 0.017 | tr |
| 17 | Cl | Chlorine | 0.03 | 0.33 | tr |
| 19 | K | Potassium | 0.06 | 0.006 | 2.5 |
| 20 | Ca | Calcium | 0.31 | 0.006 | 3.5 |
| 25 | Mn | Manganese | tr | tr | tr |
| 26 | Fe | Iron | tr | tr | 4.5 |
| 29 | Cu | Copper | tr | tr | tr |
| 30 | Zn | Zinc | tr | tr | tr |
| 42 | Mo | Molybdenum | tr | tr | tr |
| 53 | I | Iodine | tr | tr | tr |

[a] Elements comprising the four major classes of macromolecules (the other elements are frequently found as ions or salts).
[b] tr = trace amounts only.

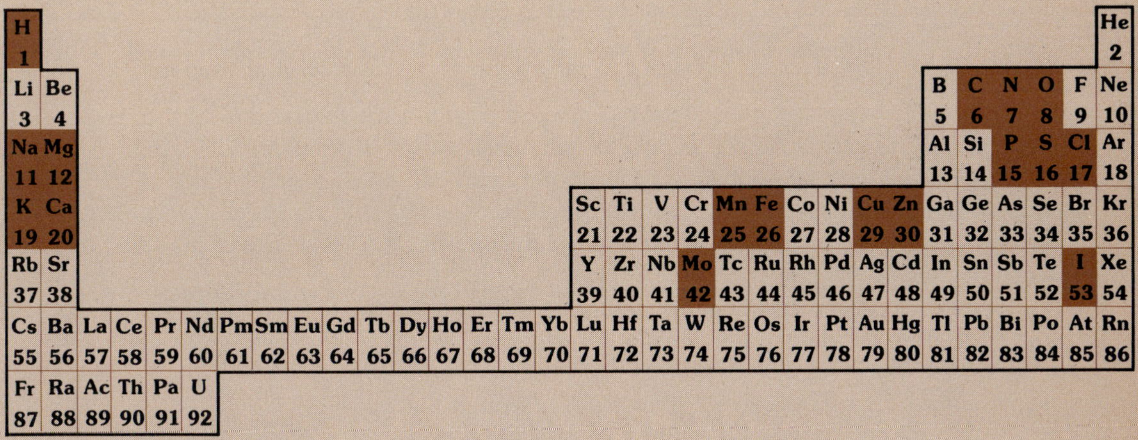

Fig. 2.1 **The periodic table of elements. Spacing is based on the subatomic organization of the atoms of each element. Elements in white boxes are known to be essential to human life.**

selves are essentially the same as those required by all living organisms.) Several points can be made from this table:

1 Only four elements are found in excess of 1 percent of total number of atoms in our bodies
2 These four elements have very small atoms (as indicated by their low atomic number)
3 Of the other required elements, those required in substantial amounts have an atomic number below 20, and only one (iodine) has an atomic number larger than 50
4 The elements that comprise the human body are more similar in their percentage composition to the elements found in seawater that to the elements that compose the land

From these points we can infer that atomic size is a more important criterion for the inclusion of elements in living systems than is availability in nature (e.g., both carbon and nitrogen are relatively rare), although availability is also important, given the overall similarity of the composition of the human body with that of the oceans. From point four, we can also infer that it is likely that life began in the oceans as opposed to the land.

**MOLECULES AND COMPOUNDS**  Elements may combine with each other to form **compounds** in the same way that letters may combine to form words (Fig. 2.2d). Water is a compound, composed of the elements hydrogen and oxygen. Table salt is another compound, made up of the elements sodium and chlorine. The smallest unit of a compound is a **molecule,** that is, two atoms of hydrogen and one atom of oxygen form one molecule of water. To attempt to split this molecule any further would result in reestablishing the individual atoms that formed it, and the properties of the substance we know as water would be lost.

**Bonds and ions**  Just as a random array of letters does not necessarily form a word, molecules are not necessarily formed merely by randomly combining atoms. Rather, the particular atoms must be capable of forming a **chemical bond** between them.

There are several different types of chemical bonds, each having different strengths. As you may know, if you dissolve table salt in water, the individual components of the salt molecule—the elements sodium and chlorine—separate from each other. Therefore, simply dissolving some compounds may be sufficient to cause **dissociation**—that is, the separation of the atoms that comprise the molecules of the substance.

Conversely, some chemical bonds are very difficult to break. The bonds that hold oxygen and hydrogen together in the water molecule are certainly not easily broken, and, in fact, can be broken in the laboratory only through such drastic means as passing a large current of electricity through the water.

Atoms are comprised of a number of smaller particles, some of which possess electrical charges. However, by definition, atoms are always electrically neutral (i.e., the number of positive charges equals the number of negative charges). In the breaking of the weaker bonds, such as those holding sodium and chlorine together in a table salt molecule, the bond is not broken equally. Thus, rather than reestablishing the atoms sodium and chlorine, each of which is electrically neutral, there is a shift

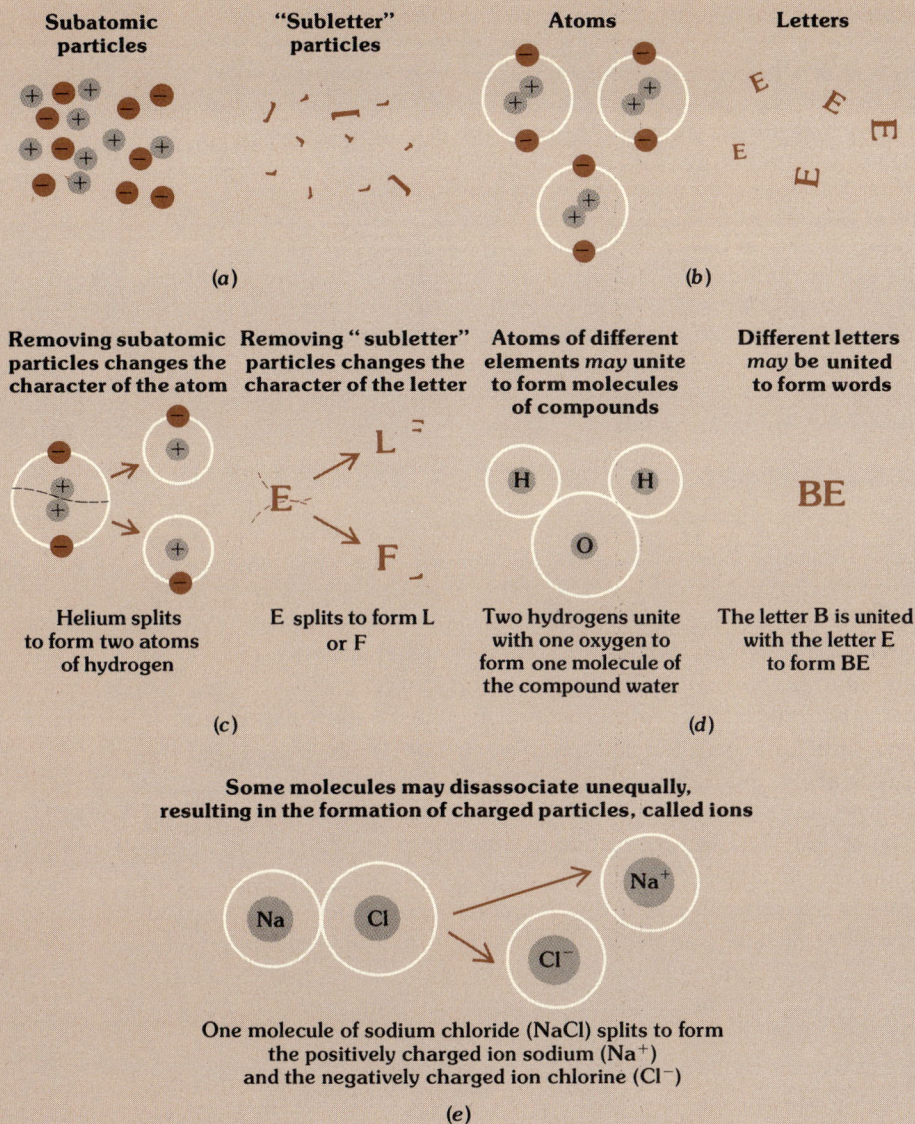

**Subatomic particles**

**"Subletter" particles**

**Atoms**

**Letters**

(a)

(b)

Removing subatomic particles changes the character of the atom

Removing "subletter" particles changes the character of the letter

Atoms of different elements *may* unite to form molecules of compounds

Different letters *may* be united to form words

L

E

F

H  H

O

BE

Helium splits to form two atoms of hydrogen

E splits to form L or F

Two hydrogens unite with one oxygen to form one molecule of the compound water

The letter B is united with the letter E to form BE

(c)

(d)

Some molecules may disassociate unequally, resulting in the formation of charged particles, called ions

Na  Cl

Na⁺

Cl⁻

One molecule of sodium chloride (NaCl) splits to form the positively charged ion sodium (Na⁺) and the negatively charged ion chlorine (Cl⁻)

(e)

**Fig. 2.2** **The organization of atoms and molecules as compared with the organization of letters and words.**

of charges such that the sodium will have a positive and the chlorine a negative charge (Fig. 2.2*e*).

Carbon is not the only element to have the capacity to form bonds with four other atoms. The element silicon, which occurs just below carbon in the Periodic Table (Fig. 2.1), also has this capacity, and it is one of the most abundant of all elements. Could life elsewhere substitute silicon for carbon? A strong argument against such a possibility is the fact that silicon has atoms that are considerably larger than are carbon atoms. Not only are large atoms generally slower to interact with other atoms, but, because of their heaviness, compounds formed from large atoms tend to be solids, rather than gases or liquids. For example, one atom of carbon combines with two atoms of oxygen to form the gas carbon dioxide, the primary waste gas of the body (see Chapter 6). Similarly, one atom of silicon can also combine with two atoms of oxygen to form one molecule of the solid silicon dioxide, also known as **quartz,** the primary component of sand. It is difficult to imagine a living system that could utilize quartz as a respiratory waste product—but such an organism would be formidable indeed, were it to sneeze!

The reason for this shift has to do with the nature of subatomic organization, a topic beyond the scope of this book, but suffice it to say, a dissociation into charged particles is very common. These particles possess most of the same properties as the original atoms, yet are different by virtue of being charged. How can they be distinguished? By convention, such charged particles are called **ions,** and they are designated by using a superscript $+$ or $-$ sign. Thus, a sodium atom is represented by Na; a sodium ion is $Na^+$. Similarly, a chlorine atom is designated Cl; a chlorine ion is $Cl^-$. Ions are of extraordinary importance in living organisms, especially in the function of nerves, muscles, and the kidneys, as you will see shortly.

**Macromolecules**   The element carbon is capable of forming a very large number of compounds because of its capacity both to form long chains with other carbon atoms and to form chemical bonds with as many as four other atoms simultaneously. In the earliest days of chemistry, it was thought that compounds based on carbon chains could only be formed by living organisms (*see box*). Hence, molecules containing carbon were designated **organic,** as opposed to **inorganic** compounds such as water or table salt. In fact, the designation is highly arbitrary, because many molecules produced by living systems do not contain carbon. Moreover, for many years, organic compounds have been synthesized outside a living organism. It has, of course, been pointed out that the designation is still accurate, insofar as synthesis outside a living organism is still being performed by a living organism (that is, the scientist himself, most of whom have been found to be living). Regardless, the distinction is still in active use.

Of particular importance to organisms are the very large organic molecules, known as **macromolecules,** of which there are four major categories. These are **carbohydrates, proteins, lipids,** and **nucleic acids.**

**Carbohydrates**   As the name would imply, carbohydrates consist of carbon atoms and water molecules stuck together—thus, with few exceptions, all carbohydrate molecules have equal numbers of carbon and oxygen atoms, and twice that number of hydrogen atoms. We know them more familiarly as the **sugars** and **starches.**

The simple sugars contain 3 to 10 carbon atoms. These sugars are called **monosaccharides** (from the Greek *mono,* meaning single, and *saccharide,* meaning sugar. It is the peculiar property of scientists to avoid the simple term if a more complex one is available—especially if the more complex term is derived from Greek or Latin.) An example of a monosaccharide is **glucose,** a six-carbon sugar which is extremely important to the body as the basic fuel of cells. (For this reason, patients being fed intraveneously are given solutions containing glucose.)

However, these simple sugars possess the capacity to be linked into longer chains. Thus, a sugar consisting of two units is called a **disaccharide.** An example is ordinary table sugar **(sucrose),** which consists of a glucose molecule attached to another six-carbon sugar **(fructose).**

Still longer chains are possible, and these are lumped under the general term **polysaccharides.** (If you analyze the way biologists count—*mono* (one), *di* (two), *poly* (many)—you will realize that behind their love for Greek terms lies a fundamental problem in counting.) There are many examples of polysaccharides, but some of the biologically more important ones are **starch, cellulose** (the material in plant cell walls from which paper is manufactured), **glycogen** (which is simply animal starch), and **chitin** (the hard shell, or exoskeleton, of insects). The primary uses of carbohydrates are as energy reserves and as structural molecules.

**Proteins**    Proteins are extremely long molecules consisting of chains of smaller unit called **amino acids.** There are 20 different amino acids, which have the same relationship to proteins as monosaccharides do to polysaccharides. Proteins differ from carbohydrates in that all amino acids contain at least one atom of nitrogen, an element never found in carbohydrates.

Proteins are extremely important as structural components of the body because they are so versatile in their composition and shape. In fact, proteins may have as many as three discernible structures (Fig. 2.3). The **primary structure** of proteins consists simply of the sequence of the amino acids which comprise it. However, these amino acids are not laid out arrow-straight, but rather are coiled like a spring (biologists prefer the word **helix**). This spiral arrangement is referred to as the **secondary structure** of proteins. Many proteins have only these two structures and as such are basically elongated strands. They are therefore referred to as **linear,** or **fibrous** proteins. An example of a linear protein is **keratin,** the protein found in hair and nails.

Another group of proteins are the **globular** proteins, so named because of their lumpy shape. This shape is attributable to their **tertiary structure,** which comes about because of the establishment of very weak chemical bonds as the protein doubles back on itself.

This capacity of proteins as a group to assume virtually any shape has led to the use of proteins as **enzymes.** Enzymes are substances that alter (usually, but not

necessarily, by increasing) the rate of a biochemical reaction, but that are not themselves used up in the process.

The significance of enzymes to life is difficult to overstate. The fact is, however,
that many of the essential chemical reactions in the body, which occur virtually instantaneously in the presence of enzymes, occur so slowly in the absence of enzymes
as to be useless. This is because in order for two molecules to engage in a chemical reaction, it is necessary that the reactive portions of the two molecules come
into close proximity. Because the macromolecules commonly found in the body
are so very large, the reactive portions are relatively very small, and the chance of
their coming into close proximity unaided is very slight. Imagine two bowling balls,
each with a white spot the size of a dime. Think how long it would take, if you
were to roll these around your living room randomly, before the white spot on one

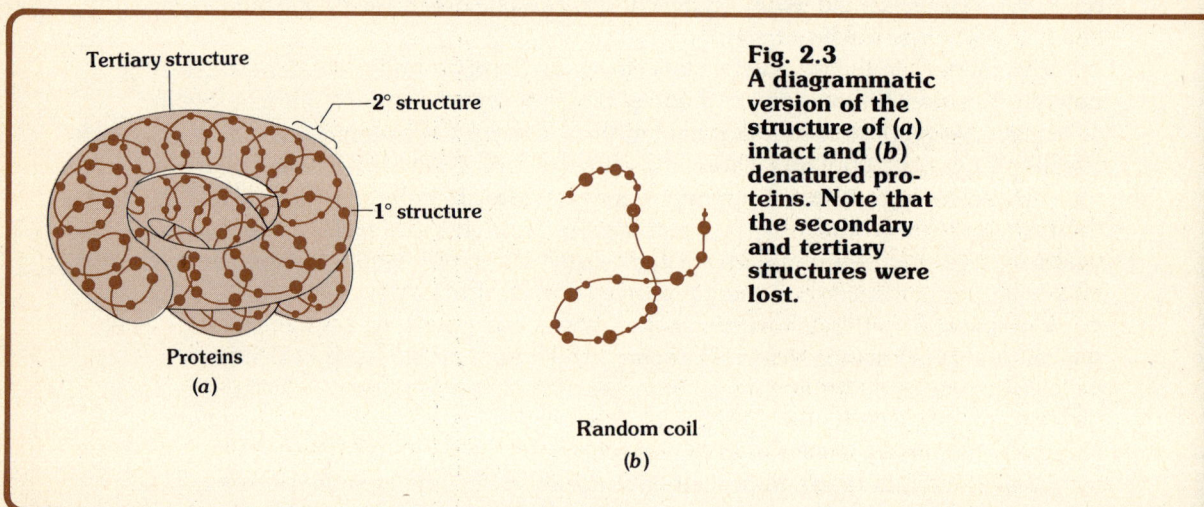

**Fig. 2.3**
**A diagrammatic
version of the
structure of (*a*)
intact and (*b*)
denatured pro-
teins. Note that
the secondary
and tertiary
structures were
lost.**

Tertiary structure

2° structure

1° structure

Proteins
(*a*)

Random coil
(*b*)

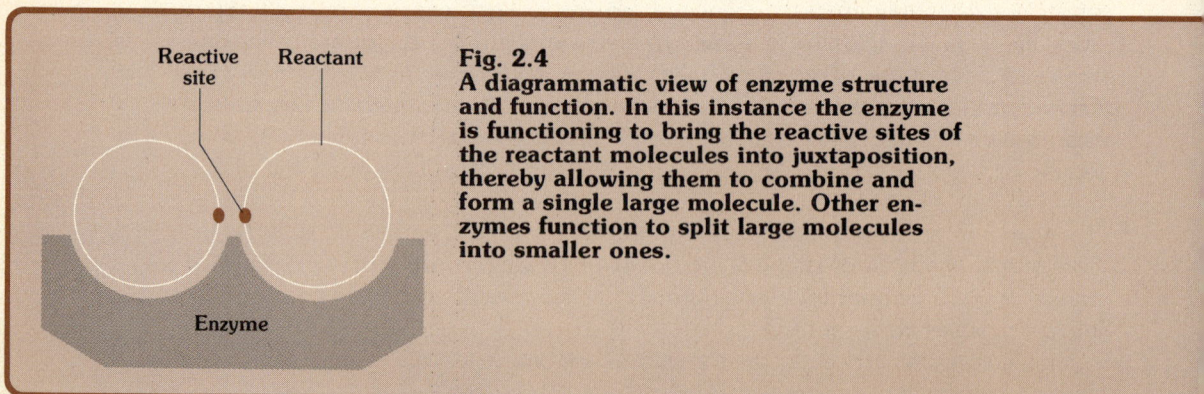

Reactive
site

Reactant

Enzyme

**Fig. 2.4**
**A diagrammatic view of enzyme structure
and function. In this instance the enzyme
is functioning to bring the reactive sites of
the reactant molecules into juxtaposition,
thereby allowing them to combine and
form a single large molecule. Other enzymes function to split large molecules
into smaller ones.**

bowling ball touched the white spot on the other bowling ball. Now suppose you had a bowling ball holder that picked up both bowling balls, and which also oriented them such that the white spots were touching (Fig. 2.4). Essentially, that is how an enzyme behaves. By virtue of its specific shape, it can "recognize" molecules with a given form and either cause that molecule to be broken into two smaller pieces or bring together two smaller molecules to form one larger one.

Most enzymes are capable of effecting only one reaction, and it is the shape of the enzymes that determines which reaction they will effect. Obviously, anything which changes the shape of an enzyme to the point at which it can no longer engage in the reaction will essentially prevent the reaction from occurring at all. Recall that it was the tertiary structure which gave the enzyme its shape. Recall, too, that the tertiary structure was established by very weak chemical bonds—bonds which are very easily broken. Moreover, once the bonds are broken, they can never be reestablished—the enzyme (protein) is said to be **denatured** (Fig. 2.3). Heat is an extremely potent force for denaturation in most proteins (*see box*). In fact, the reason that fevers much more than 105° (41° C) can be fatal is because even this slight elevation above normal body temperature is sufficient to initiate the denaturation of enzymes. So important are enzymes to life that we very quickly die once they become denatured. Less dramatically, denaturation is also involved in such things as the solidification of egg whites during cooking and in pasteurization. (Pasteurization is the process whereby dairy products or beer are heated to about 66° C, which is sufficient to kill most of the bacteria without affecting the taste of the product. The bacteria die because their protective protein coats are denatured by the high temperatures.)

**Lipids**  Lipids are the third major class of macromolecules found in the body. They are the most heterogeneous, because they are defined as anything that can be extracted with an organic solvent (e.g., alcohol, ether, etc.). Lipids generally consist only of carbon, hydrogen, and oxygen atoms (although some have **phosphorus** as well), but unlike carbohydrates, oxygen is present in much lower quantities than is carbon. There are several subcategories of lipids, which include the following.

*Fats.* Numerically, fats comprise the largest category of lipids and, in fact, the two terms are frequently (although incorrectly) used interchangeably. Fats consist of an alcohol to which are attached a number of fatty acids. Because they tend to repel water, fats are used extensively in structural areas, notably as a part of the membrane that surrounds cells. Fats are also storage molecules for energy.

*Carotenoids.* Carotenoids are the plant pigments that give the yellow and orange colors to such vegetables as beans and carrots. The group's name, in fact, derives from the plant pigment found in carrots (as well as in many other vegetables) called **carotene.** This substance is important because it is the precursor of vitamin A, an important molecule needed in vision. Thus, the old story about carrots being good for your eyes does have some validity.

*Steroids.* The steroids are an immensely important group in the body. Each consists of a series of linked rings with differing side chains (see Fig. 2.5). This group includes, among many others, the male and female sex hormones, **cortisone,** (a natural product of the body but one that is also extensively used in the treatment of inflammation), and **cholesterol,** also a natural body product but one that is highly suspect in the formation of fat deposits in arteries **(atherosclerosis).**

**Nucleic acids**   The last of the major macromolecular groups is the nucleic acids. Despite the fact they are much the rarest of the four groups, they are of enormous importance in that they are the molecules of inheritance and of regulation of protein formation.

We are concerned with two main kinds: RNA and DNA.[2] RNA is the acronym for ribonucleic acid and DNA is the acronym for deoxyribonucleic acid. If you analyze the differences in the names, you might correctly conclude that DNA is missing an atom of oxygen relative to RNA. This oxygen is located (or not located, as the case may be) in the sugar portion of the nucleic acids. It is a five-carbon sugar called **ribose** (or **deoxyribose** in DNA). Other than the oxygen atom, the sugar portion of the molecule is the same in both DNA and RNA.

Linking the sugars in a long chain are groups of inorganic material called **phosphate** groups (because phosphorus is the distinctive element). The phosphate groups are also identical in both DNA and RNA.

Attached to each sugar molecule is a **nitrogenous base.** Between them, RNA and DNA have five different bases, but each uses a total of only four. Thus, one nitrogenous base is unique to DNA, one is unique to RNA, and three are used by both.

**Vitamins**   Vitamins might well be considered a fifth group of macromolecules important to the body, yet they are generally not given that status. The reason is that the other four groups are all chemically distinct categories. Vitamins are functionally distinct, but in terms of chemical structure, they are very diverse. Some, such as vitamins A, D, E, and K are lipids; various of the B vitamins are

---

[2] Actually, there are a series of different kinds of RNA, differing principally in the length of the molecule, found in virtually all cells. For the purpose of distinguishing DNA from RNA, however, we shall treat RNA as a single entity.

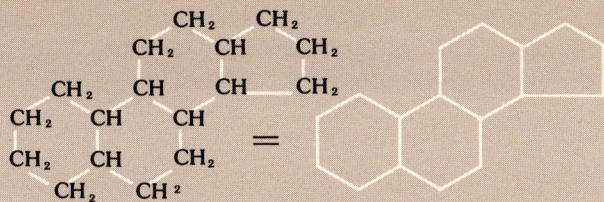

Steroid ring structure

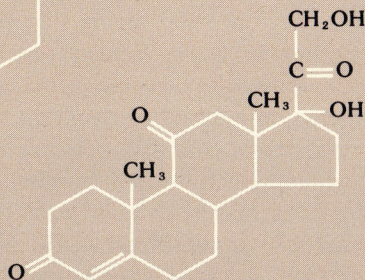

Cortisone

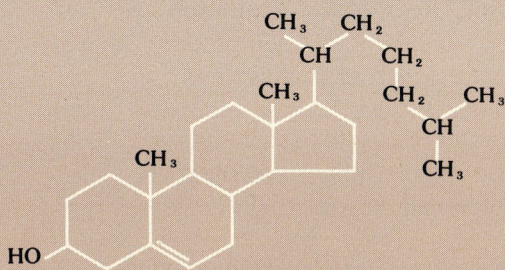

Cholesterol

**Fig. 2.5
Various types of steroids. Note the characteristic four-ring formation.**

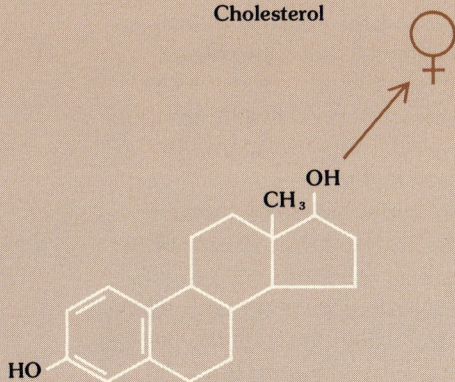

Estradiol, one of the estrogens

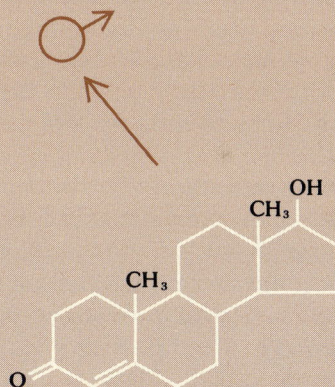

Testosterone

derivatives of the nitrogenous bases which make up the nucleic acids. Because all the vitamins belong chemically to one or another of the previously mentioned four groups of macromolecules, they are generally not afforded the status of a separate group of macromolecules.

**THE CELL**   The cell is the basic unit of all living systems, and it is the small-est naturally occurring unit that can be said to be alive. Therefore, there is ob-viously a huge gap between talking about macromolecules and talking about cells.

As it happens, cells are exceedingly diverse. A **generalized** cell is shown in Fig. 2.6; do not assume it to be a **typical** cell, for such a cell does not exist. Cells may be round (red blood cell), cuboidal (many skin cells), or irregular (white blood cell). They may be as small as 0.05 mm in length (sperm cells) or as long as 1 m (some nerve cells).

Structurally, cells share some things in common. All have a **cell membrane** of proteins and lipids, although plant cells also have an outer cell wall of cellulose; vir-tually all cells have a **nucleus,** which houses the **chromosomes** and DNA, although the red blood cells of mammals have even dispensed with a nucleus. All cells also have a certain amount of **cytoplasm**—the "soup" of the cell—but its com-position may vary greatly from cell type to cell type.

A number of formed structures, called **organelles,** may be found in most cells. These include the following:

**Mitochondria**   Mitochondria function in the production of energy (see Chapter 6), and are consequently most abundant in cells requiring large amounts of energy, such as muscle cells.

**The Golgi apparatus**   The Golgi apparatus serves to synthesize complex car-bohydrates, which are then packaged in sacs that gradually migrate to the surface of the cell where their contents are discharged.

**Lysosomes**   Lysosomes are packages of enzymes, probably products of the Golgi apparatus, and which are transported to the surface of the cell. They serve as the disposal unit of the cell and are particularly common in cells that specialize in the destruction of other cells (e.g., white blood cells).

**Centrioles**   Centrioles are a set of paired structures of importance in cell divi-sion (Chapter 12).

**Endoplasmic reticulum**   This is a labyrinth of tubes and sacs, apparently con-tinuous with the cell membrane and with the membrane surrounding the nucleus. It is extremely extensive within the cytoplasm and comes in two forms, based on its appearance under the electron microscope: **rough,** which is studded with **ribo-somes,** which are very small packages of protein and RNA that function to produce proteins; and **smooth,** which possesses no ribosomes and which functions to pro-duce lipids. The proportion of these two forms of the endoplasmic reticulum (ER) is a function of the role of the cell itself, smooth ER being very common in such cells as those of the sex glands, in which steroid hormones are produced, and rough ER being very common in muscle cells, for example, wherein large amounts of protein are produced.

**TISSUES**   In multicellular organisms such as ourselves, there are a large series of different types of cells. However, these are not randomly arrayed throughout the body but rather tend to be grouped such that like cells are joined together. A group of similar cells performing a similar function is called a **tissue,** and there are four basic types within the body, each of which contains a number of subcategories. These are elucidated in Table 2.2.

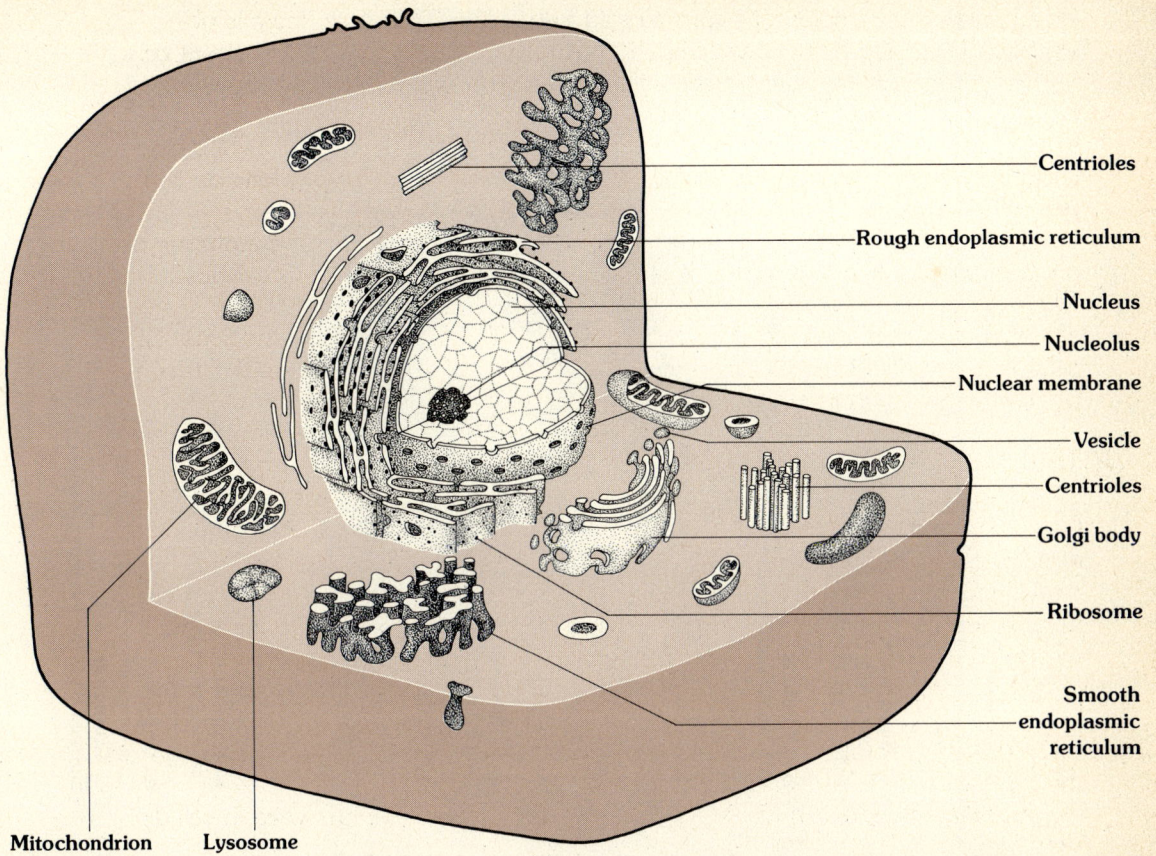

Centrioles

Rough endoplasmic reticulum

Nucleus

Nucleolus

Nuclear membrane

Vesicle

Centrioles

Golgi body

Ribosome

Smooth endoplasmic reticulum

Mitochondrion    Lysosome

(a)

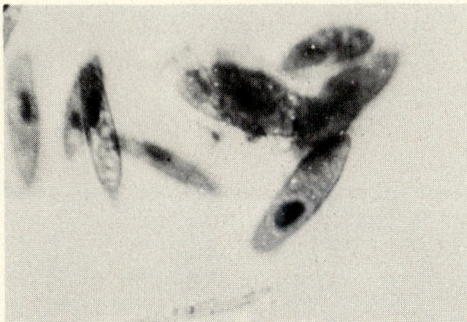

(b)

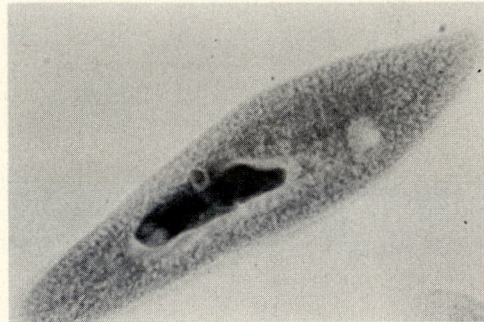

(c)

**Fig. 2.6** (a) **Diagrammatic view of a generalized animal cell;** (b) **and** (c) **light and scanning electron microscopic photographs of a** *Paramecium,* **a single-celled protozoan (about 150 microns in length). (Photos: Lester Bergman and Assoc.)**

# Table 2.2
## Cell tissue types

| Major categories | Subcategories | Where found | Appearance |
|---|---|---|---|
| **I. EPITHELIAL**<br>Broad layers of cells used in protection, absorption, or secretion. | | | |
| **A. Lining and covering**<br>(May have cilia) | **i. Simple**<br>(one layer) | **a. Squamous**<br>Flattened cells lining blood vessels | |
| | | **b. Cuboidal**<br>Cube-shaped cells found in collecting tubules of kidney | |
| | | **c. Columnar**<br>Cells taller than broad lining uterus | |
| | **ii. Stratified**<br>(many layers) | **a. Squamous**<br>Flattened cells many layers thick forming epidermis | |
| | | **b. Cuboidal**<br>Cubed cells lining ducts of sweat glands | |
| | | **c. Columnar**<br>Elongate cells found in lining of urethra | |

Cilia

**B. Glandular**
(Secretory)

   **i. Exocrine**

      **a. Simple**
      Each gland with own duct as in sweat glands
      **b. Compound**
      Many glands using common duct as in salivary glands

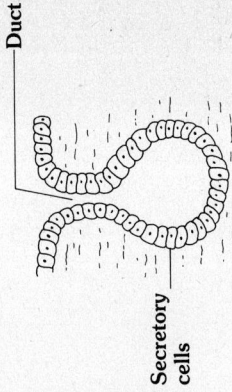

Duct

Secretory cells

   **ii. Endocrine**
   (no ducts—secretion into blood vessels)

   Thyroid gland

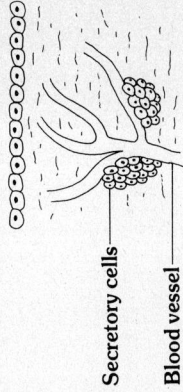

Secretory cells

Blood vessel

## II. CONNECTIVE
Contains many types of cells, some of which lay down large amounts of nonliving intercellular substance in which the cells become imbedded

**A. General**

   **i. Loose**
   (intercellular fibers are widely spaced)
   **ii. Dense**
   (fibers closely packed either at random or in parallel)

   Forms subcutaneous fat

      **a. Irregular**
      Random arrangement found in the dermis
      **b. Regular**
      Parallel arrangement found in tendons

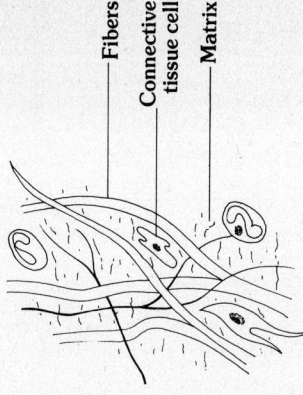

Fibers

Connective tissue cell

Matrix

**B. Special**

   **i. Cartilage**

   Tip of nose

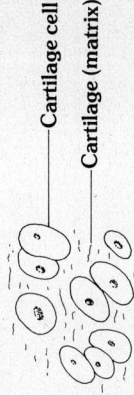

Cartilage cell

Cartilage (matrix)

(Continued)

**Table 2.2 (Continued)**
Cell tissue types

| Major categories | Subcategories | Where found | Appearance |
|---|---|---|---|
| | ii. **Bone** | Skeletal system | |
| | iii. **Blood**<br>iv. **Lymph** | Circulatory system<br>Lymphatic system | |
| III. **MUSCLE**<br>Cells contain contractile proteins which enable them to shorten | i. **Skeletal**<br>(elongate cells typified by regular transverse bands) | Muscles of arm and leg | |

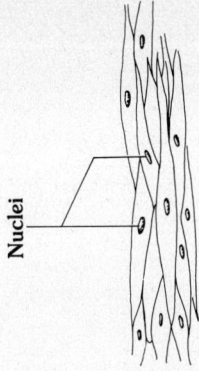

Nuclei

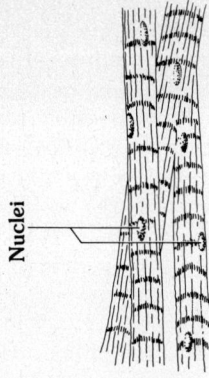

Nuclei

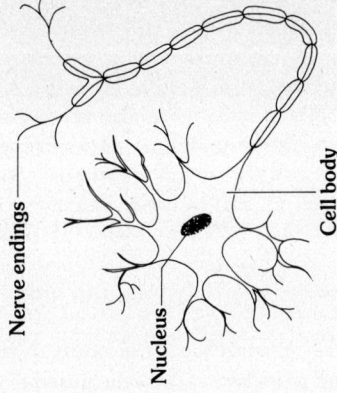

Nerve endings

Nucleus

Cell body

**ii. Visceral**
(spindle-shaped cells with no apparent bands)

Muscles of the stomach and intestine

**iii. Cardiac**
(branching cells with bands; contract spontaneously)

Heart

## IV. NERVOUS

Cells possessing the capacity to transmit information electrochemically

**i. Central nervous**
(brain and spinal cord)

  **a. Gray**
  Uninsulated cells of the brain

**ii. Peripheral nervous**
(throughout the rest of the body)

  **a. Nerves**
  Long cellular extensions
  **b. Ganglia**
  The nerve cell bodies

**iii. Special receptors**
(the special senses)

Eye, ear

A strong argument can be made that only carbon can serve as the backbone for the molecules of life. Can the same argument be made for water? Life processes, as we know them, occur in a water medium. Are there other molecules that would serve as well? A commonly suggested alternative is ammonia ($NH_3$), which has the advantage of being a small molecule (virtually the same size and weight as a water molecule) and one composed of elements that are common. However, ammonia is a gas above the temperature $-34°$ C, as are most molecules of that size and weight. It is possible to speculate that living systems could evolve at temperatures of liquid ammonia, but the rate of chemical reactions would be so slow at these temperatures that only the simplest organisms could have evolved.

**Must life be water-based?**

The molecular structure of water is such that water molecules are highly attractive to one another, and this mutual attraction results in a highly dense substance—and one that is therefore a liquid at temperatures where others are gases. Moreover, water is a highly effective solvent and it is extremely stable, retaining its chemical integrity while acting as the medium in which a great many biochemical reactions take place.

Could other life forms use another molecule? Perhaps, but on a theoretical basis, water seems the ideal choice.

**FROM ORGANS TO INDIVIDUALS**   Groups of different tissues may be grouped together into structural and functional units called **organs.** The stomach is such a structure. Typically, organs do not perform their tasks in isolation, but are linked in some fashion to other related organs. A series of such organs forms an **organ system** (sometimes just called a **system**), and there are 10 of these in all but the simplest animals. (See Table 2.3.) Collectively, of course, these organ systems comprise the individual.

**Table 2.3**
**Organ systems of animals**

| | |
|---|---|
| Integumentary (skin) | Digestive |
| Skeletal | Excretory |
| Muscular | Nervous |
| Circulatory | Endocrine (hormones) |
| Respiratory | Reproductive |

**FROM POPULATIONS TO BIOSPHERE**   Individuals do not live in isolation—at least not if they are reproducing sexually—but rather are found in some form of association with individuals of their own kind. Such groupings are called **populations.** This term is used to refer to the local unit of a single **species,** which is defined as a group of organisms that potentially or actually interbreed, producing fertile offspring. Mixtures of populations are called **communities.** Thus, the red squirrels in a forest would form a population; squirrels, in conjunction with the other forest animals and plants, would constitute a forest community.

On a still larger scale, groups of communities, together with their nonliving environment (i.e., soil, air, water) form **ecosystems,** and all the ecosystems of the world are united to form a single **biosphere.**

The next eight chapters (3–10) are devoted to an examination of the organ systems as they relate to each other in keeping the individual alive. Following that, there are four chapters (11–14), which deal with biological organization at or above the level of the individual.

## HOMEOSTASIS—THE GOVERNING PRINCIPLE OF ORGAN SYSTEMS

It is often the case that those things which are the most fundamental are the last to be discovered or elucidated. Hence, the concept of **homeostasis,** which is actually the focal point for all the dynamic interactions in which a living organism engages with its environment, was not fully developed until about 40 years ago. Yet like so many great ideas, once stated it is so logical as to be immediately apparent.

The concept simply states that interactions of a living organism with its environment are such as to maintain variables within very narrow limits. The organism is said to be in a **dynamic equilibrium** with its environment, or existing in a **steady state.** Different organisms have different capacities for control, but in humans the organ systems operate integrally to ensure that such potential variables as internal temperature, salt and water levels, oxygen and carbon dioxide levels, amount of acid present, do not, in fact, become variables but are held essentially as constants. This tight control is not mere whim—maintenance of constancy is nothing short of being essential for life. Failure to control any one of these potential variables, even for a short time, would be fatal.

Once expressed, the concept of homeostasis seems intuitively obvious—what other function could there be for the totality of our organ systems? Indeed, a primary thrust of this text is to examine organ systems not merely as a set of things to be described and labeled, but rather to analyze the contributions of each toward the maintenance of the homeostatic state of the organism.

In terms of operation, there are basically four distinct methods which the body employs to achieve homeostasis. These are **negative feedback, buffering, storage, and elimination.**

**Negative feedback** Negative feedback, despite its inherent connotation, is a primary method for maintaining constancy. It is perhaps most easily exemplified by considering the operation of a thermostat,[3] whether it be regulating a furnace, an air conditioner, or a refrigerator. Thermostats differ in their design, but functionally they are the same. For example, when the temperature of a house drops below a certain point in winter, the thermostat switch closes and the furnace comes on. De-

---

[3] It is precisely the fact that there are biological and engineering parallels in control systems that caused Norbert Wiener to write his famous book *Cybernetics* in 1948 and simultaneously found a new branch of science. It seems incredible retrospectively, that this parallel should have been elucidated so recently.

Compounds that dissociate in such a way as to release hydrogen ions ($H^+$) are called **acids**; substances that dissociate such that they release hydroxyl ions ($OH^-$) are called **bases** (or **alkalis**). The measurement of acid—base levels is on the pH scale, which ranges from 0 to 14. Low pH indicates acidity; high pH indicates alkalinity. Pure water is neutral (same number of $H^+$ and $OH^-$ ions), and has a pH of 7.

Just like the decibel scale for sound, or the Richter scale for earth tremors, the pH scale is logarithmic, meaning that there is a 10-fold difference for every change in one whole number. Thus, a cola drink, with a pH of about 3.0, is more than 10 times more acidic than tomato juice, which has a pH of 4.3 (Table 2.4).

**Acids, bases, and pH**

**Table 2.4**
**pH values for various solutions**

| Solution | pH | Solution | pH |
|---|---|---|---|
| Stomach secretions | 0.9 | Urine | 4.8–7.5 |
| Cola drink | 2.5–3.0 | Distilled water | 7.0 |
| Vinegar | 3.0 | Human blood | 7.4 |
| Orange juice | 2.6–4.4 | Intestinal secretions | 7.0–8.0 |
| Tomato juice | 4.3 | Seawater | 8.0 |

pending on the sensitivity of the thermostat, the furnace may stay on until the temperature of the house rises 2–4 degrees, at which time the thermostat switch opens and the furnace cuts off. Negative feedback, them, involves mutual control—as output (heat from the furnace) increases, input (the thermostat switch) decreases, and vice versa. Positive feedback, on the other hand, involves a spiralling effect, as occurs in a nuclear explosion—the splitting of a single atom releases particles that split more atoms, and so forth, in a chain reaction, until an explosion occurs.

There are many examples of negative feedback in biological systems, most notably in the functioning of hormones. An excellent example in ourselves, however, is the control of body temperature. A full discussion of this subject is given in Chapter 4. Briefly, a drop in the body temperature causes shivering, which is nothing more than the rapid contraction of muscles, resulting in the generation of heat. A return in body temperature to normal levels causes the cessation of shivering. An above-normal body temperature increases perspiration rate, and the evaporation of the perspiration from the skin has the effect of cooling the body. Once the body temperature has returned to normal, the rate of perspiration production drops markedly.

Interestingly, because our constant body temperature is maintained through the development of heat by cells (heat is a waste product in the generation of energy by cells) and because chemical reactions are temperature-dependent (i.e., they take place more quickly at high temperatures), there is a constant danger of positive feedback taking over in temperature regulation. That is, if the body temperature drops too low, the chemical reactions involved in muscle contractions will not occur quickly enough to generate enough heat to return the temperature to normal levels. Thus, below a certain point, the temperature of the body cannot be controlled and will quickly drop to the temperature of the environment. Except in extraordinary circumstances, death will ensue.

Similarly, once the body temperature reaches approximately 105° (41° C) chemical reactions are taking place so quickly that the heat they produce is greater than the capacity of the body to lose through perspiration, and the individual's temperature will quickly rise to the fatal level—about 108° (42° C)—at which point enzymes denature. For that reason, very high temperatures are not treated with aspirin and bed rest; rather, the individual is given alcohol rubs (alcohol evaporates very quickly and increases the cooling effect on the skin) or even ice baths in order to reestablish a normal body temperature.

**Buffering**   Proteins, including enzymes, are extremely sensitive to the level of acidity in the solution around them—and we have already discussed the vital necessity of the proper functioning of enzymes for life. Many of the foods we ingest either contain acids, or, once digested, produce acids (*see box*). Many of the chemical reactions which the cells undergo also produce acids. Because most enzymes operate only within a very narrow range of acidity, the body must in some way prevent the production of these acids from changing the acid levels of the body (or, more specifically, the acid levels of the blood), which would have the effect of denaturing enzymes—a fatal event. Put more graphically, drinking a bottle of cola would be fatal (because of the change in blood acidity), if the body were not able to neutralize the acid contained in the carbonated water.

As it happens, there are many compounds in the body which have the capacity to take up or release hydrogen ions (the constituent of acids), depending on the acid level. These compounds are called **buffers.** By taking hydrogen ions out of solution, they operate to ensure that the acid levels of the blood are unchanged, even if you drink several carbonated drinks at one time.

**Storage**   Storage is a very important part of homeostasis. Suppose you were to eat a pound of jelly beans on an empty stomach. Jelly beans, which are composed almost exclusively of sugar, would be digested very quickly and the sugar would pass into the blood. Does that mean that your blood sugar level rises and falls with your snacks? If you have ever had a glucose-tolerance test[4] you know the answer is no (unless you have a disease such as diabetes). Almost as quickly as the sugar enters the blood, any amount above the normal blood level is converted by the liver to the storage molecule glycogen.

Storage is also used for certain unusual or potentially toxic molecules, or both. Lead may be stored in the hair. DDT may be stored in fat cells (only to be liberated in large amounts should the individual later decide to go on a diet).

---

[4] A glucose-tolerance test involves the ingestion of a measured amount of glucose on an empty somach, followed a certain time later by the taking of a blood sample. If the blood sample shows a high blood glucose level, it indicates insulin production is low, a possible sign of diabetes (see Chapter 9).

Storage may therefore be temporary, as with essential nutrients such as glucose, or it may be virtually permanent, as with such potentially toxic materials as certain insecticides.

**Elimination**   In a sense, this is the most straightforward of the mechanisms used by the body to maintain homeostasis—anything present in excess is simply passed out of the body. (However, the mechanics of this process are rather sophisticated.) Excess salts and water may be lost in the urine, feces, or perspiration. The molecules responsible for the odor of onions and garlic are eliminated via the lungs (as anyone who has stood downwind of a pepperoni pizza-eater already knows); a certain amount of alcohol is eliminated in the same way. The two areas of difficulty in this process are first, possessing the capacity for removing the molecule (e.g., lead is not easily eliminated by the body and must therefore be stored) and second, regulating how much of the required materials is to be eliminated.

The body uses a number of other mechanisms for maintaining homeostasis, but these four are much the most important. We shall be considering detailed examples of each in the forthcoming chapters. However, bear in mind the concept of homeostasis as you read the next section.

**THE LIMITS TO GROWTH**   It was mentioned earlier that one characteristic of living systems is the capacity to grow. Yet growth brings problems. Consider a single-celled organism, floating freely in the sea. This organism is literally living in its food—all the requirements needed for its continued existence are contained in the ocean medium in which it lives. Exchange between the interior environment of the cell and the external environment is constantly occurring across the cell membrane. The fundamental process involved in this exchange is **diffusion,** which is simply the net random movement of molecules from areas of high concentration to areas of low concentration.

Diffusion depends on random movements of molecules and is not a particularly fast process. You can demonstrate this by adding a single drop of food coloring to a glass of water and seeing how long it takes for the coloring to spread evenly throughout the glass.

The fact is, for most life forms, diffusion is too inefficient a process to be effective at distances much greater than 1 mm. Thus, the average cell is constructed such that no point within it is greater than 1 mm from its cell membrane, the point at which diffusion takes place (*see box*).

Given that restruction, how is a single cell to grow? Must it die once its size is such that portions of it are more than 1 mm distant from its cell membrane? Or must it grow linearly, like a pencil, to keep inside this limit? As it happens, cells may attain a considerable size in one or at most two dimensions, but never in all three dimensions. Moreover, there is another confounding problem—that of surface area/volume relationships.

**Surface area/volume relationships**   Consider a cell in the shape of a cube, 2 mm in each dimension. (That is still within our limits, as the center of the cell is still just 1 mm from its membrane.) The surface area of this cell is

The maximum distance for effective diffusion in a living cell is generally accepted to be about 1 mm, which obviously places severe limits on the potential growth and shape of cells. Some of you may be aware, however, that bird eggs are thought of as single cells, and certainly the egg of an ostrich greatly exceeds the limits we have just put on maximum cell sizes. How is this possible? An ostrich egg can be considered a single cell only in the loosest sense. Most of the egg consists of yolk, the storage material responsible for growth of the developing bird embryo. The actual area of embryonic development is confined to a tiny area at top of the yolk—and very near the shell surface, for diffusion purposes. Moreover, rather early in development, the bird embryo develops a sort of egg lung, supplied with its own circulatory system, to carry oxygen back to the embryo from the surface of the egg.

**Diffusion and cell size**

Furthermore, the relative metabolic activities of a given cell determine oxygen demand, such that very inactive organisms may be able to survive even with diffusion differences considerably greater than 1 mm. The fact is, however, that some limit exists for all organisms, and in no case is that limit very high. Therefore, you can afford to laugh with impunity the next time you see an elephant-sized amoeba slopping toward the heroine in a science fiction movie—the laws of diffusion prohibit the existence of such creatures.

$6(2 \times 2) = 24$ mm$^2$. Its volume is $(2 \times 2 \times 2) = 8$ mm$^3$. Let us now double the linear dimensions of the cell, such that it is now 4 mm in each direction. The surface area is $6(4 \times 4) = 96$ mm$^2$. Its volume is $(4 \times 4 \times 4) = 64$ mm$^3$. Note that even though we only doubled the linear dimensions of the cells, we quadrupled the surface area and octupled its volume (Fig. 2.7).

Suppose, for the sake of argument, that the original 2-mm cell requires all 24 mm$^2$ of surface area to serve as the site of diffusion for its 8 mm$^3$ of contents—in other words, each cubic millimeter of cytoplasm requires 3 mm$^2$ of surface area. By doubling the size of the cell, we now have only 1.5 mm$^2$ of surface area for each cubic millimeter of cytoplasm—obviously not nearly enough, under the hypothetical restrictions just given.

The point is an increase in volume must be compensated for by an equivalently much larger increase in surface area. This principle not only accounts for the observation that cells are large only in one or at most two dimensions, but it also accounts for the relatively huge surface areas of our intestines and lungs (see Chapters 4 and 6, respectively). In short, an understanding of surface area/volume relationships is a basis for understanding the shape and size limitations placed both on cells and organs.

Consider the following example. The human thigh bone (**femur**) is capable of sustaining only about 10 times the normal body weight. If we were to triple the dimensions of a person, so he now stood 5.5 m (18 ft) tall, the weight (volume) of the individual would have increased so much more quickly than the cross-sectional area of the femur that the individual would be unable to stand up. A very famous biologist

**Problem: How does this cell, with a requirement of 3 mm² surface area for every 1 mm³ of volume, increase in size?**

Surface area = 24 mm²
Volume = 8 mm³
Vol/s.a. ratio = 1:3

2 mm

**Fig. 2.7 Surface area/ volume relationships.**

**Alternative I: Unacceptable to increase equally in all directions— ratio is insufficient**

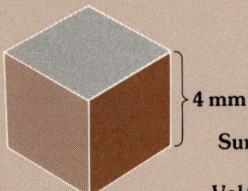

4 mm

Surface area = 96 mm²
Volume = 64 mm³
Vol/s.a. ratio = 1:1.5

**Alternative II: Ratio acceptable, but too much versatility lost by linear shape**

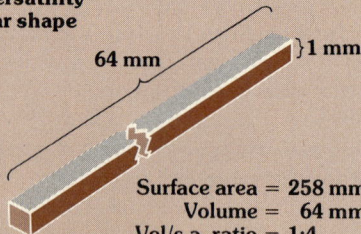

64 mm
1 mm

Surface area = 258 mm²
Volume = 64 mm³
Vol/s.a. ratio = 1:4

**Alternative III: Cell division preserves required ratio and offers the most potential for versatility in shape**

2 mm

Surface area = 192 mm²
Volume = 64 mm³
Vol/s.a. ratio = 1:3

named Haldane once noted that in John Bunyan's *Pilgrim's Progress,* the 12 m (40-ft) giants that the protagonist, Christian, met were always sitting down. Haldane pointed out that they had to be sitting down because if they stood up their legs would break. For this reason, the thigh bones of different mammals are not simply varying lengths of the same basic plan, but show considerable changes in width as a function of body volume (i.e., weight).

**Multicellularity**  The earliest life forms were single-celled organisms of two types: **autotrophs** (meaning, ''self-feeding,'' a reference to the fact that most were capable of photosynthesis) or **heterotrophs** (meaning ''other feeding,'' i.e., organisms that fed on the autotrophs and other heterotrophs. Green plants are autotrophs; humans are heterotrophs.) In such a world, there was a premium in size, particularly for the heterotrophs, for the larger they were, the greater the kinds of other organisms they could eat—and, by virtue of their size, they were protected from being eaten themselves.

Yet, as we have seen, there was a limit to how large they could grow because of the relative inefficiency of diffusion and because of the need to provide enough surface area to serve as the diffusive surface for their growing volume. Moreover, there would seem to be little advantage in simply increasing in one dimension. Thus, considerable benefit was afforded an organism that could divide into a series of cells that did not separate to form new organisms, but rather which remained together to form a single organism in which the cells cooperated and depended on each other—a **multicellular** organism. The presumed evolutionary relationships of living organisms are diagrammed in Fig. 2.8.

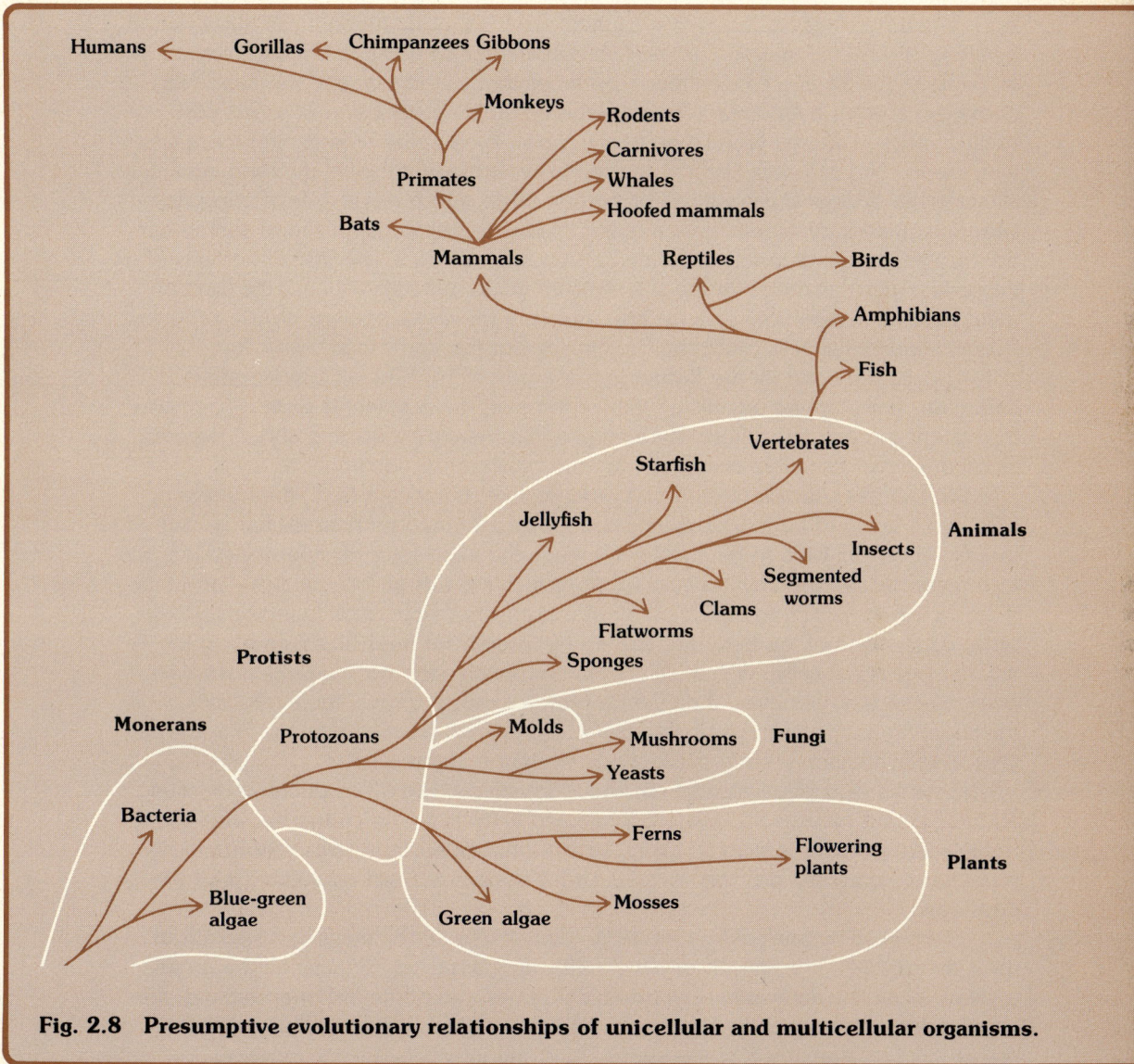

Fig. 2.8 **Presumptive evolutionary relationships of unicellular and multicellular organisms.**

**Deducing the need for organ systems** An added benefit of multicellularity was the potential for **specialization**—the development of different types of cells within a single organism—a potential which was quickly realized and which led to the development of tissues, organs, and finally organ systems. The advantages of specialization are readily apparent when you realize that all except the most primitive animals possess the same set of 10 organ systems that you do.

The development of specialized organ systems freed organisms from total dependence on the sea and allowed them to colonize harsher environments, such as land. The advantages of land life were twofold. First, at that time it was unexploited by living organisms, making competition minimal for some time. Second, oxygen is available in the air at a concentration 35 times that of the sea, and the availability of oxygen at such high levels allowed a more complete utilization of foods with vastly increased energy production (see Chapter 6). Yet the land environment was very hostile. Not only was water not instantly available, but the air environment actually acted to dry out organisms. A protective shield which resisted dessication, the **skin** (integumentary system) was needed. However, the construction of such a barrier also prevented diffusion of oxygen and nutrients. Thus, specialized portions of the surface were left moist, for the absorption of these materials—hence the development of **lungs** (respiratory system) and **intestine** (digestive system). Waste products from the individual cells could not be diffused out through the impermeable skin—hence the need for the **kidney** (excretory system). The air was not dense enough to buoy up the organism—hence the need for a **skeletal** system—nor was movement through this diffuse medium possible merely by floating about, necessitating the development of **muscles** (muscular system) for locomotion.

With all this specialization, it was obvious that many cells (e.g., the muscle cells) were going to be too far away from the oxygen and nutrient exchange surfaces to allow diffusion to be effective—hence the need for a **circulatory** system. In a sense, this was the critical system, since it represents an internal sea—and this is the fluid that must be so closely monitored to ensure that its constituents remain constant. All the cells of the body are bathed by **interstitial fluid** (the fluid filling the tiny gaps between cells), which is in direct communication with the blood; moreover, no cell is farther away than 1 mm from a capillary. Thus, the availability of all the necessary nutrients in this closely monitored internal sea allows for the extreme degree of specialization seen in the individual organ systems.

In the more sophisticated multicellular organisms, two kinds of communicating and integrating systems allows for coordinated activities—the **endocrine** (hormones) and **nervous** systems. Finally, such sophisticated organisms could no longer reproduce merely by dividing in two, and a specialized **reproductive** system was developed.

The preceding analysis is admittedly a bit simplistic, but it serves to point out three things. First, it is difficult to have a little specialization. Virtually all the organ systems would be needed in an land animal, even a hypothetical one. Second, the organ systems are mutually dependent, and the malfunction of any one of them places the entire organism in jeopardy. Third, the interdependence of organ

**Table 2.5**
Comparison of problems and solutions of amoeba and human

| Problem | Solution | |
| --- | --- | --- |
| | Amoeba | Human |
| Food | Engulfed | Digestive system; circulatory system |
| $O_2$ and $CO_2$ exchange | Diffusion | Respiratory and circulatory systems |
| N wastes | Diffusion | Excretory system |
| Support | Buoyed up by water | Skeletal system |
| Movement | Amoeboid movement (flowing) | Muscular system |
| Internal communication | Diffusion | Nervous and endocrine systems |
| Reproduction | Divides in two | Reproductive system |
| $H_2O$ | Diffusion | Integumentary (to prevent dessication) and digestive systems |

systems is manifested through the requirement that the internal environment of the organism be maintained essentially constant—that is, each organ system makes its own contribution in maintaining the organism as a whole in homeostatic balance.

Despite their obvious complexity, the organ systems do not usurp the basic functions of the cell (as performed by a unicellular organism) but rather operate collectively to provide a suitable environment in which all of the cells of the body may perform these same cellular functions. For example, even though their solutions are vastly different, the environmental demands of organisms as diverse as the unicellular amoeba and the human are essentially the same (Table 2.5). This sameness provides the justification for using a single species of our own, to illustrate biological principles, as is done in this book.

## SUMMARY

For well over 100 years it has been recognized that the cell is the basic unit of living systems. Not only are all living systems composed of one or more cells, but the individual cell is the smallest biological unit to which we can ascribe the properties of life.

Nevertheless, the cell does not possess magical properties, but is nothing more than a totality of sophisticated chemical reactions. Thus, in order to understand how the cell works (and, hopefully, to understand the nature of life itself), it is necessary to examine the levels of organization below the cell, most specifically the molecules and macromolecules which are the building blocks of the cell.

As a complete organism, the single cell has very severe restrictions in size, which ultimately translates into restriction in versatility. The size restrictions are not biological restrictions, but rather are imposed by the laws of physics and mathematics, most notably the laws of diffusion and of surface area/volume relationships. The solution was multicellularity, which must rank, along with the development of the cell itself, as one of the most significant events in the history of life on earth. Multicellularity led to organ system development, which, in turn, led to a much more complete colonization of the earth's surface than was possible by unicellular organisms. However, for the most part, the individual cells of a multicellular organism carry on virtually the same set of functions that a unicellular organism does—the organ systems merely provide the environment in which these cellular activities can take place.

An 83-year-old woman makes a misstep on the edge of the sidewalk, falls, and breaks her hip. She may never walk again. Her 5-year-old greatgrandson falls from the window of his third-story bedroom, suffers a greenstick fracture of his leg, and is running again in 4 weeks. Why the difference?

A scientist discovers the thighbone of an extinct dinosaur and is able to reconstruct a model of the whole animal from this one bone. Is it pure speculation, or is there validity to his method?

Most of us could throw a baseball more than 96 km/hr (60 mph) or lift 34 kg (75 lb) off the floor. If we are that strong, why do we have so much trouble opening a stuck window?

Because of their prominence, we all recognize the roles played by the skeletal and muscular systems—yet, ironically, for the same reason, we seem disinclined to appreciate the remarkable degree of interrelationship between these two systems, and even less to question the limitations these systems place on us. As you shall see, however, these systems are subject to the same physical laws as are all systems, and it is these laws which place limits on how our bones and muscles serve us.

chapter **III**

# chapter III
# Support and locomotion—skin, bone, and muscle

In response to the physical laws of gravity and inertia, most animals have evolved mechanisms for support and locomotion. To be sure, you can point to such species as coral and barnacles which have dispensed with locomotion (at least during their adult stages) and to species such as jellyfish and octopi, which have only limited support structures, gaining much of what they require from the water in which they live. But among the larger land-dwelling animals, support and locomotion systems are virtually indispensible.

When you think of these phenomena in terms of the organ systems responsible, you think first of the skeletal and muscular systems (Fig. 3.1). Yet these systems provide both more and less than the complete answer. As we shall see in subsequent chapters, bones are used not only for support, but also as storage sites for calcium (one of the essential elements), for the production of blood cells, and for fat storage. The muscles not only function in moving parts of the body but also in moving blood (through contraction of the heart muscles and muscles contained in the walls of blood vessels), and in moving food along the digestive tract. Breathing, eating, defecation, and all expressive behavior, from talking to gestures, also require muscles. Thus, support and locomotion form only a portion of the tasks performed by these two systems.

Similarly, support and locomotion are not entirely accomplished by these two systems alone. The skin and underlying connective tissue are responsible for the maintenance of form on the surface of the organism. (If this is not intuitively obvious, pinch up a fold of skin and notice how quickly and completely the skin returns to its normal shape once you release the fold.) Similarly, there could be no sustained locomotion without the circulatory system to transport food and oxygen to the muscles, and without the respiratory and digestive systems, which provide the body with these materials in the first place. Our concern here, however, is with the three organ systems primarily involved—the integumentary, skeletal, and muscular systems.

**SKIN AND SUPPORT**   The skin is deceptively complex, given that it is never more than 8 mm thick. Structurally, it consists of two reasonably distinct layers. Externally, there is an **epidermis,** which is composed of many layers of dead cells, all flattened together to provide a relatively impervious layer. These cells contain large amounts of the fibrous protein **keratin,** which is also found in hair

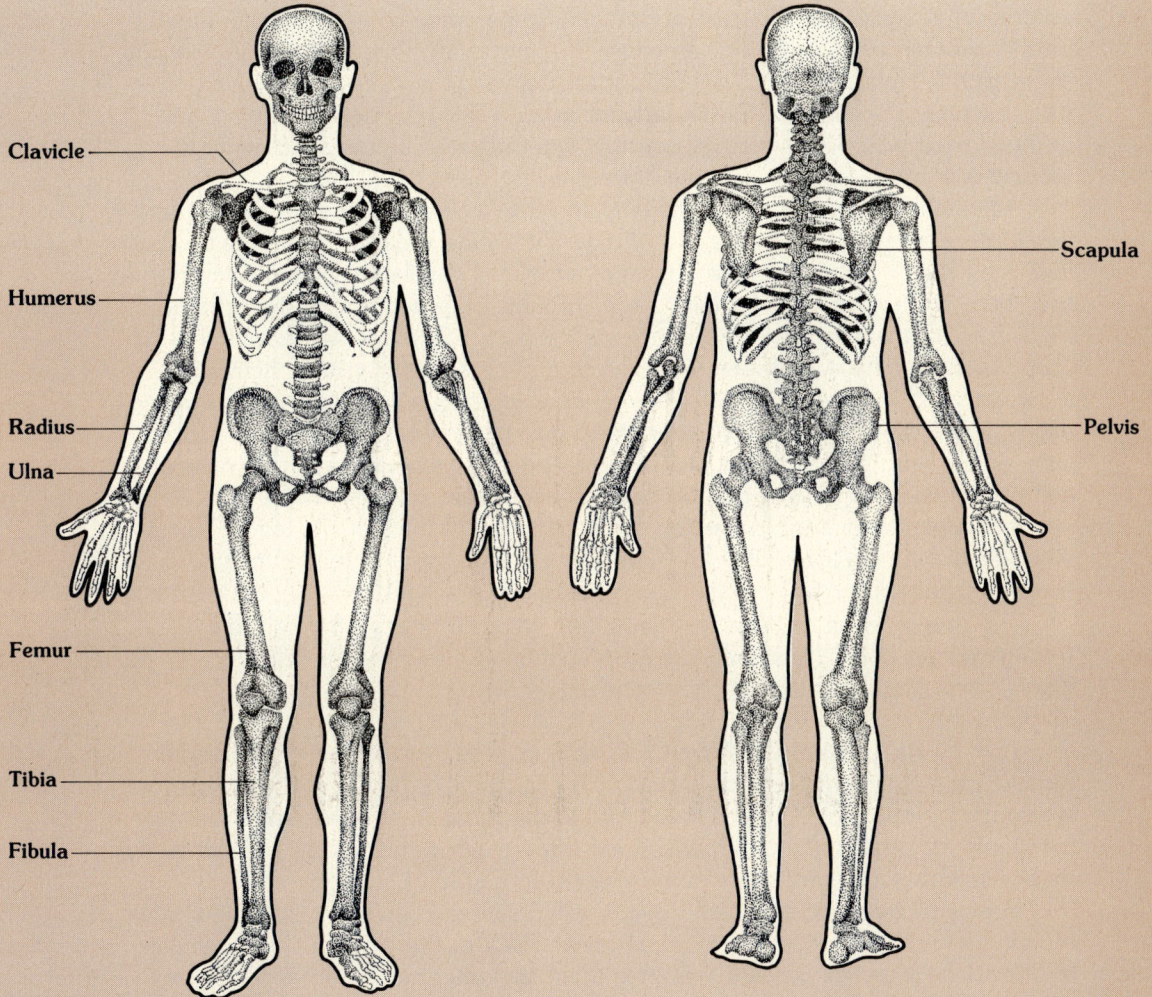

Fig. 3.1   **The human skeletal system.**

and nails, and which serves to give the epidermis its characteristic toughness. The outermost portions of the epidermis are constantly being sloughed off in small amounts (or in sheets, if you happen to have a sunburn), only to be replaced by new generations of cells being produced in the more interior reaches of the skin. During our lifetime, we shed about 23 kg (50 lb) of skin in this manner. The skin is particularly subject to changing conditions, and a sudden decision by its possessor to become a manual laborer is reflected by the generation of extra layers of epidermis in high-wear areas, a manifestation we call **callouses.**

The epidermis is laminated onto the **dermis,** in rather the same way that the outer layer of a piece of plywood is attached to the core. Too many sets of tennis on the first warm day of spring can cause this lamination to work loose in areas of stress, with the formation of a fluid-filled space between the layers. We call such structures **blisters.** The fact that the underlying layer of skin, the dermis, is alive and well endowed with nerve endings is amply demonstrated by our reaction to the seeping of perspiration into a broken blister.

Although the epidermis consists of a single type of epithelial cell, and is therefore a tissue, the dermis is a complex of different tissues and must therefore be classified as an organ. Other than the requisite number of capillaries, nerves, glands, and hair shafts, the dermis also contains large amounts of two important types of fibrous proteins that are of enormous importance in the role of the skin as an aid to support. The first of these, **collagen,** acts as a kind of organic baling wire, holding everything in place. It is the toughness of collagen which keeps our skin from becoming baggy at the knees and ankles, like a cheap pair of pantyhose. However, nothing lasts forever, and the gradual stretching of collagen is responsible for the development of wrinkles later in life. **Elastin,** the second protein, gives skin its unique property of snapping back into place when stretched, since, as the name of the protein implies, it has a rubberlike quality and will return to its original length after having been stretched.

The properties of the skin are many, and we shall be considering them throughout the book. However, its role in support is by no means insignificant, and it is a logical place to introduce this fascinating system.

**THE STRUCTURE OF MUSCLE AND BONE**   As we mentioned earlier, muscle tissue is not only used in altering the alignment of bones to each other, but is also used in the propulsion of blood and in the movement of materials through many internal organs. As it happens, these three major functions are performed by three structurally different types of muscles tissue (Fig. 3.2).

Muscle tissue associated with bone is called **skeletal** muscle; this is the type of muscle that normally comes to mind when we think of the tissue. On the cellular level, skeletal muscle is striated, because of the alignment of certain types of proteins which play an important part in the actual contraction of the muscle fibers. In addition, the various skeletal muscle cells are joined together such that it is frequently very difficult to determine just where one cell ends and another begins.

Muscle tissue associated with the internal organs is called **visceral** muscle (viscera being the name given to the internal organs). Unlike skeletal muscle, visceral

**Fig. 3.2
Skeletal,
visceral, and
cardiac muscles.**

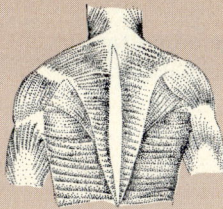

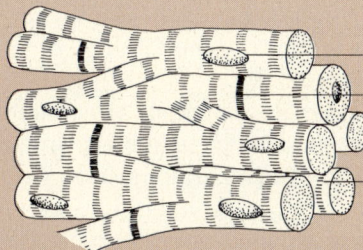

Skeletal muscle

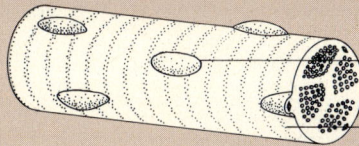

Nucleus

Striation

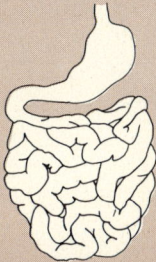

Visceral muscle

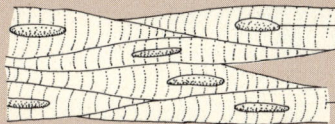

Nucleus

Smooth muscle cell

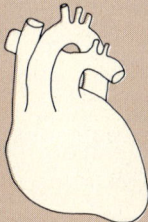

Cardiac muscle

Nucleus

Disc

Striation

muscle cells are easily separated one from another, and they do not possess the striations of skeletal muscle. Functionally, visceral muscle is used for the typically mild, but continual, contractions occurring, for example, as food is passed slowly along the digestive tract. In contrast, skeletal muscle is used for very rapid and powerful contractions, but of rather short duration.

The muscle of the heart is the third type of muscle tissue. **Cardiac** muscle, as it is termed, is found only in the heart, and it appears as something of a hybrid between the other two types of muscle. The cells are reasonably separable, but are striated. Moreover, cardiac muscle is capable of continual activity (in fact, it *must* be capable of continual activity if we are to survive), but can also contract very powerfully, as in times of vigorous activity. Cardiac cells also possess the unique capacity of spontaneous contraction, at a rate of about 72 times/min. Within the heart, the

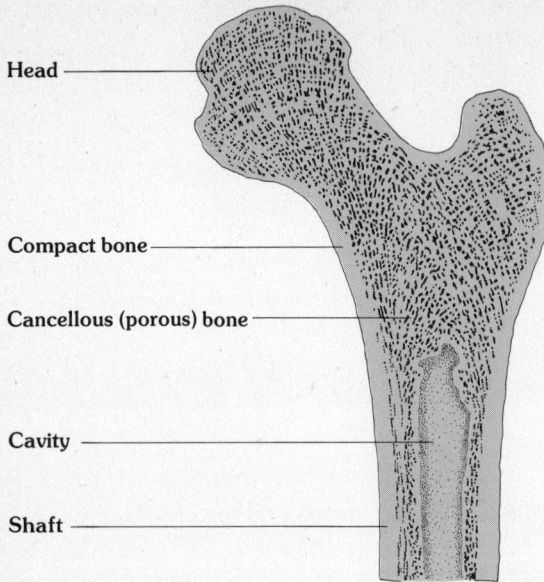

**Fig. 3.3**
**Stress lines in a long bone (femur).**

Head

Compact bone

Cancellous (porous) bone

Cavity

Shaft

actual rate is affected by nerves, but this is simply a modification of the spontaneous rate of the individual cells themselves.

Bone has a fascinating histological structure. Think for a moment how you might design bone tissue. You require a tissue that is extremely strong, one capable of retaining its rigidity during the contraction of powerful muscles. No problem—just use a modified version of tooth enamel, right? The difficulty with this solution is that bony tissue must be capable of growth in length and in thickness so that the organism itself can grow; moreover, ideally, bony tissue should be able to repair itself should it be damaged. Tooth enamel just will not do.

The actual solution is a nonliving matrix of calcium and phosphate salts, which give strength, interwoven by strands of **collagen,** the same linear protein we saw in the dermis. Collagen gives flexibility to the bone. Throughout, there is a series of canals which contain tiny blood vessels, supplying not only nutrients to the living bone cells which manufacture collagen, but also providing a method of adding or removing calcium and phosphate salts. The actual arrangement of the inorganic salts is along lines of stress (see Fig. 3.3), which act to ensure maximum strength and maximum resistance against collapse during muscle contraction. Maintenance of these stress lines, and of the bone itself, depends on the continued action of muscles pulling against the bone. Paralysis, prolonged bed rest, or space flights (no gravity, hence much-reduced muscle activity) all operate to reduce the amount, and thus the strength, of the bone. The typical long bone is hollow (Fig. 3.3), thereby providing lightness without an appreciable loss of strength.

A related tissue is **cartilage,** which forms the entire skeletal system of sharks and their relatives, but which is much more limited in land vertebrates presumably because it is not as strong as bone. Nonetheless, it plays a most important role in our own skeleton, in that most of the long bones of the skeleton (such as those of the limbs, and in contrast to the flat bones of the skull) are preformed in cartilage in the fetus, and only gradually is this cartilage replaced by bone. During this period of replacement, which is not completed until adulthood, the cartilage continues to grow, even as it is being replaced by bone. The net effect is to cause the ends of the bone to grow farther apart. At maturity, these cartilage zones of growth are finally replaced by bone, and bone growth ceases (Fig. 3.4a, b).

However, the fact that the bone no longer increases in size does not mean that the potential for new bone production has been lost. In the case of fracture repair, once again cartilage is produced, initially around the area of fracture, and, over a period of weeks, this cartilage collar is replaced by bone, until the original configuration of the bone is restored. This process may not occur so successfully if the broken ends of the bone are not adjacent and aligned. Moreover, the whole process is dependent on an adequate blood supply; thus, if fragments of bone occur at the site of fracture, they may degenerate, causing the reformed bone to be shorter than it was originally (Fig. 3.5).

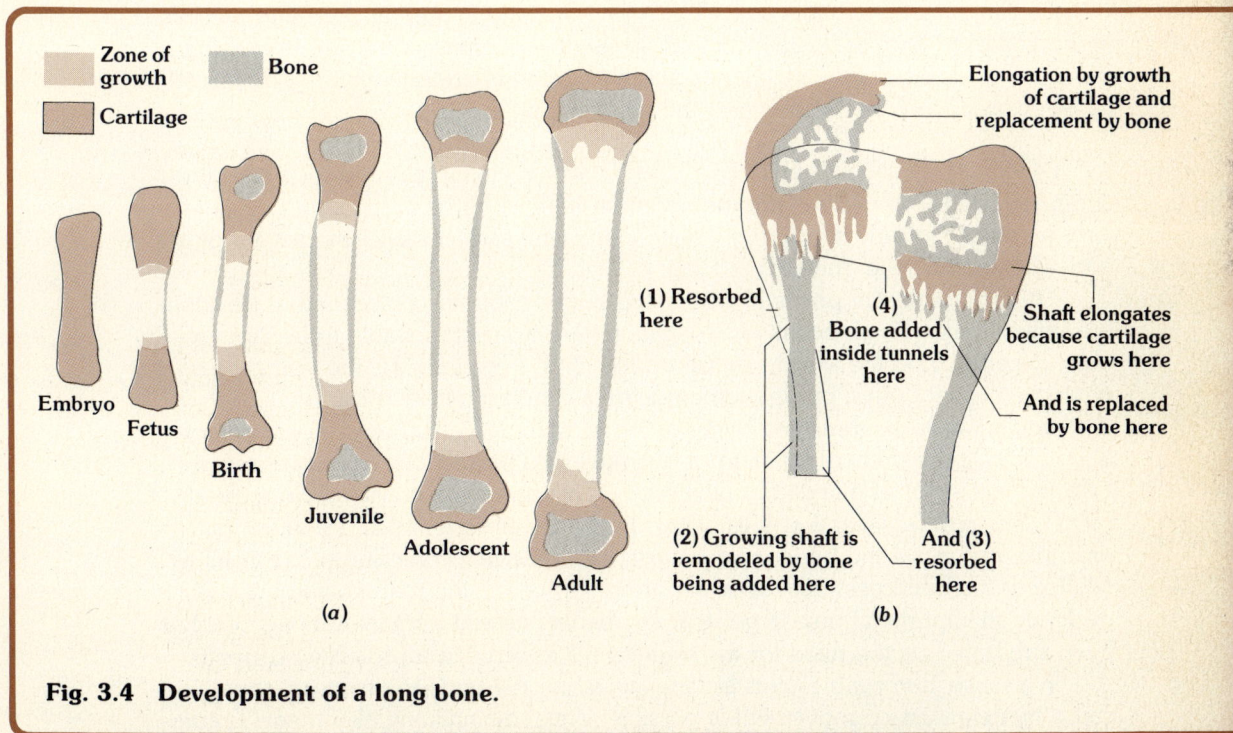

Zone of growth    Bone

Cartilage

Elongation by growth of cartilage and replacement by bone

(1) Resorbed here

(4) Bone added inside tunnels here

Shaft elongates because cartilage grows here

And is replaced by bone here

Embryo

Fetus

Birth

Juvenile

Adolescent

Adult

(2) Growing shaft is remodeled by bone being added here

And (3) resorbed here

(a)

(b)

Fig. 3.4 **Development of a long bone.**

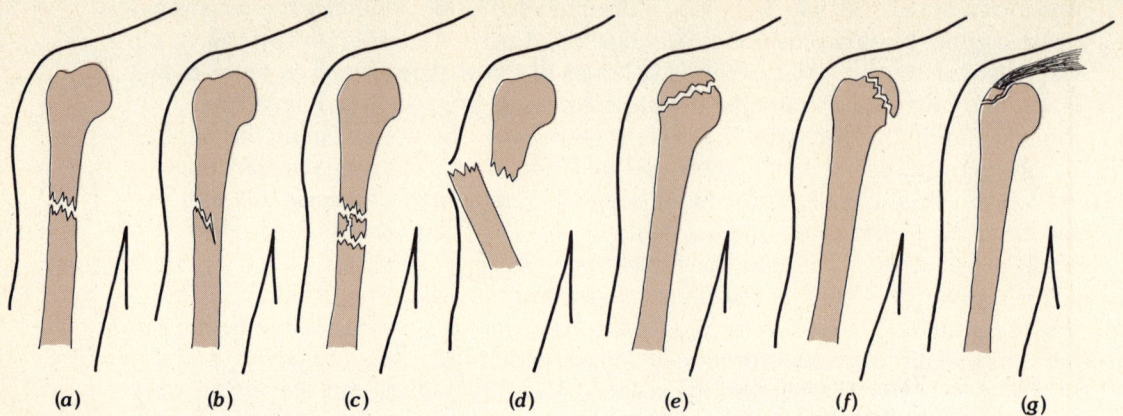

**Fig. 3.5   Types of bone fractures:** (*a*) **simple;** (*b*) **greenstick (incomplete break);** (*c*) **comminuted (multiple breaks);** (*d*) **compound (skin pierced);** (*e*) **articular;** (*f*) **epiphyseal (break in zone of bone growth);** (*g*) **evulsion (severe pull on tendon breaks off a bone fragment).**

Cartilage has other roles as well. It is found as intervertebral discs and as pads in the knees. In both of these instances, it serves as a shock absorber, because it is more compressible than bone. Cartilage is also found in areas in which great flexibility is required, such as the ears and the tip of the nose.

**BONE AND MUSCLE ASSOCIATIONS**   We have already seen how the competing demands of strength and lightness were resolved in the structure of the long bones. A similarly conflicting duality exists when we consider the nature of bone–bone and bone–muscle associations.

To begin with, the points at which two bones come in contact with each other are called **joints.** These points of association vary from immovable to freely movable, depending on the joint in question (Fig. 3.6). For example, the human skull is composed of 29 bones, but movement is possible only between the jaw and the skull and between the three tiny bones of the middle ear—all other joints are immovable, because the primary function of our skull is to protect our brain and sense organs. This is not as mandatory as it may at first seem, for considerable movement may occur at several of these joints in the lower vertebrates. (Snakes, for example, can move the bones of the upper jaw relative to one another, as well as relative to the braincase, in swallowing prey.) However, the mammalian brain is evidently so important (or so large) that any benefit gained from joint movement has been subsumed by the need for the skull bones to serve in a protective capacity.

In a sense, immovable joints are the easiest for the body to deal with, simply because by definition no movement is possible. Thus, maintaining alignments is auto-

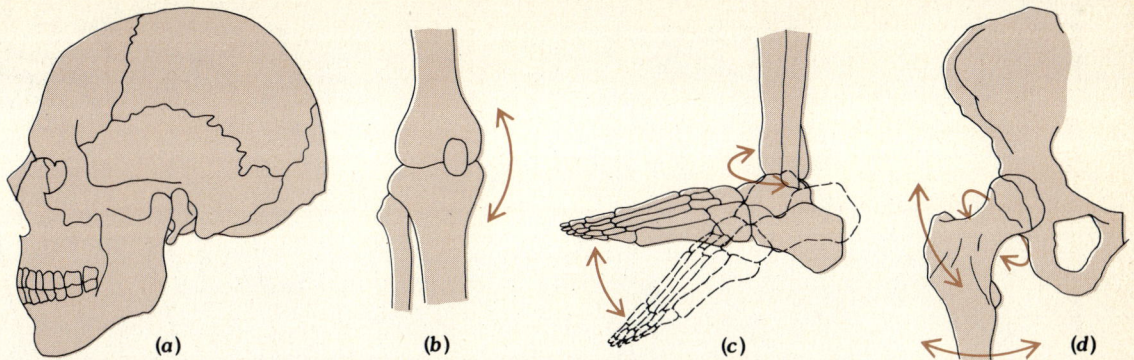

**Fig. 3.6   Four types of articulation:** (*a*) immovable; (*b*) hinge (one plane of movement); (*c*) pivot (two planes of movement); (*d*) ball-and-socket (multiple planes of movement).

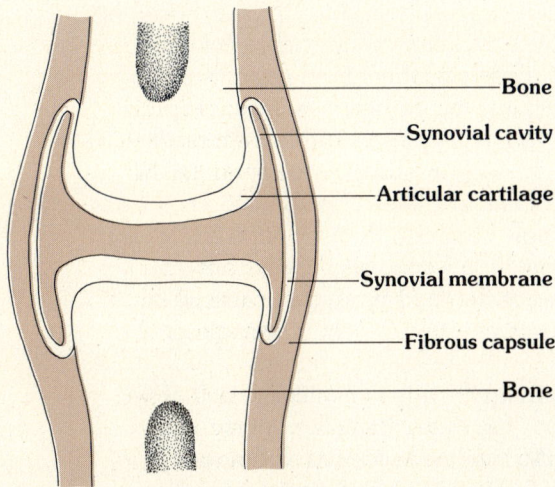

— Bone

— Synovial cavity

— Articular cartilage

— Synovial membrane

— Fibrous capsule

— Bone

**Fig. 3.7
Diagrammatic representation of a movable joint. The actual distance between the bones is exaggerated here for purposes of illustration.**

matically accomplished. Indeed, in ourselves, many of the points of juncture between the skull bones have all but been obliterated, so complete has the association become. Problems exist, however, where one bone must move relative to another. The major problems are friction and limitations of movement.

Friction must be avoided, and to achieve this end, the tips of the bones are covered with smooth cartilage. In addition, the cartilage produces a lubricating fluid which is prevented from escaping by the presence of a **synovial** membrane which surrounds the entire structure (Fig. 3.7). Thus, the bones themselves never really

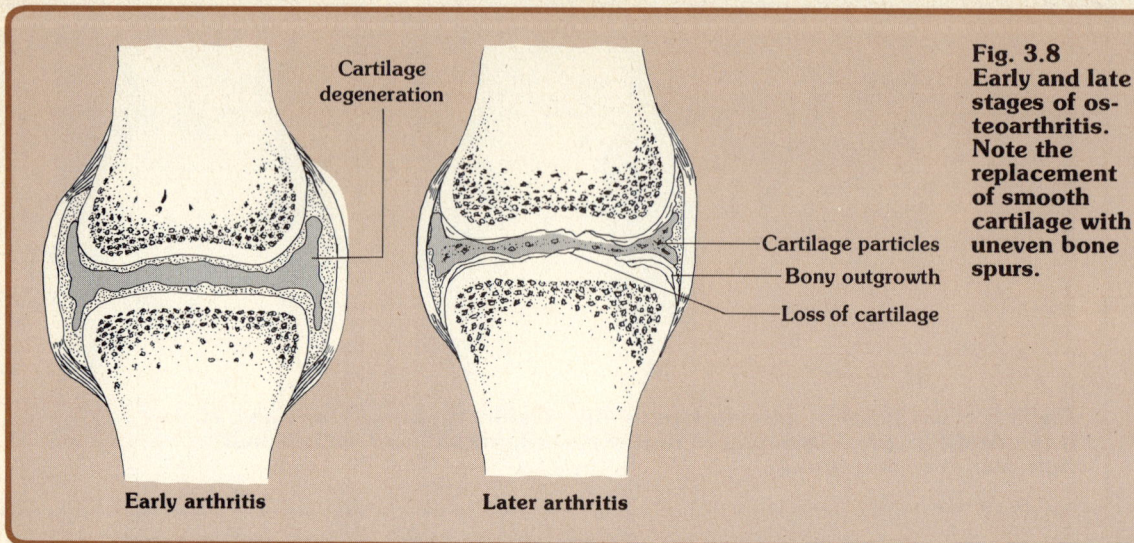

Cartilage
degeneration

Cartilage particles
Bony outgrowth
Loss of cartilage

Early arthritis

Later arthritis

Fig. 3.8
Early and late
stages of os-
teoarthritis.
Note the
replacement
of smooth
cartilage with
uneven bone
spurs.

touch, but slide over one another, separated by a thin film of lubricating fluid, in the same way that oil prevents the bearings of a bicycle wheel from directly coming in contact with another metal surface. However, just as the bearings begin to squeak and burn out without sufficient oil, so, too, is it possible for the lubricating fluid between bones to cease being produced. In such instances, movement of the joints is painful, and the joints become inflamed and swollen. This condition is known as **osteoarthritis,** the type of arthritis most common in older people (Fig. 3.8). Arthritis, which simply means "joint inflammation," may also result from a variety of causes other than simply the wearing out of the fluid-producing capacity of bones, and we shall consider these other causes at various points throughout the book.

Arthritis should not be confused with **bursitis,** which is the inflammation of the fluid-filled sacs of lubricants, called **bursae** (from the Latin, meaning "purse"), which are found surrounding the straps of connective tissue known as **tendons** (and which serve to connect muscle to bone) at joints (Fig. 3.9). This type of inflammation, which typically comes about because of too much exercise (i.e., tennis elbow), is normally temporary, and recovery is complete following rest. The condition of osteoarthritis, just described, is permanent, for all intents and purposes.

The second major problem existing at joints is to ensure unhindered movement within a set range, while preventing movement outside this range. For example, you can bend your leg backward at the knee, but (hopefully!) not forward. Three different methods are used to ensure that movements at a given joint do not exceed certain limits. These include:

1 The shape of the articulating surfaces themselves may be largely responsible for determining which way the bones will move. Little lateral movement is possible at the elbow,

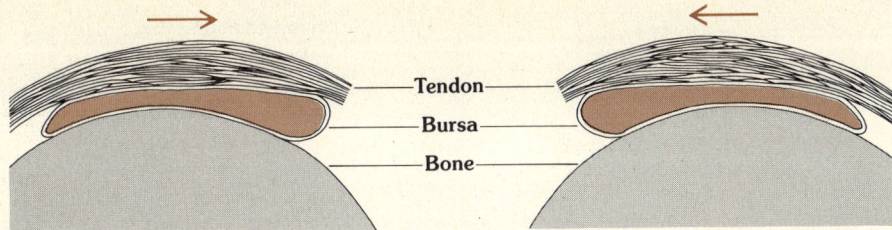

**Fig. 3.9  Relationships among tendon, bursa, and bone. The bursa acts as a flexible bearing in allowing frictionless movement of the tendon relative to the bone.**

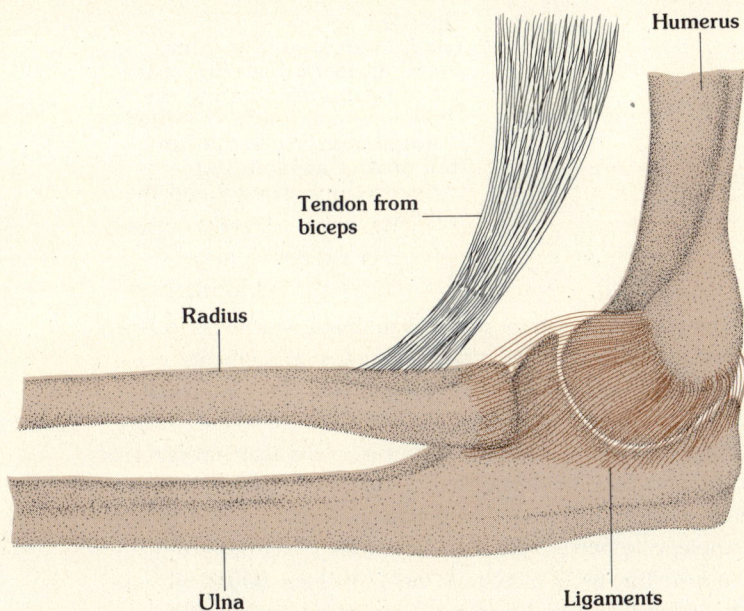

**Fig. 3.10
The human elbow joint. Muscles and other tissues have been re-moved except for the ligaments holding the bones together.**

for example, because of the nature of the association of the bones of the upper and lower arm (Fig. 3.10). This situation contrasts with that at the shoulder joint where there are few limitations placed on movement by the structure of the bones themselves (Fig. 3.11).

2 Muscles may operate to hold bones in place. The muscles associated with the shoulder girdle, for example, are of critical importance in confining movement of the arm relative to the body within acceptable limits. The shoulder blade is entirely held in place along the back by muscles.

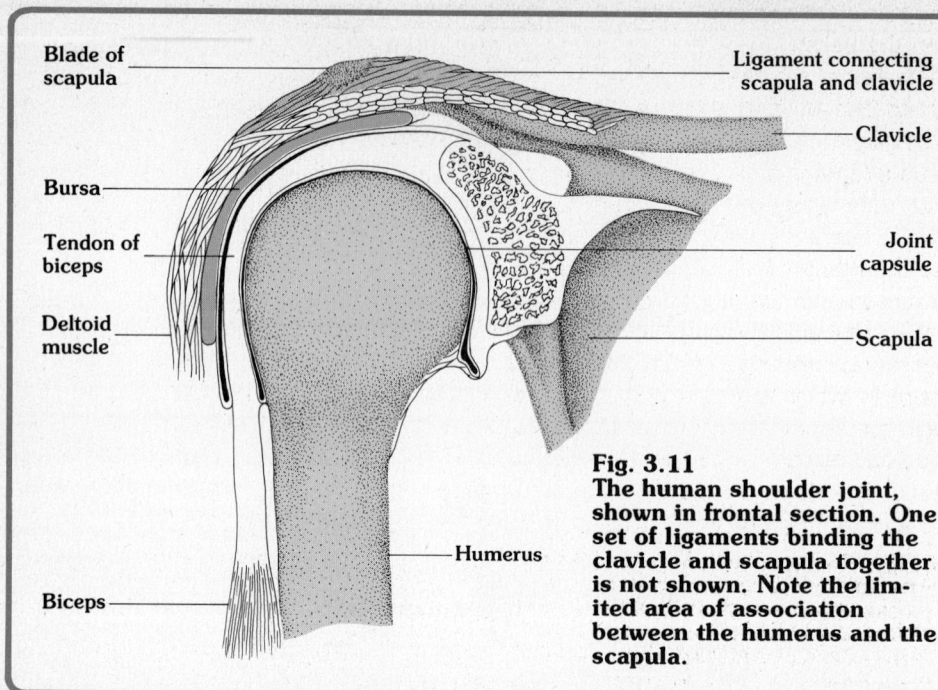

Blade of scapula

Bursa

Tendon of biceps

Deltoid muscle

Biceps

Ligament connecting scapula and clavicle

Clavicle

Joint capsule

Scapula

Humerus

Fig. 3.11
The human shoulder joint, shown in frontal section. One set of ligaments binding the clavicle and scapula together is not shown. Note the limited area of association between the humerus and the scapula.

3 There is a specialized set of connective tissue straps, called **ligaments,** which connect one bone to another (recall that tendons, although structurally similar to ligaments, connect muscle to bone), and which serve to maintain certain alignments. For example, it is not possible to move the bone of the upper leg **(femur)** backward relative to the pelvic girdle, not because of the nature of the bone alignments, but because a ligament prevents this movement (Fig. 3.12). Ligaments lack the flexibility of response of muscles, but represent a considerable gain in terms of weight and energy conservation. The function of the ligament just mentioned, for example, is to prevent the body from falling over backward at the hip. If this task were to be solved by using muscles, it would be necessary to increase the size of the musculature in the hip region and to invest a continuing supply of energy to operate these muscles. Because this movement is unnecessary anyway, the easier solution is to foreclose any possibility of such a movement through use of a tendon.

In most joints, limitations on movement are achieved by an interplay of all three methods, although the relative importance of each varies from joint to joint. Movement of the arm at the shoulder and of the leg at the hip are in many ways similar, but there is a far more complete articulation between the femur and the pelvis than between the humerus and the scapula (see Fig. 3.1 for locations of these bones). Thus, dislocations of the hip are less common than are dislocations of the arm (a necessary hazard of the extreme freedom of movement of the arm). However, fractures of the head of the femur are far more common than are fractures of the head of the humerus, a consequence of the more limited range of movement of the femur on the pelvis.

The ultimate test of the body in providing for the hazards of having bones moving relative to one another comes at the most complex joint of the body, the knee. This is actually a triple joint, as contact occurs between two points of the femur and the tibia/fibula, as well as between the kneecap and the femur/tibia (Fig. 3.13). The knee is intended to operate essentially in a single plane, as a hinge, in much the same way as the elbow, but the bony associations are less confining in the knee than they are in the elbow. Thus, considerable reliance must be placed on the presence of numerous ligaments which strap the knee in place, preventing any substantial amount of lateral movement. In addition, the chamber containing the lubricating fluid which separates tibia from femur is divided in half by a pair of cartilage pads, which also serve to aid in shock absorption. Despite these elaborate precautions, the knee remains susceptible to injury, particularly when lateral movement is forced (as when a football player is tackled from the side and his football cleats prevent any sideways movement of the foot). Torn ligaments tend to remain weaker and more susceptible to being stretched; damaged cartilages can be removed and new ones will grow, but not to the same size or thickness. Thus, once damaged, a knee seldom recovers fully.

This rather extensive knee damage is actually a **sprain.** Oddly enough, we tend to think of sprains as being minor, in contrast to a fracture, but sprains have a way of becoming chronic. Sprains are defined as injuries to a joint capsule, typically involving a stretching or tearing of tendons or ligaments. Unfortunately, both

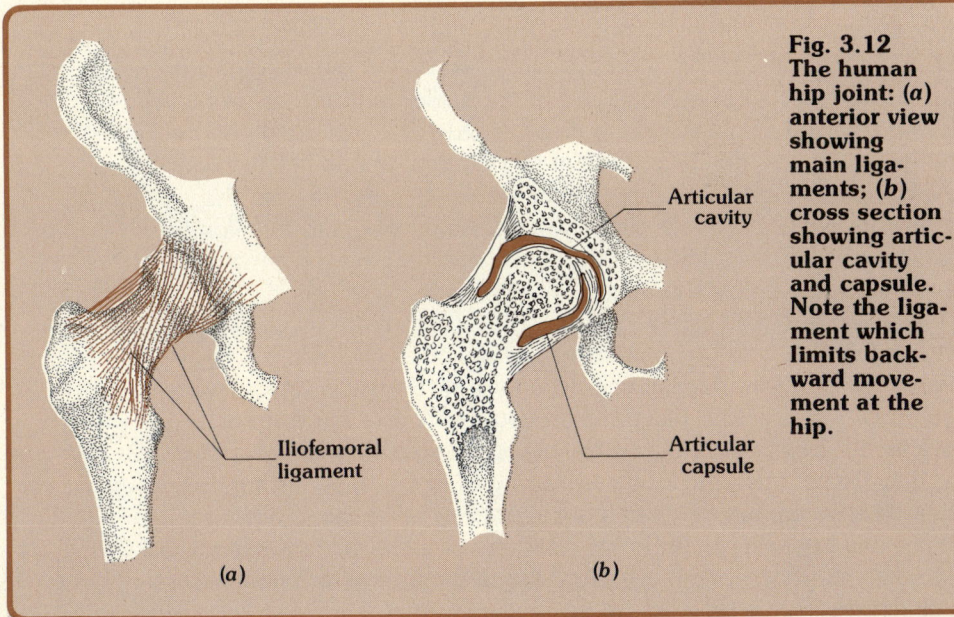

Fig. 3.12 The human hip joint: (a) anterior view showing main ligaments; (b) cross section showing articular cavity and capsule. Note the ligament which limits backward movement at the hip.

Articular cavity

Articular capsule

Iliofemoral ligament

(a)          (b)

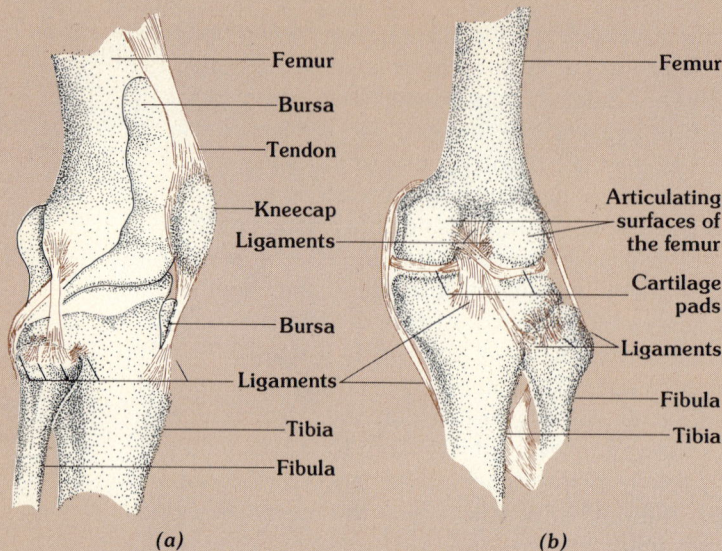

Fig. 3.13
The human
knee joint:
(a) lateral view;
(b) posterior view.

Labels in figure (a): Femur, Bursa, Tendon, Kneecap, Ligaments, Bursa, Ligaments, Tibia, Fibula

Labels in figure (b): Femur, Articulating surfaces of the femur, Cartilage pads, Ligaments, Fibula, Tibia

these structures have much poorer regenerative properties than does bone, and, once stretched, often remain weak. A **dislocation** is actually a severe sprain, and again, tends to become chronic (e.g., a "trick" knee).

**HOW MUSCLES FUNCTION**   It is possible to consider how muscles function both at the whole muscle (gross) level and at the single-cell (microscopic) level. As to the functioning of the whole muscle, a given muscle can, depending on the particular alignment of the muscle fibers, shorten by one-third to one-half its relaxed length during a complete contraction. It should be obvious, therefore, that the larger skeletal muscles (which, by necessity, always cross over a joint) function by pulling one bone toward or away from another bone and that this movement is achieved as a consequence of the shortening of the muscle.

Another fact which is almost as obvious, but which sometimes escapes notice, is that muscles can function only by contracting and never by expanding. For example, a primary function of the **biceps** muscle is to draw the lower arm toward the upper arm. Thus, you use your biceps when you touch your hand to your shoulder. However, the biceps are not responsible for straightening the arm out again—that is, they do not push the lower arm away from the upper arm by expanding. Rather, muscles are typically arranged in antagonistic pairs to allow for both flexing and extending (Fig. 3.14). In this case, when the arm is straightened, it is the **triceps** muscle, running along the back of the upper arm, which contracts while the biceps relaxes (Fig. 3.15).

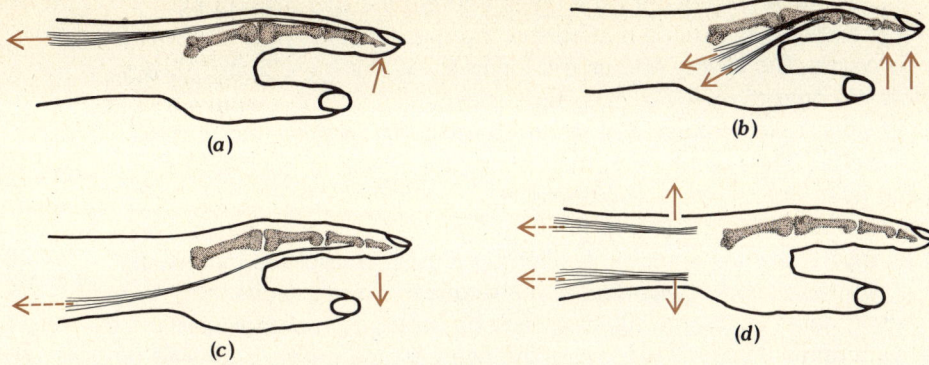

Fig. 3.14 **Muscles of the hand and forearm: (a) the muscle which extends the index finger; (b) associated muscles which assist in extending the index finger; (c) the antagonistic muscle, which flexes the index finger; (d) associated muscles which stabilize the extension of the index finger.**

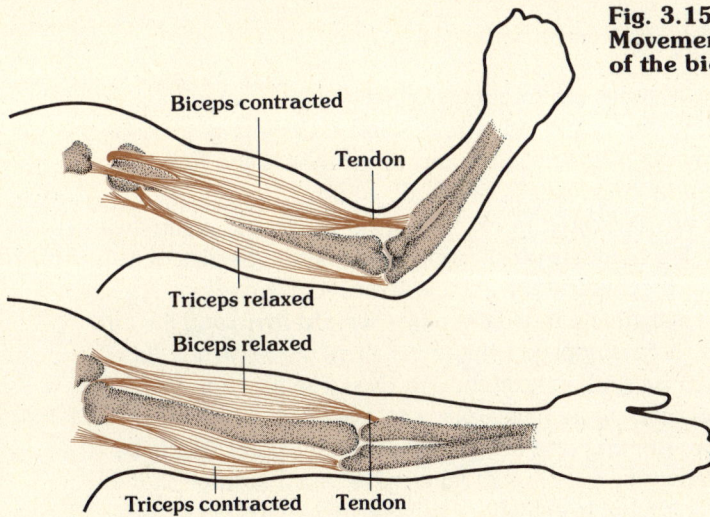

Fig. 3.15
**Movement of the forearm as a function of the biceps and triceps.**

Biceps contracted

Tendon

Triceps relaxed

Biceps relaxed

Triceps contracted    Tendon

Another component of muscle function relates to the strength of the contraction. Although a given muscle unit either contracts fully or not at all, the muscle as a whole is capable of enormous variation in the intensity of a given contraction. For example, it is quite apparent that you would use a less intense contraction to hold an egg between your thumb and forefinger than you would if you were bending bottle caps—and so much for the egg if you get the two mixed up!

The point is simply that varying numbers of the muscle fibers of a given muscle may be contracting at any one time. Moreover, all the muscles are always in a partial state of contraction—total relaxation is achieved only in death. This partially contracted state, called **tonus,** is necessary to maintain posture and form to the body. The specific fibers involved in maintaining tonus are, of course, constantly changing over time, so as to avoid fatigue.

At the other extreme, a muscle in a prolonged state of complete contraction may go into **tetany.** This is the situation typified by muscle cramps, as may occur after an unaccustomed period of exercise, and a return to normal tonus levels may take hours or even days. This same situation may be brought on pathologically, by such diseases as **tetanus** (lockjaw), which, as the name would imply, is typified by this condition of tetany.

At the microscopic level, the actual method of muscle shortening results from the sliding of two linear proteins over each other (Fig. 10.14). Contraction is achieved by increasing the amount of overlap between these two proteins. Because there are many such units all arranged in parallel, there is a summation of shortenings, which culminate in the ultimate shortening of the entire cell and finally of the muscle itself. The physiology underlying this molecular sliding is discussed in Chapter 10.

**MUSCLES AS LEVERS**   Muscles are frequently classified on a functional basis, depending on whether they flex or extend a limb, close an opening, and so forth. It is also possible to use mechanical analogies and to categorize the functioning of muscles in relation to the three classes of levers.

The three classes of levers are illustrated in Fig. 3.16a–d. In class I levers, the type used by cartoon coyotes to flip boulders off cliffs in a never-ending series of attempts to crush roadrunners, the mechanical advantage achieved depends on the ratio of the distance between the weight and the fulcrum (the **weight arm**) and the fulcrum and the point at which the force is applied (the **power arm**). Archimedes illustrated this point graphically when he said, in reference to class I levers, that he would be able to move the world if only he had a place to stand (he would also need a fulcrum, of course, to say nothing of a rather extraordinary pole). Because of the obvious mechanical efficiency of this class of lever, you might assume that all the muscles of the body were of this type. In actuality, this class of levers is poorly represented in bodily musculature, although one example is the muscle which runs up the back of the neck and functions to keep the head raised.

Class II levers are exemplified by a wheelbarrow—the weight is between the fulcrum and the point of application of the force. Again, mechanical advantage depends on the ratios of weight arm to power arm—thus, it is possible to carry more weight in a wheelbarrow with long handles, although a maximum is quickly reached because of the inherent clumsiness of this contrivance. Once again, this class of lever is poorly represented in the body, one example being the calf (**gastro-**

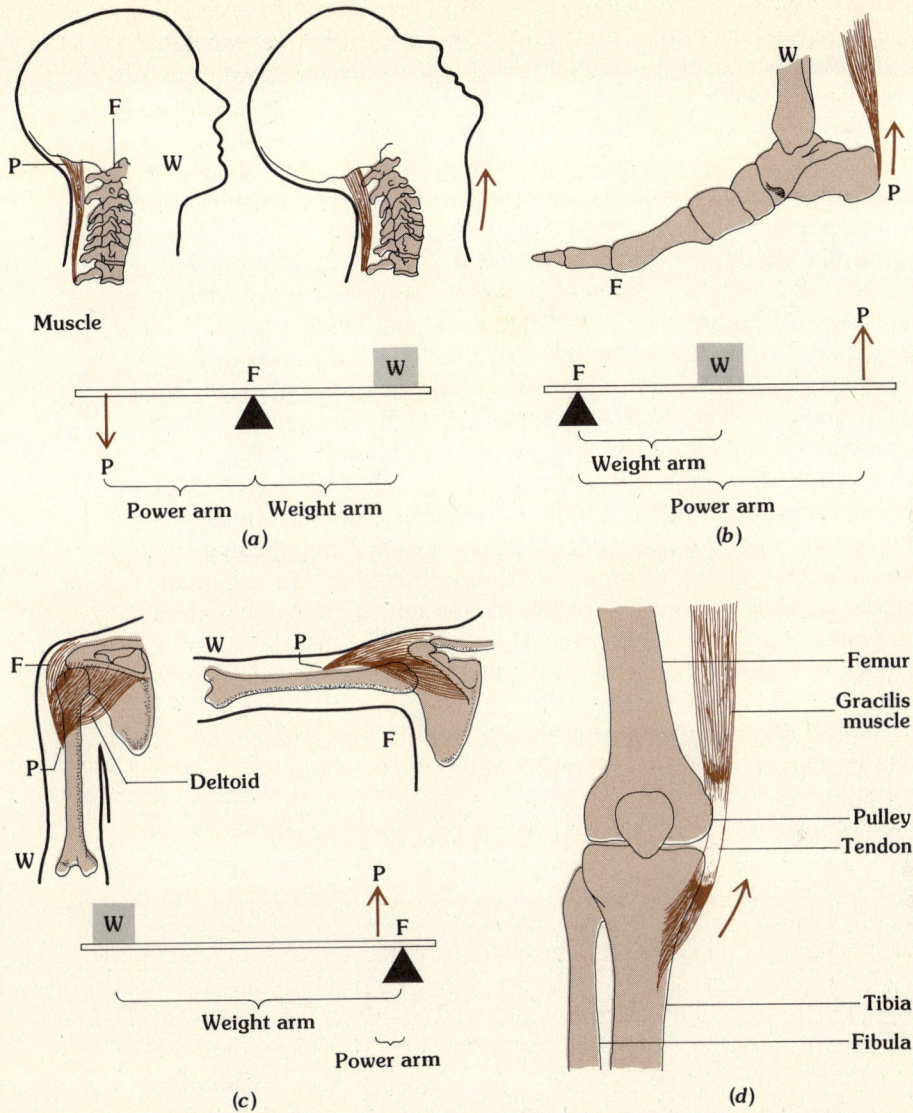

**Fig. 3.16** Muscles and bones as lever systems: (*a*) class I levers—the power arm generally is longer than the weight arm; (*b*) class II levers—the power arm is always longer than (and includes) the weight arm; (*c*) class III levers—the power arm is always shorter than (and is included by) the weight arm; (*d*) the articular ends of the long bones may be modified into pulleys which serve to determine the direction of movement when the muscle contracts.

**cnemius**) muscle. When this muscle contracts, the heel is raised upward, causing the whole weight of the body to be lifted—the person is standing on his toes.

Most of the muscles of the body exemplify class III lever systems in which no mechanical advantage can occur since, by definition, the weight arm is always longer than the power arm. An example of this class is provided by a fishing pole, where a slight jerk on the handle causes the tip to fly back suddenly through an arc of several feet.

Somehow it seems contradictory that the muscles of the body would be designed to sacrifice mechanical advantage so regularly, given that "mechanical advantage" has such a desirable ring to it. Yet the reasons for this dependency on class III levers are two-fold and are rather obvious, once elucidated.

First, because muscles cannot contract to more than one-half their relaxed length, even the longest muscle is capable of contracting only a few inches at most. In order to gain mechanical advantage, the power arm must move through a wider arc than the weight arm—thus, in a class I or class II lever system, the bone being moved will move an even shorter distance than the muscle did as it contracted. Therefore, such a system would work very well if what was intended were a series of powerful twitches, but it will not work if extensive movement is desired.

Perhaps an example is in order. Stand up with your right arm at your side, palm facing forward. Now touch the tips of your fingers to your right shoulder. Note that the fingers (or weight) swing through an arc covering the distance roughly between your knee and your shoulder (about 60 cm, or 2 ft), whereas your biceps muscle, which is responsible for this movement, shortens only about 5 cm (2 in.). Vertebrates as a group are characterized by speed, either because they must catch their prey or because they must avoid becoming prey. Speed, in turn, requires the sacrifice of mechanical advantage for the ability to move limbs through large arcs with relatively short (but powerful) muscle contractions. Hence, we might not be capable of picking up an anvil with one hand, but we can throw a baseball 160 km/hr (100 mph).

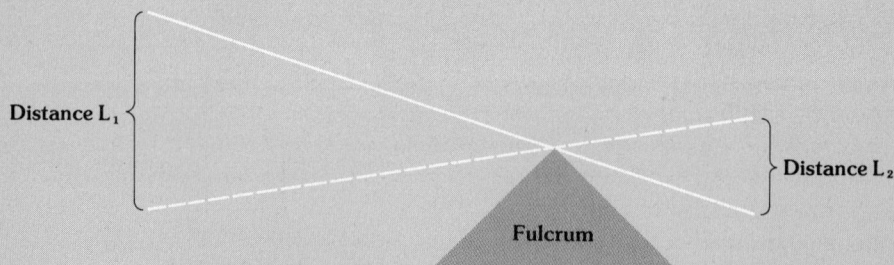

Distance $L_1$

Distance $L_2$

Fulcrum

**Fig. 3.17   The relationship of distance and speed in a lever. Even though the unequal arms travel through arcs of different lengths they do so in the same span of time. Therefore, the longer arm moves faster than does the shorter arm.**

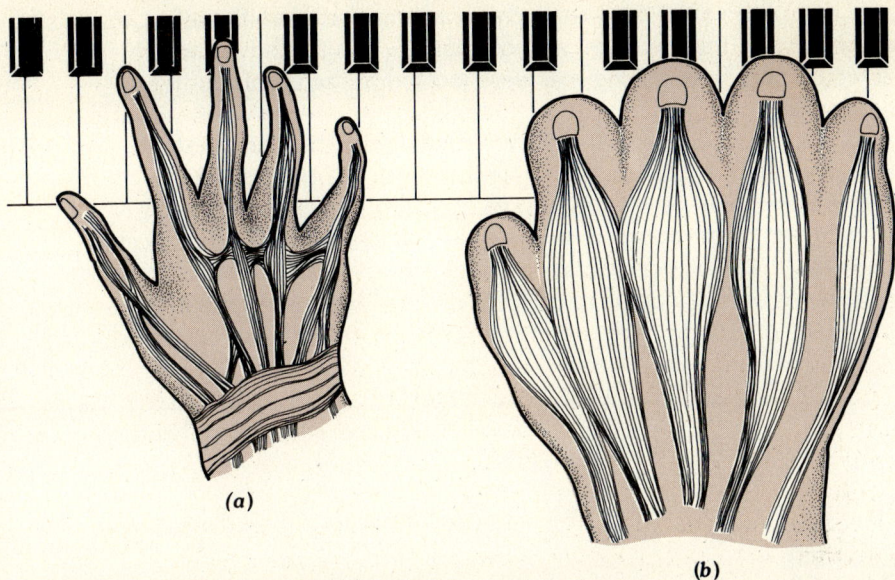

**Fig. 3.18**  (*a*) **Movement of the fingers of the hand is accomplished primarily by muscles located in the forearm. This arrangement is efficient both because the weight of the end of the limb is reduced and because dexterity is increased.** (*b*) **Note the weight and clumsiness of a hypothetical hand containing all the muscles required to move the fingers.**

The second reason for a class III lever dependency is that both the power arm and the weight arm of any lever must move through their respective arcs in the same period of time (Fig. 3.17). Thus, in the example just given, in the same time the biceps was contracting 5 cm, the fingertips moved 60 cm. Not only can we move our limbs through large arcs, therefore, but we can move them very rapidly as well—certainly far more rapidly than the muscles themselves can contract. Vertebrates have consistently sacrificed mechanical advantage for speed of movement and amount of movement, and both are achieved through a class III lever system.

Because the effectiveness of this design for maximizing speed and range of movement at the expense of strength is itself maximized if the weight is kept to a minimum, most of the large and powerful muscles are concentrated near the center of the body. Thus, our upper arms and legs are much more heavily muscled than our lower arms and legs (Alley Oop's arms notwithstanding). Most of the muscles used to move the fingers are actually located in the arm, above the wrist, and long tendons connect the muscles to the bones of the finger. This is clearly advantageous from the standpoint of increasing dexterity (Fig. 3.18), but equally important is the necessity for concentrating as much muscle as possible near the body and away from the end of the arm, because such an arrangement has the effect of shortening the weight arm of the lever. By way of analogy, imagine how much more awkward it would be if your fishing reel were located at the tip of the rod, rather than at the handle.

Of course, all these generalizations are relative. One finds, for example, that the points of insertion of the muscles on the forelimb bones of a badger, which needs strength rather than speed, are farther down the bone (thereby increasing the length of the force arm) than occurs in the cheetah, which is adapted for speed (Fig. 3.19). Nonetheless, the basic generalization remains true—vertebrates have utilized class III lever systems because of the need for speed and because of the inherent contractile limitations of muscle fibers.

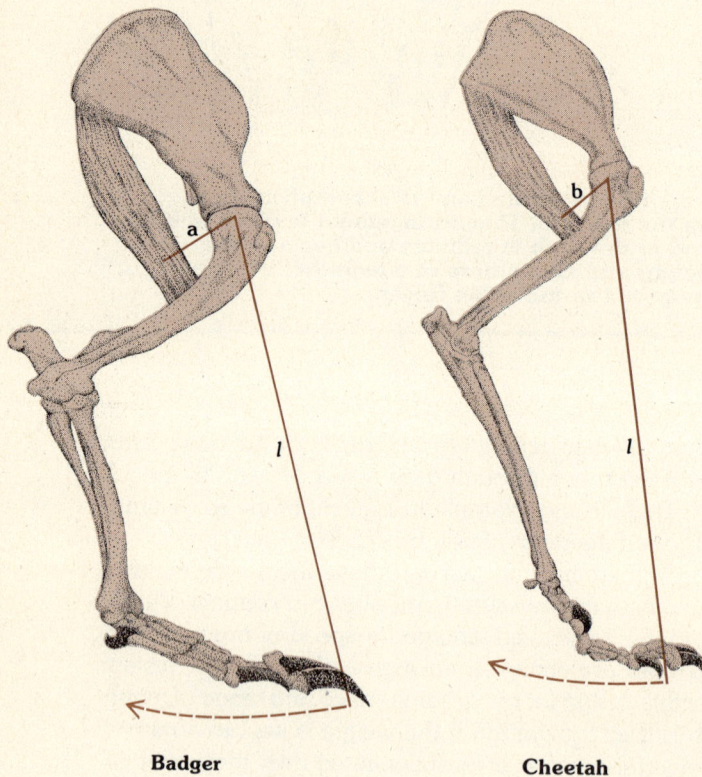

**Badger**　　　　　　**Cheetah**

**Fig. 3.19  Power and speed in a badger and a cheetah as a function of muscle placement.**

Place your fingertips on the back part of your cheek near the jaw angle and clench your teeth tightly. You should be able to feel the sudden bulging of the contracting **masseter** muscles. Try the same thing, but hold your fingertips on your temple. This muscle is the **temporalis.** Why are two different muscles needed to close the jaw? As it happens, the structure of the jaw in different mammals is closely linked with the type of food they eat—and the differences in mammalian jaw structures are at least partially attributable to differential reliance on either the masseter or the temporalis muscles. Consider the jaw types illustrated below.

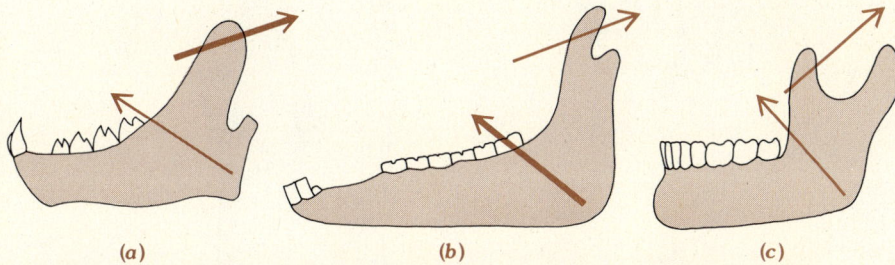

(a)    (b)    (c)

Jaw types of the (*a*) carnivore, (*b*) herbivore, and (*c*) omnivore (human). The direction of pull of the temporalis muscle is up and to the right; of the masseter, up and to the left. The relative size of each muscle is indicated by the length and width of the arrows.

The task of the **carnivore** jaw is to withstand the struggle of its prey which is trying to escape. As such, reliance is placed on the temporalis muscle which is positioned directly in line with the tip of the jaws. Thus, when contracted, the temporalis muscle is serving not only to close the jaws but also to resist the struggles of the prey pulling directly against it. Many carnivores, including the large dog breeds, have a crest running along the midline of the skull that serves to increase the area of attachment of the temporalis muscle.

**Shape and function of the human jaw**

By contrast, the jaws of the **herbivore** are designed for maximum reliance on the masseter muscle (since grass rarely struggles as it is eaten!) which does a more efficient job of effecting the grinding action of the molars than does the temporalis.

We are **omnivores** and thus rely on both sets of muscles for the different tasks necessitated by our varied diet. Thus, our jaw, like our diet, is something of a combination of typical carnivore and herbivore types.

**MODIFICATIONS FOR BIPEDALITY**   One of the most characteristic features of humans is our bipedal method of locomotion, which serves to distinguish us from most other mammalian species. To be sure, many mammals are capable of standing or even moving about using only their hind limbs—you need only think of elephants, bears, or dogs in the circus, or of pictures of gorillas standing

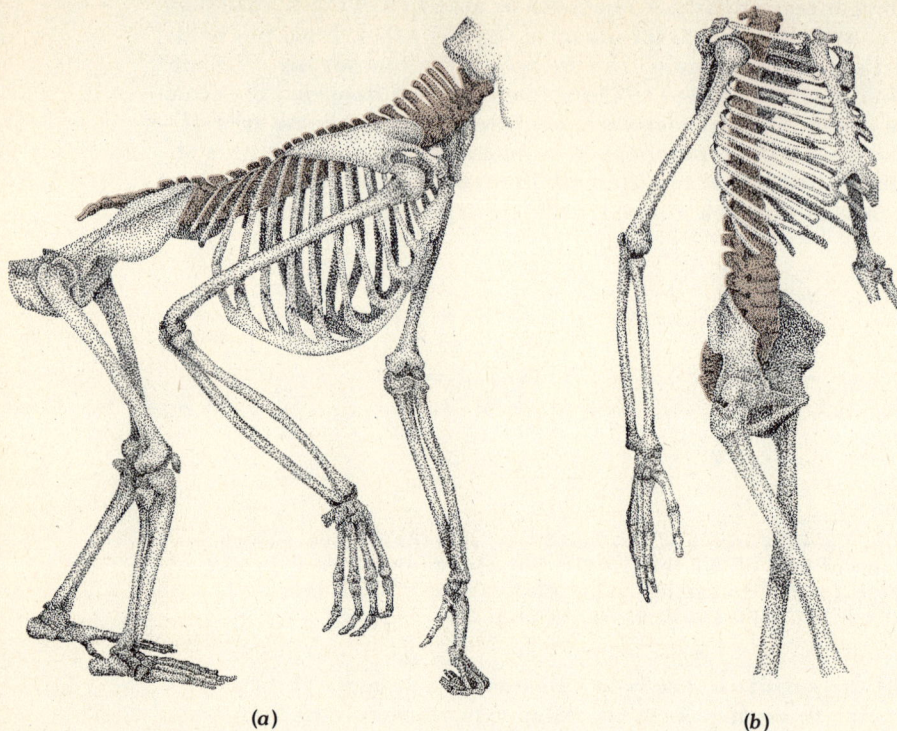

(a)                                    (b)

**Fig. 3.20   The vertebral columns of (a) the gorilla and (b) the human. Note
how much flatter the gorilla's vertebral column is in comparison to the
drawn-out "S" shape of the human's column shown in diagram (c).**

erect to thump their chests—but none of these species uses bipedality as its usual
method of locomotion. It should not be surprising, therefore, to note that there
have been a number of modifications in the muscular and skeletal systems that facilitate this method of locomotion in humans. A review of these modifications will
serve not only to demonstrate the nature of evolutionary change but also to exemplify the relationship between structure and function in the skeletal and muscular
systems.

    **The vertebral column**   A cursory look at our skeleton (Fig. 3.20) shows a dis-
tinct shape of the **vertebral column** (spine). The vertebral column of apes, like
those of most vertebrates, is slightly bowed (the extent of this bowing varies considerably among mammals). The human spine, however, is shaped like a drawn-out S.
Although this arrangement limits our ability to flex the spine (try duplicating your

Cervical curve

1st cervical or atlas
2nd cervical or axis
3
4
5
Spinous process
6
7

1st thoracic
2
3
4
5
Transverse process
6
7
Thoracic curve
8
9
10
Intervertebral disc
11
12

1st lumbar
2

3
Lumbar curve
Intervertebral foramen
4

5

Sacrum (5)
Sacral curve

(c)
Coccyx (4)

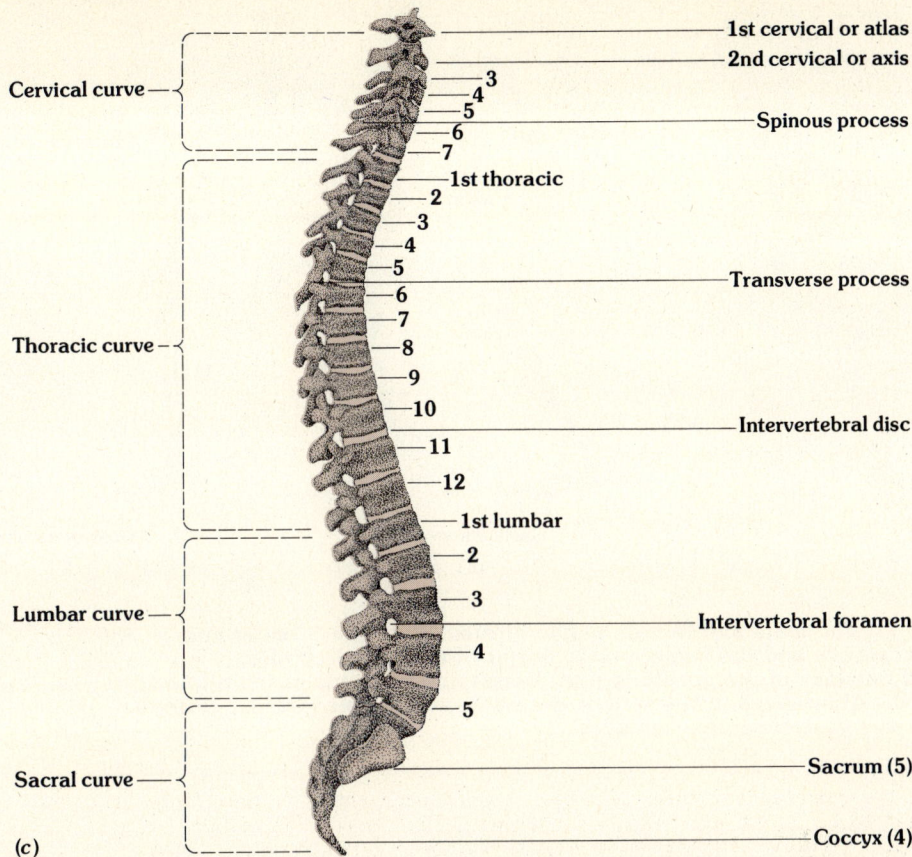

dog's ability to catch its tail or to lick the inside of its hind leg some time when you have access to a good osteopath), this shape enhances our abilities to walk erect by positioning the feet, pelvis, and skull in a vertical line. Thus, the act of remaining erect becomes, for humans, less of the balancing act that it is for the apes, and the amount of muscular energy needed to remain erect is also considerably less. Moreover, our skull is centered on our spine, rather than being thrust forward as is the case in the apes and in most other mammals. Thus, we save on the muscular effort needed to keep the head erect (note the huge necks of such animals as cows in comparison with our own.)

**The pelvis** There are also considerable differences in the shape of our pelvis as compared with that of the gorilla. This difference is not of the same direct significance as the difference in spinal curves, but it is of enormous importance insofar as

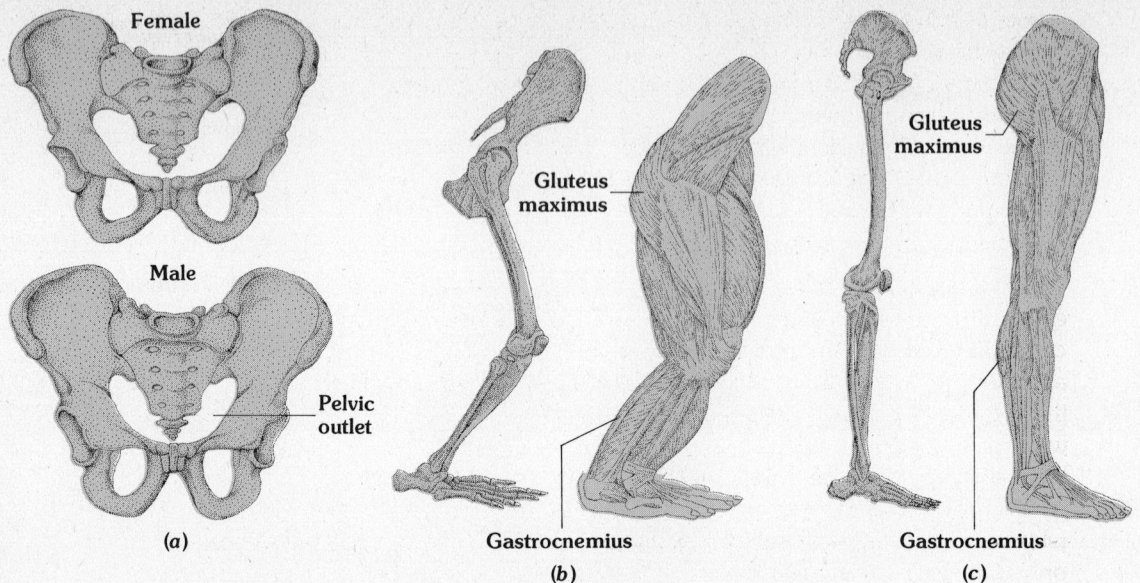

**Fig. 3.21** (*a*) The male and female pelvic girdles. Note the broader and flatter appearance and the larger pelvic outlet of the female girdle. (*b*) Bones and muscles of the gorilla leg. (*c*) Bones and muscles of the human leg. Note the differences in the size of the respective pelvic girdles, gluteus maximus, and gastrocnemius muscles.

it mirrors differences in muscular development in this region. The shorter, dish-shaped human pelvis allows for the development of a very large, but relatively compact **gluteus maximus** muscle (Fig. 3.21), a muscle which is rather small and elongate in the gorilla. The significance of this difference is perhaps most easily recognized by an example. Stand on your left leg, with your right leg extended in front of you, as if you had just kicked a football. Now bring the right leg back, keeping it straight, until the toes of your right foot are pointed well behind you. At this point, you should be able to feel your right gluteus maximus muscle contracting. The function of the gluteus maximus muscle, then, is to pull back strongly on the femur, which, of course, is a necessary movement in walking up stairs or in running. (Ordinary walking is accomplished without much effort by the gluteus maximus.) Lacking a well-defined gluteus maximus muscle, gorillas are incapable of effective bipedal walking, let alone running.

**The foot**  There are obvious and significant differences between the human foot and that of an ape (Fig. 3.22), and these are all correlated with the development of bipedality. The three most significant features are:

1 The development of the ball of the foot, which really amounts to an alignment of the toes, with a corresponding loss of opposability in the big toe (the big toe in apes is similar in its relationship to the other toes as the thumb is to the fingers);

2 The development of a foot arch, which allows a rotation forward on the ball of the foot during walking or running;

3 The increased development and refinement of a heel, which provides a site for insertion of the greatly enlarged **gastrocnemius** (calf) muscle. This muscle functions to lift the body up and forward, causing the weight of the body to be shifted forward onto the ball of the foot, as occurs during walking and running.

In essence, the human foot is part way between the largely unmodified and even primitive mammalian foot as represented by the feet of apes, and the highly specialized feet of such animals as dogs and horses, which are characterized by a lengthening of the foot to the point that, in the horse, only a single toe is in contact with the ground (the others having been all but lost). The effect in these species, both of which are highly modified for running, is to increase the relative length of the entire limb, and to utilize the ankle joint as fully as the hip or knee joint (Fig. 3.23)—that is, the leg as a whole has greater potential in terms of length of stride. Thus, whereas the distance from the ball of the foot to the ankle in ourselves represents only one-sixth of the total length of our hind limb, this fraction is one-fourth in a dog and one-third in a horse. It is partially because of the rather primitive structure of our hind limb that we are the relatively slow-moving animals that we are. [To be

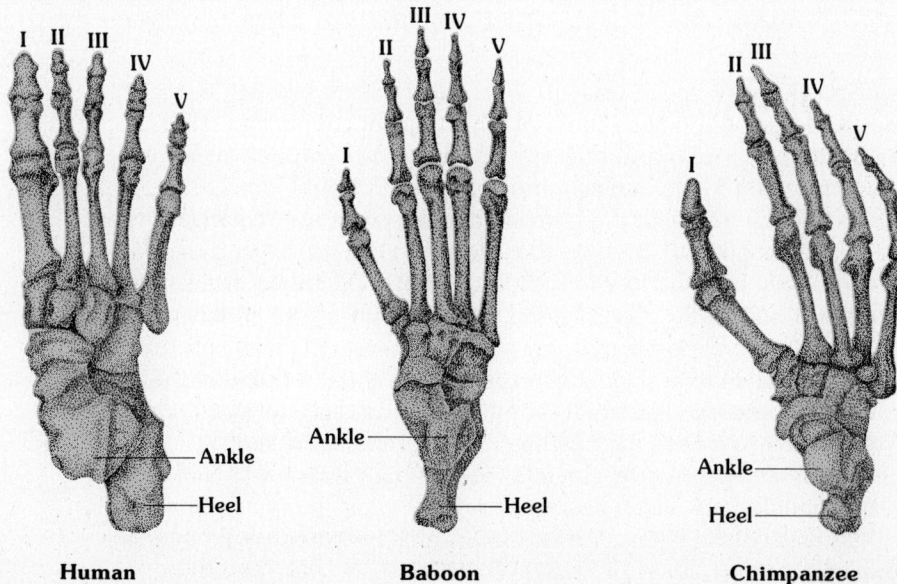

Fig. 3.22
The foot of a human contrasted with that of a baboon and chimpanzee.

Human

Baboon

Chimpanzee

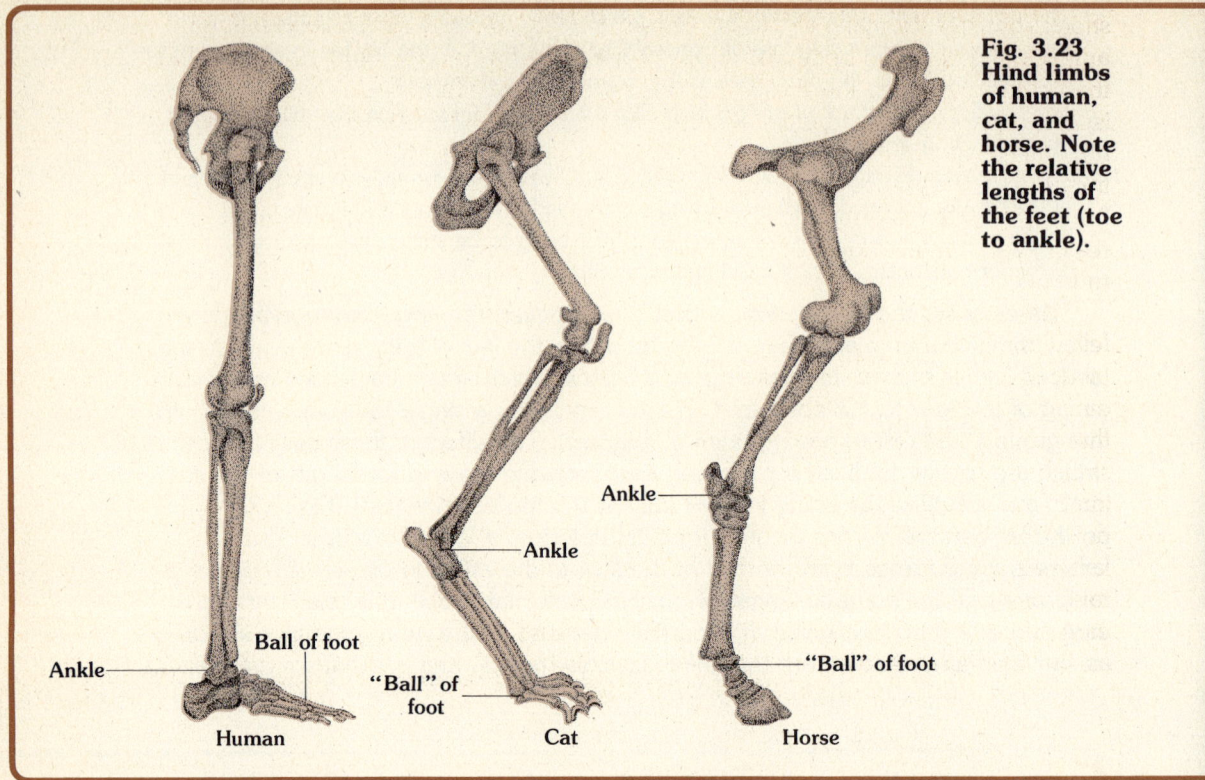

**Fig. 3.23 Hind limbs of human, cat, and horse. Note the relative lengths of the feet (toe to ankle).**

Ankle

Ankle

Ball of foot

Ankle

"Ball" of foot

Ankle

"Ball" of foot

Human

Cat

Horse

sure, much of our potential for speed was lost when we became bipedal, but such birds as the roadrunner and the ostrich (both of which are of course bipedal) are much faster runners than we are—and both have utilized the device of an elongated foot and a consequent "high" ankle joint.]

**The shoulder**  Modifications of the human shoulder were not required by the evolution of bipedal locomotion but, because the forelimbs were no longer directly involved in locomotion, they became free for other functions. Of course, the facility of the human hand has frequently been stressed [overstressed, as it happens—although the degree of thumb opposability is greater in humans than in other primates, the magnitude of structural change (Fig. 3.24) is much less than in many of the other structures we have already discussed], but there are modifications of the arm and shoulder which greatly increase the potential utility of the hand itself.

To begin, note our well-developed **clavicle** (collarbone), which assists in the maintenance of the shoulder in a lateral position; it is largely lost in many of the rapidly running animals, such as dogs and horses, because its presence would interfere with their locomotion.

To illustrate this point, consider the following. Some of you may, at one time, have broken your clavicle. Because of its rigidity, bone is not a particularly good shock absorber. In trying to protect oneself from a sideways fall by thrusting the arm out ahead of the body, the shock of the fall is transmitted up the arm and has the tendency to push the shoulder toward the midline of the body. This movement is, of course, prevented by the clavicle—but if the strain is sufficient, the clavicle may break. Rapidly running animals catch their entire weight on their forelimbs as they run; the presence of clavicles would simply increase the likelihood of their breaking. Contracted muscle is a much better shock absorber, and this is one reason why the shoulders of most large, rapidly moving mammals are so well muscled.

Another significant difference between our shoulders and those of most of our fellow mammals (including many primates) is the presence of the shoulder blade (**scapula**) lying almost flat along the back, such that the edges are pointed toward each other (Fig. 3.1). Check your pet dog or cat and note that the scapulae of these animals are along the side of the rib cage, such that the edges point almost straight up, rather than toward each other. Again, in these and other running animals, the scapulae are arranged so as to assist in shock absorption by providing a point of attachment for the shock-absorbing muscles (*see box*). In humans, the scapular rotations allow for much greater lateral movement of the arms. Stand up and raise your arms, elbows straight, thumbs up, with the palms of the hands facing each other. Now, holding the arms at shoulder level, swing them backward as far as they will go. Assuming that you are in reasonably good shape, you will find that

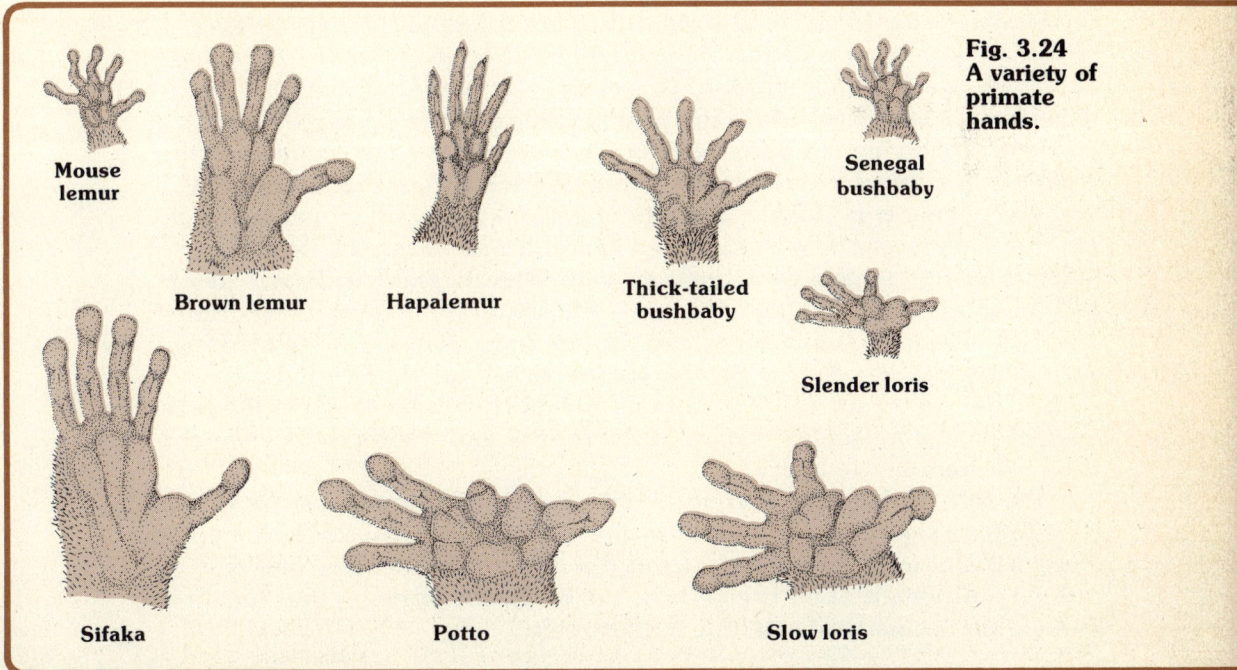

Fig. 3.24
A variety of primate hands.

Mouse lemur

Brown lemur

Hapalemur

Thick-tailed bushbaby

Senegal bushbaby

Slender loris

Sifaka

Potto

Slow loris

**Stand on a chair and jump off to the floor. Unless you are totally inept, you will land with bent knees and on the balls of your feet, taking advantage of the enormous strength and size—and, hence, shock-absorbing abilities—of the calf and thigh muscles. Try doing the same thing, but this time do it stiff-legged. The chances are extremely good that the impact of even this slight drop will be sufficient, as the shock of contact passes largely undiminished up the body skeleton, to create a first-class headache and perhaps shatter a few teeth.**

**Shock absorption**

    **The point is simply that providing the force necessary for movement is only the beginning of the role of the muscular system, as these very muscles also serve to absorb the shock of landing, as an animal jumps or runs. The skeletal system, lacking compressibility in any but a minor degree, is incapable of serving this role adequately, as the above demonstration indicates.**

the angle formed by your arms exceeds 180 degrees. Should you try that same movement on your pet dog or cat, you will hear a snapping sound once the angle exceeds about 140 degrees. The point is simply that the rotation of the scapulae in humans has greatly increased the mobility of the arm, which, of course, helps maximize the manipulative abilities of the hand.

## SUMMARY

**There are more than 200 bones and more than 600 separate muscles in the human body. In size and weight they are the dominant systems of the body. Indeed, along with the skin, the bones and muscles *are* the body, at least as we generally think of it. Yet in spite of stating the obvious, we nevertheless lose sight of it. The role played by these systems is precisely that of giving our body form and movement. We are apt to describe a person in a deep coma as a vegetable. We infer from that statement that the person is no longer capable of normal functioning, yet we might find his circulation, respiration, and excretion all excellent, and his nervous system largely functional as well. What is really different about this person is that he does not *move*—his skeletal muscles do not function and he does not communicate. In this light, the vegetable analogy, though unkind, is rather accurate, for the single most profound difference between multicellular animals and multicellular plants is the presence of a huge muscular system in the former and the total absence of such a system in the latter.**

    **The role of the skin, skeleton, and muscles in homeostasis, a role we tend to lose sight of, is to maintain the physical integrity of the organism (support) and to facilitate movement (locomotion). Because food is a prime requirement of all the cells of the body, the systems that locate and obtain it clearly play a homeostatic role.**

*A snake handler can "milk" a poisonous snake and drink the venom with impunity. Yet if the same snake were to bite him, the handler might die. Why the difference? Why can children eat garden soil without ill effects but suffer infection if the same soil enters a cut or scrape?*

*King Nebuchadnezzar II of Babylonia reportedly went mad and spent his time going about on all fours, eating grass. Had you been his advisor in nutrition, would you have recommended a dietary supplement? Would vitamin pills be enough? In short, could he survive on grass alone?*

*Why do victims of radiation poisoning frequently die of massive ulceration of the intestine, even though they ingested no radioactive material?*

*Our digestive system is one of our most interesting, yet most abused, systems. It provides a doorway to the body that is far more substantial than its delicate structure would suggest, yet drug manufacturers make millions of dollars every year selling medicines for indigestion and constipation. Are these medicines really necessary?*

chapter **IV**

# chapter IV
# Obtaining nutrients—
# the digestive system

**INTRODUCTION**   In multicellular organisms, such as ourselves, it is the function of the digestive system (Fig. 4.1) to provide the individual cells of the body with molecules small enough and structurally simple enough to be utilized by the cells for their own purposes (maintenance and growth). To achieve this end, the digestive system functions first to break large molecules into small ones, and second to absorb these small molecules selectively and pass them on to the circulatory system (Fig. 4.2).

Although the role of the system, as just elucidated, may seem straightforward enough, the fact is that there are fundamental misconceptions about what the digestive system actually does. In this chapter, we shall examine not only the specific workings of the system but also the larger question of how the digestive system serves the needs of all the cells of the body. On this latter point, you may recall from Chapter 2 that organ systems tend not to usurp the traditional activities carried on by unicellular organisms. Rather, they perform a new level of activities beyond the capabilities of an individual cell. This concept is particularly well exemplified by the digestive system. For example, unicellular organisms are, with few exceptions, limited in their choice of food items to things of a single cell or smaller, and in every case the food item is smaller than the unicellular organism that devours it. Our digestive system allows us to live on food of a much larger size, including organisms that greatly exceed the size of our own body.

To begin with, it is important to recognize that the digestive system, for all its loops and expansions, is really nothing more than a tunnel through the organism. As such, the contents of the digestive tract are not inside the organism. This may well seem paradoxical—how can a hamburger that you have just eaten not be inside you? The answer is that, strictly speaking, in order for something to be considered inside the body, that something must first pass across a cell membrane, and as long as that hamburger remains within the digestive tract itself, it has not met this criterion.

This may seem a needless semantic exercise, but, as you shall soon see, it is much more than that. The reason that we can eat food contaminated with bacteria and not become ill, or the reason that a showman can "milk" the venom from a rattlesnake and drink it without ill effects, is because those materials are not inside the body. Digestive enzymes may destroy them or they may be passed out with the feces—it does not really matter because until they cross a cell membrane, they pose no threat to the body. However, were that same showman to attempt to drink rattlesnake venom with cracked lips or a hemorrhaging ulcer, the way would be clear for the venom to be introduced inside the body (i.e., into the bloodstream) and the effects would be as dramatic as if the snake had bitten him.

The digestive system, then, is a highly specialized tube in which complex molecules too large to cross a cell membrane and/or too large for a cell to be able to use are first broken into smaller pieces by various secretions poured into the tube. These resulting small molecules are then selectively absorbed and transported, via the circulatory system, to the individual cells of the body. It is important to note that radical chemical transformations of these food items are not taking place in the digestive system—such transformations remain the province of the individual cells,

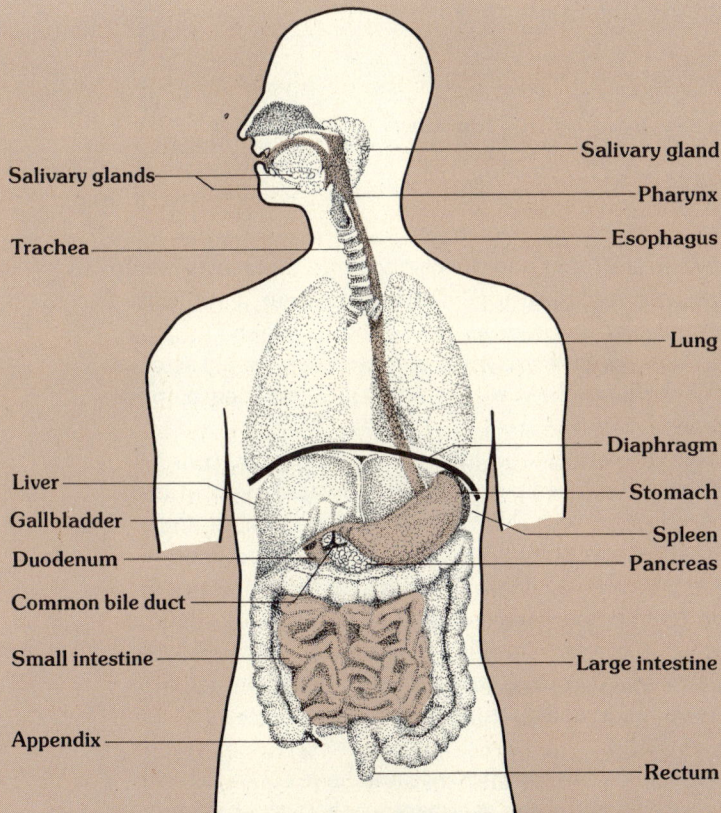

**Fig. 4.1
The human
digestive system.**

Salivary gland
Pharynx
Esophagus

Lung

Diaphragm
Stomach
Spleen
Pancreas

Large intestine

Rectum

Salivary glands
Trachea

Liver
Gallbladder
Duodenum
Common bile duct
Small intestine
Appendix

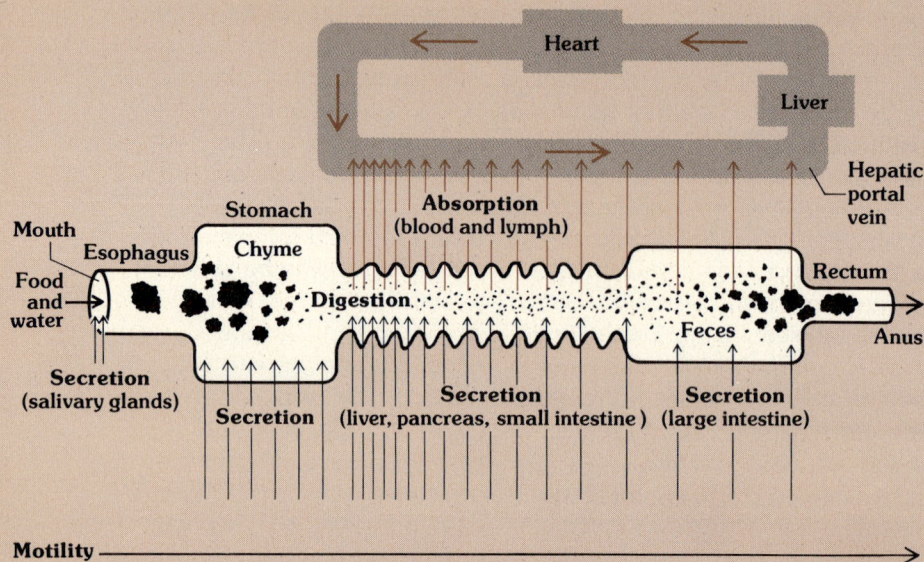

**Fig. 4.2  Diagrammatic representation of the functions of the various organs of the digestive system.**

for the most part—rather, the process is simply one of splitting large molecules (such as proteins) into their component units (amino acids). Thus, just as our teeth function to break up food items mechanically, in order to expose more surface area to the digestive enzymes, so, too, do the digestive enzymes operate like a highly selective set of teeth in that they break the long-chain macromolecules into their component simple small organic molecules.

In ourselves, the digestive system consists of a tube of approximately 6 m (20 ft) in length, running from mouth to anus. Oddly enough, the actual length of the digestive tract is a matter of some dispute, with different authorities quoting anywhere from 3 to 7.5 m (10 to 25 ft). Apparently, the reason for the uncertainty stems from the fact that measurements are based largely on cadavers, as it is not fitting to operate on a living person for the sole purpose of measuring his digestive tract; because the muscles in the walls of the tract relax at death, a substantial error is introduced if that length is automatically applied to the living person. In actuality, it is likely that there is considerable variation in the length of the digestive tracts of different individuals.

Various parts of the tract are specialized in specific ways and have come to be labeled as separate structures—the mouth, pharynx, esophagus, stomach, small intestine, and large intestine—and attached to the tract are various other organs which contribute to digestion—the salivary glands, liver, and pancreas being the most notable. We shall deal with each of these in turn, and examine the role of each in the digestive process.

**DIETARY NECESSITIES**   The digestive system is geared to digest most of the typical food items in the human diet and to absorb sufficient quantities of those things which the cells of the body require to function. Although all the plant and animal species, taken collectively, have a remarkable range of capacities for converting one molecule to another, the capabilities of such chemical transformations are somewhat more limited for any individual species. Hence, whereas green plants can convert carbon dioxide and water into glucose, or manufacture amino acids *de novo,* animals possess no such capabilities.

The implication in the previous statement is that, compared with plant cells, the cells of our bodies are somewhat lacking, given their inability to perform these same chemical transformations. This lack does not stem from any inherent weakness in animal cells as opposed to plant cells, but rather it reflects the nature of the environment in which the cells exist. We would gain no particular advantage if we had the ability to manufacture glucose from carbon dioxide and water because, since life first originated, our diet has been rich in preformed glucose. It is much better, therefore, that we use the cell's capacity for synthesis (which, of course, is not infinite) for other purposes. Thus, the nature of the chemical transformations performed by our cells has evolved as a function of our diet.

This point is made particularly clear when we examine the list of required dietary components (Table 4.1). These include four major categories: (1) water, (2) minerals, (3) an energy source, and (4) vitamins.

Because life is water-based, it stands to reason that water is a necessary dietary component. However, water is also released during the breakdown of food, and some organisms, notably certain desert mammals, require no free water in their diet but can survive on their **metabolic water.** Such organisms have evolved very sophisticated methods of conserving water loss by limiting evaporation from the lungs and by excreting a very concentrated urine. In most environments, however, water is generally plentiful and it is not surprising that it is required in some degree in the diets of most organisms, not excluding humans.

As we discussed in Chapter 2, a certain number of elements beyond those generally found in organic molecules are necessary for life. For the most part, these are elements that are relatively common in the earth's crust, which makes them easily accessible to plants. From that point, they are passed from one organism to another, as one organism ingests another, and essentially the same set of elements is required by all organisms, both plant and animal. Occasionally, our diet may be lacking in one or more of these (iodine being the most common; see Chapter 9), but typically such a lack reflects a poor dietary selection on our part, as may result from a reliance on highly processed foods (e.g., white sugar, white flour, and so forth).

Carbohydrates, proteins, and lipids all contribute to the cellular production of energy, and within broad limits these compounds are interconvertible by the human cell. Certain fatty acids are required to be present fully formed in the diet, however,

**Table 4.1**
Essential components of human diet

| | Typical daily requirements (adult human) | Typical sources | Examples of action |
|---|---|---|---|
| **Water** | 2000 ml | All food and drink | Universal solvent in biological systems |
| **Minerals** | | | |
| Sodium | } Adequate amount in ordinary diet | } Most foods | } Essential for nerve cell function |
| Chlorine | | | |
| Potassium | | | |
| Iron | 10 mg | Liver; spinach | Oxygen transport, as part of hemoglobin |
| Iodine | trace | Sea foods; iodized salt | Metabolism control, as part of the thyroid hormone thyroxin |
| Calcium | 0.5 g | Milk, cheese | Bone formation; nerve and muscle action |
| Phosphorus | 1 g | Milk, cheese | Bone formation |
| Trace elements: magnesium, manganese, molybdenum, zinc, fluorine, copper | Adequate amounts in ordinary diet | | Often as part of enzymes |
| **Carbohydrates** | Roughly half the total caloric intake | Cereals (wheat, corn, etc.), milk, fruit | Energy; structural molecules |
| **Proteins** | 50–100 g | Meats; eggs | Provides essential *amino acids*, enzyme formation |
| **Fats** | 60 g | Meats; nuts | Energy; provides essential fatty acids |
| **Vitamins** | | | |
| Water-soluble | | | |
| B vitamins | | | |
| Thiamine ($B_1$) | 1 mg | Beans, nuts, yeast | Essential in cellular respiration |
| Riboflavin ($B_2$) | 2 mg | Liver, milk | Essential in cellular respiration |
| Pantothenic acid ($B_3$) | ? | Liver, yeast, peas | Essential in cellular respiration |
| Pyridoxine ($B_6$) | 1 mg | Wheat germ, milk, meat | Essential in cellular respiration |
| Niacin (nicotinic acid) | 20 mg | Meat, wheat germ, peanuts | Essential in cellular respiration |
| Folic acid | 0.5 mg | Green vegetables | Nucleic acid synthesis |
| $B_{12}$ | 0.003 mg | ? | Essential to red blood cell formation |
| C (Ascorbic acid) | 100 mg | Citrus fruits, berries | Connective tissue formation |
| Fat-soluble | | | |
| A | 2 mg | Yellow fruits and vegetables | Part of visual pigment, rhodopsin |
| D | 0.01 mg | Fish liver; formed in humans by action of sunlight on skin | Facilitates absorption of calcium by intestine |
| E (Tocopherol) | ? | Leafy vegetables | Required for male fertility |
| K | ? | Leafy vegetables | Essential for normal blood clotting |

as are about half the amino acids—for that reason, we could not long survive on an exclusively carbohydrate diet. Interestingly, even though most animals require certain amino acids in their diets (which means that they cannot be converted from other amino acids within the body), the specific ones that are required frequently are different among different species. As it happens, those which the cell lacks the capacity to synthesize tend to be precisely those found most abundantly in the typical diet of the species.

In a similar way, certain other complex molecules are required by most animals, often to assist in enzymatic functioning. We call these molecules **vitamins** (see Chapter 2), and just as for the required amino acids, vitamins differ among species, again primarily as a function of how common (or rare) they happen to be in the normal diet of the species.

The major point that emerges from an examination of dietary requirements is that there are limits to what kinds of things we can exclude from our diets. In our anthropocentric and rather egotistical way of looking at things, we frequently delude ourselves in terms of what we can do to our bodies and still survive. For example, most plants are low in certain of the required amino acids, which suggests that over evolutionary time humans came to include a certain proportion of animal protein in their diet. Therefore, vegetarianism is not simply a moral decision but must also involve a recognition of the dietary necessity of including a sufficiently diverse protein intake to ensure that one obtains all the essential amino acids. Vegetarians who eat eggs or dairy products, or both, have an easier time of things, because these foods are rich in precisely those amino acids that are generally lacking in most plants. To be fair, however, it is necessary to point out that many total vegetarians live completely healthy lives—it is not an impossible diet, just a difficult one.

There is also increasing evidence that a diet high in animal products—especially "red meat" (as opposed to fish and fowl)—also has its hazards, as animal fats apparently promote **cardiovascular** diseases (diseases of the heart and blood vessels), and there is also evidence that large amounts of animal protein correlate positively with cancer of the large intestine.

The phrase, "moderation in all things," pertains very well to the human diet.

**THE MOUTH**   We tend to take the role of the mouth in the digestive process somewhat for granted, but in fact this attitude stems largely from our anthropocentric point of view. Even though the parents of many teenagers are convinced that their children live exclusively on hamburgers and French fries, the fact is that, as species go, we have a very complex diet, not only in terms of the origin of the food (both plant and animal), but also in terms of all the various types of things we eat. Carnivores, such as your pet dog who has lived its entire life eating the same brand of dog food, or herbivores, such as cattle, which eat grass and hay seemingly without end, have a simpler time of things.

To begin with, our mouth monitors what we eat. We have taste buds on our tongue which respond to sweet, salt, sour, and bitter (see box). It was discovered relatively recently that we also have taste buds throughout the mouth, most notably on the roof of the mouth. This finding accounts for the perennial complaint of individu-

The four categories of tastes do not represent properties of the food we eat but rather the manner in which our taste buds respond to molecules of particular sizes and shapes. For example, sweetness is not an inherent property of the molecule for table sugar (sucrose)—rather, one group of taste buds is stimulated by its presence and our brain interprets this stimulation as "sweet." Because we require large amounts of glucose, it is only to be expected that we should have a sensory apparatus that responds favorably to food items possessing glucose, and we tend to think of sweet food items as being "good to eat." The fact that the four taste categories are bodily responses rather than inherent properties of the molecules is well exemplified by the way in which certain artificial sweeteners, such as saccharine, which possess no food energy and which the body cannot use, can "fool" our taste buds and be transformed into something "good to eat."

**Fooling our taste buds— artificial sweeteners**

als with false teeth (which seal off the roof of the mouth) that their food doesn't "taste as good" as it did when they had their own teeth.

As you are well aware, we may reject a potential food item because it registers too strongly (unsweetened lemon juice) or not strongly enough (unsalted oatmeal) in one of the four taste categories. Cows and dogs, with their nonvarying diets, don't have to worry about such subtleties. Moreover, we keep the food in our mouths long enough to assay it with our nose. Food odors pass up into the nasal cavities through the opening where the nose connects with the mouth—and because these odors are gases, and because hot foods convert more rapidly to gases than do cold foods, we tend to think that hot foods "taste" better than cold foods. Similarly, the reason that most food seems so bland to us when we have a cold is that our sense of smell is impaired. Hence, we say we are "tasting" our food, but we are really also smelling it. Again, animals that live on simpler diets have little need for so refined a monitoring system.

A second function of the mouth is the cutting and grinding of food by our teeth. We take it for granted that teeth are for chewing food. Yet if you have ever watched your dog demolish a heaping bowl of dog food in 29 sec, you must be aware that it is hardly chewing each mouthful 30 times. In contrast, cows never seem to stop chewing. Why the difference? The protein in the food of carnivores is relatively easily digested—they use their teeth largely for killing and ripping off hunks of food which they then swallow essentially intact. The incisors of most carnivores are small and of little use—it is the canine (eye) teeth which are large and function as described above. The molars of carnivores are also large and are used to crush bone—after all, there is a limit to what the digestive enzymes can do.

In contrast, plants have cell walls of cellulose surrounding the cytoplasm, and cellulose is an extremely difficult compound to break down. Thus, herbivores such as cows spend a considerable time mechanically fragmenting these cell walls with their well-developed molars. As grass rarely puts up much of a struggle, cows and other herbivores have dispensed with the canine teeth, although most have a complete set of incisors (cows are an exception) for nipping off the vegetation.

Because of our mixed diet, our set of teeth is complete and relatively unspecialized. We can exert over 20 kg (50 lb) of force with our incisors, which we use, for example, in taking a bite out of an apple; we can exert over 90 kg (200 lb) of force with our molars, which we use to grind food. Our canine teeth are present, but are no larger than any of the other teeth (although they generally are deeper rooted, perhaps a remnant of our ancestors who had larger canines, as do many other modern primates), and they function essentially as an additional set of incisors. Thus, just as the shape of the jaw exemplifies the type of food eaten, (see Chapter 3) so, too, does the size and number of teeth indicate the nature of food specialization among species.

A third function of the mouth is to transport the food along the rest of the digestive tract. To a certain extent, gravity helps out here (if you've ever tried swallowing anything while standing on your head, imagine the problem a giraffe has in drinking water). However, much of what we eat is coarse or dry or both, and the mucus secretions of glands lining the mouth and of the salivary glands in particular (see Fig. 4.1) function in a lubricating capacity to allow food to slide down to the stomach with relative ease. These glands produce from 1 to 1.5 liters of secretion daily, serving primarily as a lubricant.

Although the function of salivary secretions is limited exclusively to lubrication in such animals as the dog, in humans, with a high carbohydrate intake and relatively long chewing time, the salivary glands also produce an enzyme, **salivary amylase** (from the Latin *amylum,* meaning "starch"), which has the capacity to split the polysaccharide starch into its constituent monosaccharide unit, glucose. The effects of this enzyme can be demonstrated by chewing a soda cracker for 30 sec or so, after which time it will begin to taste sweet. As we do not normally chew soda crackers for 30 sec, the importance of this enzyme has long been questioned, especially because it is rendered inoperative in the acidic environment of the stomach. However, not only may it be reactivated in the more alkaline small intestine, but, during a full meal, it may remain protected (hence, active) in the stomach for a considerable period of time, because the better part of an hour is required before the acidic secretions of the stomach are completely mixed with the meal.

Swallowing is a complex matter for the obvious reason that, in the region of the pharynx, the pathways for food and air cross over. Swallowing air is no particular problem, for it is quickly belched back up again. However, "breathing" food or water can pose an immediate threat to life, especially if a piece of food lodges in such a way as to seal off the passage to the lungs. Fortunately, the coughing reflex is normally effective to clear the airway. This reflex operates poorly, at best, in one who is unconscious, which is why it is recommended that an unconscious person never be given liquids.

Swallowing normally happens so quickly that we are generally unaware of a momentary interruption in our breathing pattern. Figure 4.3 illustrates the events involved. The tongue is arched backward, thereby propelling the food into the

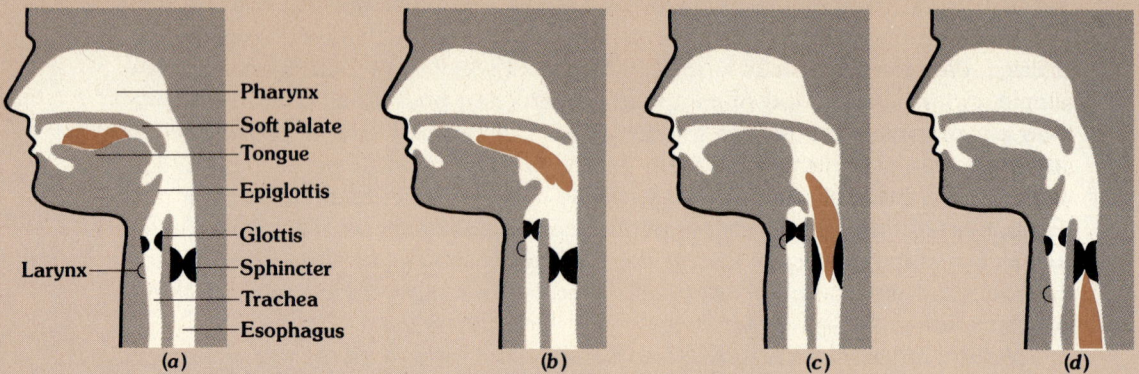

Fig. 4.3   **The events occurring during swallowing. Note the rise and fall of the larynx.**

**pharynx.** At the same time, the **soft palate** is raised, to provide greater accessibility to the pharynx and also to prevent food from backing up into the nasal passages. Then the **larynx** rises, forcing the **glottis** (entrance to the airway) closed. [You can detect this by feeling your larynx (Adam's apple) bob as you swallow.] From the pharynx, the food enters the **esophagus.**

**THE ESOPHAGUS**   The esophagus is a straight tube, approximately 25 cm (10 in.) long and 2.5 cm (1 in.) in diameter, which connects the pharynx to the stomach. It has no digestive function, as it secretes no enzymes and absorbs no materials, but rather functions solely to convey food to the stomach, the first of the specialized swellings of the digestive tract.

The muscle layer that lines the wall of the esophagus is of two types. For approximately the first third, the muscle layer is skeletal muscle and, as such, is subject to conscious control. This muscle layer contracts during the act of swallowing to aid in moving food from the pharynx. The last two-thirds of the esophagus are lined with visceral muscle, and as is typical for visceral muscle, contraction occurs spontaneously and without conscious control. The nature of this contraction is wavelike [a phenomenon known as **peristalsis** (Fig. 4.7)], with a wave requiring approximately 9 sec to move the length of the esophagus.

The presence of visceral muscle along most of the length of the esophagus accounts for our general lack of success in "swallowing" food that has become lodged in the esophagus. The general recourse is to drink some fluid, in order to wash the food down, but this solution is augmented by the fact that any distension of the esophagus promotes more powerful peristaltic waves, as well as triggering the release of increased amounts of saliva.

The esophagus traverses the diaphragm (Fig. 4.1) and continues for about 3 cm (1 in.) before entering the stomach. This last segment of the esophagus is normally closed (i.e., it acts as a **sphincter**) to prevent regurgitation of food from the stomach back into the esophagus, and opens only during the act of swallowing to allow food to pass into the stomach. The sphincter arrangement is needed because the contents of the abdominal cavity (below the diaphragm) are at a higher pressure than are the contents of the chest cavity (above the diaphragm). Therefore, without a sphincter, there would be a tendency for food to be pushed back into the esophagus. This sometimes occurs in any case, and results in a condition called **heartburn.** This condition, which takes its totally erroneous name from the fact that the esophagus enters the stomach at roughly the level of the heart, involves the irritation of the sensitive tissues of the lower esophagus by the highly acidic secretions of the stomach (*see box*).

**THE STOMACH**   It is in the stomach that the first true digestive processes take place, for it is here that the food first stops for a time. Thus, we may eat a meal in 15 or 20 min, but that food will remain in the stomach for about 4 hr (the precise time is a function of how much was eaten, what type of food was consumed, and how recently the meal was eaten relative to the previous meal).

The stomach performs four major tasks:

1  It regulates the flow of food into the small intestine—without a stomach, food would enter the small intestine too rapidly for efficient digestion;
2  It puts the food into solution, which increases the efficiency of subsequently released digestive enzymes.
3  It produces an enzyme which initiates the process of protein splitting;
4  It absorbs certain molecules and passes them into the bloodstream.

We shall examine each of these in detail.

**Motility**   The term **motility** refers to movement, in this case, of the stomach. As in the esophagus, the smooth muscle layer of the stomach wall contracts in wavelike fashion, although in a somewhat more complex manner (Fig. 4.4). These peristaltic waves have the effect not only of passing food along to the small intestine,

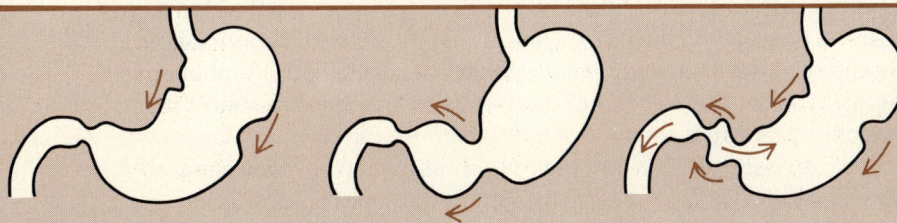

**Fig. 4.4   A wave of stomach contractions.**

**Infants may have problems both at the inlet and outlet of the stomach. They have essentially no portion of the esophagus which passes beyond the level of the diaphragm. Consequently, the sphincter is much less effective, and, as a consequence, regurgitation is much more common in infants than in adults (as any baby burper can attest).**

**A more serious problem sometimes occurs when the pyloric sphincter malfunctions. A failure to open regularly or sufficiently results in the condition called pyloric stenosis. In severe cases, essentially no food passes to the small intestine, and regurgitation and weight loss result. Typically, the infant outgrows this condition, but occasionally surgery is necessary.**

but also of mixing the food with the secretions of the stomach. Contraction waves begin rather slowly after a meal, but gradually increase in strength after the first 30 to 60 min (Fig. 4.5). The waves of contraction increase in strength and become more rapid at the end of the stomach closest to the small intestine, with the result that a powerful wave propels food in the stomach towards the small intestine. The pressure of this contraction is sufficient momentarily to open the **pyloric sphincter,** which separates the stomach from the small intestine, but before much food can escape, the wave of contraction reaches the end of the stomach and closes the sphincter again, causing most of the food to be churned back into the stomach.

A number of factors influence the strength and frequency of these stomach contractions. They tend to be stronger if the stomach contains large amounts of food, but weaker if the small intestine is distended or contains fat. This information is relayed back from the small intestine both by nerves and by hormones. Emotional factors are also involved, with contractions generally being stronger in one who is angry, and diminished in one sad or fearful.

The stomach also has the capacity to propel food back up the esophagus under certain conditions, a fact that exemplifies that motility need not be unidirectional.

**Vomiting** A discussion of this topic may at first strike you as needlessly perverse, given that our society regards it as among the most revolting of bodily functions, but the fact is that vomiting is a very important defense mechanism of the body. The complex sensory receptors (eyes, nose, and mouth) associated with assaying food quality are not infallible, and, all too frequently, toxic materials or **pathogenic** (disease-causing) organisms may inadvertently be swallowed. If the body had no recourse but to allow these materials to remain in the digestive tract for the 24 hr or more which it typically takes food to pass from mouth to anus, the length of time the absorptive surface of the body (i.e., the intestinal lining) was exposed might be sufficient to cause death.

Vomiting involves the relaxation of the sphincter at the end of the esophagus, coupled with a violent contraction of the diaphragm and abdominal muscles. These events are controlled by stimuli passing down the **vagus** nerve from a part of the brainstem called the **medulla** (see Fig. 10.18 and Chapter 10).

A variety of stimuli may result in vomiting. Stimulation in the back of the throat, excessive distension of the stomach, or intense pain all typically result in vomiting. Dizziness or motion sickness (e.g., seasickness) may cause vomiting because of reactions in the **semicircular canals** (see Chapter 10). Even more interesting is the fact that certain odors or sights can trigger vomiting (paradoxically, these include the odor of vomit or the sight of someone vomiting, conditions that can give rise to a biological application of the famous "domino theory"). In these instances, it is the conscious mind which is perceiving these stimuli, not the involuntary action of the medulla acting autonomously. What this would seem to indicate is an instance where, in humans, the thought centers of the brain have come to control the reflex centers of the brain (brainstem), a development that is discussed at length in Chapter 10.

Despite the fact that vomiting is primarily a defense mechanism, excessive vomiting can itself pose a threat to the body. What is being regurgitated are, for the most part, the contents of the stomach, which are highly acidic. A constant loss of this acid can upset the delicate pH balance of the body, and changes in the pH of the blood have disastrous consequences on the activity of enzymes (Chapter 2). This acidic fluid is also deleterious to the lining of the esophagus.

**Secretions**  The principle digestive enzyme produced by the stomach is **pepsin,** which functions to split proteins into small chains of amino acids called **polypeptides** (hence the name of the enzyme). Given that so much of a cell consists of protein, it is interesting to note how the cells of the stomach solve the problem of avoiding being digested by the enzyme they are secreting.

The dilemma faced by the stomach cells would appear to be the same as that faced by ancient alchemists (and modern science fiction writers) who purportedly spent most of their lives trying to perfect a universal solvent. However, exactly what type of container this solvent was to be stored in was never made clear.

The stomach employs four solutions. First, the enzyme is secreted in an inac-

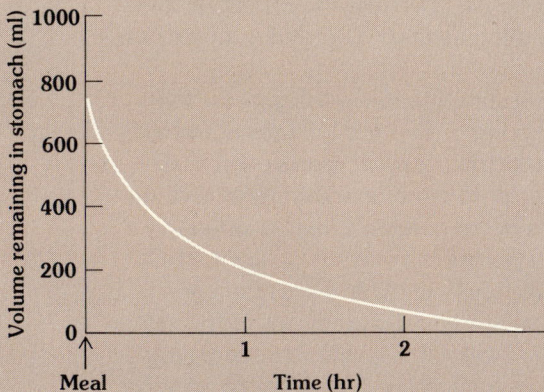

Fig. 4.5
Graphic representation showing emptying of the stomach following a typical meal. Meals high in fat may remain twice as long in the stomach.

tive form, **pepsinogen,** which is converted to pepsin (through the loss of a small end piece) only in the highly acidic medium of the stomach chamber. (Recall that protein structure may vary with the pH—and the activity of a given enzyme is generally highly dependent on the pH of the medium in which it is found.) The contents of most cells are typically somewhat alkaline, which prevents the activation of pepsinogen while it is still within the cell.

Second, another group of cells is responsible for producing large amounts of **mucus,** which coats the lining of the stomach in a layer about 1 mm thick, and thereby insulates the cells of the lining from attack by pepsin.

Third, the cells of the stomach, as are those throughout the digestive tract, are constantly being replaced at a high rate. In the stomach, a new layer of cells is formed every 1–3 days. Thus, even if damage does occur, the cells involved will soon be replaced.

Fourth, the cells of the stomach lining, like those of most epithelial tissues, do not have spaces between them, as do most internal cells, but are closely joined (a phenomenon called **tight junctions**), thereby preventing the leakage of digestive fluids.

In the stomach, conversion of pepsinogen to pepsin is achieved because a group of cells of the stomach lining produce hydrochloric acid (HCl) in sufficient quantities that the pH of the stomach is typically about 1.5. Because HCl is also a threat to the integrity of cells, the same types of precautions against destruction by pepsin also function for HCl.

**Ulcers** The fact remains, however, that, despite extensive safeguards, the lining of the stomach is all too frequently attacked and digested by pepsin and HCl. Such conditions are known as **ulcers** (sometimes called **peptic ulcers**) and may occur in the stomach **(gastric ulcers)** or in the first portion of the small intestine **(duodenal ulcers).** Gastric ulcers are about 10 times more common than duodenal ulcers, but collectively, ulcers occur in about 10 percent of adult Americans.

Exactly why some individuals develop ulcers and others do not is far from clear, but it does seem that certain individuals are genetically predisposed toward their development. With such a predisposition, a stressful life style or poor eating habits, among other things, may be sufficient to initiate the formation of an ulcer.

Usually, the pain of an ulcer is sufficiently intense (and distinctive enough) that it is treated relatively early. However, under certain circumstances, the ulcer may continue to grow. The lining of the stomach is eaten away until the underlying layer of blood vessels hemorrhage **(bleeding ulcer)** or even until the wall of the stomach is totally eaten through **(perforated ulcer).** In the latter case, food and digestive secretions are free to pass into the abdominal cavity, where infections rapidly occur. The fact that food can routinely be held in the stomach without danger yet can pose an imminent threat to life if allowed to enter the abdominal cavity is further proof that food within the digestive tract is actually **outside** the body. Such infections are lumped under the general term **peritonitis,** meaning an inflammation of the **peritoneum,** the lining of the abdominal cavity. In such instances, surgical treatment may be mandated.

Because different enzymes are active at different pH levels, one might wonder why it was that we came to use an enzyme requiring such an acidic medium, given the danger to ourselves when the system is malfunctioning. In theory, it would seem to be just as easy, and much less hazardous, to use an enzyme requiring a medium closer to pH values of the body cells (typically, mildly alkaline). Because the body can and does produce protein-digesting enzymes which are active under just such conditions, there is no theoretical objection to the presumption just raised.

The answer is that HCl has other functions beyond the simple conversion of pepsinogen to pepsin. The highly acidic medium of the stomach serves to denature most proteins, rendering them more easily digestible in the intestine. Perhaps even more important, this acidic secretion also denatures many of the proteins found in the cell walls of bacteria. In the process of denaturing these proteins, HCl serves to destroy or neutralize many bacteria before they can begin to multiply in the otherwise highly satisfactory environment of the digestive tract.

**Release of secretions**  What controls the production of stomach secretions? It would seem disadvantageous if this were a totally random process, given the potential danger to the cells of the stomach. It is rather obvious that the ideal situation would exist if secretions were timed so as to occur principally at mealtimes. As it happens, this is essentially the situation. Although there is a basal rate of secretion of about 0.5 ml/min, the secretion rate increases severalfold, to about 3 ml/min shortly after a meal.

This increase following a meal occurs because of the release of the hormone **gastrin** (Table 4.2), which stimulates the production of stomach secretions. Interestingly, gastrin not only acts on the stomach, but is in fact produced by the stomach, but, like all hormones, is dumped into the bloodstream and circulates throughout the body (see Fig. 4.6).

The release of gastrin is itself triggered by a number of factors, the most important of which is the presence of food in the stomach. Different foods have the capacity to cause differential amounts of gastrin to be released. For example, protein is more effective than carbohydrates. In addition, alcohol and caffeine are also powerful releasers of gastrin. Thus, the soup course or the predinner cocktail serve as aids to digestion, in that they maximize the output of gastrin, which in turn ensures a full release of the stomach digestive secretions. Similarly, alcohol or caffeine

**Table 4.2**
**Activities of the gastrointestinal hormones**

|  | Secretin | Cholecystokinin | Gastrin |
|---|---|---|---|
| *Secreted by:* | Duodenum | Duodenum | Stomach |
| Primary stimulus for hormone release | Acid in duodenum | Amino acids and fatty acids in duodenum | Proteins and poly-peptides in stomach; vagus nerve |
| *Effect on:* |  |  |  |
| Gastric motility | – | – | + |
| Gastric HCl secretion | – | – | ++ |
| Pancreatic secretion |  |  |  |
| Bicarbonate | ++ | + | + |
| Enzymes | + | ++ | + |
| Bile secretion | ++ | + | + |
| Gallbladder contraction | + | ++ | + |

Note the overlap in function among these hormones. Secretin and cholecystokinin are chemically very similar.

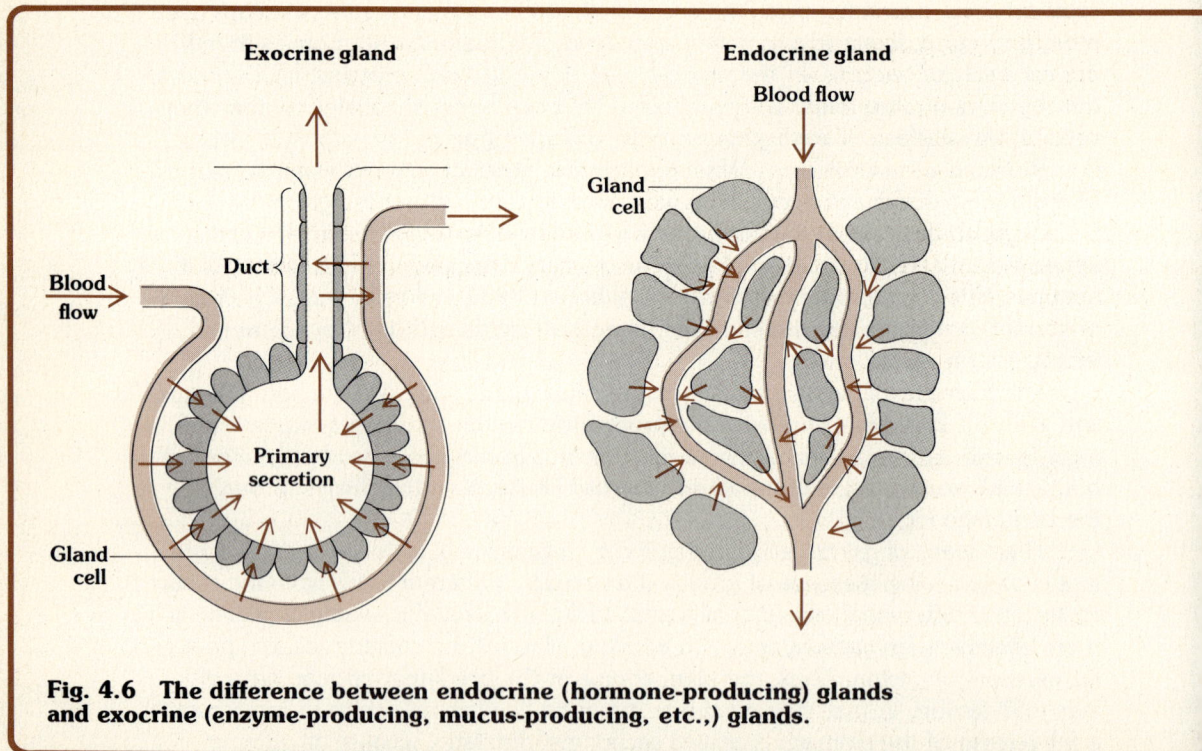

**Fig. 4.6   The difference between endocrine (hormone-producing) glands and exocrine (enzyme-producing, mucus-producing, etc.,) glands.**

ingested without an accompanying meal is counterproductive and potentially even harmful, because the consequence will be a sizable volume of secretions released without any food on which to act. It can readily be understood why individuals with stomach ulcers are advised not to drink such gastrin stimulants as alcohol and coffee, especially when not in conjunction with a regular meal. It does not take a great deal of imagination to speculate on the consequence to an ulcer of the release of stomach digestive secretions in a stomach which is empty, save for a single cup of coffee or a cocktail. In short, the efficiency of the digestive process would seem to be maximized by regular, and rather sizable, meals, as opposed to more sporadic eating habits.

A second factor favoring the release of gastrin is distension of the stomach. This seems an eminently logical cue, as a distended stomach would indicate a large meal, which in turn would require more digestive enzymes. Conversely, the presence of a high acid concentration in the stomach, indicating a large amount of gastric secretions, retards the output of gastrin. This, of course, is yet another example of negative feedback (see Chapter 2).

As it happens, only about 80 percent of the daily total of the $1\frac{1}{2}$ liters of stomach secretions is under the direct control of gastrin. Another 20 percent or so of the total is under nervous system control, specifically, the vagus nerve of the parasympathetic nervous system (Chapter 10). Therefore, mood and such senses as sight, smell, and taste are important factors in the production of stomach secretions. An atmosphere conducive to eating—soft lights, delectable odors, pleasant conversation—results in a complete release of the 20 percent factor. Conversely, an inimicable atmosphere—greasy food, standup restaurant, arguments—tends to shut down much or all of this fraction. The consequence of the latter situation is frequently indigestion, which can be attributed to an incomplete digestive process, brought about by an insufficient supply of stomach digestive secretions.

Gastrin also has other functions within the digestive system, one of which is to constrict the sphincter at the end of the esophagus. Because heartburn is a consequence of stomach contents leaking through this sphincter, it should not be surprising to learn that individuals who have had a portion of their stomach removed (and who, therefore, produce less gastrin than do individuals with intact stomachs) typically have much more frequent problems with heartburn.

**Uptake**   As was indicated earlier, a major role of the stomach is to convert food as it is ingested into a rather soupy material called **chyme.** This is achieved primarily by (1) the production of large amounts of **aqueous** (water-based) secretions coupled with rhythmical contractions and (2) the action of pepsin in breaking up proteins, the largest and therefore the most difficult molecules to digest. The stomach plays only a very minor role in the uptake of molecules (i.e., the transport of molecules from the cavity of the stomach across a cell membrane to the circulatory system), in large measure because anatomically it is not specialized to do so (unlike the small intestine, for example). That is, the surface area of the stomach lining is relatively small in comparison to the volume of the organ. A few molecules, however, are readily taken up by the stomach. These include alcohol and such weak acids as aspirin, among others. This capacity to absorb alcohol explains why it is that the effects of drinking, especially "on an empty stomach" (where there is no diluting effect from food) become apparent so quickly.

**THE SMALL INTESTINE**   With a length of about 3.6 m (12 ft), the small intestine is actually the longest single section of the digestive tract. The term "small" comes from its diameter, which, at about 4 cm ($1\frac{1}{2}$ in.), is only about one-half the diameter of the broader, but shorter, large intestine. Anatomists have rather arbitrarily divided the small intestine into three sections, but only the first, the C-shaped *duodenum*[1] (see Fig. 4.12), need be distinguished here.

More than any other section of the digestive tract, the small intestine is responsible for the breakdown of large molecules and the absorption of the resulting small ones. In the former task, the small intestine is aided by secretions from other organs, but the activity of these secretions occurs exclusively in the small intestine.

**Motility**   Movement through the small intestine is relatively slow, albeit variable, the prime stimulus for movement into the large intestine being the entry of the next meal into the duodenum. Movement of chyme is achieved by peristaltic contractions (Fig. 4.7), but, just as in the stomach, these contractions do not result in a constant unidirectional movement of chyme. Rather, there is a sloshing action, as contractions occur in a pattern called **segmentation** (Fig. 4.8). The consequences of such a pattern, of course, are a total mixing of the chyme and a complete exposure of it to the absorptive surface of the small intestine. In addition, the walls of the small intestine have fingerlike projections called **villi,** which move like giant cilia and aid in the mixing and passage of chyme through the small intestine.

Segmentation contractions occur in a definite pattern, such that they are somewhat more frequent near the beginning of the small intestine than near its end. As a consequence, there is a tendency for chyme to be moved gradually toward the large intestine.

The basic pattern of contractions is under the control of the autonomic nervous system (see Chapter 10) and, as such, can be influenced by the emotional state of the individual. Anger tends to promote motility, for example, whereas fear retards it.

---

[1] The name is derived from the Latin word for "twelve," a reference to the fact that the length of the duodenum was originally estimated as the width of 12 fingers (about 25 cm, or 10 in.).

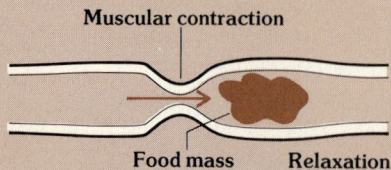

Muscular contraction

Food mass     Relaxation

**Fig. 4.7**
**Peristalsis, the force that moves food through the intestines by means of successive contractions of the intestinal walls.**

**Fig. 4.8**
**Segmentation in the small intestine.**

In addition, the mild distension of the small intestine that results from the entry of chyme from the stomach increases the intensity of contractions, but a massive distension, as might result from an injury, may lead to a complete cessation of contractions.

**Secretions** The small intestine produces a complete complement of digestive enzymes, in the sense that, collectively, they can break down all the classes of macromolecules—proteins, carbohydrates, lipids, and nucleic acids—into their component parts (Table 4.3). The small intestine produces about 2 liters of secretions every day, which includes a large mucus component, produced directly by glands, and a much smaller enzyme component, produced by cells sloughed off the lining of the small intestine. (Recall that the lining of the digestive system is completely replaced every 36 hr or so.)

The area of the human small intestine is much greater than might be predicted, given the outside dimensions of this organ. The large, fingerlike projections, the villi referred to earlier, account for a sizable increase in surface area. More importantly, each villus is covered with hundreds of microscopic folds (see Fig. 4.9) called **microvilli,** which again substantially increase the area of the small intestine to the point that, if spread flat, it would be large enough to cover a tennis court (approximately 60 m², or 2000 ft²).

Why so large? It must be remembered that virtually all the nutrients required by all the billions of cells of the body must come through this doorway—and be-

**Table 4.3**

Digestive fluids and principal digestive enzymes

| Source | Fluids | Enzymes | Substrate | End products |
|---|---|---|---|---|
| Mouth | Saliva | Amylase | Starch is split | Disaccharides |
| Stomach | Gastric juice | Pepsin | Proteins are split | Polypeptides |
| Pancreas | Pancreatic fluid | Trypsin | Undigested proteins | Polypeptides<br>Amino acids |
| | | Amylase | Starch | Disaccharides |
| | | Lipase | Fats | Glycerol<br>Fatty acids |
| Small intestine | | Peptidases | Polypeptides<br>Dipeptides | Amino acids |
| | | Lipase | Neutral fats | Glycerol<br>Fatty acids |
| | | Amylase | Carbohydrates | Disaccharides |
| | | Disaccharidases { Lactase | Lactose | Glucose<br>Galactose |
| | | Maltase | Maltose | Glucose |
| | | Sucrase | Sucrose | Glucose<br>Fructose |
| Liver | Bile | (Contains no enzymes) | Fats are emulsified by bile salts. | |

cause the chyme from which these nutrients must be drawn is constantly moving, a rather large doorway is required.

The fact remains, however, that other very large organisms (sharks, for example) have an equivalently much smaller area devoted to nutrient absorption. Thus, in addition to large body size, a second reason why the surface area of the small intestine is so great in humans is our high metabolic rate which, in turn, is intimately tied in with our endothermic ("warm-blooded") condition. This point is discussed at length at the end of Chapter 6.

How does the small intestine know when to produce its enzymes? In analogy with the stomach, the small intestine produces two hormones which collectively govern the secretory responses not only of the associated organs (liver, pancreas, and gallbladder) but also of the small intestine itself. The hormone responsible for the latter situation is called **secretin,** and its release is governed by the presence of acid in the duodenum, an event that occurs only when chyme has entered from the stomach.

The second hormone is **cholecystokinin,** the release of which is stimulated by the presence of amino acids and fatty acids in the duodenum. (Note that neither hormone is stimulated by the presence of carbohydrates.) Although cholecystokinin is not involved in the release of digestive enzymes from the small intestine, it operates to inhibit stomach contractions, which results in less chyme entering the small intestine in a given time. Therefore, the chyme already present has more time to be

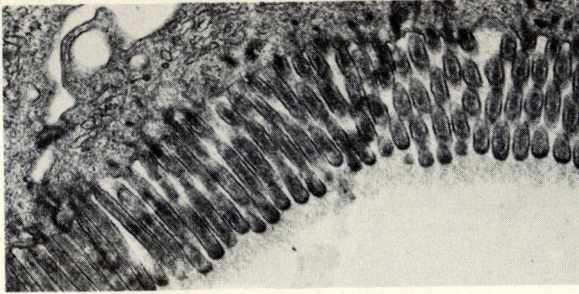

Fig. 4.9
(a) Electron micrograph of the microvilli of a single cell in the small intestine. (Rotker, Taurus)
(b) Diffusion and the semipermeable membrane.

(a)

Food coloring

Water

Time

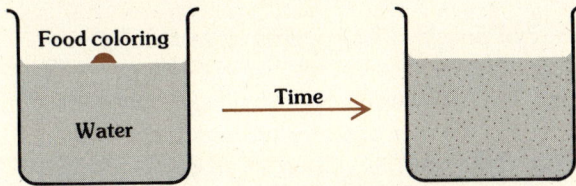

1. Simple diffusion

Membrane permeable to water but not to food coloring

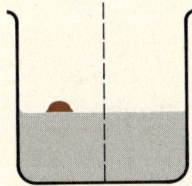

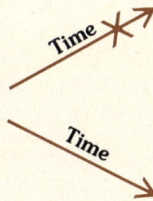

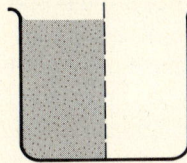

Time

No—gravity prevents this

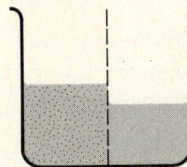

Time

Yes—some net movement to left, but countered by gravity and by increased number of water molecules at left, which tend to diffuse back

2. Semipermeable membrane

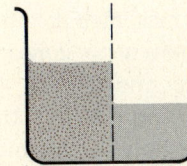

Time

Net movement continues until the number of water molecules striking the membrane at any one time is the same on both sides— more net movement to left than in 2 because fewer water molecules on left at outset, due to large amount of food coloring

3. Semipermeable membrane—additional food coloring

(b)

digested and absorbed. Consequently, meals that are high in fat tend to remain much longer in the stomach than do meals low in fat.

Secretin also affects the stomach, insofar as it inhibits the production of HCl and to a lesser extent, it also inhibits stomach contractions. Once the HCl in the duodenum is neutralized, the rate of secretin release drops, and the inhibition of the stomach is removed. The action of secretin and cholecystokinin is largely responsible for the rate at which the stomach empties into the small intestine.

**Uptake**   Most of the uptake of molecules by the digestive tract occurs in the small intestine and includes not only molecules present in the food but also most of the digestive secretions which are reabsorbed for subsequent reuse.

There are basically five ways by which food material crosses a cell membrane thereby moving from outside the body (i.e., present in the cavity of the gut tube) to inside the body.

**1. Diffusion**   Diffusion is a basic law of physics that governs all molecules in fluids (i.e., gases and liquids) and that states that molecules will become evenly spread throughout the space in which they find themselves, given enough time (see Chapter 2). A drop of food coloring placed in a beaker of water will gradually spread evenly throughout the entire beaker, although this will occur far more quickly if the water is stirred as the food color is added. This tendency for molecules to move *down a diffusion gradient* (i.e., from areas of high concentration to areas of low concentration) is therefore not a biological law, but rather a physical law, which nonetheless holds true for biological systems as well. Moreover, the law is not necessarily invalidated merely because cell membranes are present. All such membranes possess pores through which certain materials may pass, and other materials that are lipid-soluble may be able to enter the inside of a cell simply by dissolving through the cell membrane. Thus, it happens that a sizable number of types of molecules pass from the digestive tract into a cell by diffusion, the two requisites being (1) that the particular molecule is more abundant in the digestive tract than in the cell (i.e., that there is a diffusion gradient down which it can move) and (2) that it is capable of either passing through a pore or of dissolving through a cell membrane.

**2. Facilitated diffusion**   Facilitated diffusion is a specialized type of diffusion typically involving molecules which are by themselves incapable of passing through a cell membrane because of size or chemical structure, an example of which is vitamin $B_{12}$. A molecule produced by the stomach, called simply **stomach factor,** passes into the small intestine along with the vitamin $B_{12}$ in the food. The two molecules unite and stomach factor facilitates the movement of vitamin $B_{12}$ into the cells of the abdominal lining. Vitamin $B_{12}$ is required for the formation of red blood cells. In its absence, **pernicious anemia** will develop in individuals who have had a portion of their stomach removed (and who, as a consequence, may not be producing enough stomach factor), although symptoms may not occur for several years, because large amounts of vitamin $B_{12}$ are stored in the liver. Stomach factorlike molecules, termed **carrier molecules,** would seem to be effective either because they can dissolve through a cell membrane or because they possess the correct electric charge to pass through a membrane pore. Fruit sugar (fructose) is another molecule that requires facilitated diffusion.

**3. Osmosis** As used in biology, osmosis is simply the diffusion of water across a cell membrane. It may seem odd that we would single out water and dignify its movements with a special name, but the movement of water in and out of cells is so important in biology, and has so many ramifications, that the utility of giving the process its own name outstrips the inconvenience of having to explain it at length when the term is first used.

Water has the capacity to move freely across all cell membranes. (This statement is not repudiated by virtue of the fact that water cannot pass easily through our skin, because the cells that comprise the outer layers of our skin are dead and it is only their compressed remnants that create this degree of water impermeability.) Because of this capacity of water, it happens frequently in situations involving diffusion that the concentration of materials within a cell changes not because of movement of the dissolved materials, but because of the movement of the water.

To illustrate, recall the example of diffusion involving the drop of food coloring placed in a beaker of water. Suppose, for sake of argument, that the beaker was divided by a vertical barrier that was permeable to water but impermeable to the food coloring, and suppose also that a drop of food coloring was added to one side of the beaker only (Fig. 4.9b2). Although it is true that molecules of water will move freely back and forth across the membrane, because there are more water molecules per unit volume on the side to which food coloring was not added (i.e., it is pure water), there will be a tendency for a few more water molecules to move in one direction than in another. Theoretically, this movement would continue until the concentrations of water were equal on both sides—and because that would never occur (given that the food-coloring molecules will remain on one side of the barrier), in theory all the water molecules would ultimately move to the side containing the food coloring. However, in practice, this does not happen. Not only do the molecules on the food-coloring side become more numerous thus becoming exposed to a larger surface area of the membrane (which thereby increases the number of water molecules that are crossing back), but the effects of gravity are such that as water is lowered on one side and raised on the other, the pressure on the high side will have the effect of squeezing proportionately more molecules back into the pure water side.

Thus, an equilibrium is established for any given system. This does *not* mean that the water molecules cease to move across the barrier, but rather the number moving in one direction is the same as the number moving in the opposite direction. Moreover, the magnitude of the difference in heights of the two sides is a function of the relative differences in the concentrations of water. If a great deal of food coloring had been added to one side (such that, let us say, only one molecule in two was water) it is clear that a great deal more water would move from the pure water side in this second example than occurred in the first example (Fig. 4.9b3).

Osmosis is an incredibly important phenomenon in biological systems in general and our own body in particular, and we shall have reason to refer to this process frequently in discussing other biological processes throughout this book. Bear in mind that it is not a magical process, but that it simply hinges on the laws of probability as to how often water molecules will hit an opening in the barrier on one side versus the other.

Perhaps a few examples are in order to demonstrate how osmosis functions in the human body.

*Why can't we quench our thirst by drinking seawater?* The concentration of osmotically active molecules (i.e., molecules in solution) is greater in seawater than in cells of the human body. (That is, seawater is "saltier" than is the fluid of the human body.) Normally, water is absorbed by the cells of the small intestine because the chyme has fewer osmotically active molecules per unit volume than does the cell. In the case of seawater, water moves in the reverse direction—*from* the cells *into* the cavity of the intestine—thereby equalizing the concentrations on both sides of the cell membrane. Therefore, there is actually a net loss of water from the body after drinking seawater, with the result is that you become thirstier after drinking seawater than you were before.

*Why do children so often become sick after eating "junk" at carnivals?* The "junk" in this example is mostly material containing large amounts of sugar—soft drinks, candy, chocolate, and so forth. This sugar passes into solution immediately as it enters the digestive tract and has the effect of forming a highly concentrated sugar solution. Sugar is normally absorbed by the small intestine, but in a young child a large mass of sugar solution entering the small intestine over a short period of time will overwhelm the capacity of the intestine to absorb the sugar and in fact may well cause a net outflow of water from the cells (just as occurred in the seawater example discussed earlier). The consequences of this water loss may well be manifested by the nausea and diarrhea so often associated with "carnivalitis." A similar phenomenon may occur in individuals who have had all or most of their stomach removed. For this reason, such individuals are generally advised to eat several small meals each day, rather than a few large meals.

*Why can't you fool the police by drowning someone in the bathtub and then throwing him into the ocean?* Fresh water entering the lungs is much more concentrated than is the salty water of body cells. Therefore, this water will pass into the cells of the lungs, causing them to swell and frequently to rupture, with the result that some hemorrhaging typically occurs in the lungs of victims of freshwater drowning. A salt-water victim never shows this because the net movement of water is *from* the cells into the cavity of lungs. Hence, at autopsy, it is very easy to determine if a victim drowned in fresh or in salt water. So if you ever plan to use the bathtub method—put lots of salt in the water first!

In sum, water is absorbed from the digestive tract primarily in the small intestine through the process of osmosis.

**4. Active transport** Thus far, all the types of molecular movement are variations of the broad category of **passive transport**—that is, movement down a concentration gradient. However, there are many molecules required by the body that are less abundant in the food than they are in the cells of the body. Diffusion will obviously be ineffective in such instances since diffusion cannot operate *up* a concentration gradient. Through a process that is not yet completely understood, a molecular "pump" in the membrane of the cell can move specific molecules against a concentration gradient, but this process requires a considerable outlay of energy on the part of the cell in order to function. Many salts are absorbed from the small intestine by active transport, as are the monosaccharides glucose and galactose. Of course, this process does double duty, because as the salts are removed from the chyme, the water concentration of the chyme increases, thereby facilitating additional uptake of water molecules by the body through osmosis, a process requiring no expenditure of energy. As we shall see, the circulatory and excretory systems also rely on a coupling of active transport and osmosis to effect a movement of both salts and water across a cell membrane.

**5. Pinocytosis and Phagocytosis**   Many cells, including those of the lining of the small intestine, have the capacity to envelop small portions of chyme or other material and create a vacuole, which then passes into the cell for further digestion. The distinction between pinocytosis and phagocytosis is fundamentally one of degree, the latter term being used for large vacuoles, the former for small vacuoles. In both instances, the amount of material taken in is quite large, certainly much larger than in the other types of uptake mechanisms just discussed. How important these processes are in food uptake is far from clear, but it is known that they do occur frequently.

Where do the building block molecules go once they enter the cells of the microvilli? The sugars, amino acids, and some small fats enter the capillaries lining the villi (Fig. 4.10) (20 percent of the blood that leaves the heart goes to the small intestine), and from there are transported to the liver via the **hepatic portal vein.** (A portal vein is one which both begins and ends in a capillary bed. In this instance, the beginning capillary bed is in the intestine, and the ending bed is in the liver.) This nutrient-rich blood is thus exposed to a great many liver cells before being recollected and transported to the heart via the **hepatic veins** (see Fig. 4.11). In contrast, the larger fats and the fat-soluble vitamins pass into the **lacteals,** which are not vessels but rather intercellular channels, and which are part of the lymphatic system (see Chapter 5). These fats bypass the liver before ultimately entering the blood circulation just before the venous blood enters the heart, which is the point at which the lymphatic system joins the venous system.

As might be expected, the liver plays an important role in regulating both sugar and amino acid levels in the blood. Excess blood glucose is converted into the polysaccharide storage molecule **glycogen** by the liver and is stored there, to be reconverted to glucose and doled out to the body during between-meal intervals in order to ensure an essentially constant blood-glucose level.

Similarly, the liver prepares those amino acids in excess of the protein-building needs of the cells of the body for further chemical breakdown by the cells in order to generate energy. This preparation by the liver involves the removal of the nitrogen-containing portion of the amino acid molecules, and it is this nitrogenous waste product (so called because the body does not require these fragments for any other pur-

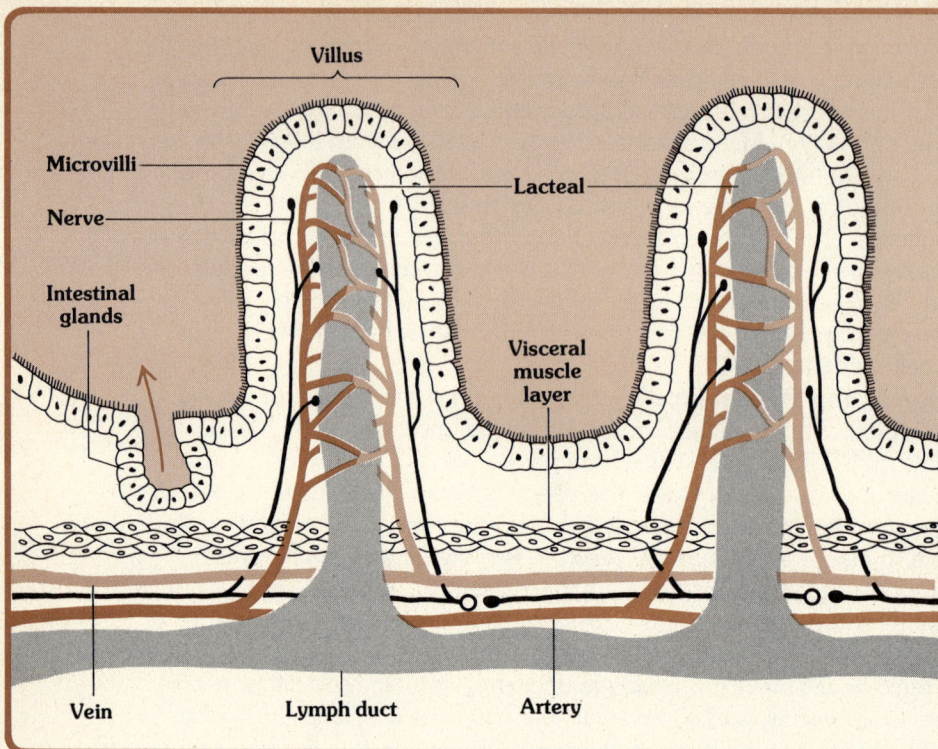

**Fig. 4.10 Diagrammatic representation of the structure of two villi.**

Villus

Microvilli

Nerve

Intestinal glands

Lacteal

Visceral muscle layer

Vein

Lymph duct

Artery

pose) which forms the basic waste product of urine, a compound called **urea** (see Chapter 7).

**ACCESSORY DIGESTIVE ORGANS** There are three organs which play an important role in the digestive process, despite the fact that no food passes through them during digestion. These are the **pancreas,** the **liver,** and the **gallbladder.**

**Pancreas** This gland is about 15 cm (6 in.) long and occupies much of the space between opposite curves of the duodenum (see Fig. 4.12). Although it is probably somewhat better known as the endocrine gland which produces **insulin** (see Chapter 9), it also plays an extremely important role in digestion. Not only does it produce a set of enzymes which are collectively capable of breaking down each of the three major food categories (a capability shared only with the small intestine, of all the digestive organs), but it also produces large amounts of **sodium bicarbonate,** which serves to neutralize the acidic chyme of the stomach as it passes to the small intestine. The pancreas produces a total of 1½–2 liters of fluids per day (mostly bicarbonate solution), which enter the duodenum via a canal located a short distance from the opening of the stomach. (It is interesting to note that, when duodenal ulcers are present, they are almost invariably located in the short space between

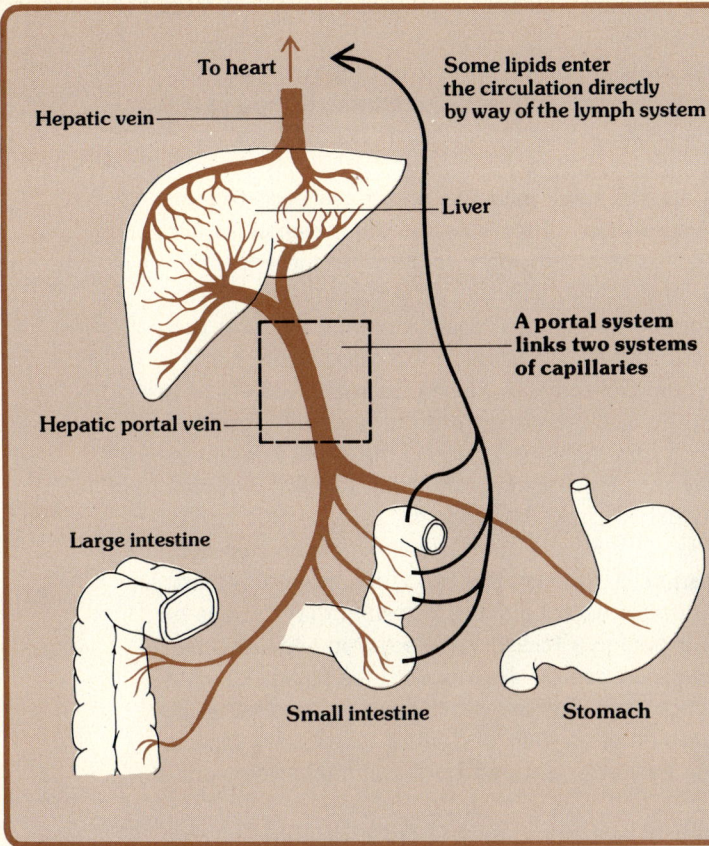

Fig. 4.11
Interrelationships of the gut tube, the hepatic portal vein, and the liver.

To heart

Hepatic vein

Some lipids enter the circulation directly by way of the lymph system

Liver

A portal system links two systems of capillaries

Hepatic portal vein

Large intestine

Small intestine

Stomach

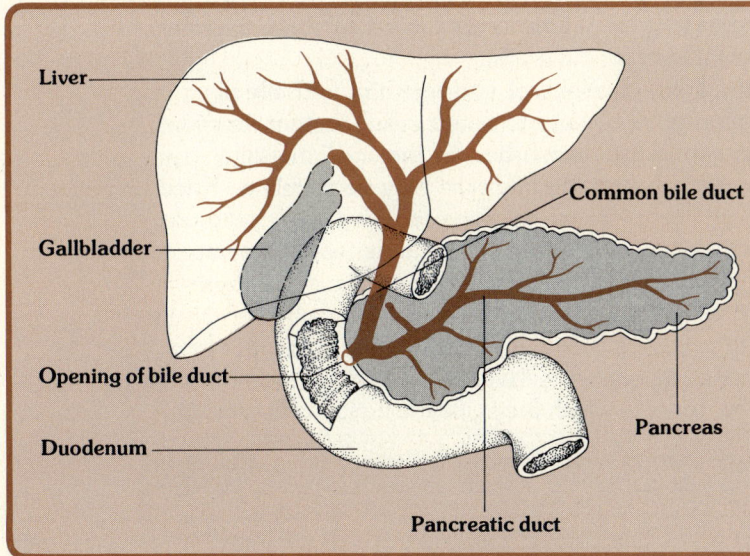

Fig. 4.12
Interrelationship of the liver, gallbladder, pancreas, and duodenum. Note that the pancreatic duct joins the bile duct before entering the duodenum.

Liver

Common bile duct

Gallbladder

Opening of bile duct

Pancreas

Duodenum

Pancreatic duct

the pyloric valve and the point where the pancreas empties—that is, in the area least well buffered by bicarbonate secretions.)

Release of the pancreatic secretions is under the control of both nerves and hormones. Just as in the case of the stomach, the sight, smell, or taste of food elicits a response. The bulk of the secretions produced by these external stimuli are enzymes, as opposed to bicarbonate.

Far more important, however, are the small intestine hormones **secretin** and **cholecystokinin**[2]. Secretin, which is produced in response to the presence of acid in the duodenum, stimulates the release of bicarbonate solution by the pancreas. (Release of bicarbonate is inhibited by nicotine, which is why individuals with duodenal ulcers are typically advised against smoking.) Cholecystokinin, which is produced as a consequence of fatty acids and amino acids in the duodenum, stimulates the release of pancreatic enzymes. Collectively, these pancreatic secretions are responsible not only for neutralizing the acidic chyme from the stomach (resulting in a pH favorable for the activation of enzymes produced by the small intestine and pancreas), but also for increasing the protein digesting capabilities of the digestive tract by 100 percent and the lipid capabilities by 200 percent.

It is interesting to note that although both the acid secretions of the stomach and the alkaline secretions of the pancreas draw on the blood for the raw materials necessary in their synthesis, the volume of each balances out such that there is no net change in the pH of the blood. Ingesting a bicarbonate solution to relieve an upset stomach actually increases acid secretion by the stomach (as gastrin production is not inhibited in such an instance) and decreases bicarbonate production by the pancreas (as the less acidic chyme entering the duodenum causes less secretin to be produced). Excess bicarbonate ions are removed by the kidneys to ensure a constant blood pH.

**Liver**   The liver is a large, roughly triangular organ of about 2 kg (4 lb) lying opposite the stomach on the right side of the body (Fig. 4.1). This organ is one of the most complex in the body, and has a long list of functions (Table 4.4). From the standpoint of digestive secretions, its role is to produce **bile.**

Bile consists of **bile salts,** the lipids **cholesterol** and **lecithin,** and bile pigments left over from the destruction of worn-out red blood cells (another function of the liver). Most of the salts and lipids are reabsorbed by the small intestine, virtually as quickly as they are released. In fact, the bile salts may be circulated twice for a single meal—as much as 8 g ($\frac{2}{7}$ oz) of these are released every meal, yet there are fewer than 4 g ($\frac{1}{7}$ oz) in the entire body. Recirculation is facilitated by the hepatic portal vein, which passes directly to the liver, where the bile salts are quickly removed and dumped back into the next slug of bile released.

By contrast, most of the bile pigments pass out with the feces giving the feces their typical brownish color. (Those bile pigments that are reabsorbed are passed to the kidneys to be excreted with the urine, to which they give the typical yellow color.)

---

[2] For many years, enzyme production by the pancreas was thought to be under the control of a third intestinal hormone, **pancreozymin.** Recent work indicates, however, that this hormone is identical to cholecystokinin.

**Table 4.4**
**Principal liver functions**

**1 Metabolism**
    Conversion of glucose to glycogen
    Breaking down glycogen to glucose
    Formation of glucose from noncarbohydrate sources
    Removal of nitrogen-containing group from amino acids
    Formation of fats from glucose, glycogen, or amino acids
    Breakdown of fats into fatty acids and glycerol
    Cholesterol synthesis and breakdown
**2 Storage**
    Glucose (stored as glycogen)
    Bile (stored in gallbladder)
    Vitamins
    Fats and fatty acids, lipids such as cholesterol
    Amino acids
    Minerals: iron and copper
**3 Secretion and synthesis**
    Secretion of bile salts into bile
    Formation of blood proteins, prothrombin, fibrinogen
    RBCs produced before birth
**4 Excretory functions**
    Breakdown of hemoglobin in old RBCs
    Ingestion of bacteria or other particles
    Detoxification of harmful acids and drugs
    Formation of urea from amino acids and uric acid from nucleic acids

Bile, and more particularly the bile salts, has the interesting property of being able to **emulsify** fats. You are aware of how difficult it is to remove oil or grease from your hands using only water. This difficulty occurs because oil and water do not mix well, as they are structurally very different molecules. We use soap to remove the oil because it has the capacity to interact both with oil and with water. Put another way, the soap serves to emulsify the oil in the water—that is, to break it up into such small clumps of oil molecules that they can be floated off the skin.

A very similar problem occurs in the digestive system. Large globs of lipid tend to remain as large globs in the digestive tract because they are in a water base (i.e., the digestive secretions). As such, the enzymes responsible for breaking down lipids can only attack the outside layer of these globs, and the efficiency of the process is much reduced. Bile acts as soap did in the previous example to break up the globs into tiny ($1 \mu$) clumps that can effectively be attacked by the lipid enzymes (Fig. 4.13). In fact, even though bile does not have any enzymatic properties itself (emulsification is a mechanical process in that no molecular bonds are broken—the principle is essentially the same as dissolving sugar in water), the effects of the bile are such that fully 25 percent more fat can be digested in the presence of bile than can occur in its absence.

The liver produces from .25 to 1 liter of bile daily at an essentially constant rate. However, the bile is typically prevented from entering the small intestine owing to the presence of a sphincter at the opening of the bile duct (see Fig. 4.12). If this sphincter is closed, the bile backs up and enters a small saclike structure

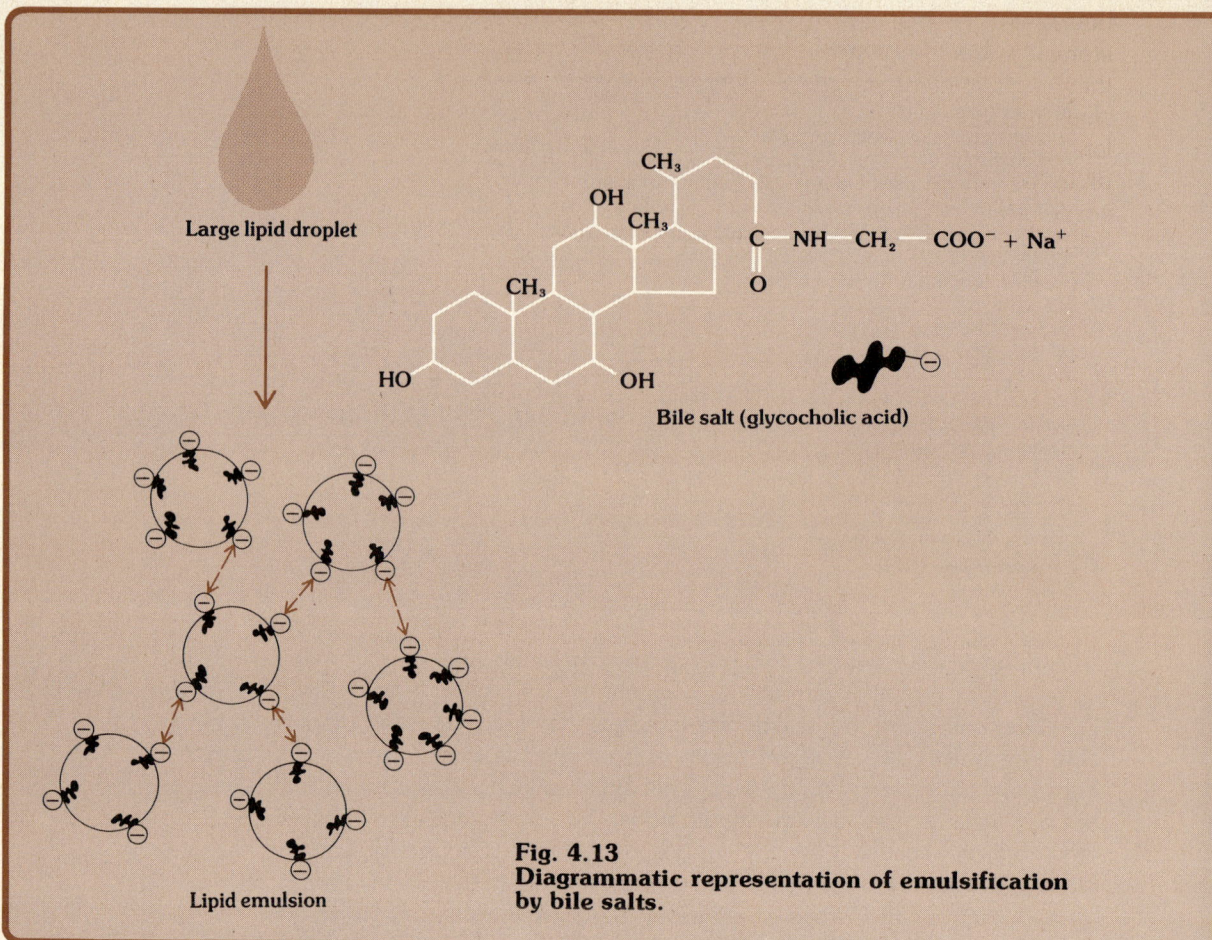

Large lipid droplet

Bile salt (glycocholic acid)

Lipid emulsion

**Fig. 4.13**
**Diagrammatic representation of emulsification by bile salts.**

about $2 \times 6$ cm ($1 \times 2.5$ in.), called the **gallbladder.**[3] The gallbladder serves to store and to concentrate the bile by removing much of the water. Cholecystokinin,[4] the small intestinal hormone which, you may recall, is produced because of the presence of fatty acids and amino acids in the duodenum, has yet another role in that it causes the sphincter at the end of the bile duct to relax and the gallbladder to contract, forcing the bile into the small intestine. (The interrelationships of these hormones are diagrammatically represented in Fig. 4.14.)

Once the bile is concentrated in the gallbladder, the cholesterol is present virtually at saturated levels. In certain individuals (especially those who are obese, dia-

[3] Some animals, such as rats and horses, have no gallbladder, and perhaps need none owing to their low-fat diets.
[4] The name of this hormone is based on its action on the gallbladder, although this is only one of its functions. It derives from the Greek words **chole** (bile), **kytis** (bladder), and **kinin** (to move).

betic, middle-aged, and female), the cholesterol may precipitate, forming **gall-stones.** These may do no harm so long as they remain in the gallbladder, but if they move into the bile duct they may block it, preventing the flow of bile into the small intestine. Instead of passing out with the feces, the yellowish bile pigments circulate back through the body, giving it a yellowish cast, a condition known as **jaundice.** Moreover, because the pancreatic duct joins the bile duct before entering the small intestine (Fig. 4.11), pancreatic secretions may also be cut off, in which case little digestion or reabsorption will occur in the small intestine.

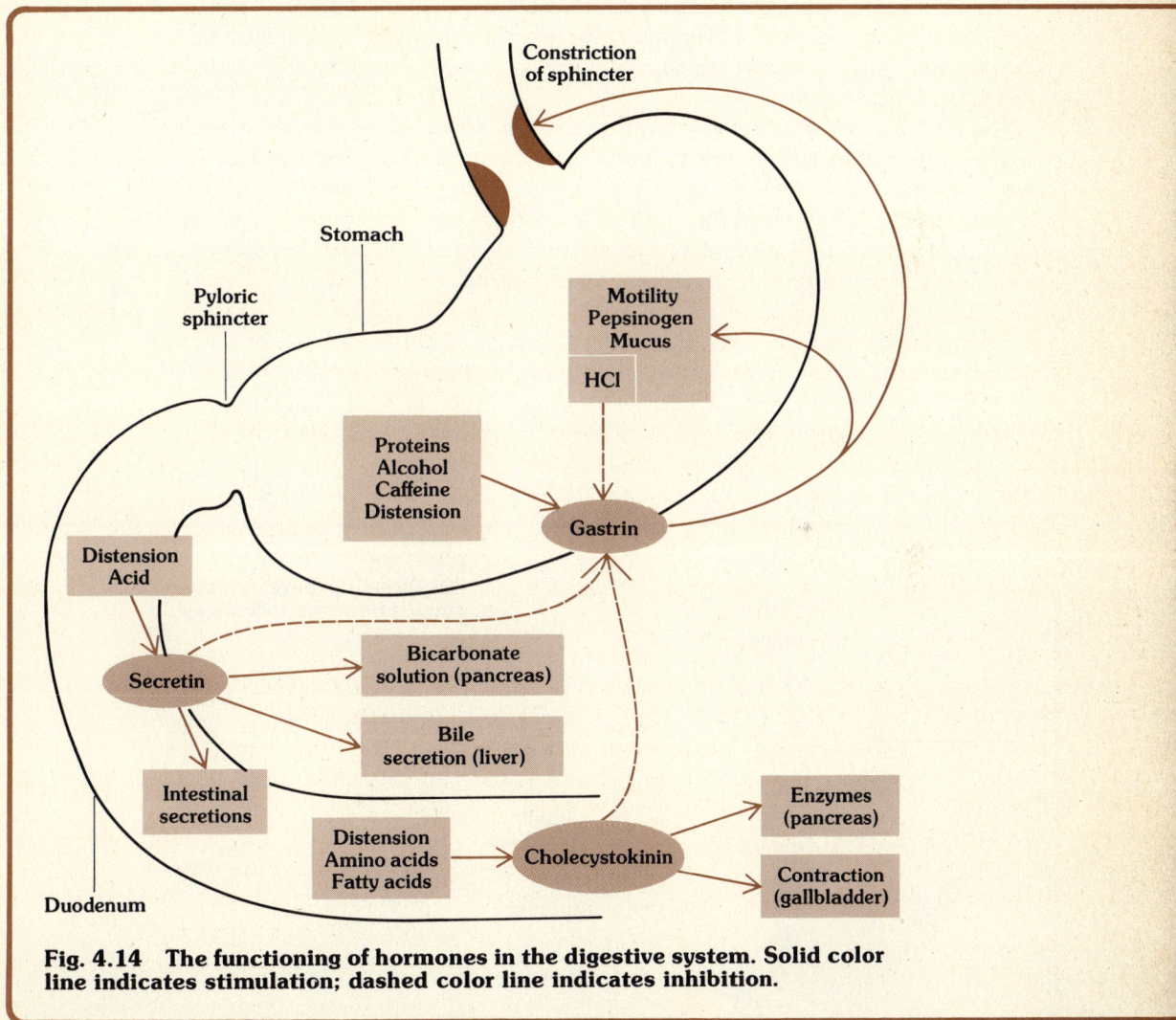

**Fig. 4.14  The functioning of hormones in the digestive system. Solid color line indicates stimulation; dashed color line indicates inhibition.**

In such instances, a **cholecystectomy** (removal of the gallbladder) is performed. An individual deprived of his gallbladder continues to produce bile but has no capability of storing or concentrating it. Thus, much of the bile enters the small intestine when little or no chyme is present, clearly reducing the efficiency of the bile. Such individuals are not as capable of digesting lipids as they were previously, and are typically placed on a low-fat diet.

Jaundice does not necessarily result only from gallstones. Liver inflammation in general (a condition known as **hepatitis**) frequently results in jaundice and, because of the myriad functions of the liver, such a disorder is far more serious than is a blocked bile duct. The danger of hepatitis is that in cases of severe damage to the liver cells they will not be replaced by normal liver cells but rather by scar tissue, a condition known as **cirrhosis.** Once a sufficient fraction of the liver becomes cirrhotic, it is no longer capable of performing adequately, and the individual dies.

Hepatitis may result from several causes. **Viral hepatitis,** as the name implies, is virally induced and may be transmitted by contaminated blood or needles (serum hepatitis) or from the feces of infected individuals. **Deficiency hepatitis** results from protein deficiency and is common in children in times of famine. **Toxic hepatitis** results when the detoxification system of the liver is overwhelmed by such heavy metals as mercury or lead, among other poisons.

**THE LARGE INTESTINE**   The large intestine, or **colon,** consists of a tube of about 1.5 m (5 ft) in length and 6 cm (2.5 in.) in diameter, oriented in roughly an inverted U shape in the lower abdominal region (see Fig. 4.1). The junction between the large intestine and the small intestine occurs in the lower right ab-

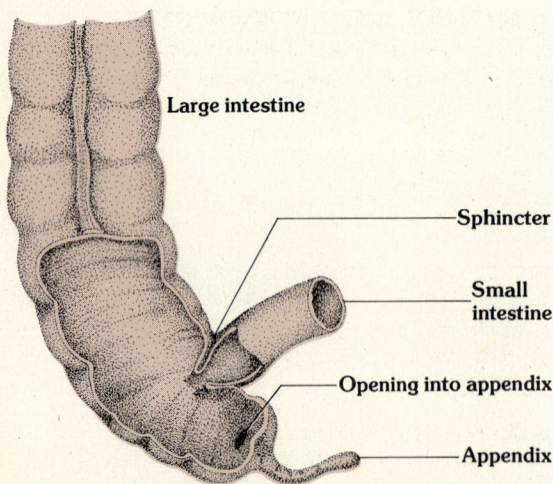

**Fig. 4.15**
**Relationship between the small and large intestines.**

Large intestine

Sphincter

Small intestine

Opening into appendix

Appendix

domen, and is marked by the appearance of a small blind sac, called the **appendix**[5] (Fig. 4.15). This structure is not known to perform any function in humans, although rather frequently it becomes infected **(appendicitis)** and must be removed. In fact, so common is this condition, that any severe pain in the abdominal region, particularly if on the right side and if accompanied by tender swelling. is suspect as appendicitis. Untreated, the inflamed appendix may burst, causing temporary relief from pain but soon resulting in a spreading of the infection to the peritoneum **(peritonitis),** the shiny tissue that lines the abdominal cavity. Peritonitis is an extremely serious disease and was frequently fatal until the development of antibiotics 40 years ago. This condition once again illustrates the consequences of a connection developing between the external body surface (i.e., the cavity of the gut tube) and the interior of the body (the abdominal cavity itself).

Movement in the large intestine is slow, with contractions occurring as infrequently as one every 30 min. Chyme may remain in the large intestine for as much as 18–24 hr, with movement occurring usually after a meal. A sphincter between the small and large intestine is normally closed, but it opens when a wave of contractions reaches the end of the small intestine. Conversely, distension of the large intestine, because of the entry of chyme, operates to close the sphincter. The result, of course, is a system that causes the flow of chyme into the large intestine to be carefully regulated.

The principal function of the large intestine is the concentration of chyme, through the reabsorption of water, and its consequent transformation into the waste material known as **feces.** Even so, most of the water in the chyme is reabsorbed by the small intestine (Fig. 4.16), for, despite the greater diameter of the large intestine, the absence of villi or microvilli in this organ results in its surface area being only one-thirtieth that of the small intestine.

The large intestine makes no significant contribution to digestion, as it produces no enzymes. However, in addition to water reabsorption, it also absorbs quantities of vitamin K and some of the B vitamins, which are byproducts of the bacteria living harmlessly in the large intestine. These vitamins are also contained in a variety of foods (see Table 4.1), but the intestinal bacteria are a sufficiently important source that doctors may recommend a vitamin supplement when prescribing oral penicillin, as penicillin has the effect of killing off the intestinal bacteria, or they may recommend foods such as yogurt to help reestablish the bacteria. The relative roles played by each digestive organ in the digestive process are diagrammatically represented in Fig. 4.17.

In addition, certain excess salts are dumped into the cavity of the large intestine to be passed out with the feces.

**DEFECATION**    At periodic intervals, typically following a meal when food from the small intestine moves into the large intestine, fecal material moves from the **colon** into the **rectum** (see Fig. 4.1). This movement is spontaneous, and we

[5] The appendix is frequently found at McBurney's point, which is midway on a line drawn from the umbilicus (belly button) to the crest of the pelvic girdle (hip point).

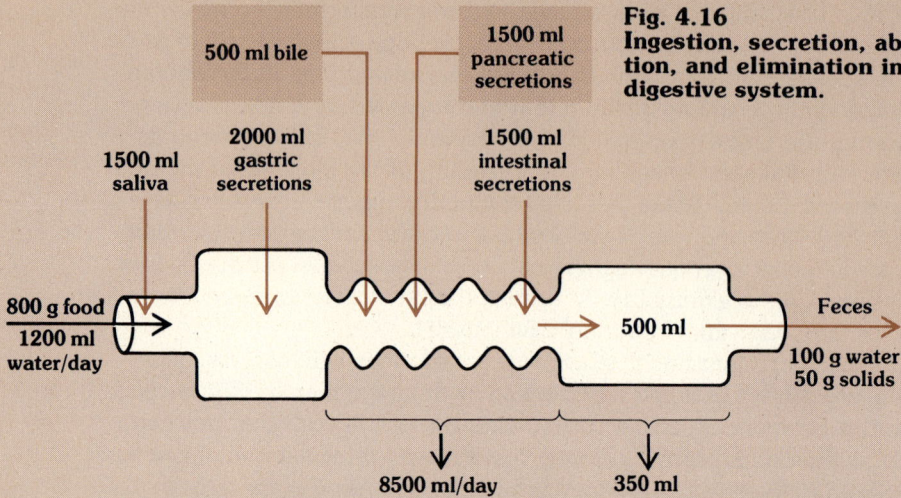

**Fig. 4.16**
Ingestion, secretion, absorption, and elimination in the digestive system.

500 ml bile

1500 ml pancreatic secretions

1500 ml saliva

2000 ml gastric secretions

1500 ml intestinal secretions

800 g food
1200 ml water/day

Feces
100 g water
50 g solids

500 ml

8500 ml/day

350 ml

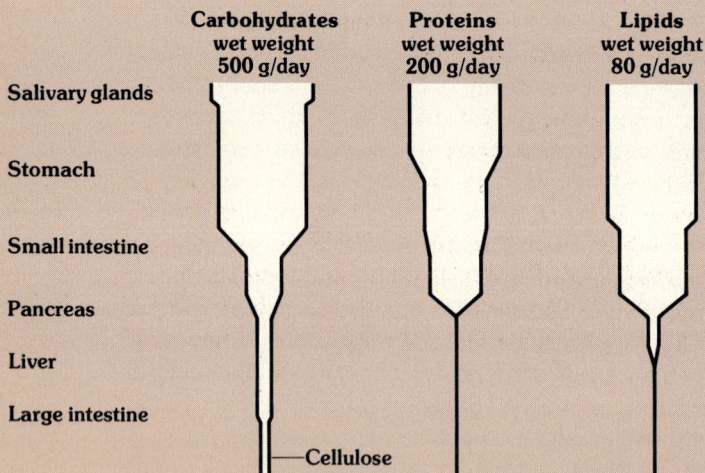

**Fig. 4.17**
The relative significance of each digestive organ in the breakdown of each major class of macromolecules. The relative narrowing of the columns indicates the significance of each organ in the digestive process.

Carbohydrates
wet weight
500 g/day

Proteins
wet weight
200 g/day

Lipids
wet weight
80 g/day

Salivary glands

Stomach

Small intestine

Pancreas

Liver

Large intestine

Cellulose

have no conscious control over it. The subsequent swelling of the rectum initiates the urge to defecate, an urge that can be resisted by the constriction of the sphincter muscle surrounding the **anal** opening. The development of this constricting ability is a maturation process usually not developed until the second year of life (as those of you who have diapered young infants are well aware).

Defecation itself involves a sharp inhalation plus contraction of the abdominal muscles. Both these events increase abdominal pressure which not only causes the feces to pass out but also causes a momentary slowing down of venous blood returning to the heart. The sudden rush of this blood, which occurs as the abdominal muscles relax, leads to a momentary increase in blood pressure which, in elderly people, may be sufficient to initiate a heart attack or stroke.

The frequency of defecation is highly variable, depending largely on the quantity and quality of food eaten. Young children may defecate several times every day; a little old lady living largely on tea and crumpets, perhaps only once a week. The significant point is that individuals vary, and there is no optimum frequency (despite overt suggestions to the contrary by laxative manufacturers). Moreover, retention of fecal material within the body poses no special danger (the commonly held belief that feces are "poisonous" is simply not valid). In fact, under certain pathological conditions, individuals have gone a year or more without defecating, and without suffering more serious consequences than the addition of rather a lot of weight. However, willful resistence to the urge to defecate is ill advised for more than short periods because water reabsorption from the feces continues and the fecal material becomes hard and painful to pass.

Diarrhea and constipation    These are two of the most common malfunctions of the digestive tract and, as such, deserve some mention here. Diarrhea may result from a number of causes, although it is perhaps most frequently associated with some intestinal infection. In essence, diarrhea is the consequence of a speeding up of peristalsis, such that the chyme does not remain in the large intestine long enough for all the water to be reabsorbed. This increase in peristaltic rate is frequently a part of the body's defense mechanism to disease, and is often coupled with vomiting, the two processes serving to clear out the digestive tract and thereby minimize the length of time the body's absorptive surface (the intestinal lining) is exposed to **pathogenic** (disease-causing) organisms.

In conditions such as intestinal "flu," diarrhea is nothing more than a nuisance. However, in such diseases as **amoebic dysentery** or **cholera**, death may result because of loss of salt (especially potassium), acidosis (because of the loss of bicarbonate ions from the pancreas), and dehydration. (The threat of dehydration is obvious when one realizes that some 7 liters of fluid are secreted into the digestive tract daily, yet the blood volume is only about 5 liters. Obviously, most of these secretions must be reabsorbed for reuse.)

Constipation is the opposite phenomenon, of course, but is more frequently a result of improper diet than of some disease condition. Much of the volume of fecal material consists of indigestible material such as cellulose from plant cell walls. A diet high in meats and processed foods and low in vegetables and fruit may result in constipation for the simple reason that the volume of indigestible material is sufficiently small that the urge to defecate is delayed, resulting in an excess amount of water being reabsorbed from the feces.

Doctors are loathe to prescribe laxatives for anything other than immediate relief, because not only can the body become accustomed to laxatives, but the problem can usually be corrected simply by altering the diet. Nonetheless, laxatives constitute one of the most sold groups of over-the-counter medications. They fall into four main groups:

1 Bulking laxatives such as bran, which has the capacity to absorb water, which in turn leads to an increased volume of soft fecal material.

2 Lubricants, such as mineral oil, which operate just as the name would imply. However, mineral oil interferes with the absorption of certain salts (such as phosphate and calcium) and fat-soluble vitamins as well. Therefore, its prolonged use is contraindicated.

3 Mineral salts, such as magnesium salts which, because they are not needed by the body, are not absorbed and thereby increase the osmotic pressure of the fecal material, resulting in less water uptake by the large intestine.

4 Irritants, such as castor oil, which serve to speed up peristaltic rates, in an analogous fashion to certain pathogenic organisms. Again, the result is that the fecal material does not remain in the large intestine long enough for more than a fraction of the water to be removed. Certain natural foods, such as prunes, also fall in this category.

## SUMMARY

The digestive system is a tube through our bodies, which is designed to break down food and make it available, in terms of both size and chemical structure, to the cells. It is also an optimum habitat for many bacteria, some of which could pose a severe threat to the body as a whole. Owing to pH changes and the presence of various enzymes, however, most organic compounds, including the cell walls of bacteria, are fragmented during their trip through the digestive tract. Nevertheless, the system is delicate and rather easily disrupted. Vomiting and diarrhea are common responses to invasion by those bacteria and viruses which are immune to digestion, but even more dangerous are breaks in the integrity of the intestinal lining, for these can lead to invasion of the body itself. Fortunately, antibiotics were developed not long after intestinal ulcers became common, the advantages and disadvantages of modern society being balanced rather evenly in this instance.

In the summer of 1974, both before and after he resigned the presidency, Richard Nixon suffered from phlebitis, an inflammation of the veins of his legs. The threat to his life, however, was from the possibility of blood clots in his lungs. What does a leg disease have to do with lung problems?

Four of the top 15 leading causes of death in the United States in 1970 were diseases of the circulatory system (heart disease, #1; stroke, #3; hardening of the arteries, #7; high blood pressure, #14). What factors of our life-style contribute so significantly to deterioration and malfunction of our circulatory system?

Why does your physician take a white blood cell count before prescribing medication for your illness? Why do some people have allergies and others do not—and what do allergies have to do with the circulatory system? Why do organ transplant patients so frequently reject the transplanted organ?

These are the types of questions considered in the discussion of the circulatory system.

chapter V

# chapter V
# Ebb and flow of
# the internal sea–
# the circulatory system

Unlike the other organ systems, most of which are functionally rather discrete, the circulatory system (Fig. 5.1) is functionally very diverse. In one sense, the circulatory system has no discrete function of its own at all, but rather acts in a support capacity for all the other systems. The circulatory system is, after all, the multicellular organism's equivalent of the primordial sea in which life presumably first arose. The other organ systems act on this internal sea either by monitoring its composition, to ensure the environmental stability (homeostasis) essential for normal cell function, or by adding materials required by the individual cells. In addition, all of the fluid occupying the tiny spaces between the cells (the **interstitial fluid**) is also in direct and dynamic association with the blood of the circulatory system. Thus, the circulatory system can be thought of as a kind of self-perpetuating conveyer belt, carrying essential materials to all of the body's cells and being systematically altered, in turn, by the various organ systems—but altered in such a way as to ensure a fundamental sameness in composition over time. The blood is the conveyor belt itself, with the heart and vessels functioning as the framework and motor supporting the belt and causing it to move.

The circulatory system functions, first and foremost, in transport. Among other things, the system transports the necessary requirements of each cell, namely food and water from the digestive system and oxygen from the lungs, from their point of entry into the body to all other parts of the body. In addition, the waste products of every cell are transported by the circulatory system to those organs, most notably the lungs (carbon dioxide) and the kidneys (nitrogenous wastes), which are responsible for expelling them from the body.

The circulatory system also carries chemical messengers, called hormones (discussed fully in Chapter 9) from their sites of production to the affected organs. As we shall see presently, certain cells of the blood also have the task of warding off foreign invaders, such as bacteria, which manage to gain access to the interior of the body. Movement of the blood is controlled by the anatomy of the vessels and by the pumping activity of the heart. However, in one sense the blood regulates its own movement, in that, by forming clots, it does not flow gradually out of the body, should one of the vessels rupture.

**BLOOD**  Blood consists of several cell types or cell derivatives, a large number of proteins including those involved in clotting, and a series of small mole-

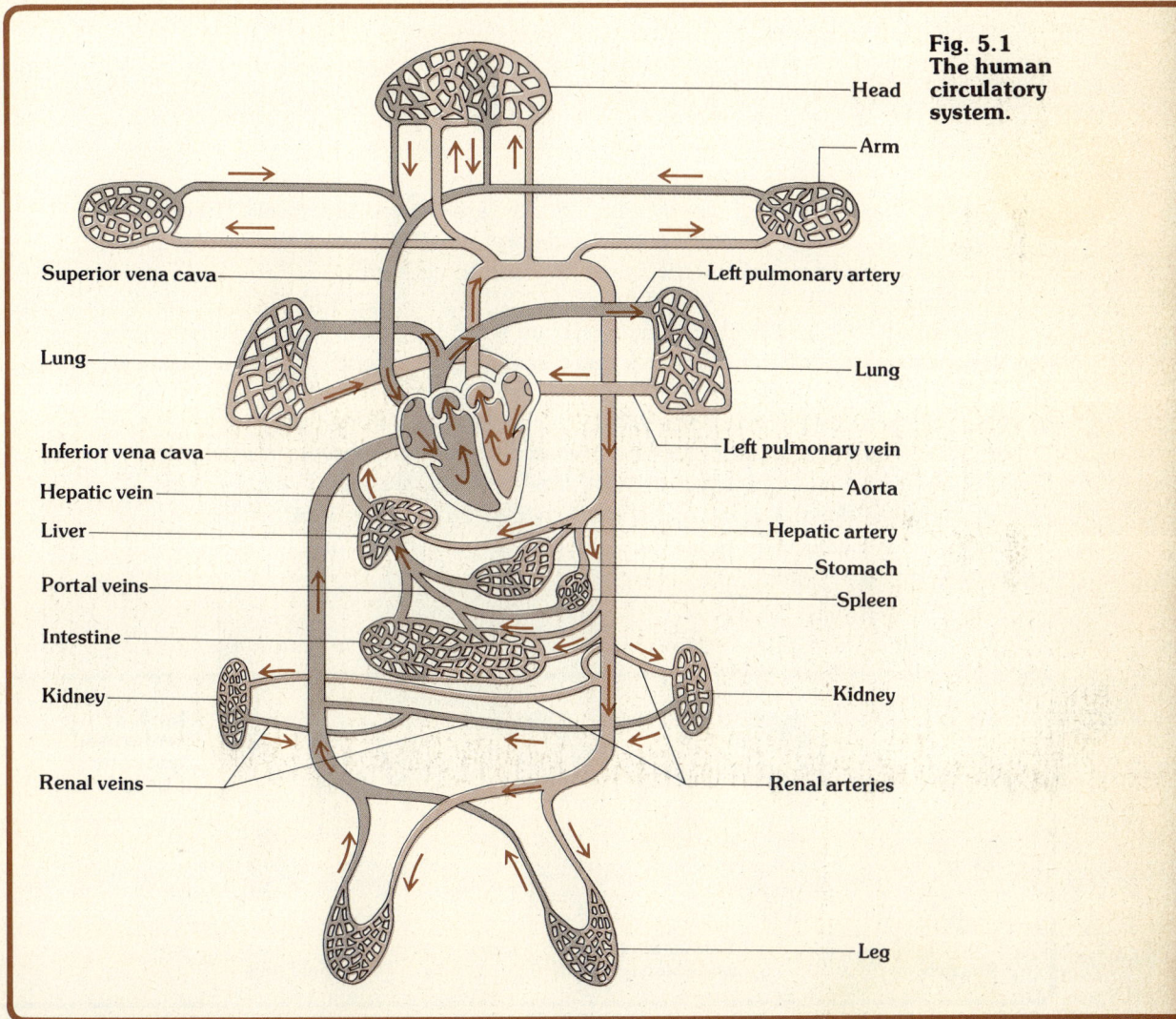

Head
Arm
Superior vena cava
Left pulmonary artery
Lung
Lung
Inferior vena cava
Left pulmonary vein
Hepatic vein
Aorta
Liver
Hepatic artery
Stomach
Portal veins
Spleen
Intestine
Kidney
Kidney
Renal veins
Renal arteries
Leg

Fig. 5.1
The human circulatory system.

cules and ions all in a water solution. Blood with all these components (see the scheme below) is called **whole blood;** blood minus the cellular portion is **plasma;** blood minus both cells and the clotting proteins is **serum.** About 8 percent of the body weight is blood. Normally, about 45 percent of whole blood consists of cells.

**Components of whole blood**

Whole blood

Plasma ← → **Cellular components**

1. Fibrinogen
2. Serum
   a. Plasma proteins
   b. Glucose and other nutrients
   c. Salts
   d. Dissolved gases
   e. Hormones
   f. Water
   g. Wastes

1. Red blood cells
2. White blood cells
3. Platelets

The two major classes of blood cells (Fig. 5.2) are the red blood cells (RBCs) and the white blood cells (WBCs). A third category, **platelets,** which are involved in clotting, are classed as cell fragments, as opposed to cells proper.

**RED BLOOD CELLS**   Red blood cells are small (7/1000*th* mm in diameter) and very numerous (about 5 billion per cubic centimeter of blood). Because they are short-lived (a given cell lasts only 3–4 months) approximately 3 million must be produced every second in order to maintain a constant level in the body.

Red blood cells are produced in the liver and spleen of embryos, in the long bones of children, and in the cavities of the ribs, sternum, and vertebrae of adults. As the cells break down, the fragments are removed by the liver. Most of the components are salvaged for reuse, but some is excreted in the form of bile (see Chapter 3).

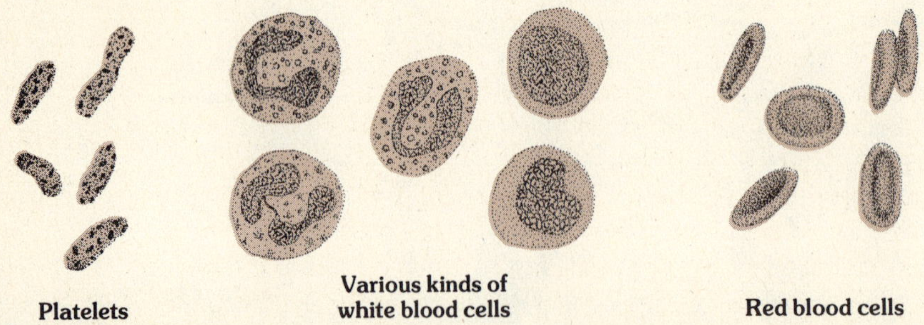

**Fig. 5.2
The formed
elements
of blood.**

Platelets

Various kinds of
white blood cells

Red blood cells

Red blood cells appear red because of the presence of iron in the protein **hemoglobin,** more than 250 million molecules of which are packed into every red blood cell. Each hemoglobin molecule has the ability to pick up as many as four molecules of oxygen in the lungs and transport them through the body.

Hemoglobin is essential for life. Although oxygen is abundant in air (21 percent) it dissolves very poorly in water. For example, 100 ml of plasma (92 percent of which is water) will dissolve only 0.3 ml oxygen, whereas the hemoglobin in 100 ml of whole blood will hold 19.7 ml oxygen. Hemoglobin therefore represents a mechanism for vastly increasing the oxygen-carrying capacity of blood beyond the level possible simply by being dissolved in plasma. As the oxygen-rich red blood cells reach tissues that are low in oxygen, the weakly attached oxygen breaks free and diffuses to these tissues. In return, carbon dioxide is given up by the tissues, but most of it does not attach directly to the hemoglobin molecule. Rather, the carbon dioxide largely combines with and is dissolved in the water portion of the blood. Carbon monoxide, on the other hand, does attach to hemoglobin and does so both much more tightly and 250 times more readily than does oxygen. It is for this reason that a person suffering from carbon monoxide inhalation is by no means instantly safe once he has been removed from the carbon monoxide-laden atmosphere into the outside air for it is only gradually that the hemoglobin releases the carbon monoxide. If too many of the hemoglobin sites have been occupied by carbon monoxide, the individual may still die, even if given pure oxygen.

**Sickle-cell anemia**   The normal shape of the red blood cell is biconcave— that is, it is rather like a donut, but without a complete hole through the center. The shape of the RBC is apparently determined in large measure by the shape of the hemoglobin molecule because the substitution of even a single amino acid in this long-chain protein may be sufficient to change the shape of the hemoglobin molecule and ultimately the shape of the RBC itself. This is precisely what occurs in sickle-cell anemia, so named because of the tendency of the RBCs to be sickle-shaped (Fig. 5.3), a trait which is maximized as oxygen leaves the hemoglobin. Although public awareness of this disease has increased recently because of its relative frequency in black Americans, essentially the same disease (although involving a different amino acid) is found in other populations as well, notably in people of Mediterranean descent. However, because the incidence of the disease is much lower in these populations, it has not gained the same recognition as in the black population.

We shall examine the genetics of this disease in Chapter 13, but for the present it is sufficient to note that as they pass through the narrow capillary beds, the RBCs normally must change shape in order to effect a passage. Sickle-shaped cells are evidently less able to undergo such shape changes and instead tend to rupture. The effect of such a rupture is not only a block of the capillary vessels but also a locally sharp increase in the osmotic pressure of the blood because of the liberation of the hemoglobin molecules. Localized swelling occurs because of osmotic differences, severe pain may also occur and, in areas where the capillary bed is serving critical tissue, as in the kidneys and the retinas of the eyes, permanent

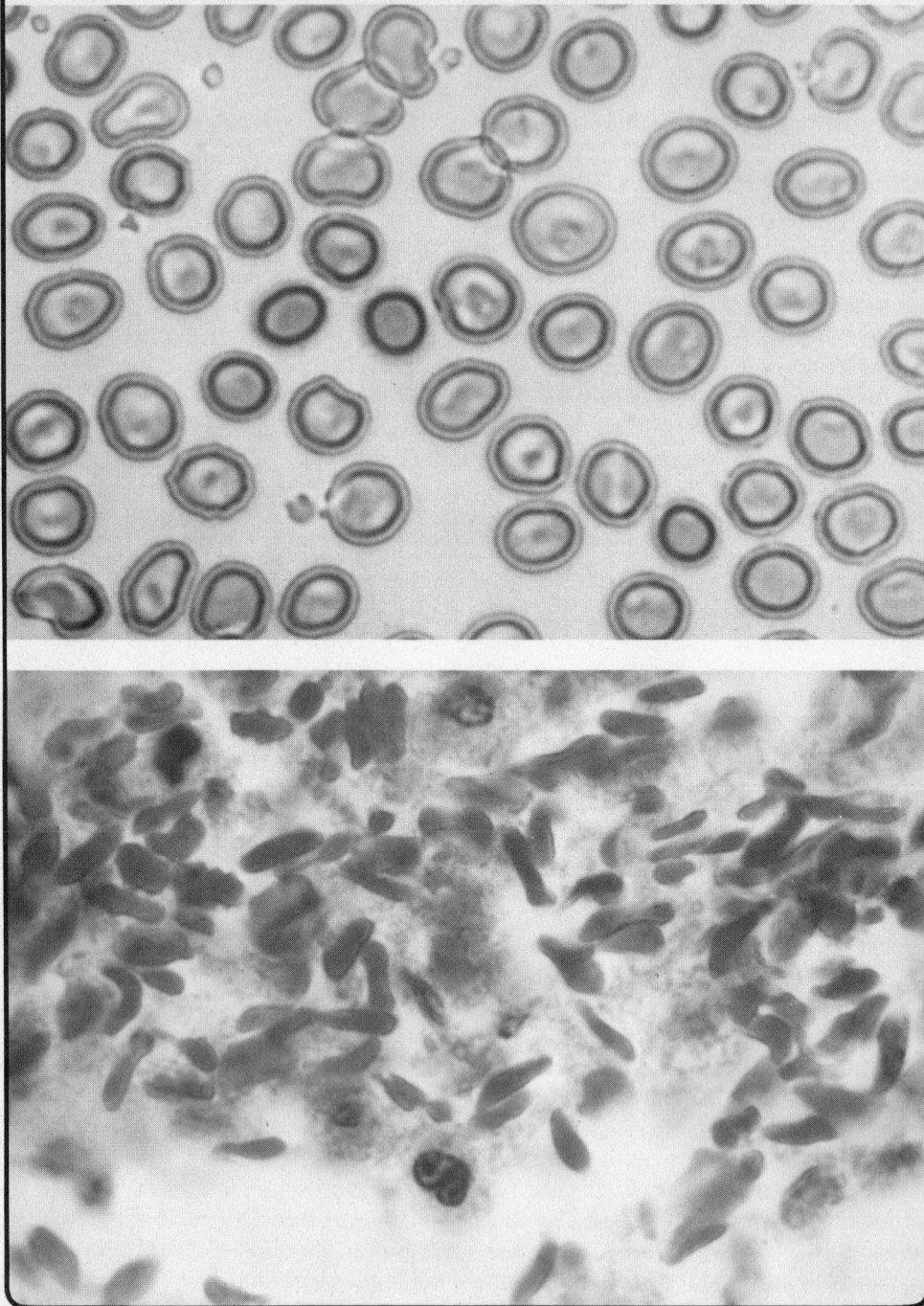

Fig. 5.3 (*Top*) normal and (*bottom*) sickled red blood cells in a person with sickle-cell anemia. (Rotker, Taurus)

tissue damage may result. The consequences of this rupturing indirectly illustrates the necessity of packaging hemoglobin molecules inside a cell membrane in the first place. If all the hemoglobin molecules were simply dissolved in the plasma, the consistency of the blood would be like molasses—and the osmotic gradient established would cause large amounts of water to be drawn from the surrounding tissues. The net result would be a huge increase in blood volume, without any increase in oxygen-carrying capacity of the blood. (Recall that it is the number—not the size—of the dissolved particles that determines osmotic concentration.)

**WHITE BLOOD CELLS**   White blood cells (WBCs) are less than 1/500*th* as common as red blood cells, numbering about 7,500,000 per cubic centimeter of blood. There are several different types, each with its own name, but it is sufficient for our purposes merely to distinguish their two major functions: engulfing of bacteria, and producing antibodies. Each of these functions is performed by a different type of WBC.

At one time or another, all of you have had slivers and you will have noted that, in addition to the irritation caused by its presence, there is a localized redness and swelling caused by increased blood flow to this area of invasion. If the area becomes "infected"—that is, if the body is unable to eliminate the bacteria present on the sliver before they begin multiplying and expanding—a great many WBCs may sacrifice themselves in attempting to repell the invaders. The collection of dead WBCs, damaged tissue, and bacteria is called **pus.** This also occurs in pimples, which are actually localized reactions to bacteria.

Invasion by bacterial cells results in a multiplication of the engulfing form of WBCs, although the same result does not generally follow from a viral infection. Therefore, when you have some vague illness and the doctor takes a blood sample, one of the things he checks is the WBC count. A higher WBC count (up to twice the normal level) suggests that the cause of the disease is bacterial, and some antibacterial drug, such as **penicillin,** may be prescribed. If there is no elevation in WBC count, the disease is probably viral in nature and is treated symtomatically if at all, because antibacterial drugs are ineffective against viruses.

The significance of WBCs is made graphically evident by the following two examples. The first example involves the manner of death following exposure to atomic radiation. As a result of the exposure, the high-energy rays cause a shutdown of WBC production. Death follows in about two weeks, as the bacteria, which are always present but are normally held in check, begin multiplying, creating respiratory infections and open sores in mucous membranes.

The second example is a consequence of the uncontrolled production of WBCs, a condition known as **leukemia.** Leukemia may involve a proliferation of engulfing WBCs (from bone marrow) or the antibody-producing WBCs (from lymph tissue). It may also be described as either **chronic** (wherein death may result from the metabolic drain of maintaining an abnormally high level of WBCs—up to 500,000,000 per cubic centimeter) or **acute** (wherein death may result because the WBCs cause a spreading of the disease to the lymphatic system proper, even though the WBC level never reaches that of chronic leukemia). Additional problems arise from the crowding out of RBC production sites in the bone marrow and from the fact that these abundant WBCs do not function properly.

**Immune response**   Another type of WBC functions in the immune response. The nature of an immune response involves the combining of one protein called an **antibody,** normally produced by WBCs within the body, with a foreign protein, called an **antigen** (from "**anti**body **gen**eration"). The mechanics of the immune response are exceedingly complex and well beyond the scope of an elementary text, but a brief consideration is essential because of the importance of this phenomenon not only in establishing immunity to disease, but also in creating problems in tissue and organ transplants.

The immune response begins with the entry of foreign material into the bloodstream. In order to act as an antigen and thereby trigger antibody production, the foreign material must consist of very large molecules, and, because the largest molecules of the body are primarily proteins, it is generally found that antigens themselves are proteins.

The presence of foreign protein in the blood ultimately triggers production of antibodies by a specialized type of WBC in the lymphatic system. A given WBC may divide five or six times in as many days and, with the continued presence of antigens, each generation of WBC produces larger amounts of antibody. This time lag accounts for the fact that many diseases last from 3 to 7 days—the time it takes for sufficient antibodies to be produced to destroy the antigen associated with the disease. During this tooling-up process the individual in whom it is taking place is said to be ill.

Antibodies, like enzymes, tend to be highly specific for particular antigens. Therefore, the fact you have antibodies for (i.e., are immune to) measles provides you with no capacity for resisting chicken pox. Like enzymes, antibodies appear to operate by binding with an active site on the antigen. If the antigen is on a bacterial cell wall, this binding may cause the wall to rupture and the bacterium to die. In some cases, the combining antibody causes bacteria to clump where they are susceptible to being engulfed by WBCs or of being destroyed in lymph nodes (*vide infra*).

Some of the WBCs retain the mold, so to speak, for the antibody in question. The cells retaining this mold may pass it along with future generations of cells and thus retain a lasting immunity. For this reason, they are called **memory cells.** The duration of this memory varies with the disease. For example, in most individuals, true measles, once caught, conveys lifelong immunity—that is, the body retains the capacity to produce antibodies effective against the pathogen on an indefinite basis. For other diseases, such as diphtheria or typhoid fever, immunity lasts for a period of years, but not necessarily for life. Still other diseases, such as most upper respiratory ailments (colds and the flu), generate only a short-term immunity.

One reason why immunity may be short-term is related to the specificity of the response. Although antigens and antibodies are generally very large molecules, their reactive sites usually involve only a small portion of the molecule. Thus, if through mutation, the viruses or bacteria alter the precise structure of the reactive portion of their surface protein, which acts as an antigen, the antibody that was formerly effective in combining with the antigen may no longer be capable of doing so and, as a consequence, there will be no resistance to this newly mutated strain. The viruses responsible for the diseases we call "colds" and "the flu" are very quick to mutate, which explains not only why vaccines to fight them are of limited effectiveness, but also why it is that you may become ill with a cold several times a year.

**Active and passive immunity**  Immunity itself can be subdivided into two major categories labeled **active** and **passive.** Active immunity occurs when the cells of the individual's own body produce the effective antibodies. This need not result only from contracting the disease, however. As you are well aware, for a number of serious diseases (smallpox, diphtheria, polio, whooping cough) most children in this country are immunized by being given a vaccination or "shot." For these diseases, which are both common and serious, vaccines have been prepared that consist of dead or severely weakened **pathogens** (disease agents, such as bacteria), or portions of pathogens, which, although incapable of generating the full-blown disease, are sufficient to provoke the production of antibodies by the cells of the body of the person being vaccinated. Note, however, that although an individual vaccinated with a weakened or dead pathogen can gain immunity to the pathogen without ever having contracted the disease, the effective antibodies are nonetheless produced by his own body. For some vaccines, a series of injections ("booster shots") may be necessary, because the initial amount of antibody produced is frequently small. Repeated exposures to the antigen promote a stronger antibody production and thus increase the resistance to the actual disease organism.

Active immunity is in contrast with passive immunity, which occurs when the antibodies themselves, which have been manufactured outside the recipient's body, are injected into the body of a person who has already been exposed to a potentially dangerous antigen. Passive immunization is used for such things as snake bites, rabies, or botulism poisoning. Antibodies are prepared by injecting rabbits, horses, duck eggs, and so forth, with a weakened strain of the antigen, and these antibodies are then removed and injected into the recipient. Immunity is passive in that the recipient's body does not itself produce the antibodies that nullify the invading antigen. To be sure, for most of these antigens it would be possible to have every person build up his own active immunity by injecting him with weakened or dead pathogens, but such a vaccination program is only carried out for the most common serious diseases. As the overwhelming majority of you will never be bitten by a snake or a rabid animal, there is little point in being vaccinated in advance on the chance it might one day happen. (People who frequently handle wild animals do receive rabies vaccinations just as your pet dog or cat does.)

Passive immunity sounds ideal—a short-cut to immunity, so to speak. However, there are problems with this procedure. First of all, because there is no continuing internal production of the antibodies, repeated injections may be necessary to ensure that a sufficient quantity of antibody is present at all times. (Repeated injections may not be necessary for such things as snake bite, wherein all of the antigens are injected with the bite, but are required for such diseases as rabies, for which the antigens continue to be produced as long as the disease agent remains alive.)

A second problem is that the body has no way of knowing that the injected antibodies are "good" proteins, but only that they are **foreign** proteins—and the body's natural response to a foreign protein is to produce antibodies to destroy it. Hence, we have the paradoxical situation of the body's producing antibodies not only to destroy foreign antigens, which may be life-threatening, but also to destroy foreign antibodies, which may be life-saving.

A third problem involves the mechanics of extracting the antibodies from the organism that produced it. For many years, antibodies for rabies were produced in horses, and then a small portion of the horse's serum (containing the antibodies) was removed and injected into the individual who had been exposed to rabies. However, the other proteins present in the horse's serum often caused a violent reaction in the individual being injected, a reaction that gave rise to the horror stories connected with rabies injections. In recent years, however, antibodies from duck eggs have been used rather than from horses, the violent reaction has been eliminated, and the whole process, although not yet pleasurable, is now more easily tolerated.

A second type of passive immunity is exemplified by the transfer of antibodies to the infants of nursing mothers in their milk. After a few weeks, the infant's own immune system begins to function, and the maternal antibodies are no longer required. Of course, bottle-fed babies do not receive this benefit.

**Allergies**  A second area of interest involving the immune response concerns the development of allergies. Allergies seem to result from an incomplete immunity— that is, there has been a recognition of a foreign antigen, but for some reason antibody formation has not ensued correctly. Normally, as antibodies are produced by the WBC, they are liberated into the bloodstream itself. In the case of allergic reactions, the antibodies are retained on the surface of the WBC, or even within its cytoplasm. Thus, the reaction between antigen and antibody, rather than occurring in the bloodstream occurs on the surface of the WBC, frequently resulting in a rupturing of the WBC. Rupturing of the WBC causes the release of the protein **histamine,** which has the effect of increasing the size of the small arteries, but not the veins. Thus, in the region of the body in which these WBCs have been destroyed and histamine has been released, there is an increase in blood flow to the area because of the larger arterial size, but blood flow away is not increased because the veins are not affected. As such, fluid accumulates in these tissues and the tissues become swollen. If the antigen involved is pollen, the areas affected are the mucous membranes of the eyes and nose—which, of course, as anyone who has hay fever can attest, results in swollen, reddened eyes and nose. Treatment may consist of receiving a series of desensitization injections designed to convey greater immunity (although these may have to be repeated at frequent intervals) or of taking a pill containing an **antihistamine,** which, as the name would imply, nullifies the histamine and hence its effects.

A more serious variation on this same theme occurs when the exposure to the allergy-inducing antigen is not limited to just a small area of the body such as the nose and eyes. Sometimes the allergic reaction is triggered farther down in the respiratory tract. If the tissue surrounding the tubes leading to the lungs becomes swollen, then the tubes themselves become compressed, and movement of air through them is impeded. This condition is known as **asthma;** again, the treatment is an antihistamine.

Even more serious consequences ensue when the allergy-causing antigen is injected into the bloodstream as the reaction is frequently body-wide. Such a situation may occur from bee stings or, more commonly, from a reaction to drugs such as penicillin. This body-wide reaction is called **anaphylactic shock,** and it can be fatal. In this case, the diameter of the small arteries increases throughout the body, and the blood tends to pool in abdominal capillary beds, thereby depriving the brain of blood. The consequences are virtually the same as if the person had lost a great deal of blood from a severe wound. It is out of concern for the dangers of anaphylactic shock that a doctor who has given you an injection of penicillin will ask you to remain in his office for a few minutes if you are not sure if you are allergic to the drug.

**Tissue transplants**   The third consequence of the immune system is that the body possesses the capacity to distinguish self from nonself. A simple example of this occurs in blood transfusions. Blood transfusion is the oldest form of tissue transplant, and it was recognized early in this century that not everyone's blood was the same. (Recognition of this fact came from frequent failures in transfusion, that is, the recipient died.) If of an incompatible blood type, the RBCs of the donor's blood form clumps because of the presence of plasma antibodies sensitive to blood types other than the individual's own. These antibodies are present from birth—no prior exposure to a foreign blood type is necessary. Because the amount of antigen introduced in a transfusion is very large, the reaction is also extreme. The clumps of RBCs ultimately lodge in capillary beds, such as those of the kidney. If there are enough such clumps, kidney function is impaired, and the individual may die.

On a more complex scale, the cells of any individual contain some proteins that are different from those of all other individuals (see the discussion of polymorphism in Chapter 13). By the presence of these differing proteins, the body recognizes foreign materials and produces antibodies to destroy the tissue. Therefore, in any organ transplant, the ultimate problem to be overcome is not the mechanics of the transplant itself, but the likelihood of rejection of the transplant. To be sure, some organs are more likely to be rejected than are others. Kidney transplants have a higher success rate than do heart transplants because the number of junctions between donor organ and recipient body is fewer in the kidney than in the heart, meaning that there are fewer interfaces where problems can arise. In any case, tissue typing is essential before a transplant can be undertaken. This process involves a determination as to how different—that is, how many proteins are involved—are the tissues of the host and the would-be recipient. Close relatives are frequently used in kidney transplants because tissue similarity is generally the greatest. (Because their tissue proteins are identical, rejection is not a problem when a transplant is made between identical twins.)

Regardless, a certain amount of drug therapy accompanies any transplant procedure. The problem is that drugs that tend to reduce the incidence of antibody formation to the newly transplanted tissue naturally also reduce antibody formation to pathogens. Hence, the doctor must walk the narrow line between too little medication, which would result in rejection of the transplant, and too much, which would open the door for death by secondary infection.

**PLATELETS**   Platelets, which are derived from cellular fragments, are roughly 35 times as common as WBCs in the bloodstream, numbering about 250,000,000 per cubic centimeter of blood. Their entire function is to prevent the blood from escaping from the vessels of the system by forming a plug or clot, in the same way that self-sealing automobile tires prevent total air loss from the tire.

**Clotting**   Clot formation is exceedingly complex (Fig. 5.4), but the essentials of it are as follows. Cells that are damaged (as by a cut) release a substance called **thromboplastin;** this substance is also present in platelets that rupture in the vicinity of cell damage. Thromboplastin acts as an enzyme to convert the plasma pro-

**When bleeding occurs–two processes automatically minimize blood loss**

(1) Constriction of blood vessels at site of bleeding   (2) Blood clot formation

**Fig. 5.4
Blood clotting.**

These two processes are initiated by breakdown of

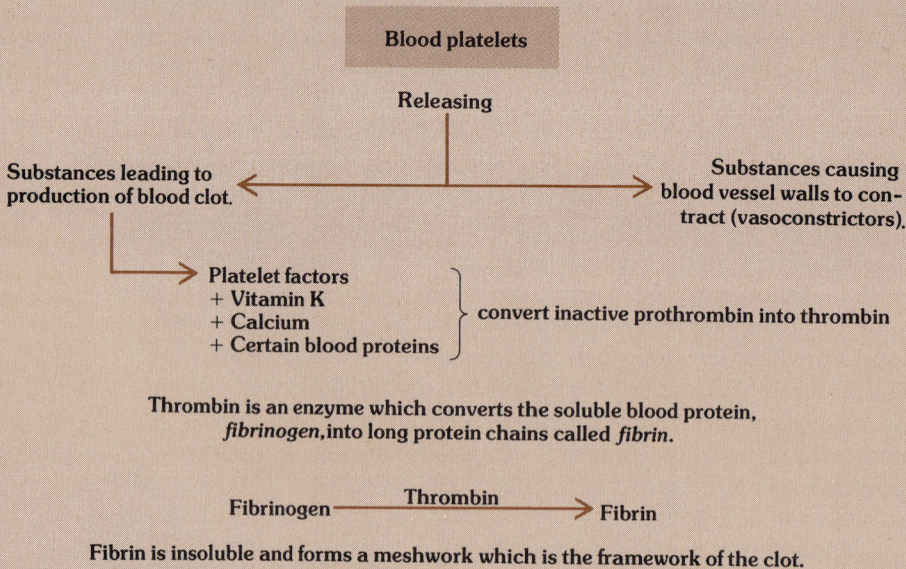

**Blood platelets**

Releasing

Substances leading to production of blood clot.

Substances causing blood vessel walls to contract (vasoconstrictors).

**Platelet factors
+ Vitamin K
+ Calcium
+ Certain blood proteins**

convert inactive prothrombin into thrombin

Thrombin is an enzyme which converts the soluble blood protein, *fibrinogen,* into long protein chains called *fibrin.*

Fibrinogen ——— Thrombin ———→ Fibrin

Fibrin is insoluble and forms a meshwork which is the framework of the clot.

Thus, all of the elements necessary to produce a clot are in the blood at all times, waiting for a sequence of reactions triggered by platelet breakdown at a damage site–resulting in the formation of an insoluble meshwork of fibrin.

tein **prothrombin** into its active form, **thrombin.** In turn, thrombin acts as an enzyme to convert the very abundant, but inactive, plasma protein **fibrinogen** into **fibrin,** which consists of long-chain, "sticky" molecules that are pulled over to the wound and gradually close it, trapping blood cells in the process, and thus forming a clot. Because sharp cuts, as from a razor blade, damage relatively few cells, not much tissue thromboplastin is released, and, as a consequence, such cuts take longer to clot.

Note that it is important that clot formation be restricted in the body to sites of actual vessel rupture—if this reaction occurred body-wide, all the blood would immediately form one giant clot, which would, of course, render the blood as useless as if it had all escaped from the original rupture.

**Abnormal clotting**  Unfortunately, clot formation is not always restricted to events as outlined above. Prolonged bed rest, which leads to a rather sluggish rate of blood flow, seems to promote clot formation. For this reason, people hospitalized for long periods are frequently given anticlotting drugs to prevent the random formation of clots inside vessels. Similarly, postoperative patients are urged to walk about as soon as possible after an operation. Abnormal clot formation may also occur in middle-aged and elderly people because of the development of uneven fat deposits inside the blood vessels themselves, which serve as sites for clot formation. Finally, such inflammatory diseases of the veins as **phlebitis,** with which former President Nixon was stricken, may cause the spontaneous formation of clots in the affected vessels. The danger of such clot formation is that the clots may break loose and lodge in a critical vessel. Typically, the clot will lodge as it nears the first capillary bed it encounters. Thus, it will pass through the veins as they increase in diameter, through the heart and thence into the arteries leading to the lungs. Should the clot lodge here—where it would be termed a **pulmonary embolism,** embolism being the name given to any object which blocks blood flow—death may result almost instantaneously.

**Anticoagulants**  The drug frequently given to minimize the likelihood of random clot formation is **heparin,** which is also added to the blood stored in a blood bank. If you have ever had a clotting test, you know that it consists of drawing your blood into a thin glass capillary tube and timing how long it takes to clot. (This is also a classic example of how tissue damage is not necessary to promote clotting, only the cessation of blood flow.) Think of the problems facing a mosquito, which must draw up blood through a much narrower tube. A clot in the middle of her proboscis would put Ms. mosquito out of business permanently. Thus, like all bloodsucking beasts, mosquitoes must produce an **anticoagulant,** which is any compound which interferes with one of the steps prior to fibrin formation.

**Other factors**  Two other related points deserve mention. First, vitamin K is necessary for the formation of prothrombin, without which the chain of reactions mentioned above cannot function. In the absence of vitamin K, excessive blood loss occurs through the spontaneous and frequent (but normally not serious) internal hemorrhages that occur in all of us. The second point has to do with **hemophilia,** an inherited trait manifested in a virtual absence of clotting ability. In this disease, the platelets do not release thromboplastin, which is, of course, the necessary first step in clot formation.

## HEART AND CIRCULATION
### Structure and function of the heart
The human heart (Fig. 5.5) is actually a double pump in the sense that, although left and right halves beat simultaneously, normally there is no flow between halves. Blood returns from the body low in oxygen and high in carbon dioxide and enters the right **atrium** through two major vessels—the **superior vena cava** (precava) from the arms and head and the **inferior vena cava** (postcava) from the body and legs. This blood then passes through the **tricuspid** valve into the right **ventricle,** as the atrium contracts. When the ventricle contracts (Fig. 5.6), the tricuspid valve prevents backflow into the atrium and the blood instead flows through the **pulmonary artery,** which quickly branches and carries this blood, still in a deoxygenated state, to both lungs. The blood passes through the capillary beds of the lungs where carbon dioxide is given off and oxygen taken up, and then returns to the left atrium of the heart via the **pulmonary veins.** The blood, now oxygenated, then passes through the **bicuspid** or **mitral**[1]

---

[1] The term refers to a fanciful similarity to the two-cusped hat, or mitre, worn by bishops. T. H. Huxley, the nineteenth century British biologist who coined the word "agnostic," provided anatomy students with a method of remembering that the mitral valve was located on the left side of the heart, by noting that, at least in his opinion, bishops were never right.

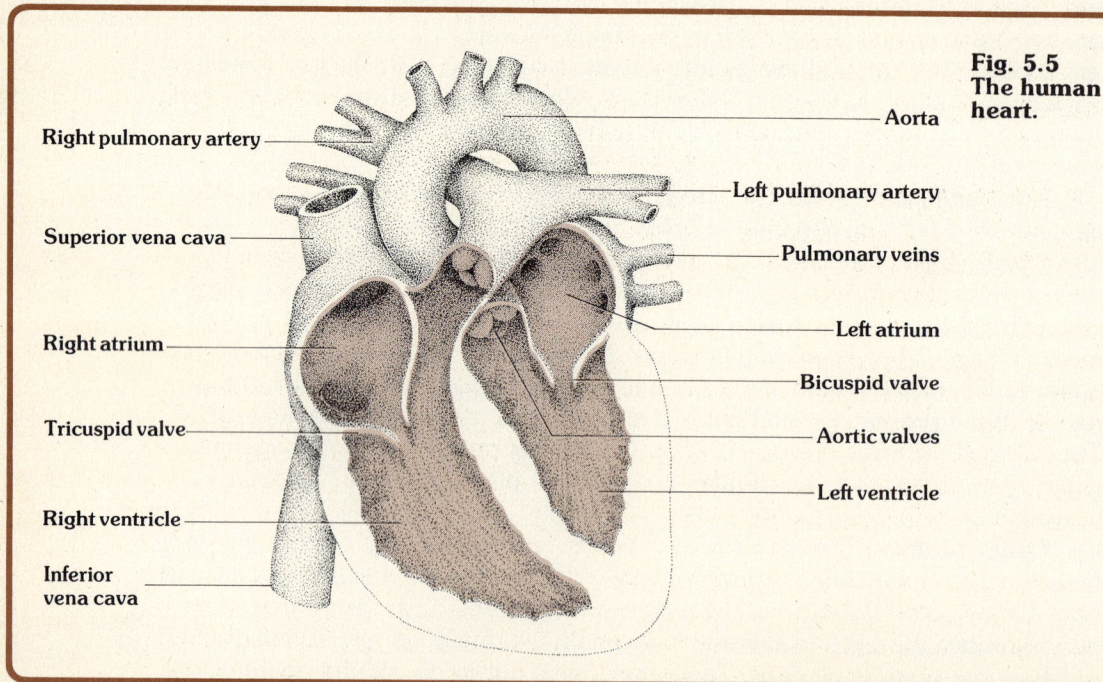

**Fig. 5.5 The human heart.**

Right pulmonary artery

Superior vena cava

Right atrium

Tricuspid valve

Right ventricle

Inferior vena cava

Aorta

Left pulmonary artery

Pulmonary veins

Left atrium

Bicuspid valve

Aortic valves

Left ventricle

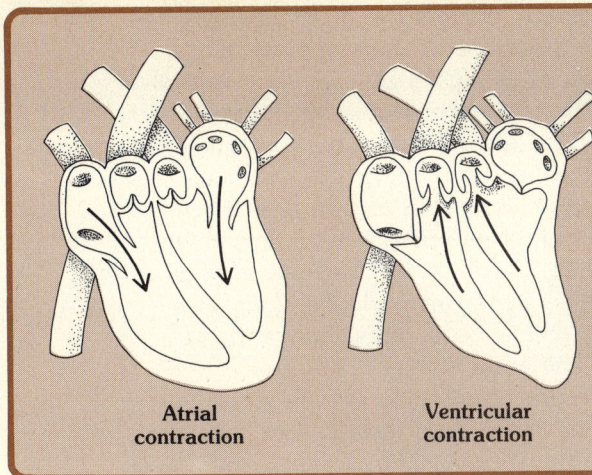

**Fig. 5.6**
**Heartbeat.**

Atrial
contraction

Ventricular
contraction

**valve** to the left ventricle as the atrium contracts. When the ventricle contracts, this valve prevents backflow into the atrium, and instead the blood passes into the **aorta,** which sends branches to all parts of the body.

Obviously, leakage of these valves poses severe problems to the proper functioning of the heart—and the valves are, in fact, subject to damage because of such illnesses as **rheumatic fever,** once a frequent consequence of untreated strep throats. Such damage is detectable by the presence of a modified heart sound called a **murmur.** A murmur, itself, is not necessary serious; the danger is that when valve damage is severe, the threat of **congestive heart failure** is markedly increased. In this condition, the blood is forced back into the atria as the ventricles contract. The usual consequence is for more blood to enter the pulmonary circulation (as it is usually the bicuspid valve which fails), and fluid may be forced into the lung tissue, causing the individual literally to drown in his own fluids.

Three points deserve special emphasis. First, note that the distinction between arteries and veins hinges on whether the blood is flowing toward or away from the heart, not on its oxygen content. Second, a heartbeat actually consists of a double beat, with both atria contracting simultaneously and just prior to the contraction of both ventricles. Third, the circulatory pattern consists of a double circuit, with the right half of the heart responsible for collecting deoxygenated blood from the body and conveying it to the lungs, and the left half responsible for receiving oxygenated blood from the lungs and sending it to all parts of the body—barring abnormal circumstances, there is no interchange of blood between these two circuits except in the capillary beds.

**Physiology of beat**  Although cardiac muscle has the ability to contract spontaneously, as can be shown from tissue cultures of free-floating heart muscle cells in an appropriate medium, in the intact heart there are two sets of controls that coordinate the muscle cells of the heart to ensure a smooth and powerful contraction. One of these controls is the **autonomic nervous system,** discussed in detail in Chapter 10. Suffice it to say that this is the system that is responsible, among other things, for speeding up the heartbeat when you are angry or frightened. The other controlling mechanism, and the one responsible for the normal rate of contraction, is the **sino-atrial node (S-A node)** so called because it apparently represents the remnants in ourselves and other mammals of the **sinus venosus,** the chamber that collects blood and empties into the atrium in the lower vertebrates. Normally, this node of tissue sends out impulses at the rate of 60–72/min to the heart muscle. The atria receive the impulse first because they are nearer the source and therefore are first to contract. The ventricle receives the impulse via the **atrioventricular node (A-V node),** which is connected to the S-A node by special tracts of tissue, and the ventricles contract a split second later. Obviously, the sequence is all important, because otherwise the ventricles would be contracting before they were filled with blood. In individuals who have suffered damage to the heart muscle, most typically because of a heart attack, and who have **arrhythmia** (irregular beat), an artificial stimulating device called a **pacemaker,** may be installed, which serves, as the name would suggest, to regulate the heartbeat.

**Blood pressure**   It would seem intuitively obvious that the blood must be at considerable pressure, as the ventricles contract, in order to be propelled through the maze of vessels that collectively make up the circulatory system. This blood pressure is at a maximum as the ventricle contracts, and at a minimum just before the next beat. The reason that there is a falling off of blood pressure between beats is both because of the friction of the blood along the walls of the vessels and because the vessels have the capacity to stretch somewhat. The energy required to stretch the vessel walls naturally comes from the pressure of the blood itself, and such an energy expenditure results in a reduction in pressure.

**How is blood pressure measured?**   In a typical person at rest, the blood pressure at its maximum (i.e., during ventricular contraction) is about 120 mm of mercury (mm Hg). (This method of expressing fluid pressures is standard—what it means is that the heart is exerting enough pressure to raise a 1-mm-wide column of mercury to a height of 120 mm.) The minimum (interbeat) pressure falls off to about 80 mm Hg. Under emotional or physical stress, where the demands of parts of the body for additional blood are increased, the heart not only beats more frequently, but more strongly, and consequently the blood pressure is increased. However, taking blood pressure by medical personnel remains a standard diagnostic aid because substantial and consistent deviation from the "normal" figures given above not only implies some imbalance or disease but may, in itself, pose a threat to the body because of the strain on the blood vessels. The cuff placed around your upper arm is inflated to a point where all blood circulation in your arm is cut off. The air is then let out of the cuff gradually, until such time as blood begins to spurt through the main artery of the arm, propelled by the beating heart. Frequently, a stethoscope is placed directly on the artery at a point close to the surface of the skin (typically the inside of the elbow), in order to aid the observer in detecting the beginning of flow. The pressure at which blood flow begins again represents the maximum output of the heart, and as we noted above, a figure of about 120 is typical. Additional air is let out of the cuff until a point is reached whereupon blood flow through the artery is continuous. This represents the minimum, or interbeat, blood pressure of which a typical value is 80. The full blood pressure reading includes both values and is written 120/80.

**Why is blood pressure measured?** Why is so much importance placed on blood pressure? Why is it necessary to record both figures? The answer to the first question is simply that high blood pressure **(hypertension)** poses the same dangers to the circulatory system (hence to the body as a whole) as would result from trying to force too much air into your bicycle tire—the possibility of a blow-out is dramatically increased.

The vessels of the circulatory system are, not surprisingly, best adapted to carry blood to the cells of the body at "normal" pressures of about 120 mm Hg. To be sure, there is a substantial margin of safety to allow for the higher blood pressures that normally accompany vigorous physical activity. The arteries and **arterioles** (small arteries) are thick walled and heavily muscled; under control of the nervous system, they have the capacity to expand and contract, causing a diminishing of blood pressure such that blood does not enter the thin-walled capillaries at a pressure so high as to cause them to rupture. A consistently high blood pressure, however, suggests that this safety system either is overloaded or is not working properly, with the result that the blood is entering the capillary beds at a pressure which may endanger them. Ruptures of these blood vessels may cause blindness, if the rupture occurs in the eye, kidney failure if in the kidney, or temporary or permanent paralysis if it occurs in the brain (where such ruptures are frequently called **"strokes"** or **cerebrovascular accidents**).

**Why does blood pressure fall?** The reason that both beat and interbeat pressures are recorded is that the difference between them (normally about 40 mm Hg) is a measure of the efficiency of the damping action (Fig. 5.7) of the arteries and arterioles. As long as these vessels retain their elasticity, there should be a substantial difference between beat and interbeat values. The absence of such difference therefore, suggests that the elasticity of the arteries may be substantially lost and that the individual may be suffering from "hardening of the arteries" **(arteriosclerosis).** Obviously, the dangers posed by even momentary rises in blood pressure are substantial in people with arteriosclerosis, and it is therefore most important that people suffering from hypertension because of arteriosclerosis be diagnosed as such and treated accordingly.

Arteriosclerosis of the arteries that supply the heart muscle itself (the **coronary** arteries) leads to a condition known as **angina pectoris** (literally "pain in the chest"). No longer capable of appreciable stretching, the arteries are unable to carry extra blood to the heart muscle during times of stress when the heart is beating more strenuously. Deprived of oxygen, the heart muscle "protests" in the form of a constricting pain. Fortunately, this is not a fatal condition, and the pain passes relatively quickly, especially if medication (in the form of nitroglycerin pills) is available.

**Atherosclerosis** Another condition which can lead to hypertension is **atherosclerosis,** which is a narrowing of the arteries due to the deposition of fats (Fig. 5.8),

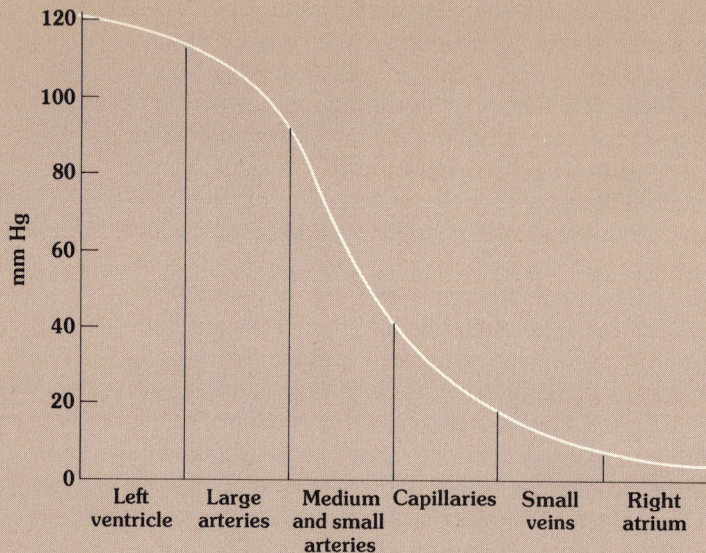

**Fig. 5.7
Changes in blood pressure in the different vessels.**

*y-axis:* mm Hg (0, 20, 40, 60, 80, 100, 120)

*x-axis:* Left ventricle | Large arteries | Medium and small arteries | Capillaries | Small veins | Right atrium

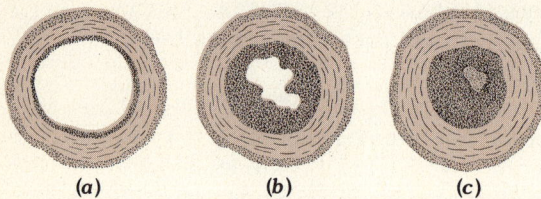

(a)    (b)    (c)

**Fig. 5.8
Cross section of an artery showing the development of atherosclerosis, and a coronary thrombosis. (a) Fatty material (dark area) is deposited around the inner lining of a vessel; (b) fatty material now clogs much of the passageway; (c) a thrombus has formed and the vessel is completely blocked.**

including the now infamous **cholesterol.** This condition is frequently accompanied by arteriosclerosis, and there is some tendency to treat both as variants of the same condition. Not only does a narrowing of the artery result in higher blood pressure (since the same amount of blood is being pushed through a narrower tube) but the irregular surface which results may initiate clot formation.

**Functions of blood pressure**    Blood pressure serves a number of functions, two of which are of particular interest here. The first function is obvious—the fact that the blood is under rather high, but stable, pressure ensures that the nutrients carried by the blood will reach the cells of the body not only in large amounts, but also in a very predictable manner. A situation could be conceived in which the blood was not under pressure but simply sloshed around the body—in large measure the insects, a highly successful group, employ this approach—but it would be unsatisfactory in large warm-blooded organisms such as ourselves.

The second function of blood pressure is somewhat less obvious. As we have already seen, a major task of the blood is to deliver oxygen and nutrients to all the cells of the body, and to carry away cellular metabolites such as carbon dioxide and nitrogenous wastes (Fig. 5.9). This exchange is effected only in capillary beds, which are comprised of vessels with walls only one cell thick and which are just wide enough to allow blood cells to pass through. Even given that our bodies contain an estimated 100,000 km (60,000 miles) of capillaries and that no cell is more than the width of one or two cells from a capillary, the problem of actually transferring these substances remains. That is to say, it is one thing to have materials needed by the cell in the proximity of the cell; it is quite another to get them from the blood across the capillary wall and through the cell membrane. In theory, there are three possible methods. First, we could rely on simple diffusion—but this is a relatively slow process in liquids, and the diffusion gradients are not large for most of the substances. Moreover, except for substances actually being used up by the cell, such as glucose and oxygen, we could expect to transfer no more than 50 percent of any given material from the capillary to the cell, given that, by definition, the laws of diffusion would predict equal distribution of materials on both sides of the capillary wall. In addition, using diffusion assumes, of necessity, that all these materials can successfully diffuse across a cell membrane (i.e., the wall of the capillary). Diffusion does, in fact, occur with glucose and oxygen, but cannot account for movement of many of the other blood components, such as salts. A second possibility is osmosis, but as we have seen, that is by definition limited to the movement of water only. A third possibility is active transport, but the energy drain on the body would

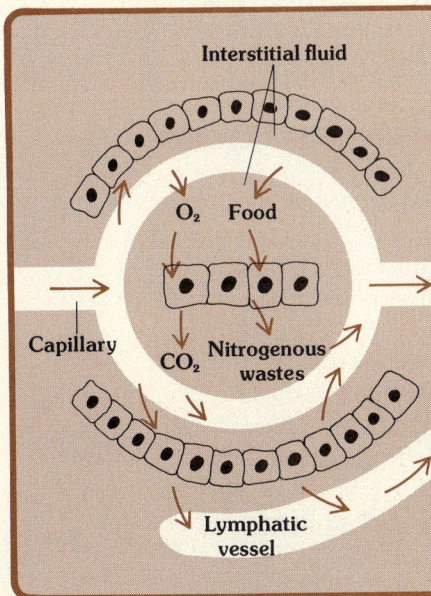

**Interstitial fluid**

**Fig. 5.9**
**Capillary-cell interchange.**

$O_2$   Food

**Capillary**

$CO_2$   **Nitrogenous wastes**

**Lymphatic vessel**

The formation of a clot, or **thrombus,** in a vessel is dangerous anywhere, of course, but particularly so in the coronary arteries. The thick muscular walls of the ventricles are not supplied directly from the blood which enters and leaves them with every beat as might be supposed, but rather from a separate pair of arteries that branches off the aorta. These are relatively small arteries (the heart, after all, is a small organ, weighing only about 400 g, or 9 oz), and are therefore particularly susceptible to blockage by a thrombus. Moreover, unlike the situation in many organs of the body, there are few auxilliary routes for the blood. Therefore, a thrombus in a coronary artery effectively prevents blood from reaching that portion of the heart muscle "downstream" of the clot. Because of the immense demands for oxygen and glucose on the part of heart muscle, the muscle cells, once deprived of these essentials, die rather quickly.

There are two principal consequences of the death of these heart muscle cells. First, if the area involved is not large, the "heart attack" (**coronary thrombosis** or **myocardial infarction**) may not be fatal. The muscle cells will not regenerate, but will be replaced by scar tissue. However, if the area deprived of blood is large, enough of the heart tissue may die that effective contraction is no longer possible and death ensues immediately.

The second consequence of a coronary thrombosis is caused by the death of the heart muscle cells deprived of blood. The dying cells interfere with the passage of impulses from the S-A and A-V nodes with the result that the ventricle may begin to twitch at a high frequency, as opposed to contracting regularly at a low frequency. This is called **ventricular fibrillation,** and death is imminent unless a normal heartbeat can be restored. If the individual is being monitored in a hospital, the "paddles" (which seem to be used at least once a week on the doctor shows on television) may be employed to restore the heartbeat electrically. The drug **lidocaine** is also frequently used as soon as a heart attack is diagnosed, as it acts to prevent ventricular fibrillation.

**Heart attacks**

seem prohibitive, given the sheer volume of material that travels between capillary and cell.

Having disposed of our three alternatives, what is left? The answer is blood pressure. Tiny gaps between the cells making up the capillary wall allow all the small molecules to be squeezed out of the capillary by blood pressure, in the same way that water is forced out of a perforated hose when you water your lawn. Left inside the capillary are the plasma proteins, which are too large to pass through these openings, and the blood cells themselves, along with a portion of small molecules, including water. (The capillary should not be thought of as containing only randomly scattered cells and proteins, left high and dry like a series of tiny beached whales. Not all the water and small molecules leave, for reasons that will presently become apparent.)

However, getting the substances needed by the cells to them is only half the battle—and, in fact, that is the easy half. How are cellular wastes returned to the blood? Equally important, how are all the water and excess small molecules returned? The body obviously does not have the reserves to allow these materials to take one-way trips only. The answer lies in the plasma proteins. Since they are free molecules, they exert enormous osmotic pressure. Not only are they left in a highly concentrated form, given the loss of water to the tissues, but the interstitial fluid bathing the tissues is now rather dilute (from the standpoint of dissolved materials) because of the influx of water from the capillaries. The increased viscosity of the blood together with its movement through the tortuous passages of the capillary bed, has resulted in a sharp drop in blood pressure, whereas the osmotic differences between the blood and the interstitial fluid has actually risen somewhat. The net result is that, at the venous end of the capillary bed, most of the water returns to the capillary down an osmotic gradient—and this water brings with it the excess salts, plus the waste product of the cells. Thus, the body takes advantage of the necessity of having to have blood pressure to ensure a constant source of fresh blood to the cells by using this same factor to push water and nutrients out of the capillaries to the cells—and the presence of the plasma proteins ensures that sufficient osmotic pressure will develop to allow a return of these materials plus the cellular metabolites.

**Movement in veins**   One problem still remains. Given that it is necessary for the blood pressure to be rather low as the blood enters the capillary beds (otherwise the capillaries would rupture), and even lower as the blood leaves the capillary beds (so as not to nullify the osmotic gradient which is responsible for the return of materials to the blood), how do we get this venous blood, at very low pressure, back to the heart? Why do the effects of gravity not cause all the blood to settle to the feet?

Because all the blood does not settle to the feet, obviously there is a solution to this dilemma. As it happens, the veins are equipped with a series of one-way valves along their length, which allow blood to move in one direction only—toward the heart. In a sense, these valves operate like a bucket brigade, with the blood being "passed on" from one valve to the next.

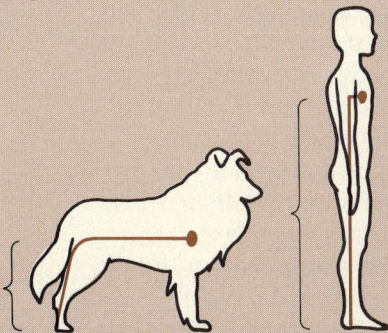

Fig. 5.10
**Differences in the vertical height venous blood must be moved in quadrupedal versus bipedal organisms.**

Have you ever felt momentarily dizzy when you stood up suddenly? While you are lying down, the effects of gravity on circulation are minimal, because no point of your body is more than a few cm below your heart. When you are standing, the blood in your feet may be a meter or more below your heart. Obviously, there must be some change in circulatory pattern to maintain a sufficient blood supply to the head, once the body is subject to the effects of gravity.

The veins are highly distensible vessels, normally possessing a considerably larger inner diameter than do the arteries. Indeed, more than 60 percent of the blood is normally found in the veins. In order to prevent pooling of the blood in the legs in an individual who is standing erect, the skeletal muscles in the legs function to squeeze the veins and thereby decrease their diameters. Because smaller vessels can hold less blood, there is less pooling of the blood in the veins and greater return to the heart.

**The dangers of standing up**

There are limits to this method of compensation, however—limits which we exceed, in certain situations. For example, soldiers often faint when forced to stand at attention for long periods, because their leg muscles are not active and their leg veins are not compressed. For that reason, soldiers are instructed to wiggle their toes while at attention, an activity that will involve some contraction of leg muscles. Similarly, test pilots and astronauts may black out once gravitational (*g*) forces become sufficiently great so as to force blood back into the legs and away from the brain.

But even given that the blood can flow in one direction only, how does the blood move at all, given the very low pressure that is present in veins? The muscles of the body play an important role. Many of the veins are surrounded by skeletal muscle, and, as these muscles alternately contract and relax, blood is forced through the veins. Normally, this is an efficient process even though, because of our rather recent bipedal locomotion, the human heart is $1-1\frac{1}{4}$ m ($3\frac{1}{2}-4$ ft) above our feet (our quadrupedal ancestors had to force blood only the length of the hind limb— see Fig. 5.10). However, under certain circumstances, problems may arise. A number of leg veins run very close to the surface of the skin. As long as the skin is taut and the muscles below are active, blood continues to flow rapidly through these veins. As the skin becomes less taut because of advancing age, or because of a sudden weight loss, or if the individual is standing a great deal but not moving about (i.e., the leg muscles are not in fact alternately contracting and relaxing), blood may tend to pool just above each valve, giving a swollen and knotted appearance to the veins. (This is the same phenomenon so eagerly courted by teenage boys with reference to the veins of their arms; the appearance of knotted veins bulging through the skin as the fist is clenched is typically used as an index of strength and maturity.) Ultimately, the veins may become permanently stretched and the individual is said to have developed **varicose veins.** These are not only unsightly, but because of their exposed position, are relatively easily injured, and the individual may elect to undergo a "stripping" operation, in which the vein is surgically removed; returning blood is transported through other veins, for one of the important features of the circulatory system is the presence of secondary routes of blood flow.

Gravity continues to take its toll on venous blood pressure as the blood moves from the legs to the abdominal region and, in fact, the blood pressure may be negative (with respect to atmospheric pressure) in the postcaval vein just before it empties into the right atrium. Movement over this last short distance is aided by the process of breathing. As we shall see shortly, inhalation involves the formation of a partial vacuum, which not only causes air to rush into the lungs, but also acts to assist the blood to pass from the postcava into the heart.

During late pregnancy, the fetus is so large that the visceral organs become packed together, permitting less room for the diaphragm to move during inhalation, and causing breathing to be more shallow. Both the packing of the organs and the interference with the diaphragm act to decrease the efficiency of venous return from the legs. Partly for this reason, varicose veins often arise as a result of pregnancy, a fact which helps to explain why varicose veins are far more common in women than in men.

**THE LYMPHATIC SYSTEM**   We have spoken earlier of the methods by which fluids pass from the capillaries to the tissues of the body and back again. It is easy to see that if this process were less than 100 percent efficient, there would be progressive fluid loss from the circulatory system. It is a primary task of the lymphatic system to forestall this possibility. The lymphatic system (Fig. 5.11) consists of a series of vessels which penetrate most of the organs of the body (the central nervous system is a significant exception) and which provide an alternate route by which interstitial fluid can be routed back to the circulatory system proper. The process by which this is accomplished parallels that of the venous system. Within the lymphatic vessels there is abundant protein, which has the effect of creating a sizable osmotic gradient running between the tissues and the vessels. Movement of the lymph itself is largely by one-way valves and muscular activity—there is no lymphatic pressure to speak of. The fluid is emptied into the circulatory system proper at the point of lowest blood pressure (to prevent movement the wrong way)—that is, into the large veins just before they enter the right atrium.

The significance of the role played by the lymphatic system can perhaps best be understood by examining the effects of two diseases. **Elephantiasis** is a condition resulting from the blockage of lymphatic channels by a parasitic worm. Typically, this condition affects an arm or leg, but other areas may be involved (e.g., the scrotum). In these areas of the body there is no way for fluid lost from the circulatory system to be returned. As a consequence, the build-up of fluid in these areas frequently causes massive swelling of the affected region (Fig. 5.12).

Similarly, the disease **kwashiorkor,** caused by severe protein deficiency (see Chapter 4), has an effect on the efficiency of the lymphatic system. As an individual starves, his carbohydrate and then his fat reserves are utilized first, to provide energy for the rest of the body. Ultimately, protein is broken down for energy and

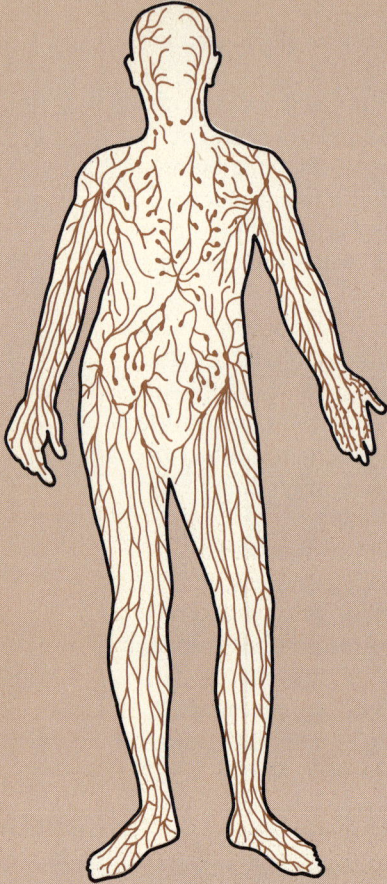

Fig. 5.11
The lymphatic system. Spots
indicate lymph nodes.

victims of kwashiorkor are typified by matchsticklike arms and legs (Fig. 5.13).
They also possess greatly distended abdomens, giving the false impression that
they have just engorged an enormous meal. In reality, this swollen abdomen results
from the pooling of interstitial fluid in this region. Among those proteins broken
down for energy are those in the lymphatic system itself; without the proteins to provide the necessary osmotic gradient, the system ceases to work efficiently.

The lymphatic system has another, and equally important role. At intervals
along the system of ducts are arrayed numerous lymph nodes that have the function of monitoring the fluid passing through the canals. The presence of bacteria or
viruses triggers the production of those white blood cells responsible for engulfing
and for antibody production, for these cells are abundant in the lymphatic system
as well as the circulatory system proper. It is the swelling of these nodes, caused by

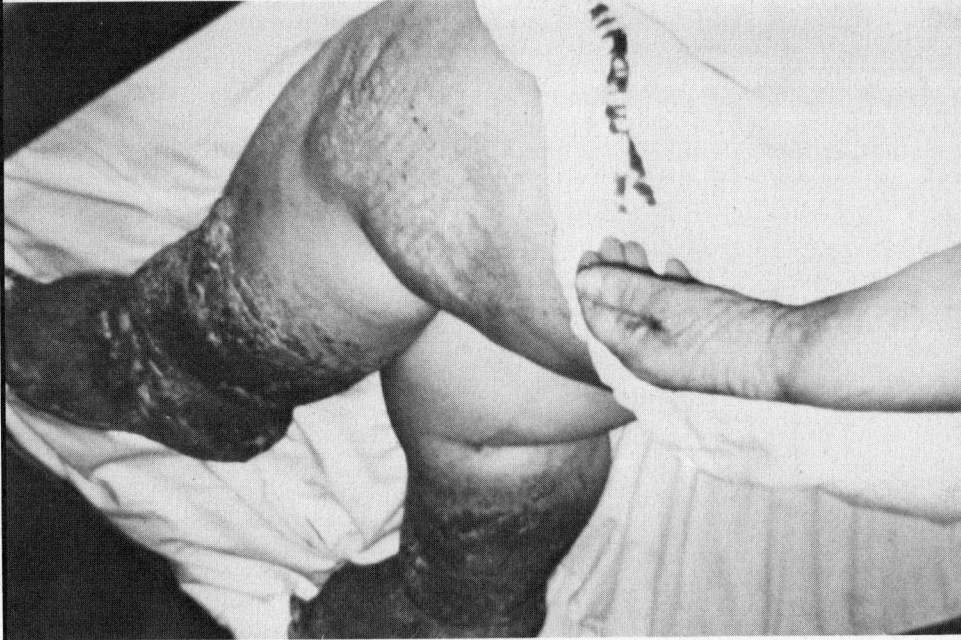

Fig. 5.12 Elephantiasis, which results from the blockage of the lymphatic system by a parasitic worm and causes grotesque swelling below the point of blockage. (Lester Bergman and Assoc.)

Fig. 5.13 Child with kwashiorkor, caused by a diet deficient in proteins. The word is Ghanese, meaning literally "red boy." (Lester Bergman and Assoc.)

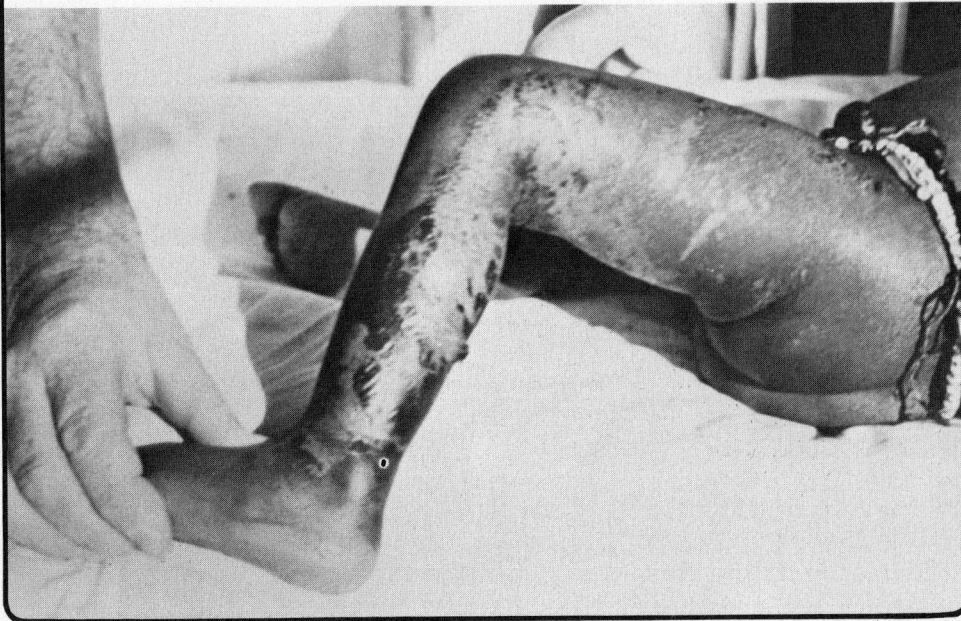

the rapid generation of additional WBCs (or sometimes, as in the case of mumps, due to infections of the nodes themselves) particularly in the region of the neck and armpit, which causes the swollen "glands," so characteristic of many diseases.

## SUMMARY

The average person tends to think of the circulatory system as a pump and a set of sealed tubes through which the blood must course as it transports oxygen to the cells and carbon dioxide back to the lungs. As we have seen, the circulatory system is much more complex than the simple model just given.

To begin with, such a model ignores the very significant problem of movement of materials from the blood to the cells of the body which, as we have seen, can only be partially accomplished by diffusion. The model also fails to take into consideration the dynamic interrelationship between blood and interstitial fluid, and between interstitial fluid and the lymphatic system. Hence, on purely anatomical grounds, this commonly held model of the circulatory system is seriously deficient.

The model is even more deficient on physiological grounds. Variations in the level of body activity demand blood to be supplied at different rates—therefore, the heart must be more than just a simple pump. There must be a mechanism for reducing blood pressure before the blood reaches the thin-walled capillaries. As we have seen, the blood also contains agents to combat invading organisms, as well as specialized proteins which act to plug all but the largest leaks in the vessels.

In short, the common model of the circulatory system as a set of inside plumbing is much too simplistic to be of much use. A somewhat more adequate model would be to compare the circulatory system with the water in an aquarium, and to think of the fish of the aquarium as being the cells of the body. The water is the source of food and oxygen for the fish and serves as well as their sewage system. The filter, which circulates and cleans the water, is the combined action of the heart and the kidneys; the air pump is the lungs. The model is not complete, of course— the aquarium cannot generate more water, it cannot prevent leaks, it cannot destroy invading organisms—but it is a useful model in that it forces us to think of the cells of our body as being afloat in an internal sea, rather than as a solid mass of tissue penetrated by an irrigation system.

We generally think of hiccups as nothing more than an occasional nuisance, but in the years before his death, Pope Pius XII (1876–1957) suffered repeatedly from serious attacks of hiccups, which lasted for weeks at a time and prevented him from sleeping. What are hiccups, and how do they relate to the respiratory system?

Every year a number of swimmers drown after breathing deeply several times prior to an underwater swim. How is it that increasing the oxygen level of the blood by deep breathing can lead to drowning?

In 1900, the two leading causes of death in the United States were tuberculosis and pneumonia, both of which are lung diseases. In 1970, tuberculosis was of minimal significance as a cause of death, but another lung disease, emphysema, is now the tenth leading cause of death —a disease that was unknown in 1900. Why have the lungs always been so vulnerable a pair of organs?

These are the types of questions we shall consider as we discuss the functioning of the respiratory system.

chapter **VI**

# chapter VI
# Gaseous exchange– the respiratory system

Respiration refers to the metabolic utilization of oxygen by the body. It is exclusively a cellular phenomenon—every cell in the body is capable of utilizing oxygen in its own metabolic processes. But if respiration is a **cellular** event, how can there be a respiratory **system?** The need for physiological accuracy should compel us to substitute another term for the network of tubes (Fig. 6.1) leading from the nose and mouth to, and including, the lungs ("ventilation" is more correct, for example), but the historical usage of "respiratory system" is so completely entrenched that it is unlikely the phrase will ever be replaced. This double usage of "respiration" is rather confusing, and it is important that you distinguish organ system from cellular event when you see the term used.

Moreover, it would be a mistake to dismiss this distinction in usages as a mere semantic exercise. Once again, we have an excellent example of the point made in Chapter 2, namely, that organ systems do not tend to take over cellular functions, but instead operate primarily to regulate the environment in which the cells must live. The primary role of the "respiratory system," then, is to supply oxygen to the cells and to remove carbon dioxide (a cellular waste product) from the body, with the circulatory system operating as the go-between that links the lungs with the individual cells of the body. Thus, just as the digestive system does not function by utilizing food for energy, the lungs do not "use" oxygen, but rather function simply to provide the cells of the body with this vital substance.

**MECHANICS OF BREATHING**   The chest cavity is virtually surrounded by an armor of ribs and muscles. Applied against the interior wall of the chest cavity is a shiny, slippery tissue called the **pleura.** The chest is almost entirely taken up by a pair of **lungs** which, despite their size, together weigh only about 1 kg (2.2 lb). The outer surface of the lungs is also covered with pleura, and, as the lungs are alternately filled and emptied, they are protected from abrasion against the inner wall of the chest cavity by a thin film of lubricating fluid produced by the pleura, in the same way that the wheels of your car move on the axle protected by a thin film of oil or grease. Just as your car can develop squeaks if it is not lubricated regularly, so, too, can inflammation of the pleura (a condition known as **pleurisy**) operate to interfere with its lubricating capacities, which leads to a situation whereby the very act of breathing becomes a painful experience.

Each lung is completely enclosed in its own chamber, the only opening being through the **bronchi**—the pair of tubes that carries air from the nose and mouth to the lungs. Not only are the lungs isolated from all the other body organs, but they are even isolated from each other. This is a very important point, for it means that if the wall of the chest is breached by an injury thereby creating a new opening to the chest cavity, only one lung will collapse. Once the breach is closed and healed, the collapsed lung will generally reinflate spontaneously, but the ability of the remaining lung to continue to function normally quite literally spells the difference between life and death.

**Fig. 6.1
The human
respiratory
system.**

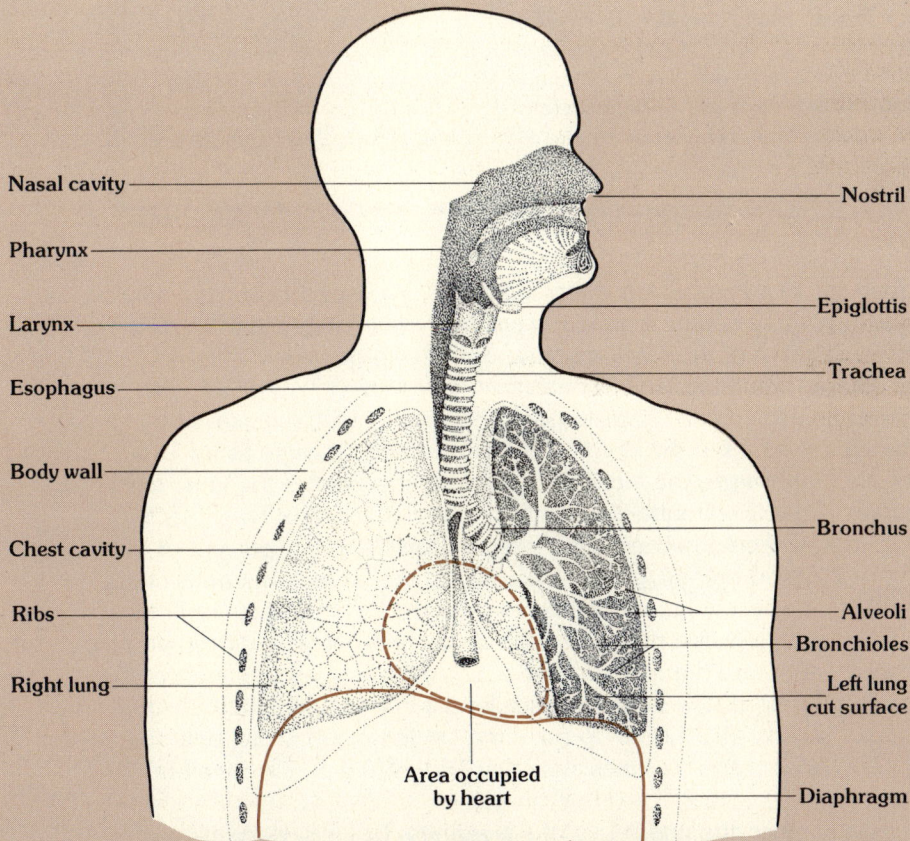

Nasal cavity

Pharynx

Larynx

Esophagus

Body wall

Chest cavity

Ribs

Right lung

Nostril

Epiglottis

Trachea

Bronchus

Alveoli

Bronchioles

Left lung
cut surface

Diaphragm

Area occupied
by heart

Perhaps the classic testimonial of the tendency of biological systems to avoid waste is the utilization, by mammals and certain other animals, of air during exhalation for the production of sounds used in communication. In ourselves, we term such sounds "speech."

The mechanics of sound production in humans and other mammals is fascinating. Between the pharynx and the trachea is the larynx, or voice-box. The larynx consists of a skeletal framework of cartilage, believed to be derived from the support structures of the gills of fish, our distant ancestors, across which are stretched a pair of tough, fibrous strands, the vocal cords. These are always folded to the side during inhalation (try talking sometime as you inhale—it cannot be done), but during exhalation, they may be engaged if their owner wishes to speak. Within broad limits, we can control the frequency the sounds produced (i.e., whether high- or low-pitched) by increasing or decreasing the tension on the vocal cords in the same way that a guitar string may vary in the note it produces depending on how much it is tightened. Overall length of the vocal cords is also an important determinant of frequency, and, just as you can effectively shorten your guitar string by pinching it between your finger and a fret to play a higher note, women tend to have higher voices than do men because their vocal cords are shorter. The longer vocal cords of men require a larger larynx, which is why the larynx is more prominent in men than in women. This prominence of the larynx in men accounts for its common reference as the Adam's apple (rather than Eve's apple).

**The larynx**

We have seen in Chapter 4 how, through the *active* processes of chewing, swallowing, and peristalsis, the essential liquids and solids are ultimately put in contact with the absorptive surfaces of the small intestine. Breathing, however, is not entirely analogous. Rather, breathing can be considered to be analogous with the operation of a bellows or an accordion. The volume of the chest cavity is increased by the action of muscles attached to the ribs that act to raise the rib cage, as well as by the action of the **diaphragm,** the large, flat muscle that separates the lungs from the digestive organs (Fig. 6.2). Contraction of these muscles is, of course, an active process (although not very energy-consuming—at rest, only about 3 percent of our energies are directed toward breathing), but the filling of the lungs is not an active process. The increase in chest volume operates to produce a partial vacuum in the lungs, and air rushes in to equalize pressures both inside and out. The difference in pressures is generally not more than 1 mm of mercury (mm Hg) (as a reference point, note that normal atmospheric pressure at sea level is 760 mm Hg), but it is sufficient to cause about 500 ml (1 pt) of air to rush into the lungs. (Actually, only about 350 ml (0.7 pt) of air enters the lungs; the other 150 ml (0.3 pt) fills the mouth, pharynx, and bronchi.)

Exhalation is very much the equivalent of squeezing a bellows. Relaxation of the chest muscles and diaphragm (a passive process) results in a diminished volume of the chest cavity, and about 500 ml (1 pt) of the air in the lungs and pharynx is squeezed out.

The picture presented, then, is one of a repeated movement of about 500 ml (1 pt) of air in and out of the lungs as you breathe. This amount is called the **tidal volume,** and it represents just a fraction of total lung capacity. At the maximum, the lungs are capable of holding about 6000 ml (12 pt) of air, but no more than 4500 ml (9 pt) can be exchanged at any one breath. Even during a forced exhalation, in which the muscles of the abdominal wall are contracted, forcing the organs of digestion up against the diaphragm, 1500 ml (3 pt) of air remains in the lungs, as they cannot collapse totally between breaths.

The fact that, of a total lung capacity of 6000 ml (12 pt) fully 25 percent is incapable of being utilized at any given time is a troublesome finding, as it suggests inefficiency, a failing we would hope not to see in the organ system design of our bodies. There is no question, however, that a "blind sac" design (utilizing the same

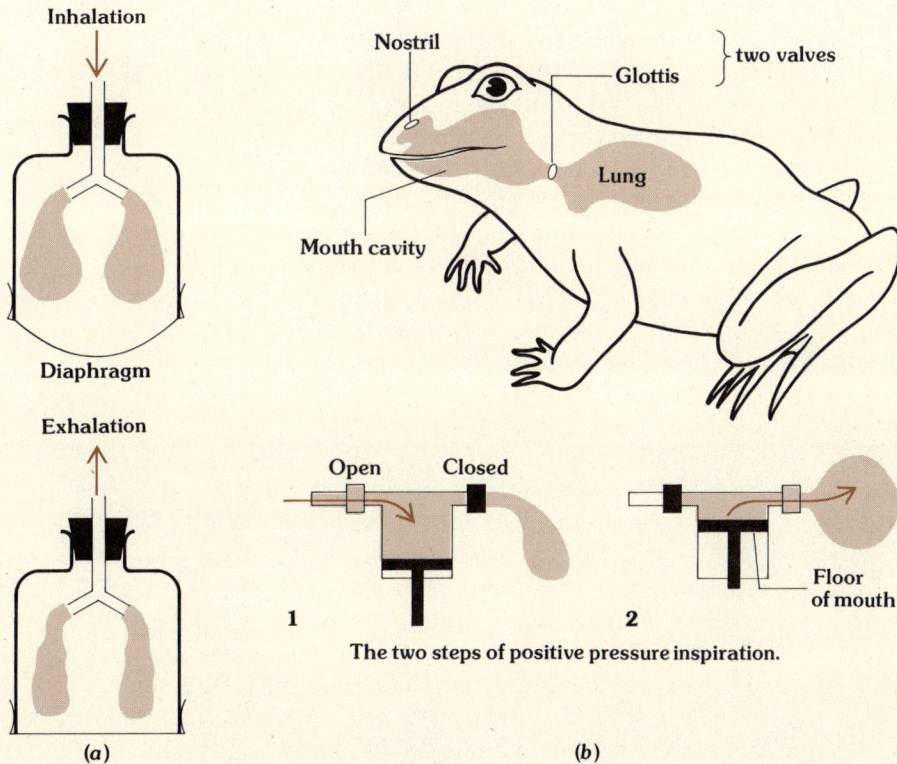

Fig. 6.2 (a) Diagrammatic view of the functioning of the diaphragm and filling of the lungs; (b) lung-filling in the frog, using positive pressure.

opening as entrance and exit) is inefficient, a conclusion bolstered by the fact that this same design was initially utilized in the digestive system of multicellular animals, but was soon abandoned by more advanced species in favor of the "tunnel" design. In contrast, the respiratory system of birds is far more efficient (Fig. 6.3). Birds require even larger amounts of oxygen per unit body weight than we do because flying requires huge amounts of energy, yet the lungs of birds are considerably smaller, relative to the lungs of mammals of the same body size. Birds utilize a system whereby the lungs are open at both ends and air is passed over them in a continuous stream, in the same way that water is passed over the gills of a fish. Thus, the entire lung surface is constantly being bathed by fully oxygenated air, and there is no need for the lungs of birds to be large enough to allow for a residual volume, as in ourselves.

**Gas exchange**  The lungs are not simply smooth-walled balloons, but, in an analogous manner with the digestive system, they possess various infoldings that serve to increase their surface area. These infoldings in the lungs take the form of some 3,000,000 small, cuplike structures called **alveoli.** Collectively, they increase the surface area of the lungs to the point that, if flattened out, the lungs would equal the area of a 24 × 30 ft living room.

The walls of the alveoli are extremely thin, and air in the lungs is separated from the blood in the capillary by only 2/10,000 mm (0.000008 in.) (see Fig. 6.4). (If you consider that the diameter of a red blood cell is about 70/10,000 mm [0.00003 in.] you will appreciate just how narrow a separation is involved.) There is no active transport system for oxygen, but because the distance is so small, diffusion operates quickly to move oxygen from the lungs to the blood.

If we consider (1) that only 21.7 percent of the air is oxygen, (2) that the average amount of air actually reaching the lungs with each breath is about 350 ml (0.7 pt), and (3) that we breathe approximately 12 times per minute, we can calculate that the volume of oxygen entering the lungs each minute is $(0.217 × 350 × 12) =$ approximately 900 ml (1.8 pt). Of this amount, only about 200 ml (0.4 pt), is actually

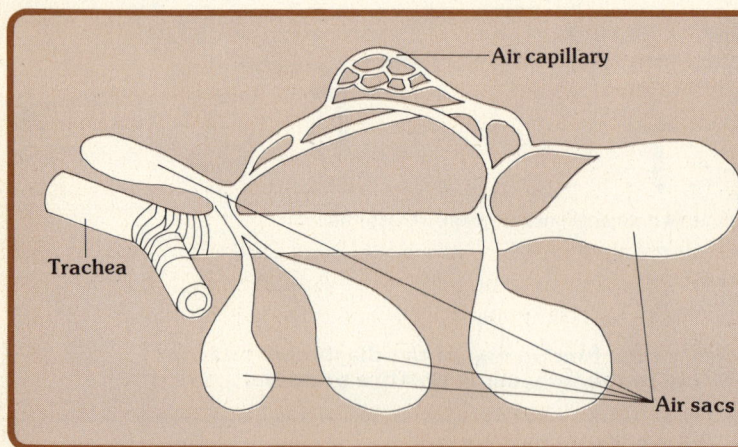

Air capillary

Trachea

Air sacs

Fig. 6.3
Respiratory system in birds.
The lung proper is limited to the area immediately surrounding the air capillaries.

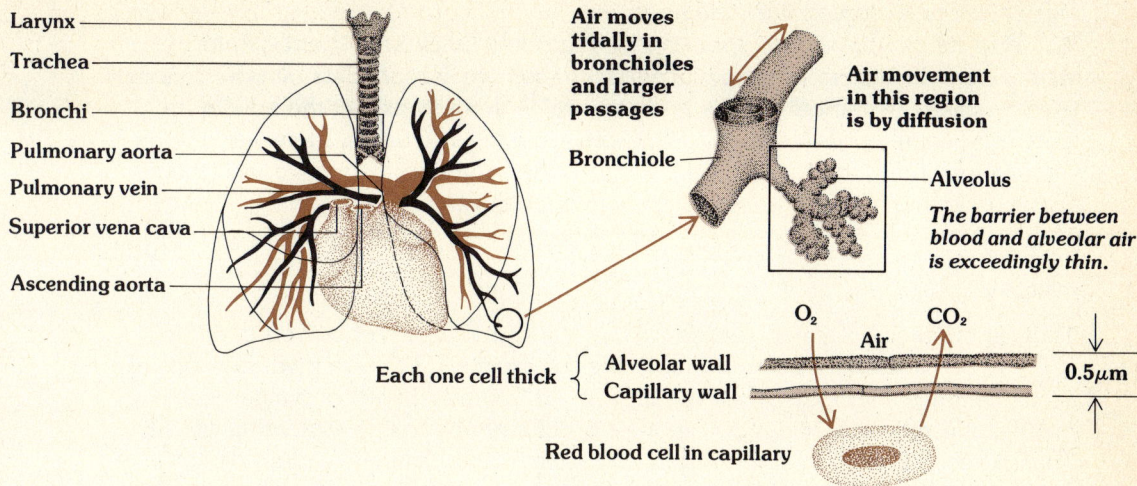

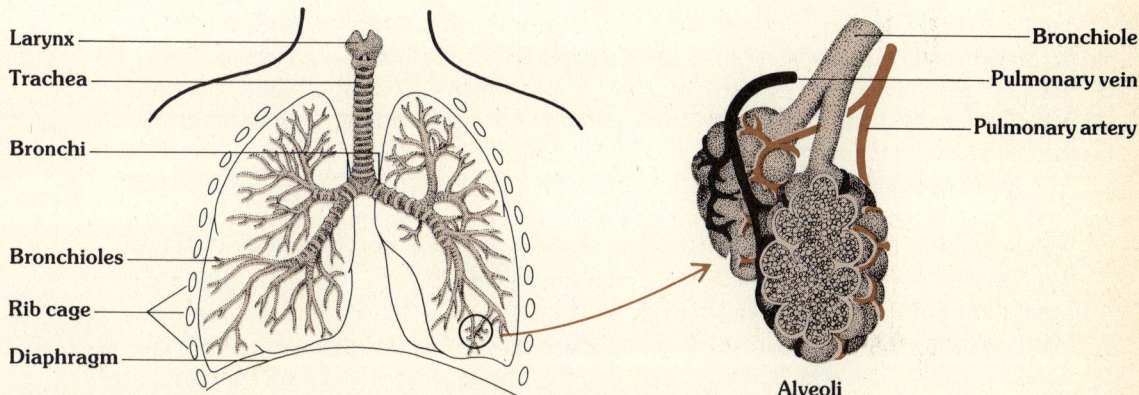

**Fig. 6.4** (*a*) Gas exchange in the lung; (*b*) close-up view of the fine structure of the lung.

taken up by the blood; the other 700 ml (1.4 pt) passes back out of the lungs with each exhalation. Again more inefficiency, right? In actuality, this is not so much inefficiency as a reserve that can be drawn on if needed. From our discussion of the circulatory system, we know that the blood entering the lungs via the pulmonary arteries is deoxygenated, but this is a relative term. In fact, this blood contains about 800 ml (1.6 pt) oxygen per liter, compared with 1000 ml (2 pt) oxygen per liter for oxygenated blood. Thus, normally, 200 ml (0.4 pt) of oxygen are absorbed from the lungs, to reconstitute the blood and raise the oxygen level back to 1000 ml/liter. If the tissues are using up oxygen at a more rapid rate, the amount of oxygen in the deoxygenated blood will be less than the normal level of 800 ml/liter and more oxygen will be absorbed from the lungs.

Thus far, we have been speaking as if the only function of the respiratory system were the absorption of oxygen. As we shall see in more detail shortly, this oxygen is utilized by the cells of the body to enable them to extract the maximum amount of energy possible from such energy sources as glucose. However, the oxygen atoms are not destroyed in the process, but rather are combined with the atoms of carbon in the glucose molecule in the following equation:

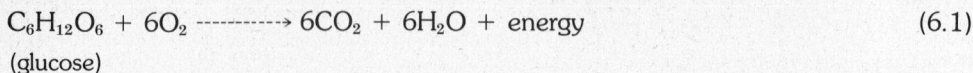

$$C_6H_{12}O_6 + 6O_2 \dashrightarrow 6CO_2 + 6H_2O + energy \tag{6.1}$$

(glucose)

Thus, in the metabolism of glucose ($C_6G_{12}O_6$), six molecules of carbon dioxide ($CO_2$) are produced for every six molecules of oxygen ($O_2$) consumed. In terms of actual volumes, the body at rest requires about 200 ml (0.4 pt) of oxygen per minute, and produces about 200 ml (0.4 pt) of carbon dioxide as a waste product. Thus, a second function of the respiratory system is to eliminate carbon dioxide through the lungs during exhalation. However, just as blood entering the lungs is not totally devoid of oxygen, so, too, is blood leaving the lungs not entirely free of carbon dioxide. In fact, blood leaving the lungs still contains about 2600 ml (5.2 pt) of carbon dioxide per liter of blood. More inefficiency? Hardly—but in order to understand the significance of this high level of carbon dioxide in the blood, we must first examine in greater detail how oxygen and carbon dioxide are transported by the blood.

**Gas transport**  Oxygen is extremely insolvent in water, and only about 1.5 percent of the oxygen carried by the blood is dissolved in the plasma. The rest of the oxygen is loosely bound to the hemoglobin molecules of the red blood cell. Carbon dioxide is somewhat more soluble in water, and about 8 percent is carried in dissolved form in the plasma. Another 25 percent is loosely attached to the hemoglobin molecule. Most of the carbon dioxide, however, combines with water to form carbonic acid ($H_2CO_3$), which, like all acids, tends to dissociate into its component ions, as illustrated by the following equation:

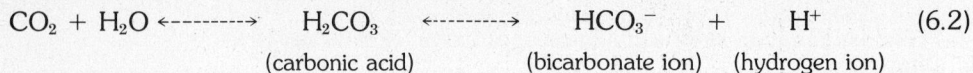

$$CO_2 + H_2O \dashleftarrow\dashrightarrow \quad H_2CO_3 \quad \dashleftarrow\dashrightarrow \quad HCO_3^- \quad + \quad H^+ \tag{6.2}$$

(carbonic acid)      (bicarbonate ion)   (hydrogen ion)

These ions are extremely soluble in water, and the plasma typically carries large amounts of each. The presence of the hydrogen ions is very important in regulating the pH of the blood. As discussed in Chapter 2, most proteins are very sensitive to changes in pH, and it is the maintenance of a certain number of hydrogen ions in the blood that compensates for the largely alkaline proteins and allows the blood pH to approximate neutrality (pH = 7.4).

The reactions just given are all reversible. As carbon dioxide leaves the lungs during exhalation, more of the carbonic acid is converted into carbon dioxide and water, and, in turn, more of the ions are reconverted into carbonic acid (i.e., the equation tends to move from right to left). As such, the volume of carbon dioxide eliminated during each exhalation determines how many hydrogen ions remain in solution in the blood. This is an extremely important point. Stated another way, the body can control blood acidity by the amount of carbon dioxide exhaled through the lungs.

**Regulating breathing rate** Thus far, we have spoken as if breathing were always occurring at the same rate. Obviously, it does not. For example, you breathe very rapidly after running for awhile. Logically, the body requires more energy during periods of activity than it does at rest—and obtaining that energy requires the utilization of large amounts of oxygen. Therefore, it is only logical that there should be some mechanism for allowing increased air flow into the lungs. As it happens, there are two methods of achieving this end. The depth of breathing may be increased, from a normal level of 500 ml (1 pt) per breath to as much as 4500 ml (9 pt). Correspondingly, the frequency of breathing can increase, from 10 or 12 breaths/min to as many as 300 breaths/min. The former method is actually more efficient, as can be demonstrated from the following comparison. Suppose you and a friend both ran for a quarter of a mile. At the end of the run, you find your rate of breathing is still the same (12 breaths/min), but the volume of each breath has increased to 4500 ml (9 pt). Allowing for 150 ml (0.3 pt) of air in the mouth and pharynx, the amount of air entering your lungs per minute is 12 × (4500 − 150) = 52,500 ml (105 pt).

In contrast, your friend's breathing pattern is unchanged with respect to depth (i.e., 500 ml (1 pt) per inhalation), but he is breathing 100 times/min. Allowing for 150 ml (0.3 pt) of air in the mouth and pharynx, the amount of air entering his lungs per minute is 100 × (500 − 150) = 35,000 ml (70 pt). Thus, because of the dead air space in the mouth and pharynx, it is advantageous to breathe deeply, rather than more frequently.

How is a change in respiratory rate or depth initiated by the body? An increase in physical activity requires additional energy production by the cells of the body. The cells require more oxygen to produce this energy and, consequently the amount of oxygen in the blood drops below the normal level. Correspondingly, the carbon dioxide level in the blood increases. Thus, the initial effect of increased activity by the body as a whole is a change in both blood oxygen and blood carbon dioxide levels.

Theoretically, the body could use either carbon dixoide or oxygen blood concentrations as the cue for altering respiration rates, because the change in concentration of the two gases is equivalent. As it happens, the body monitors *both* gases, although it appears that changes in carbon dioxide are of greater significance. In experimental situations, where only one of the blood gas concentrations was allowed to change, respiratory rates altered more quickly and more strongly to carbon dioxide changes than to oxygen changes. Carbon dioxide is not monitored directly, however, but indirectly through hydrogen ion concentrations. [See Eq. (6.2) to verify why this indirect method is nonetheless entirely effective.]

Where are the blood gases monitored? Receptors for blood oxygen and carbon dioxide are located adjacent to the receptors for blood pressure—in the aorta and in the arteries that carry blood to the head. In addition, the **medulla** portion of the brainstem is highly responsive to changes in the hydrogen ion concentra-

Changes in the rate and depth of breathing occur as an automatic response to changes in the blood concentrations of oxygen and carbon dioxide. Yet it is also true that we have considerable conscious control over our breathing patterns. As such, we can, through the exercising of such control, alter blood-gas concentrations—sometimes with unfortunate consequences.

Hyperventilation—an increase in breathing rate, not in response to changes in concentrations of blood-gas—is frequently voluntarily undertaken by athletes prior to engaging in competition, the objective being to boost blood oxygen levels. However, as we have seen, carbon dioxide is also being exhaled at a rate equivalent to oxygen intake. Thus, the increase in blood oxygen during hyperventilation is achieved only at the cost of reducing blood acid levels—Eq. (6.2) moves from right to left. This reduction in acidity may cause temporary unconsciousness, which can be a dangerous circumstance if the athlete in question is a swimmer. Every year, a number of amateur swimmers drown, as a result of hyperventilating prior to swimming underwater.

**Hyper-ventilation**

tion of the blood, and it is the medulla, operating through nerves running to the diaphragm and to the muscles of the rib cage, which is ultimately responsible for directing a change in respiratory rate (Fig. 6.5). A high blood hydrogen ion concentration promotes increased breathing, causing more carbon dioxide to be eliminated—and more oxygen to be taken up by the blood. As the carbon dioxide is eliminated, the hydrogen ion concentration in the blood falls—and so does the respiratory rate.

**RESPIRATORY DISEASES**  Despite being encased in the rings of bony armor of the ribs, and despite the evident precautions of warmed and filtered air, the lungs remain one of the more vulnerable organs of the body. This is partly because our need for oxygen is less subject to interruption than is our need for food or water—the cells of the brain will begin to die in less than 10 min from the time they cease to receive oxygen. In any case, until very recently respiratory illnesses were the leading cause of death in this country and remain the leading cause of death in many countries in which antibiotics are not widely available.

At one time or another, all of us have been affected by viral diseases of the upper respiratory tract, which we generally designate by the term "cold." Inflammation of the cells of the nasal epithelium causes excess mucus secretion, which causes sneezing and coughing. These upper respiratory infections seldom last more than a few days and are usually not serious.

Inflammation at the opening of the pharynx may be more serious, partly because the infection is deeper in the body and partly because of the likelihood of secondary infections from the ever-present bacteria in our mouths. Sore throats may become strep throats; inflammation of the **Eustachian tube,** which connects the ear and throat, may occur; and perhaps most serious, an infection in this region may pose a threat to the lungs.

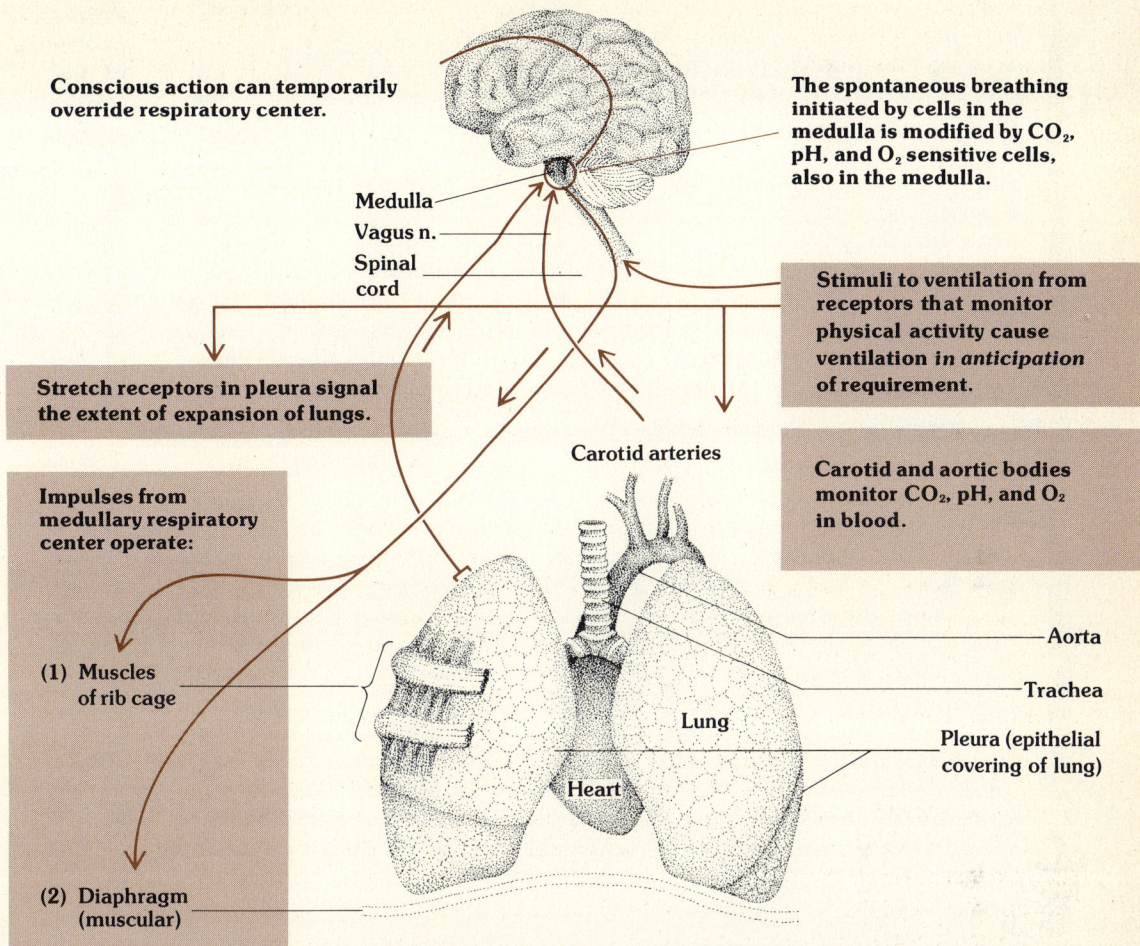

Conscious action can temporarily override respiratory center.

The spontaneous breathing initiated by cells in the medulla is modified by $CO_2$, pH, and $O_2$ sensitive cells, also in the medulla.

Medulla
Vagus n.
Spinal cord

Stimuli to ventilation from receptors that monitor physical activity cause ventilation *in anticipation* of requirement.

Stretch receptors in pleura signal the extent of expansion of lungs.

Carotid arteries

Carotid and aortic bodies monitor $CO_2$, pH, and $O_2$ in blood.

Impulses from medullary respiratory center operate:

(1) Muscles of rib cage

Aorta
Trachea
Lung
Pleura (epithelial covering of lung)
Heart

(2) Diaphragm (muscular)

Fig. 6.5  The control system in breathing.

You are probably aware that head colds sometimes become chest colds (at least this is how we designate them—the causative agents may well be different, but one can lead to the other). Inflammation of the bronchi may lead to **bronchitis;** this condition may become chronic if untreated, as the resulting coughing prompts additional sites for infection. Chronic bronchitis may also be caused by continual irritation by pollutants in the air we breathe, including the willfully introduced pollutants in cigarette smoke.

Air enters the body through the nostrils and passes across the **olfactory epithelium** (see Chapter 10), where it is warmed, moistened, assayed, and filtered before moving into the mouth cavity and pharynx. Filtering is accomplished by the trapping of particulate matter in the air by the mucus secreted by cells that line the air passages and by the movement of both mucus and particles back to the nostrils. Where the amount of particulate matter in the air is especially great (dusty room; pepper being sprinked too enthusiastically), or where excess mucus is produced, such as when you have a cold, the nasal passages become irritated, and following a deep breath ("aaahh . . ."), you exhale violently (". . . chooo"), the entire process commonly called a **sneeze.** A similar reaction, which we term a **cough,** occurs as a result of an irritation of the walls of the trachea or bronchi. Yet another form of protection occurs when we breathe air containing various types of noxious chemicals, such as acid vapors, for the response of the respiratory system is to momentarily shut down.

**Protective reactions of the respiratory system**

Another reaction of the respiratory system is not protective of the system, but it is at least as annoying as are coughs and sneezes. **Hiccups** result from a too-frequent stimulation of the diaphragm by the **phrenic nerve,** which runs from the brain to the diaphragm. Most of the "cures" recommended for hiccups (holding one's breath; breathing into a paper bag; drinking a glass of water slowly) result in an increase in the blood carbon dioxide level, which should lead to an increase in respiratory rate, a response mediated by the phrenic nerve. An increase in the activity of the phrenic nerve theoretically should operate to cancel out the hyperstimulation that exists during hiccups, but, as you are no doubt aware, such "cures" are not always effective. One method now in wide use is to swallow a spoonful of granulated sugar. Because the phrenic nerve is closely associated with the esophagus, the mechanical effects of this gritty material apparently have a direct effect in nullifying the hyperstimulation of the nerve.

**Hiccups**

In a very real sense, all these problems are minor irritations compared with the life-threatening diseases of the lung itself. The very pollutants that caused chronic bronchitis may mechanically break down the alveoli of the lungs, effecting a marked reduction of surface area available for gas exchange. If enough area is lost, the individual is said to be suffering from **emphysema** and may be constantly short of breath. Unfortunately, alveoli do not regenerate, and for that reason emphysema is a very serious disease, but one that frequently goes undetected until a large portion of the alveoli have already been destroyed. Bacterial or viral infections of the lungs may cause the lungs to fill with mucus,[1] thereby effectively preventing that por-

---

[1] These conditions should not be confused with the genetic abnormality of children known as **cystic fibrosis,** in which an excess of mucus is produced by the lungs in the apparent absence of any disease agent.

tion of the lungs from assisting in gas exchange. Such illnesses are lumped under the general heading **pneumonia.** Fortunately, bacterial pneumonia is much the more common and is now treatable with antibiotics. However, as recently as 1900 it was the leading cause of death in the United States.

Another serious lung disease, **tuberculosis,** is now considered a disease of the distant past by most Americans, although it, too, was a leading cause of death in your grandparents' day. The disease organism, a bacterium, infects a portion of a lung, which in otherwise healthy individuals walls off this portion in something called a **tubercle.** Depending on the overall health of the individual, the tuberculosis bacteria may remain isolated in this tubercle for long periods of time. However, a subsequent weakness from an injury or other disease frequently results in a freeing of these bacteria, which then can infect other areas of the body, including the bones or spinal cord.

**CELLULAR RESPIRATION**   It has been stressed previously that a principal role of the respiratory system is to supply the cells with oxygen. Similarly, the role of the digestive system is to break down complex molecules into the relatively simple ones that can be used for energy production or synthesis of new molecules by the individual cell. In both cases, the circulatory system acts as the vehicle for moving these materials from their site of origin to the cells.

Why do cells need oxygen? Cells need oxygen for the same reason that oxygen is needed to burn a log—oxygen plays a dominant role in the splitting of chemical bonds, in turn releasing energy. Why do cells need energy? Cells require energy to enable them to drive the various chemical reactions involved in the synthesis of new compounds.

The primary compound used in energy production in the cell is the monosaccharide **glucose,** a six-carbon sugar. However, glucose is to the cell what a log is to a fireplace—it has lots of potential, but some preparation is needed first. Thus, just as one might break up a log to make starting a fire easier so, too, does the cell expend some energy in breaking the glucose molecule into two three-carbon pieces. These two molecules, called phosphoglyceraldehyde (PGAL) not only have a great deal of energy trapped in their chemical bonds, but are also rather more easily broken down than was glucose. The cell converts each PGAL molecule, through the use of enzymes, into **pyruvic acid.** This process takes place in the cytoplasm of the cell and does not, in fact, require any oxygen but the amount of energy produced by this transformation is small.

The breakdown of glucose in the absence of oxygen is called **anaerobic glycolysis** or **fermentation.** A few primitive organisms, such as certain bacteria and yeast, are incapable of utilizing oxygen, and are therefore limited in the amount of energy they can obtain from the breakdown of glucose. The pyruvic acid in such organisms is ultimately converted into one molecule of ethanol (grain alcohol) and one molecule of carbon dioxide. These processes can be demonstrated rather easily. If one allows bread dough, which contains yeast, to sit in a warm place for a couple of hours, the dough rises because yeast produces carbon dioxide, which is trapped within the dough. If one leaves the dough for a few more hours, the rising will cease and the odor of alcohol becomes apparent. In a closed system such as a bowl of bread dough, the yeast organisms ultimately die, in effect drowning in their own wastes—in this instance, alcohol.

In ourselves, the pyruvic acid normally formed by the cytoplasmic breakdown of glucose is transferred over to the mitochondria for further processing (Fig. 6.6). Mitochondria are the sites in the cell at which oxygen is actually utilized. The pyruvic acid is passed into a rather complex cyclical series of reactions called the **Krebs cycle** (or citric acid cycle), and is ultimately broken down into carbon dioxide and water, with the release of a considerable amount of energy. This conversion of pyruvic acid to carbon dioxide and water, in the presence of oxygen, is called **aerobic glycolysis,** or **respiration.** The overall equation is given in Eq. (6.1).

The problem faced by the cell is in trapping the energy given off when the chemical bonds of glucose are broken. It would obviously be erroneous to think of all the molecules of the cell as warming their "hands" as the glucose "burned"—the analogy of the breakdown of glucose with the burning of a log is only partially correct. Rather, what is desired is a controlled "burn" whereby energy is released in small amounts, more easily trapped by the cell. Again, consider an analogy. A gallon of kerosene would last a whole evening in a camp stove, giving off a controlled warmth. Al-

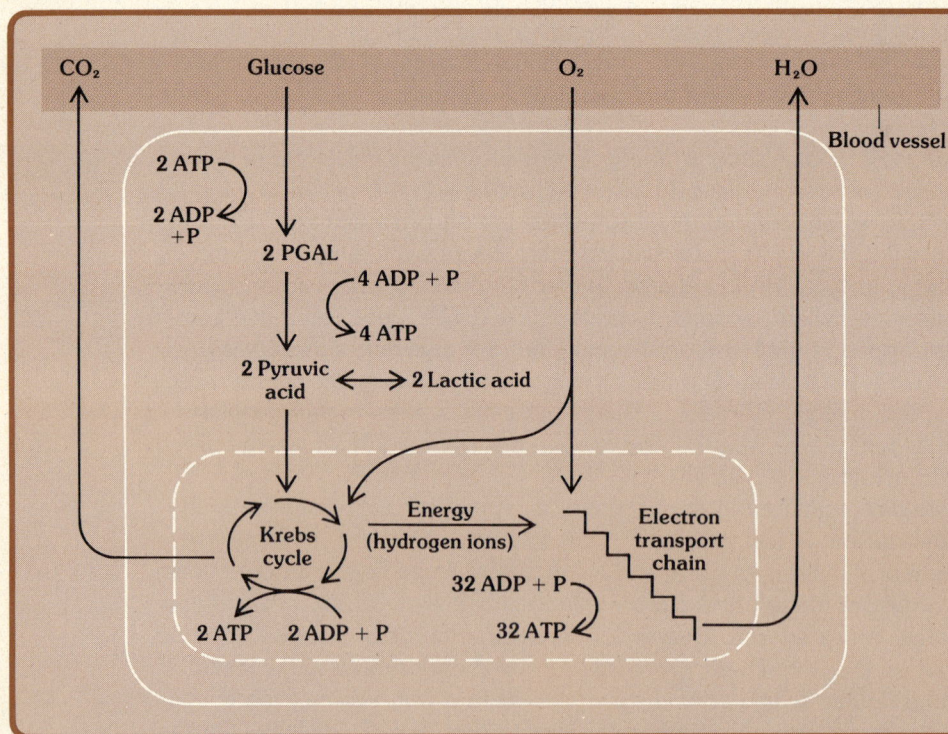

Fig. 6.6 Diagrammatic representation of cellular respiration. Solid line is the cell membrane; dotted line is a mitochondrion.

ternatively, the same gallon could be thrown into the campfire. The same amount of energy would be given off, but in dangerously high amounts and for a uselessly short period of time. Obviously, a controlled "burn" is preferable.

Energy that is trapped by our cells and becomes usable to power other reactions is trapped in the form of a molecule known as **adenosine triphosphate** (ATP)—and this same molecule is used in all plants and animals that utilize oxygen. It consists of an **adenosine** portion (this is also one of the nitrogenous bases of DNA and RNA, as well as a component of one of the vitamins—again, note the multiple duty of such molecules in the body) to which are attached three inorganic phosphate groups. The first of these is normally bound, but the second and third are high-energy bonds—that is, it takes a great deal of energy to connect the second and third phosphate groups, but similarly when these are broken off, a great deal of energy is released. A useful analogy may be made with a switchblade. Energy is required to push the blade into the handle (because of the need to stretch the spring), but this energy is released again when the switch is tripped and the blade released.

ADP (an adenosine plus two phosphates) is a common molecule in the cell as are the phosphate groups themselves; much of the energy released by the breaking of chemical bonds during respiration is stored by virtue of its being used to fix another phosphate group onto a molecule of ADP to yield a molecule of ATP. In a few instances, this is accomplished directly from a specific chemical reaction; more frequently, the energy of a particular chemical reaction is trapped in a group of enzymes collectively called the **electron-transport chain.** These enzymes are also found in the mitochondria and have the effect of parceling out the energy released from a given chemical conversion in the Krebs cycle in still smaller doses such that a maximum number of ATP molecules can be formed. In a sense, the energy drop is like a descent on a staircase, rather than a straight drop.

In comparison to industrial systems, glycolysis (including both anaerobic and aerobic portions) is an efficient process, with some 40 percent of the energy present in the chemical bonds of the glucose molecule being trapped in the form of ATP, the rest being lost as heat. Each molecule of glucose yields 36 molecules of ATP, two from fermentation, and 34 from respiration. By comparison, a well-tuned internal combustion engine will convert no more than about 20 percent of the energy stored in the chemical bonds of gasoline into kinetic energy (movement).

This situation represents the normal sequence of events in animals, including ourselves. Under certain circumstances, however, we are also capable of utilizing fermentation exclusively. If the average person were to run a mile, he would be panting profusely at the end, and his legs would feel rubbery. The individual is said to be in a state of **oxygen debt**—his muscle cells did not receive sufficient oxygen to allow for a complete breakdown of glucose and, instead, the process stopped at the fermentation level. In ourselves, and in any animal in a state of oxygen debt, the pyruvic acid is converted temporarily to **lactic acid,** which is also a three-carbon molecule; it is the presence of this molecule in the muscles which gives the legs the sensation of being rubbery and weak. In the presence of oxygen, the lactic acid is reconverted to pyruvic acid and thence on to the Krebs cycle. The fact that plants and animals have different end products for fermentation may be viewed either as a blessing or a curse—lactic acid is easily reconverted to pyruvic acid, whereas ethanol is not (as it is just a two-carbon molecule); in contrast, if we produced ethanol in our muscles as a consequence of fatigue, rather than lactic acid, one might expect to see more people jogging!

Respiration includes more than just the breakdown of glucose (glycolysis), however, as other types of molecules can also be used by the cells for energy production. If you are on a diet or are fasting, first your stored fats and then the proteins of your muscles are broken into their respective simple molecules for utilization by the cells. Amino acids and fatty acids enter the Krebs cycle directly, whereas the alcohol portion of the fat molecule (see Chapter 2) is first converted to PGAL.

**ENDOTHERMY**   It is common knowledge that some animals are "warm-blooded" and that others are "cold-blooded." In actuality, however, this distinction is somewhat erroneous—at any given time, the blood of an insect or lizard may be as warm as ours. For that reason, the nature of the distinction was for years described as **homeothermic** as opposed to **poikilothermic,** meaning "same temperature" and "variable temperature," respectively. Recently, however, it has been realized that many of the so-called poikilothermic animals are, in fact, capable of maintaining a surprisingly constant internal temperature by varying their behavior. For example, by alternately sunning on a rock in the early morning hours and seeking shade during the heat of midday, lizards can frequently succeed at maintaining their body temperatures at roughly the same level throughout the day.

The true distinction is between animals that are capable of maintaining their body temperatures constant through internal physiological regulation and those that lack this capacity. Animals in the first class are called **endotherms;** those in the second are called **exotherms.**

Endothermy poses some real advantages, but as we shall see, it is analogous to holding a tiger by the tail—once the commitment is made, there is no letting go. Only birds and mammals have achieved complete endothermy.

The primary advantage in being an endotherm is obvious—such organisms are capable not only of surviving, but of living fully active lives under environmental temperature regimes in which exotherms simply cannot thrive. Insects, which are exotherms, are easy prey if caught exposed when the temperature drops much below 10°C (50°F), because they are, as a rule, incapable of rapid locomotion at such temperatures (flight may be impossible, for example, especially if there is no direct sunlight). Birds and mammals, however, are capable of normal rates of activity virtually regardless of the outside temperature.

How can these behavioral differences be explained in physiological terms? As it happens, the rate of movement of molecules in a fluid environment is directly related to temperature; consequently, the rate of chemical reactions is also directly related to temperature. The amount by which a given chemical reaction is increased or decreased owing to a temperature change varies greatly depending on the reaction, but as a general rule, metabolic reactions in animals increase from two to three times for every 10°C (18°F) increase in temperature. Thus, on an overcast morning with the outside temperature about 4°C (40°F), the reason that a mouse can catch a grasshopper with no difficulty is that the energy-producing reactions in the cells of the mouse are operating at about 38°C (100°F), whereas those of the grasshopper are operating at about 4°C (40°F). This is a difference of almost 35°C (63°F), which is enough to ensure that those reactions are taking place about 10 times more quickly in the mouse than in the grasshopper. Incidentally, it is no accident that the normal body temperature for all endotherms is 38°C (100°F) ± 4°. Much above this temperature, proteins begin to denature (see Chapter 2).

Not only do endotherms have an advantage over exotherms in cold weather but they can also colonize cooler regions of the globe more easily. There are, for example, many more species of reptiles and amphibians in Oklahoma than there are in North Dakota, but the number of species of birds and mammals differs only slightly between those two states. There are no reptiles at all in the polar regions, but a relatively large number of bird and mammal species are found there. Increased habitat opportunities are an immense advantage in a crowded world.

Finally, growth and development is considerably more rapid in endotherms than in exotherms. In both, 60 percent of the energy in the chemical bonds of glucose is given off as waste heat. In exotherms, however, the determinant of food intake is growth rate, which tends to be low. In endotherms, it is the need to maintain a constant temperature, despite continuous heat loss, that determines metabolic rate. As a consequence, the food intake of exotherms is high in order to maintain constant body temperature, and the 40 percent of food energy trapped as ATP is enormous and can be put to use in growth. Consequently, rabbits grow more rapidly than turtles, for example.

However, there are disadvantages in being an endotherm. (That should be obvious from the fact that most of the animal species of the world remain exothermic.) Think for a moment of the source of heat of an animal. When you are cold, you shiver, a reaction that involves a rapid contraction of the muscles of the body. At the cellular level, a great deal of glucose is being metabolized to produce the ATP necessary to allow the muscle fibers to contract—and, as we noted earlier, some 60 percent of the available energy in glucose is "lost" as heat. Endotherms have put this "waste" heat to use by regulating its rate of production such that the internal temperature of the body remains constant. Shivering is a case of the tail wagging the dog—originally, muscles contracted in order to propel the organism, and the food energy that was not trapped in high-energy bonds from the breakdown of glucose was passed off as heat. In shivering, the muscles do not contract for the purpose of locomotion, but only as a consequence of the need to produce more of this "waste" heat.

The example of shivering provides a clue to the disadvantages of being endothermic. Fundamentally, the disadvantage is the rate at which heat must be produced by cellular respiration in order to maintain a constant body temperature. This fundamental disadvantage is manifested in several ways:

**1. Food requirements** Per unit weight, the endothermic animal needs up to ten times more food than does the exotherm. To be sure, under certain conditions, as described above, the endotherm may be at an advantage over the exotherm in capturing the food, but in times of scarcity, the endotherm will, as a rule, be first to die.

**2. Oxygen requirements** Per unit weight, endothermic animals require more oxygen (except at very high temperatures), in order to metabolize the increased volume of food (Fig. 6.7). Given the abundance of oxygen in the atmosphere, this requirement may seem, at first, to pose no problem, but it is surely the reason why such completely aquatic mammals as whales and porpoises have never reverted to gill respiration (oxygen being 30 times less available in water than in air), but rather have continued to put up with the hazards of having to surface several times an hour in order to breathe. The increased need for oxygen also accounts for the huge lungs and rapid rates of breathing of mammals.

**3. Specialized circulatory system** Endotherms require a highly efficient circulatory system to deliver glucose and oxygen to the individual cells. This includes a rapidly beating heart and blood circulating at high pressure, as well as complete segregation of oxygenated and deoxygenated blood. Significantly, only the birds and mammals have such features—and of course only these two groups are endotherms.

**4. Physical limitations** Endotherms have definite size and shape limitations, because of surface area/volume relationships (see Chapter 2). Obviously, the more surface area per unit volume, the more heat loss. Thus, for example, the Arctic hare is rather globular in shape, with relatively short ears and legs, as com-

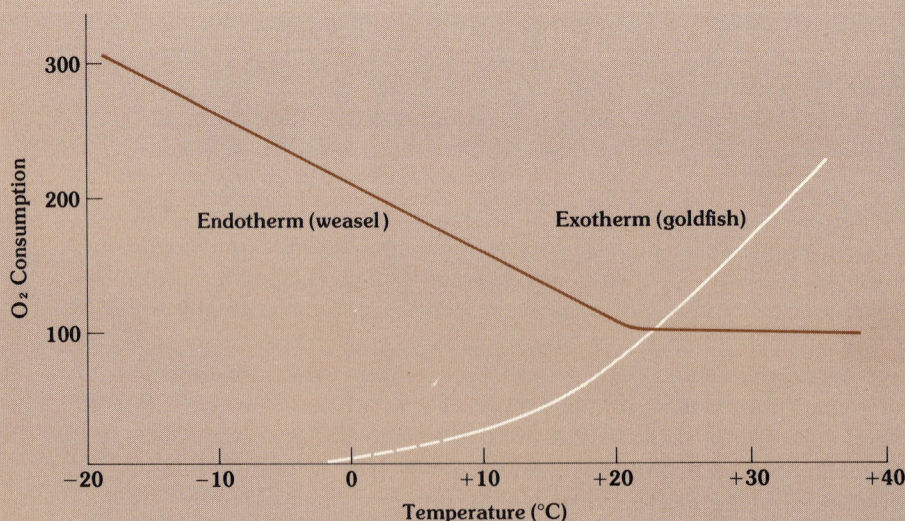

Fig. 6.7 Rate of oxygen consumption in endotherms as contrasted with exotherms.

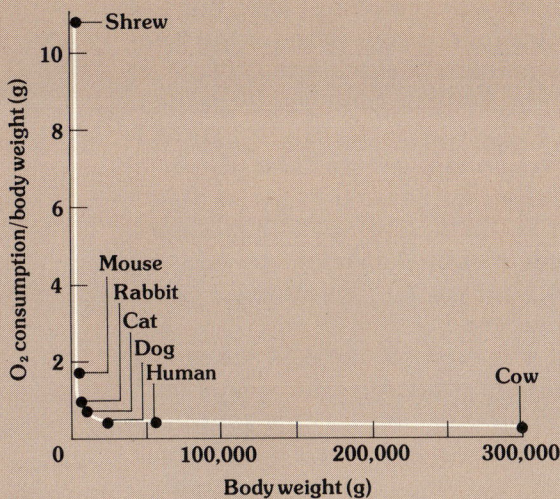

**Fig. 6.8**
**Oxygen consumption in endotherms as a function of body size.**

pared with the more rangy jackrabbit of warmer climes, where heat loss is not as significant a problem as the environmental temperature is much closer to the internal temperature of the organism. Even more significantly, the minimum possible size of endotherms is about that of hummingbirds and shrews. These organisms have incredibly rapid heart beats in order to propel blood laden with oxygen and glucose (and these organisms have enormous appetites for each) to the cells, which must break down glucose at very high rates to provide the heat necessary to maintain the temperature of the body at a constant level (Fig. 6.8). In contrast, the overwhelming majority of insects, for example, are smaller than shrews or hummingbirds—and certainly one reason why all animals are not endotherms is because endotherms are incapable of occupying the microhabitats of insects and other small exotherms.

It is interesting to note that many endotherms in temperate climates have adapted to food shortages in winter, either by migrating (birds) or hibernating (mammals). Such behavioral modifications tend to expand seemingly narrow physiological limits. Hibernation is a particularly interesting phenomenon because during hibernation, for all intents and purposes, the organism becomes physiologically an exotherm. In a sense, such animals have the best of both worlds.

**SUMMARY**

**Respiration is the metabolic utilization of oxygen, which is a cellular event, but this term is also used interchangeably with the word "breathing." This dual usage is the cause of repeated misunderstandings.**

The situation is made no easier by the fact that breathing is both a voluntary and an involuntary act. Of course, we can breathe at will, but you may be interested to know that your little brother or sister cannot kill himself by holding his or her breath, despite persistent threats to do so during a temper tantrum—ultimately the child's brainstem will force the initiation of breathing once again.

Cellular utilization of oxygen was a secondary event in the history of the earth. It was not until there were a sufficient number of green plants to produce large amounts of oxygen that animals developed capacity to use oxygen. That the overwhelming number of animal species now uses oxygen is testimony to the fact that oxygen allows a 20-fold increase in energy production from the metabolism of glucose.

Respiration is a remarkable process, but even so, most of the energy present in the chemical bonds of such energy sources as glucose is lost as heat. Such large animals as birds and mammals have come to utilize this heat by trapping it within their bodies and by raising their body temperatures to the point of maximum efficiency of chemical reactions. This ability is a mixed blessing, however, as it requires a relatively large intake of food, especially in small birds and mammals, to produce enough cellular heat to keep the body temperature constant.

If you were to drink a liter (2 pints) of water, within a short time your digestive tract would absorb the bulk of this water and add it to the blood. How does the body respond to a dilution of the blood?

Perhaps your roommate is a salt nut, always adding salt to everything he eats. Much of this salt will be absorbed by the gut and added to the blood. What prevents the blood from becoming too salty and, as a consequence, from pulling water from surrounding tissues?

Should you develop bacterial pneumonia, you may very well be treated by a "shot" of penicillin. Does this penicillin remain in the blood forever? If not, how does the body get rid of the drug? What about other recently developed medicines—how is the body able to pass off such novel molecules?

In 1976, Howard Hughes, along with 50,000 other Americans, died of kidney failure. When the kidneys "fail," what physiological factors change to such a degree that death results?

chapter **VII**

# chapter VII
# Regulation of the internal sea – the excretory system

**THE NATURE OF THE PROBLEM**    To reiterate a point made earlier, the success of any multicellular organism depends in large measure on its ability to keep its cells bathed in fluid, not only containing those things required by the cells in order to maintain life, but also containing each of these constituents in the proper quantity. Primitive water-dwelling multicellular organisms, such as sea anemones, have little problem because their watery environment also serves as the fluid that directly bathes their cells. However, life on land necessitates a different solution, for the blood (and the interstitial fluid derived from it) is a closed system that must be closely monitored.

Consider the insults that large land organisms such as ourselves inflict on our circulatory system. We are constantly losing water as we exhale (exhaled breath is almost saturated with water, as you demonstrate each time you see a cloud of vapor on a cold morning). We also lose water in feces, in perspiration, and in urine. These losses tend to raise the concentration of the blood, and therefore its osmotic value. Reciprocally, our cells produce water as an end product of metabolism, and of course we ingest water in our food and drink. These tend to dilute the blood and again disrupt the vital osmotic interactions between the cells and the interstitial fluid. Therefore, an ongoing problem faced by the body is to maintain a balance between the amount of water that enters and leaves the body (Table 7.1).

**Table 7.1**
**Water intake and loss**

| Intake | ml | Loss | ml |
|---|---|---|---|
| Drink | 1200 | Skin and lungs | 900 |
| Food | 1000 | Sweat | 50 |
| Metabolically produced | 350 | Feces | 100 |
| | 2550 | Urine | 1500 |
| | | | 2550 |

A second problem concerns mineral salts, especially the ions of potassium ($K^+$), sodium ($Na^+$), and calcium ($Ca^{2+}$). These are all present in relative abundance in our normal diet, and are generally taken up freely by the digestive tract, and from there are transported to the blood. However, as we shall see in Chapter 10, a consistent balance among these ions is essential for the normal functioning of nerves and muscles, as well as for normal osmotic interactions between cells and the interstitial fluid. Hence, the elimination of excess salts is another problem confronting the body.

Many of the foods we eat contain weak acids or, once metabolized, result in the production of weak acids (i.e., the liberation of $H^+$ ions). A point that has been reiterated several times is that the acid–alkaline balance (i.e., the pH) of the body is critical, because enzymes are so pH-sensitive. For example, the normal pH of the blood is about 7.4. If it rises above 7.8 or drops below 7.0, the individual will die. Thus, a third problem faced by the body is the control of body acidity within survivable levels.

There is no storage site for amino acids in the body as there is for sugar (glycogen in the liver) or lipids (fat in fat cells). Therefore, amino acids in the diet exceeding the amount needed for the manufacture of new proteins are metabolized by the cells to produce energy. Before they can be used for energy generation, however, the liver must remove the nitrogen-containing portion of the amino acids, because nitrogen is not used in energy production. This molecular fragment is split off in the form of **ammonia** ($NH_3$), which is highly toxic. Thus, yet another problem faced by the body is the elimination of metabolic waste products (**metabolites**).

**THE ROLE OF THE KIDNEYS** The kidneys play an important role (although it is not necessarily exclusive) in solving each of these four problems. For example, in maintaining water balance, the body can do little to regulate loss in feces or in the breath (and water lost in perspiration is a function of the need for maintaining a constant temperature, not of the need for osmotic stability), nor can it do much to regulate water intake in the food or through metabolic production. However, as we shall see, thirst is a variable over which the body does have control and, within broad (but not infinite) limits, there is also control over water lost in the urine—and it is the kidneys which exercise this control.

Similarly, although somewhat more simply, salt loss occurs in the production of tears and perspiration; although the amount lost is very small, the body has little control over this loss. Once again, the kidneys are the primary regulators of the amount of salt that leaves the body.

The control of blood pH is complex, involving both the respiratory system and a series of buffers in the blood, as well as the kidneys, but it is the kidneys alone which actually have the power to excrete $H^+$ ions.

Finally, although the respiratory system has the principal role in the removal of one by-product of metabolism, carbon dioxide ($CO_2$), it is exclusively the kidneys which handle the excretion of the various nitrogen-containing metabolites.

From this summary it should be apparent that the role of the kidneys in the maintenance of stable levels of various components of the blood (i.e., **fluid homeosta-**

sis—see Fig. 7.1) is both more extensive and more important than is generally assumed. Indeed, it has rightly been said that the composition of our body fluids is not so much a reflection of what the mouth takes in, but of what the kidneys keep.

It should be obvious that "fluid homeostasis" is a broader term than "excretion." Excretion is defined narrowly as the elimination of metabolic wastes, which would include $CO_2$, nitrogenous wastes, and metabolic water. Only the last two are handled by the kidneys, the main organs of the "excretory" system, and of course these organs handle many other tasks as well. It seems more reasonable, therefore, to use an expanded version of excretion—the elimination of any substance given off by a cell—which would include all the substances under discussion, but would still distinguish them from the passage of feces from the digestive tract (elimination). The kidneys and related structures might then best be called the **urinary system,** to emphasize that excretion is handled by several different systems.

**An historical perspective**   The three major roles played by the mammalian kidneys are (1) excretion of nitrogen-containing metabolites, (2) osmoregulation (control of water and salt balance), and (3) control of blood pH. However, these roles were not all acquired at the same time during the evolution of the vertebrates.

The kidneys of freshwater fish function primarily to excrete excess water (because there is a tendency for the fish to pick up water by osmosis, the cells of the fish being more concentrated than the surrounding water). However, ammonia, the dominant nitrogen-containing metabolite, is not excreted by the kidneys, but by the gills. Saltwater fish, which tend to lose water to the more concentrated salt water about them do not use their kidneys to any extent in salt regulation, but rather rely on salt-secreting glands in the gills.

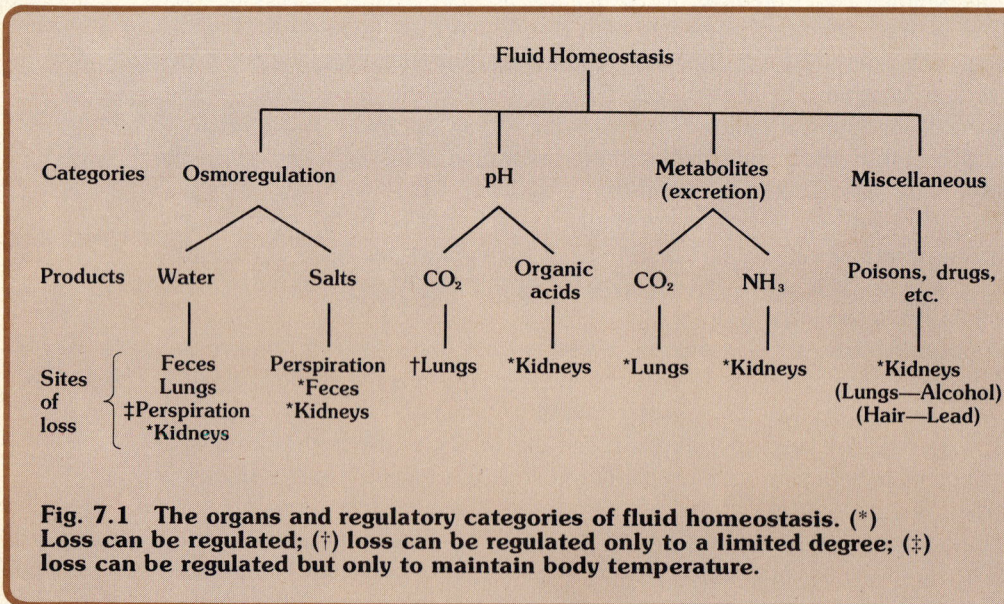

Fig. 7.1   **The organs and regulatory categories of fluid homeostasis. (\*) Loss can be regulated; (†) loss can be regulated only to a limited degree; (‡) loss can be regulated but only to maintain body temperature.**

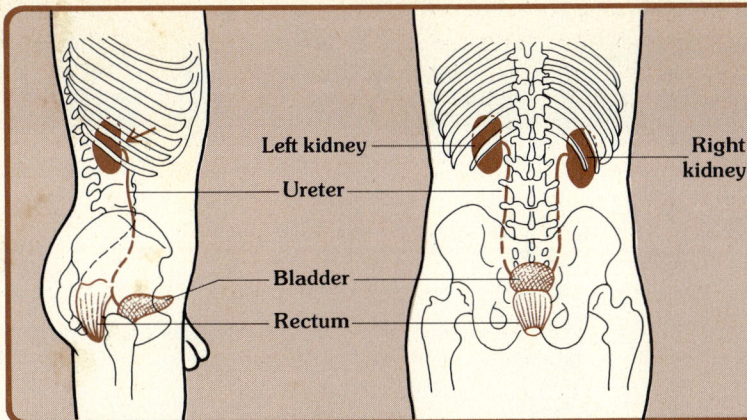

**Fig. 7.2
The human
urinary system.**

Left kidney

Ureter

Right kidney

Bladder

Rectum

Similarly, marine reptiles and birds have evolved special salt-secreting glands in the nostrils to excrete excess salt, rather than relying on the kidneys. The nasal glands of saltwater crocodiles empty into their eyes—hence the term "crocodile tears" for a false display of grief.

One might assume from the above facts that marine mammals use a similar technique, but in fact they do not. The mammalian kidneys have taken over all these tasks, but with limited efficiency, as far as marine mammals are concerned, for they are unable to drink seawater and must instead rely on the water contained in their food and that metabolically produced by their cells. Thus, ironically, as the kidneys have come to have increasingly greater importance as we move up the vertebrate ladder, limitations are imposed that did not exist earlier—marine mammals cannot drink seawater, because the ability of the kidneys to produce urine with a high salt content is limited, whereas marine fish, reptiles, and birds, all with less sophisticated kidneys, can and do drink seawater because they have evolved special salt-secreting glands to excrete the excess salt.

**THE URINARY SYSTEM** The urinary system consists of a pair of **kidneys,** a pair of tubes, the **ureters,** which lead from the kidneys to a single **urinary bladder,** and a tube leading from the bladder to the outside, the **urethra** (see Fig. 7.2).

The kidneys are a pair of bean-shaped organs,[1] rather small in size (about 10 × 5 cm, or 4 × 2 in.), located somewhat higher up alongside the vertebral column than is generally thought. A cross section of the kidney shows that it is composed of two layers, an outer **cortex** and an inner **medulla.**

**Structure of the nephron** The basic functional unit of the kidney is the **nephron,** about 1 million of which are found in each human kidney. This number is more than sufficient to meet our needs and, in fact, we can survive with as little

---

[1] As an illustration of the nature of the difference between zoologists and botanists, to a botanist a bean is kidney-shaped.

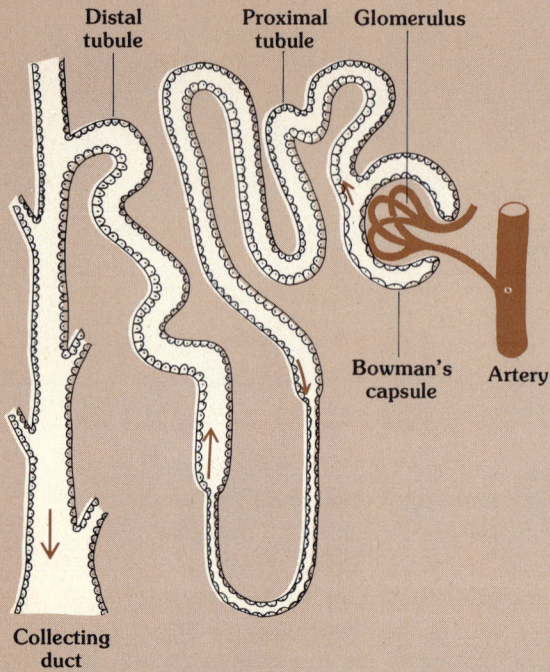

Distal tubule

Proximal tubule

Glomerulus

Bowman's capsule

Artery

Collecting duct

Fig. 7.3
Cross section of a nephron.

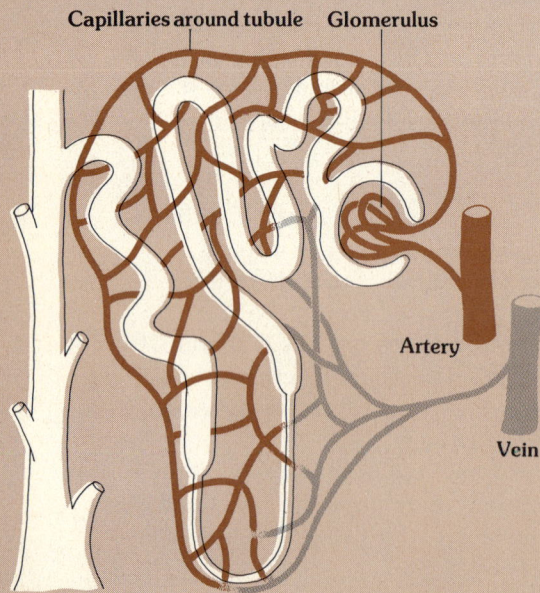

Capillaries around tubule

Glomerulus

Artery

Vein

Fig. 7.4
A nephron, with surrounding capillaries.

as two-thirds of one kidney. The nephron itself consists of several components (Fig. 7.3). First, there is a cuplike structure, called **Bowman's capsule,** which is the filtering portion of the nephron. Collectively, all of the Bowman's capsules provide the body with about 1 m² (3.3 ft²) of filtration surface. From the capsule, the nephron continues as a **convoluted tubule,** which has a descending (proximal) and an ascending (distal) arm, and which ultimately joins a **collecting duct,** leading, in turn, to the ureter.

Another portion of the nephron is in the form of a knot of arterial capillaries, called the **glomerulus,** which is embedded in the capsule much like a fist in a baseball glove. Blood leaving the glomerulus then enters a capillary bed that surrounds the convoluted tubule (see Fig. 7.4).

**Function of the nephron—general methods**   The function of the nephron is to prepare urine from blood, which is a task of a magnitude equal to spinning gold from straw. Three processes are involved in this transformation. These include **filtration, reabsorption,** and **secretion** (Fig. 7.5).

**Filtration**   At any given time, the kidneys are receiving about one-fifth of the blood in the body. Put another way, every drop of blood in the body passes through the kidneys on the average of once every fifth circuit. As the volume of whole blood is about 5 liters (1.3 gal) and a complete circuit requires about 1 min, the kidneys receive and process approximately 1 liter (.26 gal) of blood each minute or over 1500 liters (396 gal) each day.

Because the kidneys are located so close to the aorta, and the arteries supplying the kidneys are so large, blood reaches the kidneys at a rather high pressure. Moreover, the arterial capillaries comprising the glomeruli are of a unique type, inasmuch as they are called capillaries only because of their size. It is important to note that, unlike typical capillaries, no oxygen exchange occurs in the glo-

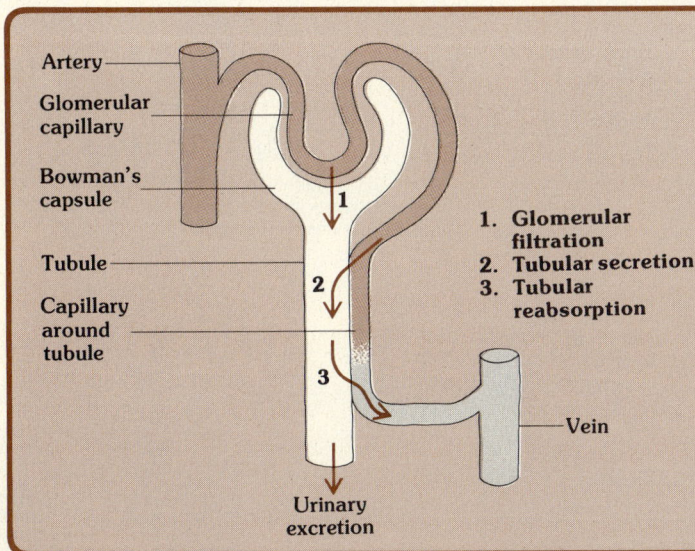

Artery
Glomerular capillary
Bowman's capsule
Tubule
Capillary around tubule

1
2
3

Urinary excretion

Vein

**Fig. 7.5**
**A summary of the physiological activities of the nephron.**

1. **Glomerular filtration**
2. **Tubular secretion**
3. **Tubular reabsorption**

meruli (i.e., the capillaries merge back into an artery), and the capillary walls are thick, like the walls of a small artery. As such, they are capable of receiving blood at high pressure without rupturing. In fact, the pressure is sufficiently high so as to force about 20 percent (see Fig. 7.6) of the blood plasma, excluding the proteins, through the pores in the walls of the glomerulus, across the wall of Bowman's capsule, and into the cavity of the nephron itself. The material that makes this journey is called **glomerular filtrate,** a perfectly reasonable name because it is filtered by the glomerulus. Some 180 liters (47.5 gal) of glomerular filtrate are produced daily.

In theory, urine formation could involve one of two routes. One could conceive of a kidney that could somehow recognize nonessentials and could allow only these to be filtered out of the blood. The difficulty with such a system is that a very great number of molecules would somehow have to be recognized as nonessential, and there is no provision in such a system for new molecules, such as newly developed drugs, which must ultimately be excreted from the body. Conversely, one could imagine a kidney that could allow almost everything to be filtered out of the blood, coupled with some mechanism whereby essential molecules would be again picked up. Because the number of essential molecules is small and presumably not subject to change, it is not surprising that this second alternative is the one that is actually used in the mammalian kidney.

Thus, the material that filters through the glomeruli is decidedly not urine, but rather has essentially the same composition as blood serum (see Chapter 4). It is

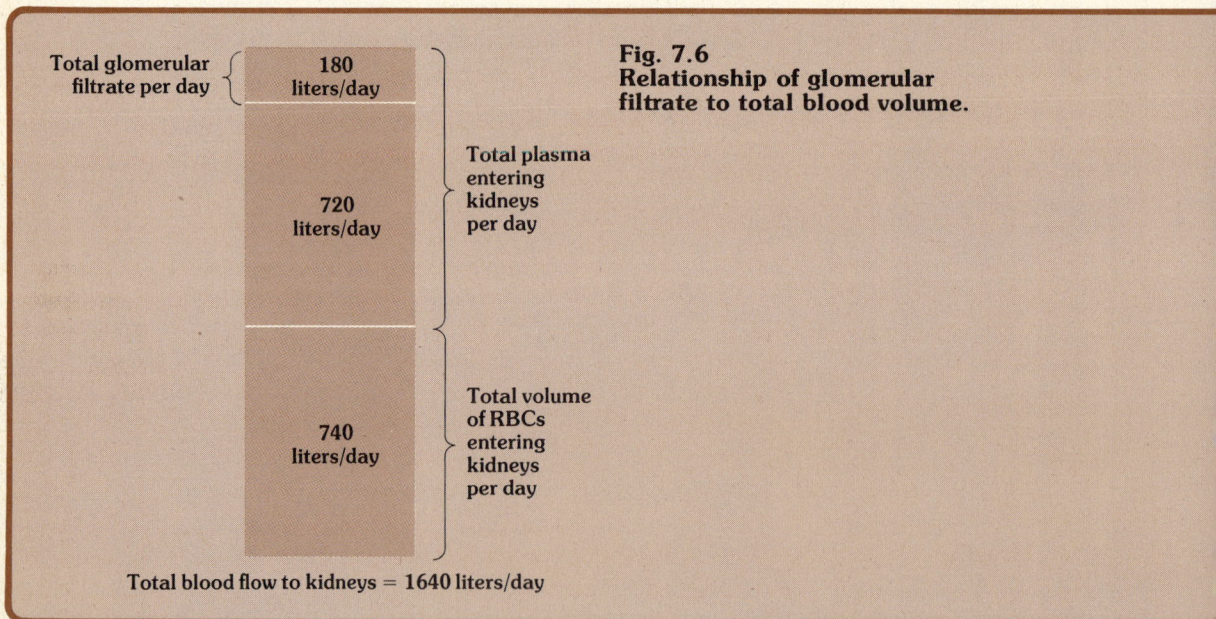

Total glomerular filtrate per day { 180 liters/day

720 liters/day — Total plasma entering kidneys per day

740 liters/day — Total volume of RBCs entering kidneys per day

Total blood flow to kidneys = 1640 liters/day

Fig. 7.6
**Relationship of glomerular filtrate to total blood volume.**

**Table 7.2**

**Filtration, excretion, and reabsorption values for several common components of glomeruler filtrate**

| Substance | Amount filtered per day | Amount excreted per day | Percent reabsorbed |
|---|---|---|---|
| Water (liters) | 180 | 1.8 | 99.0 |
| Sodium (g) | 630 | 3.2 | 99.5 |
| Glucose (g) | 180 | 0 | 100 |
| Urea (g) | 54 | 30 | 45.5 |

only *after* the glomerular filtrate has been acted on by the processes of reabsorption and secretion that it finally becomes urine.

**Reabsorption** It is the function of the convoluted tubules to recapture the molecules required by the body. This process of recapture is complex, but essentially consists of the active transport of such small molecules and ions as glucose and $Na^+$ and the subsequent passive transport of water and $Cl^-$. As a consequence, the nonessential molecules remain within the tubule and ultimately come to form urine. The osmotic gradient set up by the nonessential molecules ensures that a sufficient amount of water remains within the tubule to keep them in solution. Full details of this process are discussed under the section on osmoregulation below.

This process of reabsorption is not as perfect as it might seem at first glance. For example, because much of the tubule is permeable to water and because so much water is reabsorbed, certain other molecules that are readily dissolved in water also are reabsorbed. This includes the nitrogenous waste molecule **urea,** about 50 percent of which is reabsorbed, even though it has no necessary function in the blood (Table 7.2).

Moreover, there is a limit as to how much of a given substance can be reabsorbed. For example, if the level of blood glucose is so high that the amount entering the glomerular filtrate exceeds 375 mg/min then the reabsorptive capacities of the tubule will be exceeded, and any excess above this level will be passed out in the urine (Fig. 7.7). Although this amount is several times greater than normal blood values, glucose is so essential a molecule that it can never be considered a waste product. Other substances normally reabsorbed also have limits which, if exceeded, are voided in the urine. The point is that the plasma levels of glucose and many other molecules can be **maintained** by the kidneys but cannot be **regulated** by them. That is, below a certain limit, *all* the substance is reabsorbed; above a certain limit *none* of the excess is reabsorbed. (The blood levels of $H^+$, water, and $Na^+$ are exceptions to this statement because they are regulated by the kidney, as is explained below.)

**Secretion** Most of the small molecules in the plasma pass into the tubule during filtration, at which point a select few are reabsorbed. However, hydrogen ions enter the tubule almost exclusively by secretion, as do such foreign chemicals

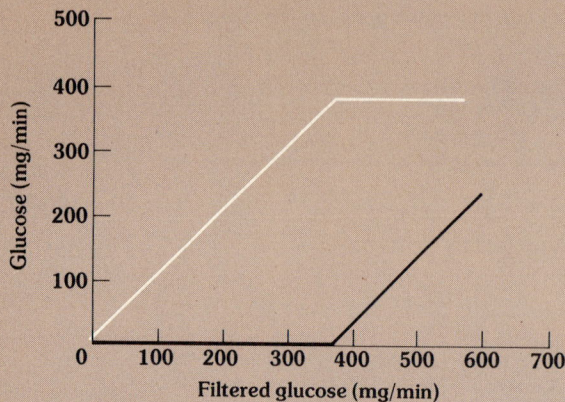

**Fig. 7.7**
**Reabsorption threshold for glucose: white line represents glucose reabsorbed; black line signifies glucose excreted.**

as penicillin and other drugs, partly because their size limits their effective filtration. In addition, potassium ($K^+$) is both filtered and reabsorbed—but the potassium that ultimately passes out in the urine is secreted. Secretion represents a kind of backup mechanism, whereby the final constitution of the urine can be altered through the active transport of select materials from the cells forming the walls of tubules into the cavity of the tubules.

### Function of the nephron—physiological events

**Removal of metabolites**   The nitrogen-bearing fragments of amino acid molecules can be split off by the liver, yielding molecules of **ammonia** ($NH_3$). Ammonia is an extremely toxic molecule, and the cells of the body can withstand only very low concentrations of it in the blood. Therefore, in order to maintain the ammonia level below toxic levels, large amounts of urine must be passed, for there is no possibility of concentrating the ammonia at any point in the body. As an example, frogs produce a volume of urine on a daily basis equal to 25 percent of their body weight; earthworms produce a volume equal to 60 percent of their body weight in order to maintain blood ammonia levels below the toxic level. (This is why earthworms dry out so quickly!)

It is intuitively obvious that large, land-dwelling organisms can ill afford such monumental water loss. Accordingly, it is to be expected that such organisms have evolved ways of converting ammonia to some less toxic molecule, which allows a higher concentration in the urine and, therefore, less water loss. In mammals, including humans, the molecule in question is **urea,** which is synthesized in the liver, and which consists of two ammonia molecules bound together with a $CO_2$ molecule. Urea is a great deal less toxic than is ammonia, with the result that we produce only 2 percent of our body weight in the form of urine every day. We also produce some ammonia as well, as anyone who has ever diapered a baby can attest. However, the reason that soiled diapers become so "fragrant" in the diaper pail is that bacteria act to split the urea molecule back into ammonia molecules. Thus, the ammonia concentration increases with time.

An even more efficient solution is employed by most birds and reptiles, which produce **uric acid.** This molecule is rather nontoxic, and can actually be excreted at a pastelike consistency, obviously an enormous saving of water. We also produce some uric acid, primarily resulting from nucleic acid breakdown. It is interesting to speculate why mammals do not use uric acid in greater amounts, however. One possible explanation is that, in reptiles and birds, the pastelike uric acid passes from the bladder into a common chamber with the rectum rather than through a narrow urethra. One could imagine the difficulty (to say nothing of the pain) of a man trying to pass a pastelike material down the length of his urethra. Freud's statement, "Anatomy is destiny" seems to have special significance here.

**Control of blood pH**   Challenges to an otherwise constant blood pH come primarily from two[2] sources—the production of $CO_2$ as a consequence of cellular metabolism (see Chapter 6) and the ingestion of food containing weak organic acids. Elimination of $CO_2$ occurs via the lungs; elimination of acids (in the form of $H^+$) occurs via the kidneys.

As was outlined in Chapter 2, excretion is only one of several methods used by the body in homeostasis. A second method is **buffering,** and this term is limited, by definition, to mechanisms that prevent changes in pH. Proteins and amino acids are efficient buffers, in that they possess the capacity to release or to take up $H^+$, depending on the pH of the surrounding fluids.

Carbon dioxide production causes an increase in the acidity of the blood in that it partially dissolves in the water of the plasma forming the weak acid $H_2CO_3$ (carbonic acid[3]), which then dissociates into the ions $H^+$ and $HCO_3^-$. This release of $H^+$ causes an increase in the breathing rate, resulting in the exhalation of larger amounts of $CO_2$, which ultimately lowers the level of $H^+$ in the blood (see Chapter 6 for details).

Not all of the $H^+$ results from the formation of carbonic acid, of course, and even though buffering can act temporarily to prevent a change in blood pH, sooner or later all the buffer molecules will be tied up and the buffering capacity of the blood will be exceeded. The lungs can act to reduce the production of $H^+$ to some extent by passing off $CO_2$ more quickly, but it is important to note that no $H^+$ itself is passed off by the lungs. Only the kidneys possess this capacity. An increase in the $H^+$ concentration in the blood triggers increased secretion of this ion by the kidney tubule cells into the forming urine. The flexibility of $H^+$ secretion is demonstrated by the fact that although urine typically has a pH of about 6.0, it can range from 8.0 down to 4.5, depending on the amount of $H^+$ that is present in the blood.

Of course, there is a limit as to how much free $H^+$ can be secreted into the cav-

---

[2] Changes may also be introduced through abnormal events, namely vomiting or diarrhea, which tend, respectively, to raise or lower blood pH (see Chapter 4).
[3] The tart taste of carbonated soft drinks comes from the presence of carbonic acid—"carbonation" is nothing more than the dissolving of $CO_2$ in water.

ity of the tubule before the acidity of the developing urine reaches the point at which it begins to damage the cells that form the walls of the tubule (which are not capable of being protected by, say, a mucus barrier, as are the cells of the stomach). As it happens ammonia ($NH_3$) has the capacity to act as a buffer and pick up $H^+$ (thus converting to $NH_4^+$). Normally, of course, most of the ammonia has been converted to urea in the liver, because free ammonia is itself a highly toxic molecule to living cells. However, when blood $H^+$ levels are high, tubule cells not only secrete $H^+$ but also secrete increased amounts of ammonia, which, because of its buffering action, prevents the urine from becoming too acidic.

**Osmoregulation**   It should be clear by now that the osmotic pressure of a solution can be elevated either by adding solutes or by removing solvents. Reciprocally, the osmotic pressure of a solution can be lowered either by removing solutes or by adding solvent. In the human body, both salts (solutes) and water (solvent) are being added to the blood at all times, primarily from the digestive system.

The kidneys maintain the osmotic pressure of the blood constant by removing appropriate amounts of both salts and water. This is an important point, because theoretically, the osmotic pressure of blood could be maintained by monitoring only one variable (i.e., either the salts or the water, but not both).

It should be noted that loss of water through the skin, the lungs, and the gut is essentially uncontrollable. Similarly, it would seem an impossible task if it were necessary for us to drink water in precisely the same amounts as it is being lost from the body. Hence, it is the kidneys exclusively which possess the capacity for *controlled loss of water* (as well as salts, although the other sources for salt loss are far less important than they are for water loss). Of course, there are limits to this controllability. Of necessity, some loss must occur in order to keep other urinary constituents in solution. However, the range of potential daily water loss through the kidney is staggering—from 400 ml (8.0 pt) to 25 liters (6.6 gal) per day.

The basic method of osmoregulation by the kidney is rather simple, although it is complicated by two overriding factors, namely, the hormones **ADH** and **aldosterone,** which will be discussed subsequently. In the basic method, both the salts and water enter the glomerular filtrate, as both are freely filterable. During the course of movement through the convoluted tubule, $Na^+$ is reabsorbed by active transport. As a consequence, $Cl^-$ follows, in order to maintain equality in electrical charge on both sides of the cell membrane (note that the movement of $Cl^-$ is by passive diffusion). Because of the drop in the level of solutes, the osmotic pressure of the glomerular filtrate falls, and water flows, by osmosis, out of the cavity of the tubule and into the tubule cells themselves. Both salts and water quickly reenter the circulatory system by way of the network of capillaries which surround the tubules (Fig. 7.4).

The preceding paragraphs describe the qualitative aspect of the system. To a large extent, the quantitative aspects are predetermined as well. Approximately 180 liters (47.5 gal) of glomerular filtrate are produced each day, but, of that amount, only about 2 liters (0.5 gal) typically leaves the body as urine. (It is just as well that the loss is small—with a body fluid volume of only 40 liters (10.6 gal), we would become a totally dehydrated mass of crystals within 6 hr, unless we were prepared to drink 7 liters (1.8 gal) of water, with the correct salt content, every hour of our life.)

Therefore, the truly critical part of urine formation involves not so much the volume but the composition of the 2 liters (0.5 gal) of urine. Volume and composition must be such as to ensure that blood osmotic values remain constant.

A drop in blood volume results in a drop in blood pressure. This is an important point, not only in the abstract but in highly pragmatic ways as well. A person suffering from extensive bleeding (drop in blood volume) is in imminent danger of death because of the drop in blood pressure, and not because of a drop in the oxygen-carrying capacity of the blood owing to the loss of red blood cells. Therefore, such individuals are immediately given transfusions of serum or plasma, which restore blood volume and raise pressure, but which make no significant contribution to the oxygen-carrying capacity of the blood. The change in blood pressure is monitored by pressure sensors in various parts of the body, which stimulate a specialized portion of the floor of the brain, called the **hypothalamus,** to initiate the thirst reflex. Thus, water is added by drinking, and the blood volume is restored.

Simultaneously, these same cells of the hypothalamus increase the production of a hormone called ADH. ADH stands for antidiuretic hormone, a **diuretic** being any substance which increases urine output. This hormone is discussed more completely in Chapter 9. ADH increases the permeability of the tubule cells to water (apparently by increasing the diameter of the pores of the cell membrane through which the water molecules must move). Thus, in the presence of ADH, water loss in the urine is decreased, water retention by the body is increased, and blood volume ultimately increases as a consequence (Figs. 7.7 and 7.8). Interestingly, thirst (water in) and urine production (water out) are governed by the same portion of the brain, and the cue used (blood pressure) has a direct relationship to the blood volume. In this manner, changes in the concentration of solvent (water) in the blood are guarded against, and the osmotic value of the blood is held constant.

It is interesting to note that several common molecules affect the operation of ADH. Alcohol, for example, apparently interferes with the production of ADH in the hypothalamus such that a lessened secretion occurs. As a consequence, less water is reabsorbed resulting in an increase in urine volume. Thus, rest rooms are frequented in bars not so much because of the **volume** of liquid being imbibed as from the nature of the constitutents of that liquid.

Similarly, the ingestion of **caffeine** (from coffee, tea, or cola) results in increased urine formation, but it seems that caffeine interferes not so much with the actual production of ADH as with its site of effectiveness—the tubule cells themselves. In a manner that is not completely understood, caffeine affects the capacity of ADH to increase the size of the pores in the membrane of the tubule cells, the pores remain small, and the reabsorption of water is less complete. The similar results of ingesting both alcohol and coffee suggests that, to the extent that a hangover may result from dehydration, ingesting a second diuretic (coffee) to "sober up" may be counterproductive in trying to get rid of a headache.

The situation with regard to salt control is somewhat different. The major ion

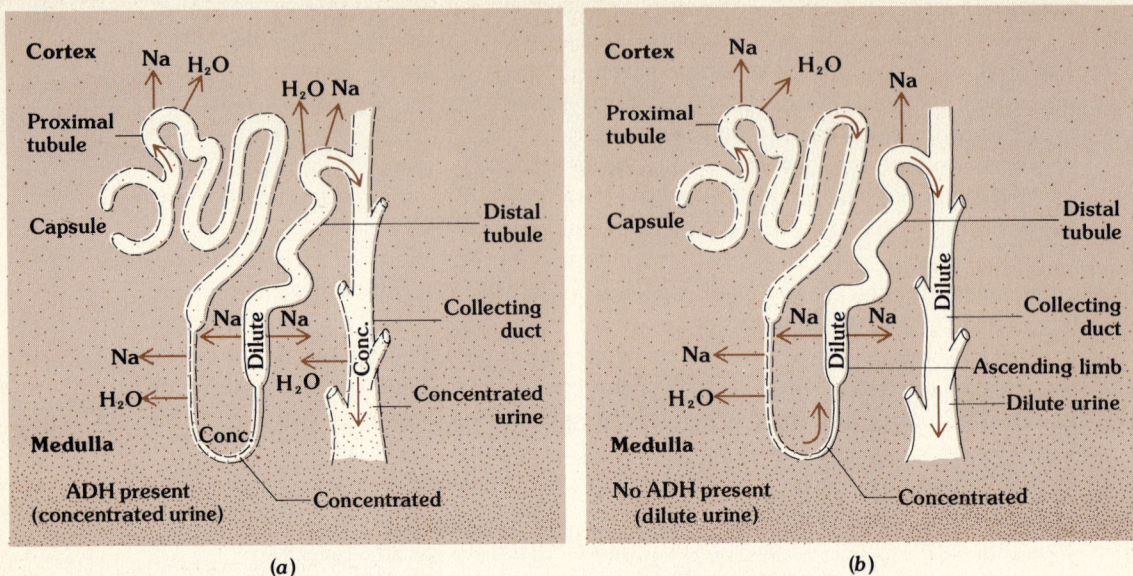

**Fig. 7.8** (a) Effect of presence of ADH on urine formation; (b) effect of absence of ADH on urine formation. Dashed lines indicate areas of tubule which are permeable to water.

in the interstitial fluid is $Na^+$ (most of the other common ions are much more abundant within cells than outside them), and therefore the amount of $Na^+$ in this fluid has a profound effect on its osmotic value, hence its total volume. The significance accorded to the regulation of this ion is indicated by the fact that its concentration never varies by more than 2 percent from the norm.

Because $Na^+$ is such a common constituent in our food, there is seldom if ever a deficiency of this ion (i.e., there is essentially no need for a salt "thirst," although such a phenomenon does exist in many other mammals). The problem, then, is reduced to that of determining the amount of $Na^+$ that will be reabsorbed by the cells of the tubules, because large amounts of $Na^+$ are present in the glomerular filtrate and, unless reabsorbed, will be lost in the urine.

A reduction in the $Na^+$ level in the interstitial fluid causes greater spontaneous excitation of nerves (for reasons discussed at length in Chapter 10). These nerves include those that go to the kidney. Increased nervous stimulation of the kidney results in the formation of a protein called **renin,** which serves as an enzyme in the conversion of the plasma protein **angiotensinogen** (produced by the liver) into **angiotensin** (Fig. 7.9). This protein, in turn, stimulates the release of aldosterone, a hormone produced by the **adrenal glands** (a set of endocrine glands lying just above the kidneys—see Chapter 9). Aldosterone, in turn, acts on the tubule cells to increase the rate of reabsorption of $Na^+$. Thus, in a circuitous fashion, the kidney controls the osmotic value and the volume of the blood and interstitial fluid by controlling the amount of $Na^+$ that is reabsorbed. Thus, in totality, the kidney achieves body fluid osmoregulation by controlling the amount of water (solvent) and sodium (the primary solute) retained by the body, and how much of each is lost in the urine.

**Urine formation**   Of the 180 liters (47.5 gal) of glomerular filtrate which enter the nephrons every day, only about 2 liters (0.5 gal) are passed off as urine. Such a reduction in volume is not only desirable, it is mandatory. If there were no reabsorption (i.e., if all the glomerular filtrate were passed off as urine), all of the water present in the plasma would be lost in less than half an hour. Perhaps even more dramatically, given such a situation, coupled with the present capacity of the human bladder, we would need to urinate every 3 min, 24 hr a day!

The point is simply that it is generally advantageous for the body to lose a minimum amount of water, for the consequences will be a reduction in the need to drink. Therefore, urination does not simply consist of voiding excess salts, acids, and metabolites, but it also involves accomplishing all that and losing the smallest possible amount of water in the process.

In theory, it would seem that urine could be no more concentrated than the blood or interstitial fluid, as the explanation of reabsorption given above has the urine remaining osmotically balanced with the blood as sodium is actively transported out of the glomerular filtrate. In practice, the human kidney is capable of producing urine as much as 3.5 times as concentrated as is the blood. The manner in which this is achieved provides an excellent example of the **countercurrent principle**.

**The countercurrent principle**   There are many examples of the application of the countercurrent principle in nature. Not only is it used in the kidneys, but also in the arrangement of arteries and veins in such diverse structures as the gills of fish and the limbs of mammal.

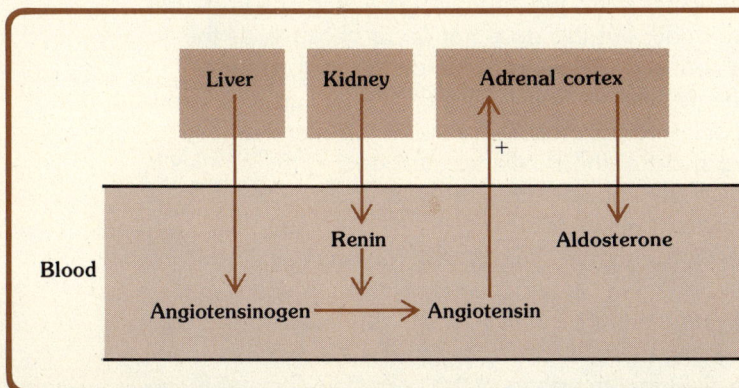

Fig. 7.9
Control of
aldosterone
production.

Consider the problem of the whale, swimming in Arctic waters, in maintaining a high body temperature. A thick layer of blubber insulates its body, but the large flippers have no such insulation. Nevertheless, the heat lost from the flippers to the surrounding water is not especially great. The arteries that enter the flippers of whales run very close to the veins, which return blood to the body, and both are deep within the flipper. Heat loss from the arterial blood does not, for the most part, pass into the water, but rather warms the venous blood before it reenters the body. Thus, the arterial blood of the flipper is at a lower temperature than is arterial blood in the body, as a consequence of the heat lost to the returning venous blood. The result is not only a minimizing of the temperature differential of the returning venous blood, but also a lowering of the temperature of the entire flipper relative to the body, which results in a smaller gradient between flipper temperature and water temperature, thereby minimizing total heat loss to the water.

Similarly, water moves across the gill of a fish in the opposite direction of blood movement within the gills. Thus, even as the water loses oxygen to the blood, it is encountering blood that contains less oxygen than does the water. In that way, virtually all the oxygen in the water can be taken up by the blood, rather than only the one-half predicted by the standard laws of diffusion.

Movement of materials differing in concentration or temperature in parallel vessels but in opposite directions exemplifies the countercurrent principle. By this arrangement, heat loss can be minimized in the flippers of whales, maximum oxygen can be extracted from the water by the gills of fishes, and urine can be maximally concentrated in the kidneys of mammals.

The countercurrent principle really is used twice in the case of the mammalian kidney—once by the tubules of the nephrons and once by the capillaries which surround these tubules. Figure 7.10 indicates that the glomerulus and Bowman's capsule are located in the cortex of the kidney, but the long loop of the nephron descends into the medulla. Because most of the $Na^+$ leaving the glomerular filtrate is actively transported by the portion of the tubule that is located in the medulla, the interstitial fluid in this region of the kidney is extremely high—in fact, some three times higher than in any other portion of the body. One would expect this salt to be carried away by the capillaries surrounding the tubules (Fig. 7.4) and to be spread equally throughout the body, but this does not occur because, as the blood in these capillaries rises to connect with veins in the cortex, it encounters increasingly lower salt concentrations, and the salt picked up by the capillaries flows out of the capillaries again and back into the interstitial fluid of the medulla (Fig. 7.11). Thus, the particular arrangement of the capillaries surrounding the tubules is such as to ensure a continuing high salt concentration in the medulla. This, then, is the first utilization of the countercurrent principle.

The second utilization occurs in the tubule itself. As the forming urine flows up the distal arm and back into the cortex, $Na^+$ is removed by active transport, and the osmotic value of the urine is quite low (Fig. 7.11). However, note that the collecting duct again proceeds through the medulla with its high interstitial fluid $Na^+$ concentration. If ADH is present (Fig. 7.7) and the pores of the collecting duct are there-

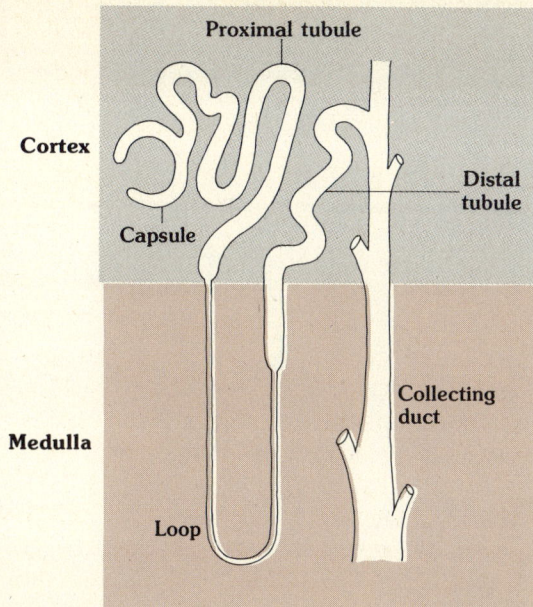

Fig. 7.10
Relationship of the nephron to the cortex and medulla of the kidney.

fore open, water moves rapidly out of the collecting duct because of the high surrounding salt concentration. As a consequence, the urine itself is concentrated, ultimately reaching a concentration about equal to the greatest concentration in the medulla (about three times greater than blood). This, then, is the second application of the countercurrent principle.

By now it should be clear that the adaptive significance of the long loop in the convoluted tubule is to allow for concentration of the urine. Moreover, it might be guessed that the longer the loop, the more concentrated the urine can become. This is, in fact, the case. The loop of the kangaroo rat, a small mammal which requires no external source of water, is proportionately much longer than is our own, with the result that the kangaroo rat can produce a much more concentrated urine. This, of course, is necessary if the rat is to live exclusively on metabolic water.

**Urination**  The collecting ducts gradually merge and come to form a single tube, the **ureter,** one of which leads from each kidney to a single, centrally located **urinary bladder** (Fig. 7.2). Movement of urine down the ureters takes place by means of peristalsis, as the ureters are lined throughout with smooth muscle.

The bladder has a capacity of between .5–1 liter (0.13–.26 gal), but when about 300 ml (0.6 pt) of urine are present, stretch receptors in the wall of the bladder respond and transmit a stimulus to the spinal cord, which in turn stimulates a contraction of the bladder. This bladder contraction pulls the **urethra** open at its bladder end (an area sometimes referred to as the **internal sphincter**), and urina-

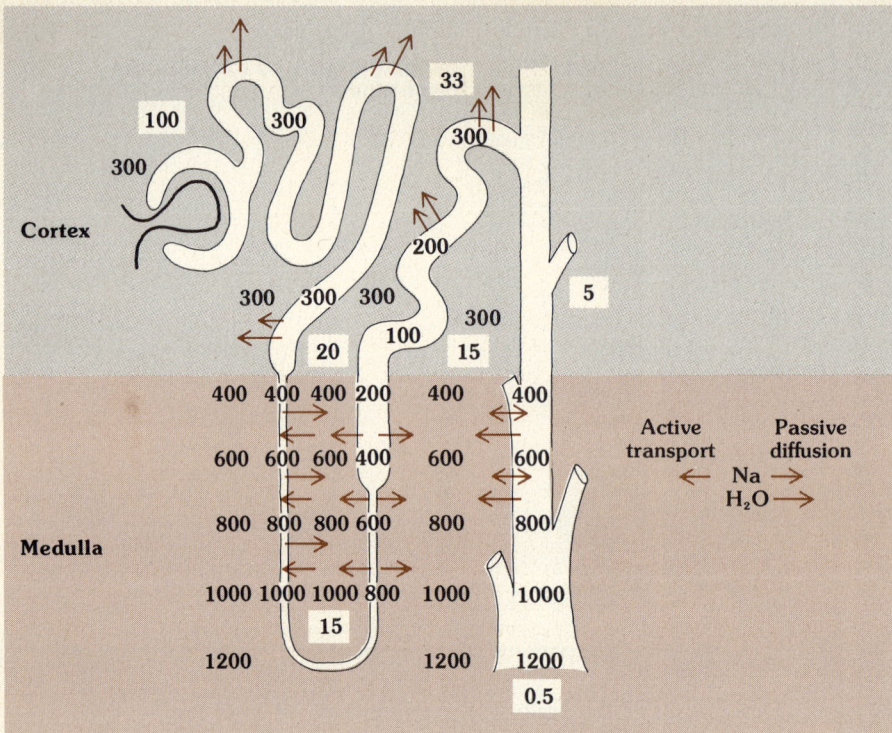

**Fig. 7.11  The countercurrent concept of urine concentration. Numbers refer to the osmotic concentration of the fluid within the tubule and the interstitial fluid. Boxed numbers indicate the estimated percentage of the original glomerular filtrate still remaining within the tubule, at various points.**

tion occurs. This whole process takes place automatically, which is not unexpected, given that visceral muscle is involved. In children past the age of about three there develops an override system whereby the child can delay urination consciously by contracting the skeletal muscle at the outer end of the urethra (an area known as the **external sphincter**). Still later, the ability to urinate at will, even in the absence of stimuli from a filled bladder, is developed. The consequences of this gradual maturation in control are twofold. First, many children are punished for bedwetting at a time when they are physiologically incapable of overriding the urination reflex. Second, the fact that "toilet trained" children may not be able to urinate at will, but suddenly need to do so 15 minutes later (just after the family car has entered the freeway) is not inconsistent with the fact that the same child no longer wets the bed at night.

The urethra is very short in women, measuring only about 2.5 cm (1 in.) in length, but because this tube runs to the end of the penis in men, the male urethra is considerably longer. It is this rather significant difference in length which accounts for the fact that bladder infection **(cystitis)** is much more common in women than in men, as invasive bacteria need traverse a much shorter distance in women.

**KIDNEY DISEASES**  Kidney diseases account for about 50,000 deaths annually in the United States. Kidney diseases may be separated into infections of the kidney itself, typically by bacteria, and mechanical interference caused by a variety of sources. In both cases, a typical symptom is **uremia**—the presence of an excessive amount of urea in the blood. Although urea is much less toxic than is ammonia, it nonetheless can poison cells at high concentrations. More typically, however, the real threat is a change in the blood pH or osmotic value because salt and acid excretion also ceases when metabolite (urea) excretion fails.

Infections of the kidney usually are of the nephron itself and are termed **nephritis.** There are about 3,000,000 cases of nephritis annually in the United States, many stemming from prior conditions of scarlet fever or strep throat. These infections may take the form of plugging up the nephrons so that only a small fraction of the urea is excreted, or, alternatively, the infection may damage the glomeruli such that larger molecules, most notably proteins, or even red blood cells, may enter the glomerular filtrate and pass out in the urine, giving it a cloudy appearance. Obviously, the seriousness of any case of nephritis depends on how much impairment of normal kidney function occurs.

Chronic nephritis may also develop as a consequence of hypertension or arteriosclerosis (see Chapter 5), both of which might be expected to play havoc with the normal functioning of the glomeruli.

Mechanical damage or obstruction may take a variety of forms. Various poisons, especially such heavy metals as lead, mercury, and cadmium, are particularly lethal to kidney cells, and obviously cellular destruction would disallow any secretion or reabsorption. Mismatched blood transfused into the body forms clumps of red blood cells; these are broken down in capillary beds where they release hemoglobin, which, in turn, clogs the glomeruli. In fact, the degree of danger posed by a mismatched transfusion is primarily a question of how much kidney damage results.

Perhaps the most common mechanical obstruction is **kidney stones,** which results primarily from the precipitation of uric acid. For whatever reason, kidney stones are three times more common in men than in women. They can be excruciatingly painful if they enter the ureters. Should they lodge, surgery may be necessary, because there is then no way for the urine to enter the bladder (although typically only one kidney is involved at a time). Generally, however, the kidney stones move slowly along the ureter and, once in the bladder, cause no further problem, as the urethra has a much larger diameter than do the ureters. However, if the ureter is torn by the passage of a kidney stone, there is a danger that scar tissue will form and the possibility that the passageway will be occluded.

A related problem is the overproduction of uric acid. Although this is the predomi-

For a variety of reasons, kidney failure is becoming an increasingly common phenomenon. The preceding pages have indicated how vital are functioning kidneys to the well-being of the organism and, indeed, death will follow in a matter of days if the kidneys cease to function. Even though the functioning of the kidneys seems complex, the fact is that the operation of the kidneys basically consists of a balancing of diffusion potentials, a simple physical phenomenon. For this reason, the "artificial kidney" was among the first of the artificial organs to be developed.

Basically, the artificial kidney consists of a salt and water chamber divided in half by a special cellophane membrane. The composition of this solution is essentially that of normal blood plasma, except that no urea is present. Blood is pumped into one side and the pores of the cellophane allow urea, excess salts, and excess $H^+$ ions to diffuse through. Then the blood is pumped back into the body.

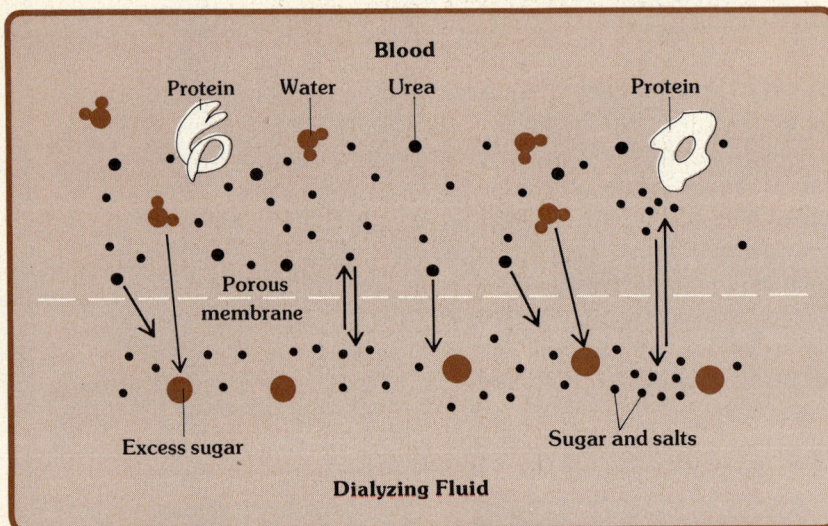

Blood

Protein    Water    Urea                    Protein

Porous membrane

Excess sugar                          Sugar and salts

Dialyzing Fluid

Diagrammatic view of dialysis, the principle upon which the artificial kidney operates. Only the small molecules can pass through the membrane pores separating blood (*top*) from dialyzing fluid (*bottom*). The dialyzing fluid is then discarded, and the blood is returned to the patient.

**The artificial kidney**

This is not a totally efficient process and certainly involves a great deal of discomfort as well as being psychologically wearing. There is also the continuing risk of the formation of blood clots. Nonetheless, twice weekly treatments have kept many individuals alive for long periods of time, when they would have otherwise died. In addition, organ transplants of kidneys have been the most successful of all organ transplants (see Chapter 5).

nant nitrogenous waste of reptiles and birds, only a small amount typically forms in humans. In individuals who produce excessive amounts, uric acid tends to settle in joints, especially at the ball of the foot at the base of the big toe, causing a swelling known as **gout.**

## SUMMARY

For most of the organ systems there is a reasonably complete overlap between anatomical entity (e.g., digestive tract) and physiological activity (e.g., digestion). Such is not the case for the excretory system and excretion. First, by definition excretion is limited to the removal of metabolites ($CO_2$, metabolic water, and ammonia) and in that sense it excludes the salt control and most of the pH control that must exist if the blood is to remain in a state of homeostasis. Second, the excretory system handles only nitrogenous metabolites—$CO_2$ is removed by the lungs.

A better term for what is typically called the "excretory system" would be the "urinary system," which suggests only that it is concerned with the formation of urine, and implies nothing as to the constituents of urine or to the physiological processes involved. Similarly, a better term for the physiological activity typically labeled "excretion" would be "fluid homeostasis," which includes all the physiological events involved in regulating the components of blood and interstitial fluid. It is unlikely that these suggested changes will supersede the existing terms, but recognition of what underlies them will aid in understanding the problems and solutions of maintaining the purity of the internal sea.

Why do girls tend to stop growing when they enter puberty, whereas boys grow rapidly?

Why are most birth control techniques initiated by women rather than by men? Which techniques are the most effective?

How serious are venereal diseases really? With all our antibiotics, why is VD on the increase?

These questions are addressed in our discussion of the reproductive system.

chapter VIII

# chapter VIII
# Creation of the next generation–the reproductive system

Thus far, we have considered a variety of systems and processes which, for the body as a whole, are primarily a means to an end. For example, the digestive and circulatory systems function to ensure the acquisition and distribution of nutrients required by the cells of the body for their continued existence; the kidneys function to maintain a balance of various components of the blood, and so on, with the end being the continued life of the cells. The reproductive system, however, is an end in itself. Biologically speaking, all other systems are subservient to the reproductive system in that they are geared to keep the organism alive until it reaches sexual maturity and is able to reproduce. Of course, we tend to take a less dispassionate view of things when looking at ourselves and usually feel that life has meaning of its own, rather than simply serving as a waiting period before (and after) reproduction.

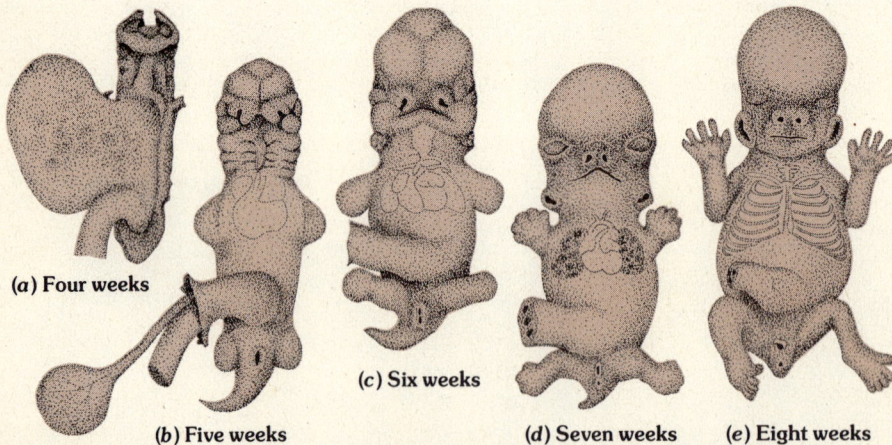

Fig. 8.1
The developing human embryo, drawn as if the body curvatures were flattened.

(a) Four weeks

(b) Five weeks

(c) Six weeks

(d) Seven weeks

(e) Eight weeks

**DEVELOPMENT OF THE HUMAN REPRODUCTIVE SYSTEM** Because the reproductive system is unique in that it is the only system to vary appreciably between male and female, we have come to think of the differences between the sexes as somehow indelibly stamped into every individual. It may therefore come as some surprise to learn that during the first 6 weeks of embryonic life, there is no structural difference between the male and female embryos (discounting chromosomal differences). The male and female **gonads** (the **testes** in the male and **ovaries** in the female) are identical under the microscope, and the external genitalia (such as they are) are also identical (Figs. 8.1 and 8.2).

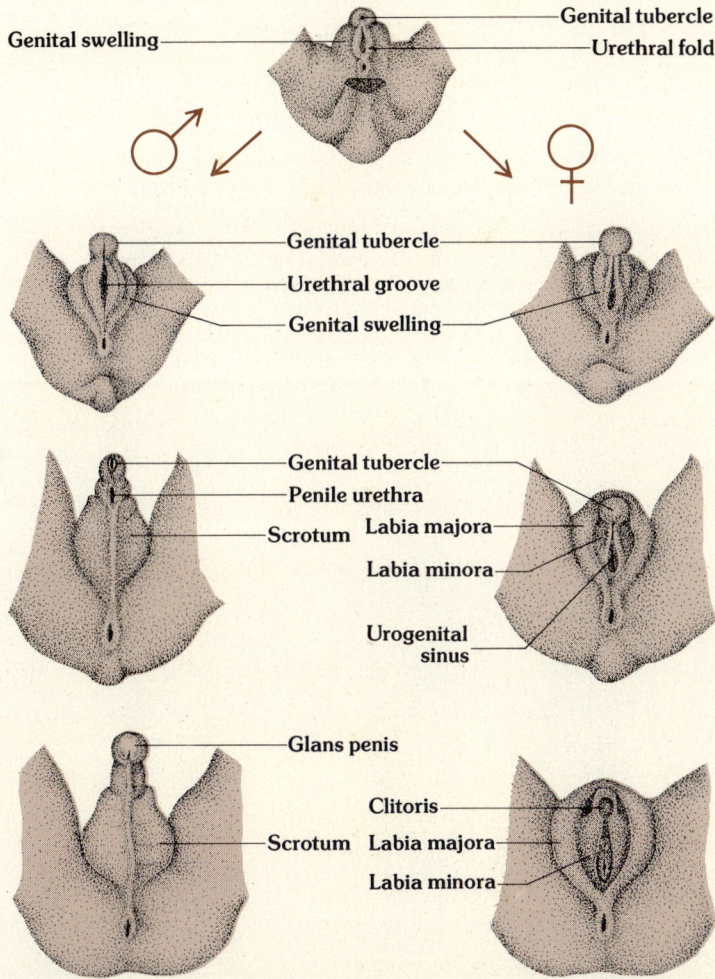

Fig. 8.2 Embryonic development of the male and female genitalia from undifferentiated precursors.

Genital swelling — Genital tubercle — Urethral fold

Genital tubercle — Urethral groove — Genital swelling

Genital tubercle — Penile urethra — Scrotum — Labia majora — Labia minora — Urogenital sinus

Glans penis — Scrotum — Clitoris — Labia majora — Labia minora

Development of male–female differences does occur during fetal development, of course, and, at birth, many of the basic anatomical differences by which we characterize males and females have already taken place (Fig. 8.3). It is important to note, however, that these differences are the result of differential growth of the same set of structures. The female structures resemble those of the undifferentiated embryo, whereas the male structures undergo considerable change. The **clitoris** of the female remains in the position of the precursor **genital tubercle,** whereas in the male it is carried forward several inches (in the adult) to become the head of the penis **(glans penis).** Because the urethra must carry to the end of the penis, it must grow along with the moving glans. That portion of the urethra within the penis therefore corresponds to the inner folds **(labia minora)** of the female. (Note that the urethra in the female does not pass through the clitoris, but rather empties just below it.) Finally, in the male, the outer folds (which correspond to the **labia majora** in the female) grow out to meet and fuse to form the **scrotum,** the sac which contains the testes.

The adult systems are illustrated in Figs. 8.4–8.6. Internal differences between male and female are somewhat greater, insofar as the same **homology**[1] of parts does not exist for all of the internal structures. For example, the basic duct of the female (consisting of the **vagina, uterus,** and **Fallopian tubes**) is not derived from

[1] Homology refers to structures having a common origin, even if differing in final form. For example, the wings of birds are homologous with the arms of humans, as both are derived from the basic vertebrate forelimb. The antithesis of homology is **analogy**—analogous structures are those that have a similar final structure but no common origin (e.g., the wings of birds and the wings of butterflies).

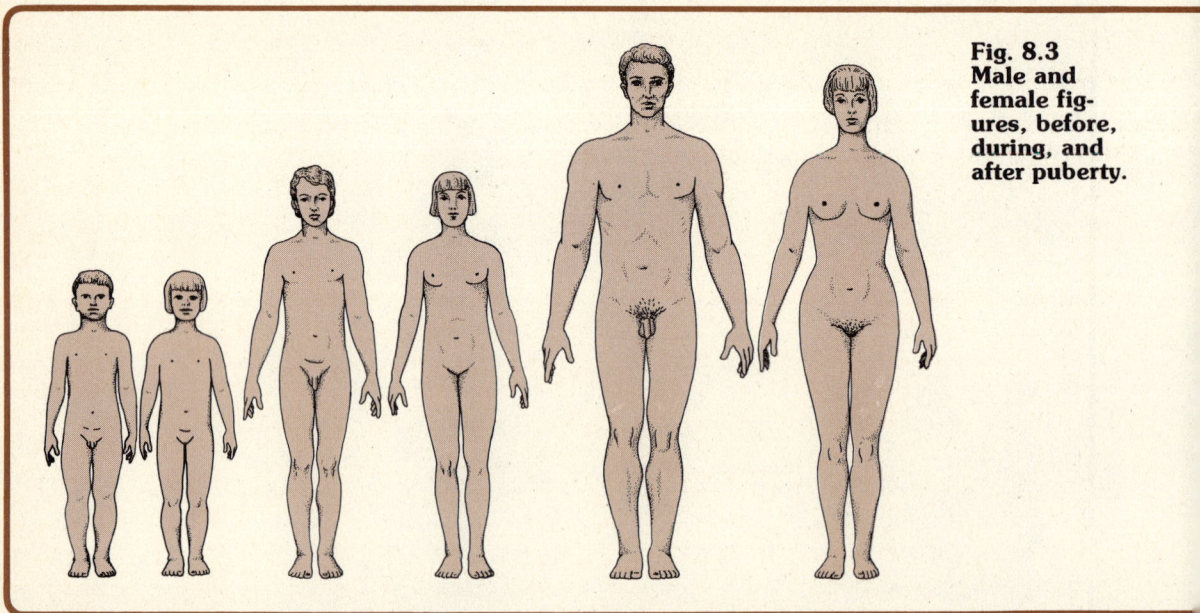

Fig. 8.3 Male and female figures, before, during, and after puberty.

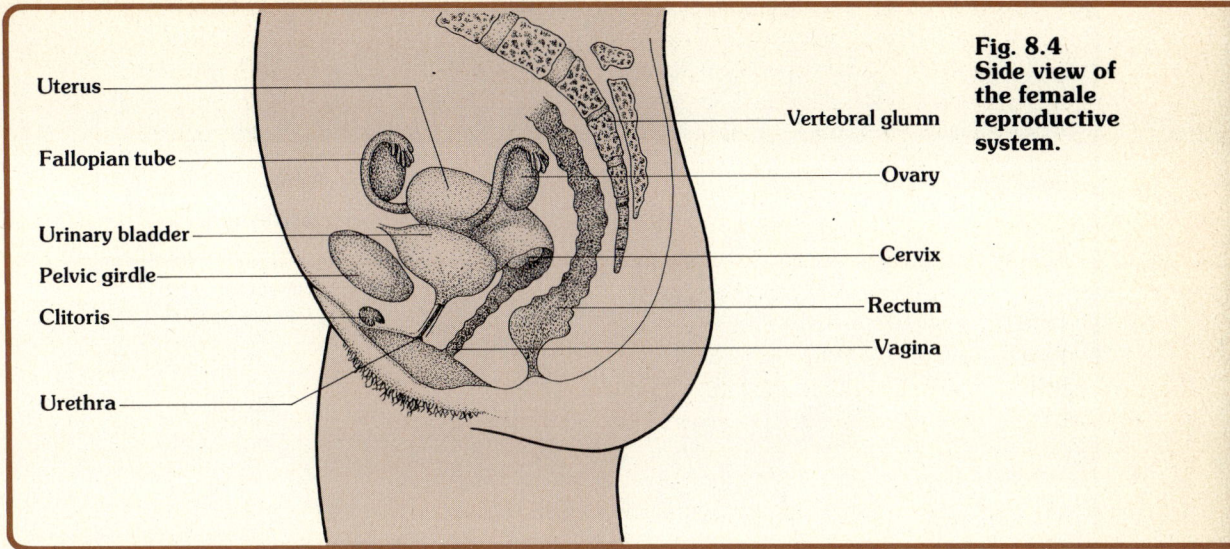

Fig. 8.4
Side view of
the female
reproductive
system.

Uterus

Fallopian tube

Urinary bladder

Pelvic girdle

Clitoris

Urethra

Vertebral glumn

Ovary

Cervix

Rectum

Vagina

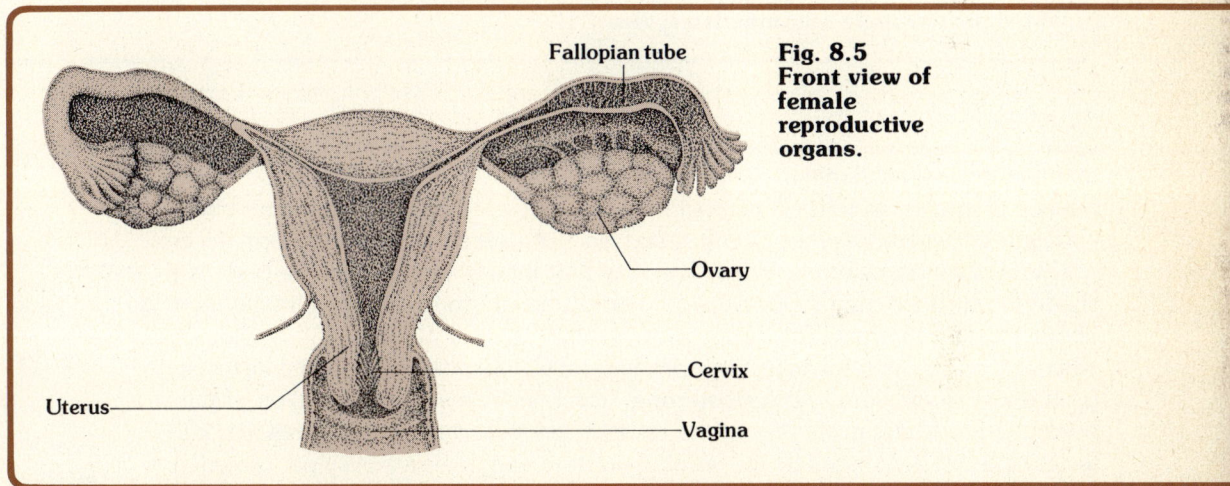

Fig. 8.5
Front view of
female
reproductive
organs.

Fallopian tube

Ovary

Cervix

Uterus

Vagina

the same embryonic structures as is the male duct **(vas deferens).** There is likewise
no female equivalent of such male glands as the **prostate** and the **seminal
vesicles.**

One of the more striking differences between the male and the female systems
is the relative position of the gonads. The ovaries of the female are located essentially in the original embryonic position, whereas the testes of the male are found
outside the main body cavity. Embryonic studies indicate that the male gonads origi-
nate at the same point as do the female gonads (recall that at the outset of their devel-

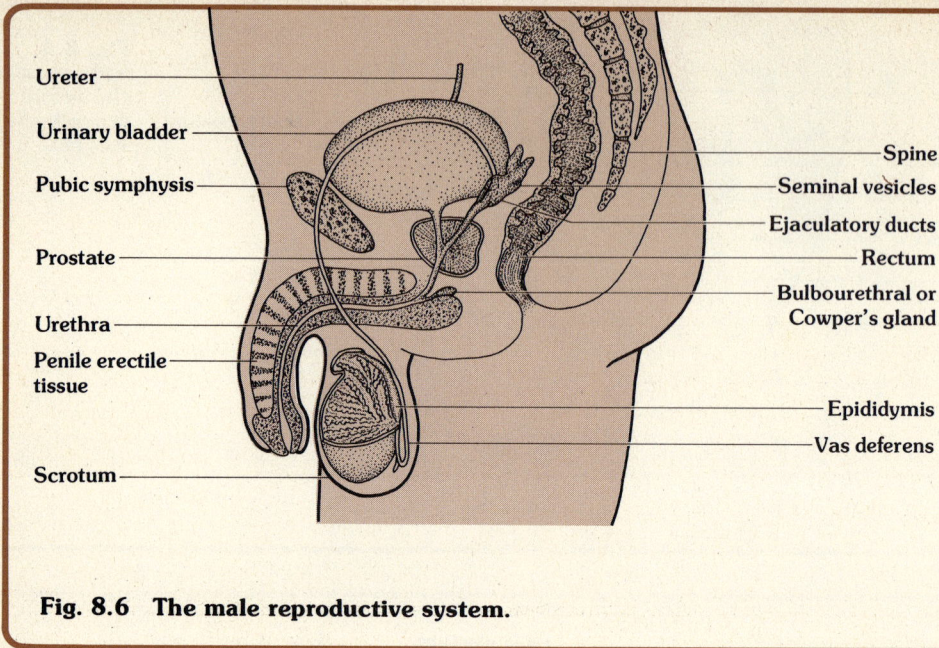

**Fig. 8.6  The male reproductive system.**

opment, there is no observable difference between male and female gonads), and that they only secondarily move downward into the developing scrotal sac. The sequence of events, which is illustrated in Fig. 8.7, is normally completed by birth, although occasionally one or both testes do not descend at the appropriate time.

Why such an elaborate ritual? It seems that the temperature of the body is a shade too high for the proper development of the sperm [the scrotal temperature is about 1°C (2°F) lower]. This supposition is supported by the fact that individuals whose testes do not descend typically produce no viable sperm (although their testes do produce the male hormone **testosterone,** which indicates that maturation of the hormone-producing properties of the testes is not thwarted). For this reason, if the testes have not descended by the onset of puberty, surgery is typically performed, not only to prevent sterility but also to prevent the development of cancer which is not uncommon in undescended testes.

It is, of course, this movement of the testes from their original point of development to the scrotum which accounts for the rather tortuous pathway followed by the vas deferens (Fig. 8.6). Moreover, this descent is not without its attendant hazards. The opening through which the testes move from the body cavity to the scrotum is called the **inguinal canal.** It essentially seals off following the descent, leaving only a small central area through which the vas deferens must pass. However, this secondary sealing off is frequently a point of chronic weakness, and a strain on the muscles in this area may cause the formation of a tear, resulting in an **inguinal hernia** (Fig. 8.8). (The adjective is important because hernias can occur in other parts of the body, most notably the diaphragm and the umbilicus.) Inguinal hernias are rather serious because a loop of the intestine may pass through the tear and become trapped in the scrotum. In severe cases, circulation may be cut off to this portion of the intestine and may result in gangrene. Fortunately, the scar tissue that forms once the tear has been sewn shut is usually stronger than the original muscle, making a second hernia (on the same side) uncommon. Of course, it is the descent of the testes and the resultant weakened area which accounts for the overwhelming prevalence of hernias in men, as opposed to women.

**ONSET OF PUBERTY IN THE FEMALE**    Prepubertal boys and girls are essentially sexless, despite the obvious differences in external genitalia, because hormonally they are virtually the same. Granted, children of each sex tend to assume different roles in our society. However, available evidence suggests that culture, and not biology, provides the major impetus for these role differences. In fact, the anatomical differences between boys and girls indicate only the **potential** for later sexual differences, a potential that is manifested at the time of **puberty.** Puberty is defined as

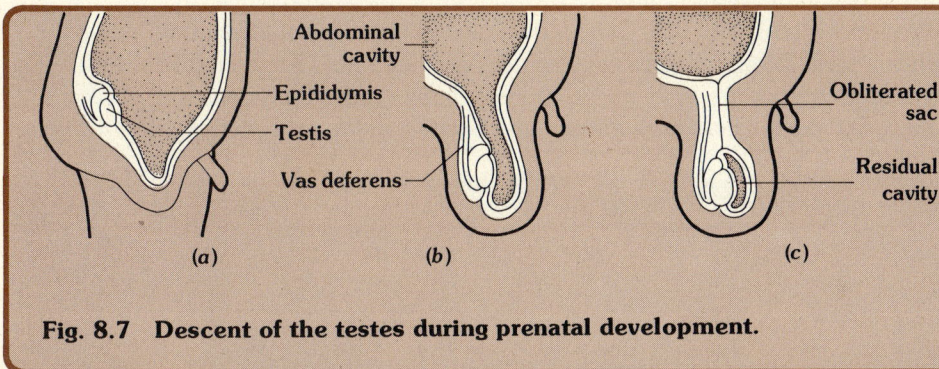

**Fig. 8.7   Descent of the testes during prenatal development.**

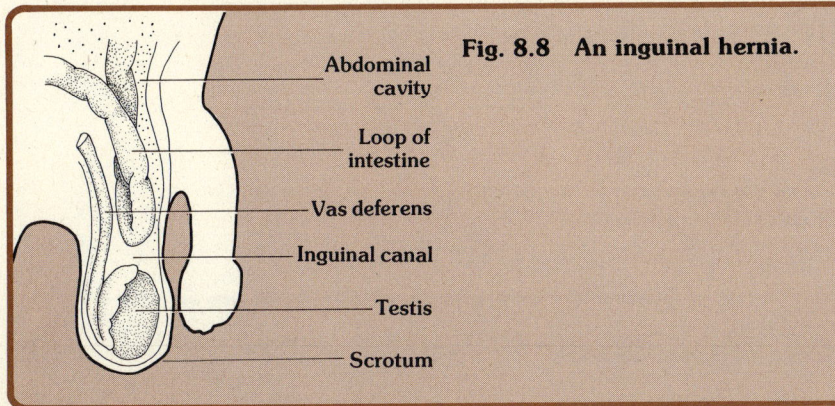

**Fig. 8.8   An inguinal hernia.**

the period during which sexual maturity occurs—that is, the time when reproduction becomes possible.

The onset of puberty in the female is triggered by factors that are far from understood. It is known that there is some correlation with overall body growth and the onset of puberty, although it is difficult to point to any one factor in growth which might be responsible. It is evident that there is no absolute correlation between body growth and the onset of puberty. Puberty can begin as early as five years of age, although such children are typically found to have brain tumors. However, in the past 100 years or so, the age of puberty has declined three or more years in most countries of the Western World, which would seem to be partly the result of improved nutrition and a consequent earlier period of body growth. The effects of an increasingly earlier sexual awakening in a society that "forbids" sexual experiences for a matter of some years (i.e., until marriage) are, unfortunately, rather obvious.

Physiologically, it is known that the onset of puberty is occasioned by the production of the hormone **estrogen,** which is produced by the ovaries. (Actually, there are a series of hormones collectively called estrogens. For the sake of simplicity, we shall consider them as a single hormone here.) Estrogen has many different functions in the body; these are summarized in Table 8-1. Among the more important of these functions, however, are the maturation of the reproductive tract and the overall development of the so-called **secondary sexual characteristics** (broadening of hips, growth of the breasts,[2] and the patterning of body hair). Estrogen also causes a cessation of growth in the long bones. Hence, women entering puberty at a young age tend to be shorter than are women who begin puberty at a later age.

---

[2] Breasts, of course, are present in both sexes but require estrogen for growth. Men treated with estrogen for certain disease conditions show considerable breast growth. Thus, the *potential* development of breast exists in both sexes, as opposed to its being an exclusively female characteristic.

## Table 8.1
### Effects of sex hormones on the body

**Effects of estrogens**

1 Growth of ovaries and follicles
2 Growth and maintenance of the linings of the entire reproductive tract
3 Growth of external genitalia
4 Growth of breasts
5 Development of female body configuration: narrow shoulders, broad hips, converging thighs, diverging arms
6 Pattern of pubic hair
7 Reduction of blood cholesterol
8 Vascular effects (deficiency → "hot flashes")
9 Feedback effects on hypothalamus and anterior pituitary

**Effects of progesterone**

1 Stimulation of secretion by endometrium
2 Stimulation of breast growth (particularly glandular tissue)
3 Elevation of body temperature
4 Feedback effects on hypothalamus and anterior pituitary

**Problems with the eggs**

Normally, only one egg follicle develops in any given month. However, from 1 to 2 percent of the time, one or more additional follicles may also develop. Fertilization of more than one egg accounts for the birth of non-identical twins, triplets, etc. "Fertility shots," now in common use for women who have difficulty conceiving, are extracts of pituitary containing large amounts of FSH.

Quite a different problem is related to the method of egg formation. Unlike a man, a woman produces no new sex cells during her life. That is, the eggs released on a monthly basis throughout the woman's reproductive life are present from birth. It is thought that the increase in such birth defects as **Down's Syndrome** (mongolism—see Chapter 13) in the children of older mothers is caused by a breakdown in egg cells held in a state of suspended animation for up to 50 years.

What initiates estrogen production? Initially, estrogen production is stimulated by the release of the hormone **FSH** (*f*ollicle *s*timulating *h*ormone) from the **pituitary gland.** ( The role of this "master gland" is discussed in considerable detail in Chapter 9.) FSH, in turn, is produced through stimulation of a region in the floor of the brain, just above the pituitary gland, called the **hypothalamus.** At present, the source of stimulation of the hypothalamus is unknown, although there are many theories. One of the more interesting of these theories involves the reception of light by the brain, possibly through activation of the **pineal gland,** a small lump on the surface of the brain apparently homologous with the "third *eye*" of certain lower vertebrates; it is well known, for example, that women who are blind from birth begin puberty significantly earlier than do women who are sighted.

Another consequence of the beginning of production of FSH is the initiation of the **menstrual cycle,** the monthly cycle which characterizes reproductively capable woman. FSH, as its name would imply, triggers the growth of a single **egg follicle** on the surface of the ovary (*see box*). The follicle consists of a maturing egg cell surrounded by follicular cells which support and nourish the egg. The follicular cells (*not the egg*) produce estrogen.

In addition to the various body-wide effects, mostly relating to the differential growth of certain tissues which we have already noted, estrogen also has several important functions in the menstrual cycle. First, the production of estrogen causes a decline in the production of FSH; this is an *excellent example of negative feedback,* as discussed in Chapter 2. Second, estrogen stimulates the growth of the cells forming the lining, or **endometrium,** of the uterus. Third, estrogen stimulates the production of a second pituitary hormone, **LH** (luteinizing hormone).

LH is not released in gradual increments, but rather as a sudden surge about 18–20 days after the beginning of FSH production. This surge of LH causes rupturing of the egg follicle and release of the egg. (Actually, ovulation is not quite so automatic and, in fact, is influenced to a considerable degree by emotional factors. This statement is evidenced by the otherwise unexpectedly high pregnancy rate in rape victims and wives of soldiers home on leave.) The ruptured egg follicle forms a small yellowish scab on the surface of the ovary called the **corpus luteum** (Latin for "yellow body"). The cells comprising the corpus luteum continue to produce estrogen, but also produce a second ovarian hormone, **progesterone.** This hormone has its main effect on the uterine lining, causing the endometrial cells to begin secreting fluids in preparation for receiving the egg, should it become fertilized.

Of course, the egg usually is not fertilized. In such instances, the body readies itself for another phase during which reproduction can take place. As it happens, progesterone also acts to inhibit the production of LH, and because LH is needed for maintenance of the corpus luteum, in its absence the corpus luteum begins to break down about ten days after ovulation. Once the corpus luteum breaks down, estrogen and progesterone production cease of necessity. Because progesterone acts to maintain the arteries supplying the endometrium, the cells of the endometrium die in its absence and are sloughed off and passed out through the vaginal opening during the period of **menstrual flow.** During this three- to six-day period, the endometrial tissue and from 50 to 100 ml (0.1–0.2 pt) of blood are lost.

How does the cycle reestablish itself? It just so happens that in the absence of estrogen the inhibition on FSH production is removed and FSH production begins spontaneously. Although it is subject to considerable variation, the length of the cycle is approximately one month. Thus, during her reproductive life, a woman is fertile for a short time every month. Such a frequent periodicity of reproductive capacity naturally promotes reproductive success.

The events as just described are summarized in Fig. 8.9.

**Pregnancy**  Whether fertilized or not, the egg travels rather slowly down the Fallopian tube, entering the uterus a few days after ovulation. During the intervening time, progesterone from the corpus luteum has acted to ensure that the endometrium is prepared to receive the fertilized egg. Upon implantation (*see box*), about seven days after fertilization, those cells of the developing embryo which contribute to the **placenta**[3] begin to secrete a hormone, chorionic gonadotropin **(CG).** "Chorion" refers to an embryonic membrane which contributes to the placenta (Chapter 12); "gonadotropin" is any hormone that directs the gonads, such as FSH or LH. Because CG production is diagnostic of pregnancy, it is the hormone analyzed for in the urine of women who suspect that they may be pregnant. The old assays using rabbits or frogs are passé as CG can now be tested for directly.

CG has properties very much like those of LH. As such, the production of this hormone preserves the corpus luteum beyond the point at which it would normally break down, thereby preventing both the sloughing of the endometrium and the liberation of any new eggs (because the estrogen produced by the still-intact corpus luteum continues to inhibit the production of FSH). For this reason, pregnant women do not have menstrual periods and produce no eggs during pregnancy. (It is exceedingly rare, but a woman may release an egg during pregnancy, which is ca-

---

[3] The placenta is formed jointly by embryonic and maternal tissue. It is discussed more fully in Chapter 12.

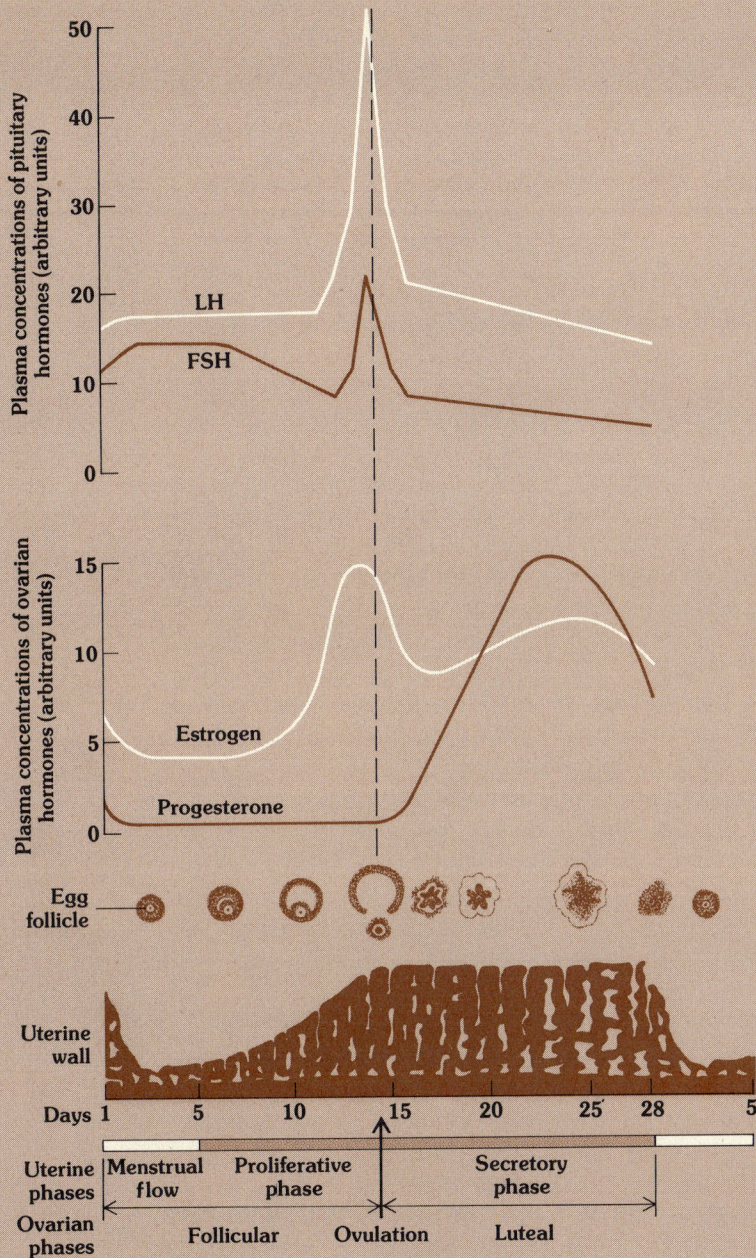

Fig. 8.9
Hormone concentrations in the blood, and ovarian and uterine changes during the menstrual cycle.

**Problems with fertilization sites**

The ovary is not completely surrounded by the top of the Fallopian tube (Fig. 8.3); therefore, very rarely, a fertilized egg may fall into the abdominal cavity and become implanted on one of the organs of the pelvic region, such as the bladder. This condition is known as an **ectopic** pregnancy. Because organs other than the uterus are incapable of sustaining a growing embryo, surgery is sometimes necessary to protect the woman.

Somewhat more frequently, a fertilized egg travels too slowly and implants in the Fallopian tube. This condition is known as a **tubular** pregnancy. The integrity of this thin-walled tube is immediately threatened and, again, surgery may be indicated to prevent the tube from rupturing and the possibility of subsequent infection.

pable of being fertilized. The result is "twins," but they are delivered a month or more apart.)

Therefore, the production of CG, which is a function of implantation of the developing embryo, in effect stops the normal monthly cycle at about the equivalent of day 21 and maintains it there throughout pregnancy. However, merely arresting the cycle is not sufficient in itself to prepare the body for birth. For example, what triggers labor? What controls milk production?

To begin with, the level of CG begins to drop after the third month of pregnancy at which time the placenta itself begins to produce progesterone and estrogen in large quantities. The level of these hormones continues to rise throughout pregnancy, and the amount of each produced by the placenta is several times that produced by the corpus luteum of a nonpregnant woman.

This increase in estrogen and progesterone to levels never reached in the nonpregnant woman serves two purposes. First, these hormones are responsible for the relatively very great increase in breast size during pregnancy (Fig. 8.10), which is a reflection of their influence both on increasing the size of the mammary glands and on increasing the rate of fat deposition in the breast. (These hormones also affect breast size to a much lesser degree during the regular monthly cycles.) Second, the high levels of these two hormones inhibit the production of **prolactin,** a hormone from the anterior pituitary, which is responsible for the production of milk by the mammary gland. This inhibition ceases at birth, for the placenta, site of estrogen and progesterone production during pregnancy, passes out of the body as the **afterbirth,** within a few minutes of delivery. Hence, in the days immediately following birth, the prolactin levels increase substantially, and the breasts of the nursing mother become even larger than during pregnancy (Fig. 8.10).

**Parturition**   The precise physiological events that initiate **parturition** (delivery) are not completely understood. An important role is played, however, by the hormone **oxytocin,** which is released from the posterior pituitary gland primarily as a result of stimuli from stretch receptors in the uterus. [The volume of the uterine canal increases from 4 ml to 6 liters (0.008 pt to 1.6 gal) during pregnancy (Fig. 8.11).] Oxytocin is a powerful stimulator of visceral muscle, which lines the uterus

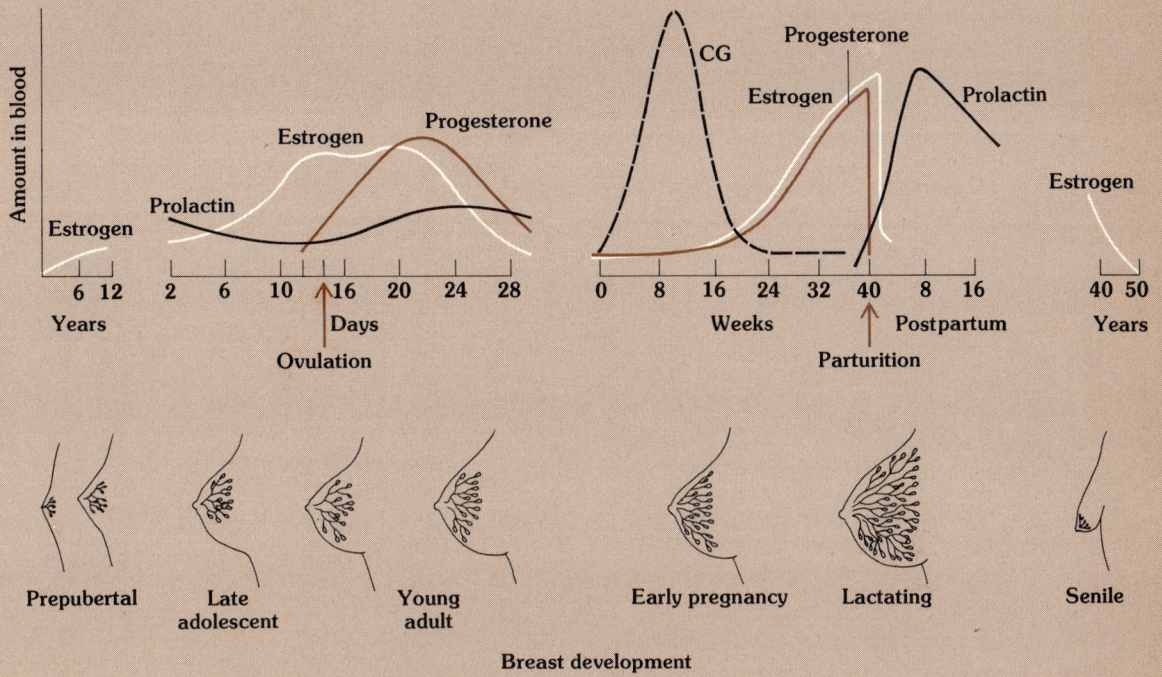

**Fig. 8.10  Interrelationship of hormonal concentrations and breast development.**

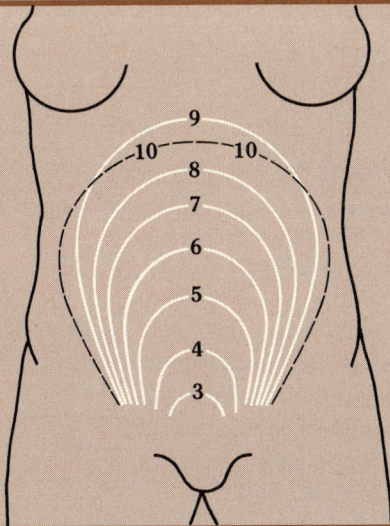

**Fig. 8.11
Increase in the size of the uterus during
the various months of pregnancy.**

**Table 8.2**
**The composition of milk**

| Component | Approximate concentration (g/100 ml) | |
| --- | --- | --- |
| | Human milk | Cow's milk |
| Water | 88.5 | 87.0 |
| Lactose (milk sugar) | 7.0 | 4.8 |
| Lipid | 3.3 | 3.5 |
| Protein | | |
|   Casein | 0.9 | 2.7 |
|   Lactalbumin and other proteins | 0.4 | 0.7 |
| Minerals | | |
|   Potassium | 0.041 | 0.150 |
|   Calcium | 0.030 | 0.120 |
|   Phosphorus | 0.013 | 0.095 |
|   Sodium | 0.011 | 0.050 |

and is responsible for the contractions associated with labor. In conjunction with oxytocin release, yet another hormone, **relaxin,** is released from the placenta. This hormone functions to dissolve part of the matrix of the ligament holding the two halves of the pelvic girdle together (see Chapter 3), which allows the ligament to stretch during delivery, thereby increasing the size of the birth canal.

**Lactation**   During pregnancy, the breasts have initially been prepared for their role as milk producers by the increase in size of the mammary glands (under the influence of progesterone and estrogen). With the loss of the placenta as the afterbirth, the inhibition on prolactin is removed and the production of milk begins usually within 24 hr after deliver. However, actual delivery of the milk (called **milk letdown**) requires the presence of oxytocin, which functions to cause contractions of the smooth muscle of the ducts within the mammary gland.

Both these hormones are initially produced at the time of parturition. However, continued production of both hormones after parturition depends on sucking stimuli (produced by the nursing infant) on the nipples of the breasts. As long as nursing continues, these hormones continue to be produced, and milk production is maintained. The composition of human milk is compared with cow's milk in Table 8.2.

**Menopause**   There is an overall decline in male sexual activity with increasing age, perhaps as a function of the gradual change in hormonal levels (Figs. 8.12 and 8.13). Nonetheless, men continue to produce sperm from puberty until death. Women, however, have a termination point, usually in their late 40s or early 50s, after which they no longer produce mature eggs and cease to undergo periods of menstruation. This is not a sudden process, but is a gradual development over from one to five years. There is no relationship between the age at which puberty begins and the age at which menopause occurs. This termination of ovarian function is also unrelated to the production of the gonadotropin FSH from the pituitary, which, in fact, continues to be produced at high levels (Figs. 8.14 and 8.15), but

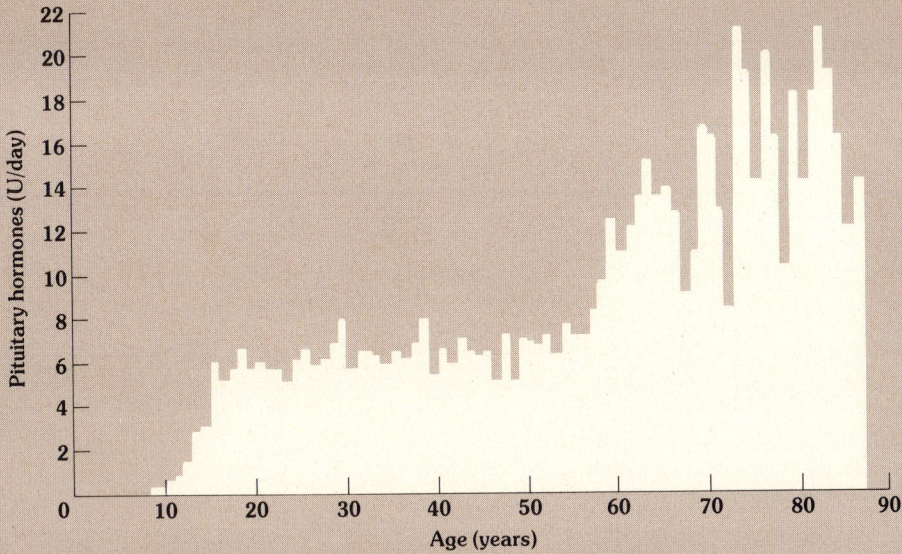

**Fig. 8.12   Excretion of the anterior pituitary hormones in the urine of men, as a relative index of amounts of hormone in the blood.**

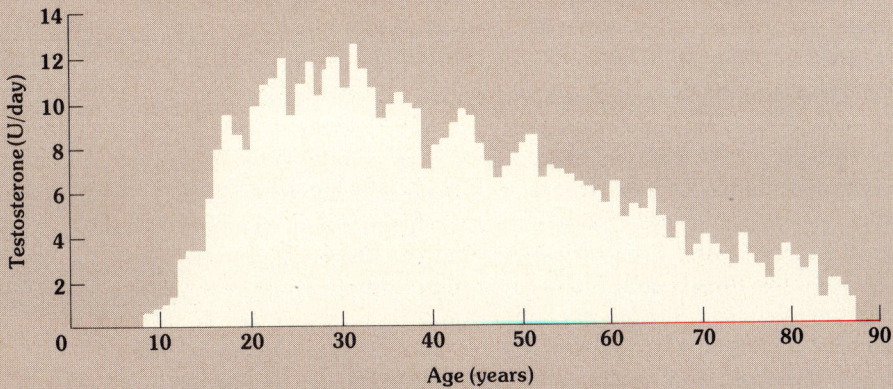

**Fig. 8.13   Excretion of testosterone in the urine of men, as an index of blood testosterone levels.**

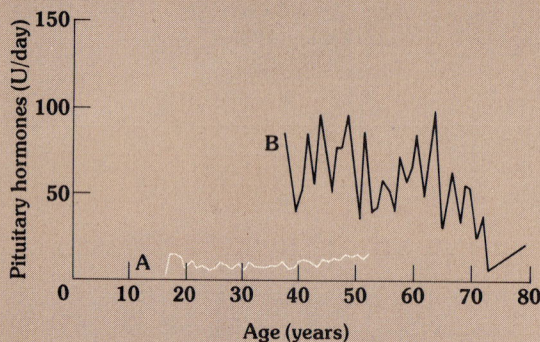

**Fig. 8.14** Excretion of anterior pituitary hormones in the urine of women, both before (A) and after (B) menopause.

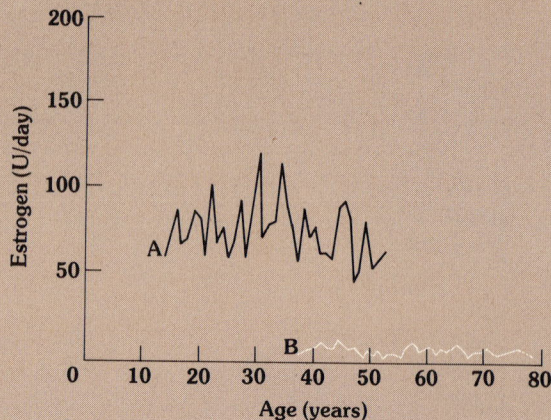

**Fig. 8.15** Excretion of estrogen in the urine of women as an indicator of blood estrogen levels, both before (A) and after (B) menopause.

rather stems from a failure of the ovaries to respond to this hormone. The decline in estrogen levels is responsible for a decrease in the size of such estrogen-dependent tissues as the genital organs and breasts. Moreover, because estrogen lowers blood cholesterol, postmenopausal women experience an increase in blood cholesterol; not surprisingly, the occurrence of circulatory diseases, including atherosclerosis and heart attacks (see Chapter 5) increases in postmenopausal women.

**ONSET OF PUBERTY IN THE MALE**   Compared to the female, the situation in the male is simplicity itself. **Testosterone,** the male equivalent of estrogen, is responsible for such secondary sexual characteristics as the broadening of the shoulders, the growth of body and facial hair, deepening of the voice, and increased development of the musculature, as well as such primary characters as increased size of the genitals. Testosterone also plays a role in baldness, as demonstrated by the fact that castrated men do not become bald. (It is unlikely, however, that such a drastic solution to the threat of baldness will ever become popular.) Moreover, unlike estrogen, testosterone causes a sudden spurt in height, which is so characteristic of the male during puberty.

Testosterone production is governed by the pituitary hormone, LH, and the feedback relationship between testosterone and LH is essentially the same as that between estrogen and FSH. Testosterone is produced by one group of cells in the testes and, in conjunction with FSH, also serves to stimulate a second group of cells in the testes to produce sperm. About 100,000,000 sperm cells are produced daily by the sexually mature male. However, there is no monthly cycle, and this rather simple set of hormonal relationships continues, essentially unchanged, from puberty to death. The hormones oxytocin and prolactin are not known to have any function in the male. CG and relaxin are not found in the male because these hormones are produced by the placenta, an exclusively female structure.

**COITUS AND FERTILIZATION**  Most land vertebrates, and certainly all mammals, utilize internal fertilization in order to achieve a successful fusion of sperm and egg. The reproductive tracts of the human provide an admirable example of how this fusion is achieved.

In the sexually excited male, the penis becomes engorged with blood as a result of increased arterial flow. This engorgement, which results from the filling of specially constructed sinuses, collapses the veins and traps the blood in the penis. As a consequence, the penis becomes erect (in as little as 5 sec) and increases substantially in size. During sexual intercourse **(coitus),** the penis is inserted into the vagina of the female. Rhythmical thrusting movements on the part of the male or the female, or both, stimulate the area just behind the glans and ultimately trigger an **emission,** which is the movement of the sperm to the urethra.

The path of sperm movement is as follows. From their site of origin, in the tubules of the testes, the sperm enter the **epididymis,** which is a tube less than 0.5 mm (.02 in.) wide, but some 6 m (20 ft) long, coiled atop each testis. Movement through the epididymis is by peristaltic activity, as the sperm are not yet motile (capable of self-propulsion). Typically, the sperm are stored here for a time, undergoing final maturation presumably under the influence of some fluid produced by the epididymis. The epididymis is continuous with the **vas deferens** and during emission the sperm move through the vas deferens, past the **seminal vesicle** and into the **ejaculatory duct,** which is located in the **prostate gland.** These glandular structures produce a variety of products that serve to nourish the sperm as well as to neutralize the acidic urethra and vagina (sperm require an alkaline environment in order to be motile).

Continued stimulation of the penis ultimately results in a powerful wave of contractions which propel the **semen** (the mixture of sperm and glandular secretions) out of the penis. This expulsion is called **ejaculation,** and is accompanied by a wave of intense pleasure; this combined event constitutes male **orgasm.** The volume of an average ejaculation is about 3 ml (0.006 pt), and contains about 300,000,000 sperm. Yet so small are the sperm cells that they collectively constitute only the volume of a pinhead; the bulk of the semen consists of fluid from the prostate and seminal vesicles. During ejaculation, the sphincter to the bladder is closed, thereby preventing sperm from entering the bladder and urine from entering the urethra.

Depending on a variety of factors, including age, time since the last ejaculation, and psychological factors, among others, the time between initial insertion of the penis into the vagina and ejaculation may range from less than 1 min to 20 min or more. Following an ejaculation, however, there is a latent period of from several minutes to several hours before another erection and ejaculation can occur (Fig. 8.16).

There is considerable variation in the size of the penis, whether erect or flaccid, partly because the amount of blood in the erectile tissues at any one time is highly variable. An average size for an erect penis, however, is about 15–18 cm (6–7 in.), with a diameter of about 4 cm ($1\frac{1}{2}$ in.). These dimensions represent close to a doubling of the flaccid measurements. Incidentally, many mammals have a bone in the penis for rigidity, as opposed to the blood erectile tissues of the human male.

The length of the vagina is only about 7 cm (3 in.), and the angle and shape of the uterus prevents the penis from entering that structure (Figs. 8.4 and 8.18). How does a 7 cm vagina accommodate an 18-cm penis? In the sexually excited female, there is a ballooning of the vagina such that it generally becomes large enough to accommodate all but the largest penis (see below).

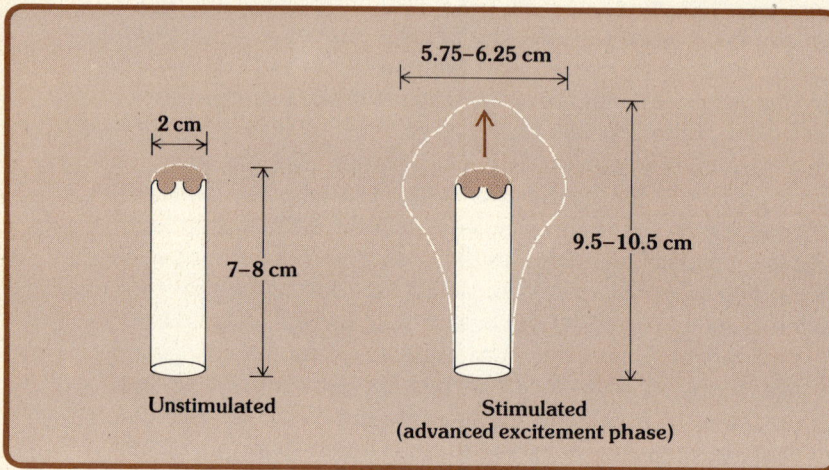

5.75–6.25 cm

2 cm

7–8 cm

9.5–10.5 cm

Unstimulated

Stimulated
(advanced excitement phase)

**The penis**

Enlargement of the vagina, caused by sexual excitation.

Failure to achieve an erection is called **impotence**. The causes of this condition are manifold, although these causes are more commonly psychological than physical. Temporary impotence may be caused by fatigue or by excess alcohol, the latter apparently acting through higher brain centers. Recall the gatekeeper's speech in *MacBeth*—alcohol "promotes the desire but takes away the response."

Sexual excitement in the female is typically somewhat slower to become manifest, but ultimately includes a swelling and firming of the breasts, engorgement of the clitoris, and the production of lubricating fluid by glands located around the orifice of the vagina. The female may respond to the thrusting movements of the penis with a **female orgasm,** which, unlike that of the male, does not involve a release of fluid. Typically, however, coital activity ends with ejaculation by the male, and depending on a variety of factors (not the least of which is the time required for ejaculation), the female may or may not experience orgasm. However, because

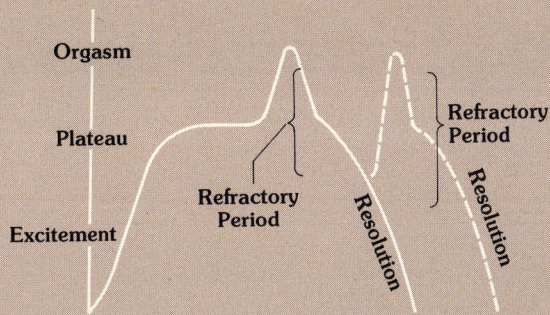

Fig. 8.16
Sexual response in the male. The refractory period is variable.

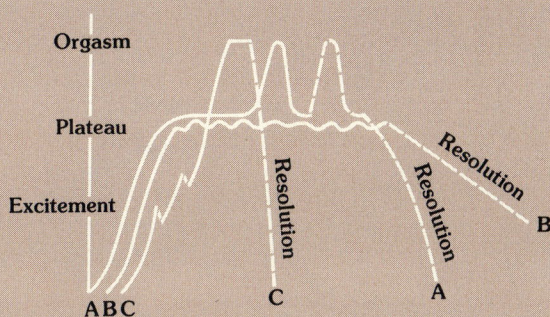

Fig. 8.17
Varied sexual responses in the female. Note the possibilities of multiple orgasms in pattern A and of remaining at a plateau of excitement without orgasm in pattern B.

the female also frequently does not have the same latent period before another orgasm can occur, she may, in fact, experience multiple orgasms during a single act of coitus (Fig. 8.17). The long debate over whether there is a difference between clitoral and vaginal orgasms now seems about over. Present evidence indicates that orgasm may occur as a result of stimulation either of the clitoris or vagina, but that the result is the same, that is, there is only one type of orgasm.

The sperm are discharged near the opening of the uterus (Fig. 8.18) and, despite their small size, can reach the Fallopian tubes and fertilize an egg (if one is present) within 30 min of ejaculation. To what extent the wave of contractions that occurs in a female experiencing orgasm assists the sperm in their migration is unclear, although it is certain that fertilization can occur even if the female does not experience orgasm. Despite the release of millions of sperm, only one sperm fertilizes the egg. Immediately after the first sperm penetrates the egg, a **fertilization membrane** is produced by the egg which effectively prevents any other sperm from penetrating.

**BIRTH CONTROL**   Depending on the circumstances, virtually everyone at one time or another has wished alternatively either for sterility or fertility. Partly because of our intelligence and partly because of the unique sexual receptivity of the

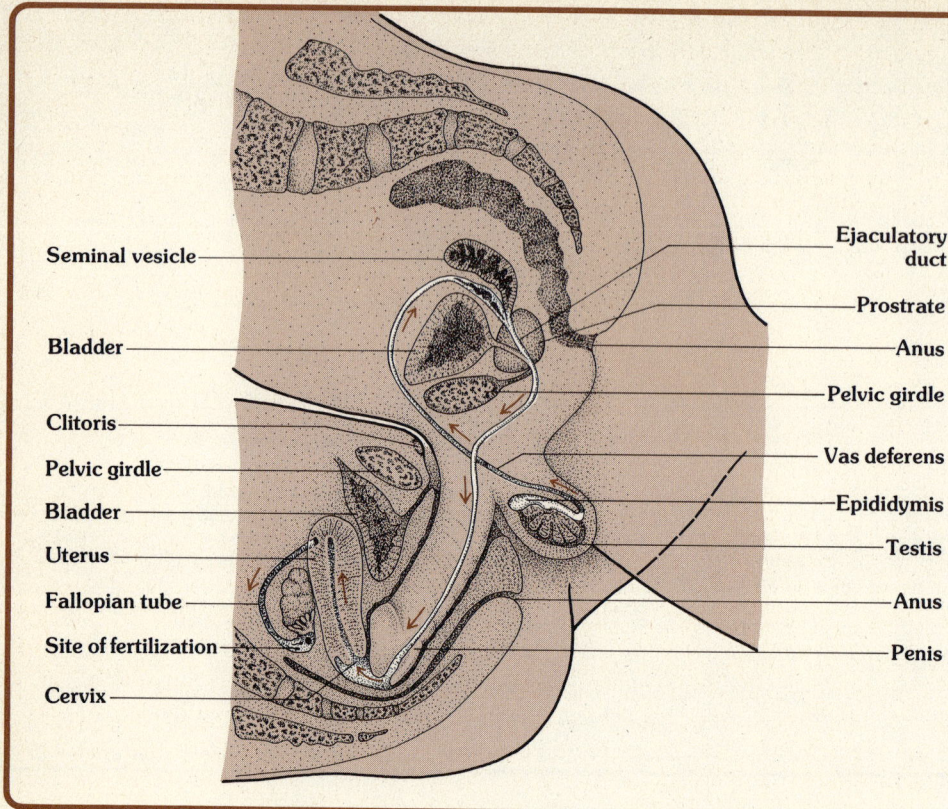

Fig. 8.18
Coitus. Note
the pathway
of the sperm
to the site of
fertilization.

Seminal vesicle

Bladder

Clitoris

Pelvic girdle

Bladder

Uterus

Fallopian tube

Site of fertilization

Cervix

Ejaculatory duct

Prostrate

Anus

Pelvic girdle

Vas deferens

Epididymis

Testis

Anus

Penis

human female,[4] we have long been interested in separating intercourse from conception.

There are a large number of birth control devices and techniques presently in use. Accurate data on their relative efficacy is difficult to come by, however, because "failures" tend to be blamed on the device or method and not on errors or lack of motivation on the part of the participants. This point will be exemplified shortly.

**Male-initiated techniques**   Perhaps the earliest form of birth control was male withdrawal **(coitus interruptus)**—there is even a Biblical admonition against this technique, so certainly it was in use over 2000 years ago. Failure of this method results from two factors. First, the male produces some lubricating fluid him-

---

[4] The menstrual cycle of the woman is very unusual and possibly unique in mammals. Most mammals are either sexually receptive only during times of ovulation (heat periods of dogs, for example) or at certain precise intervals, based on behavioral phenomena (female mice, for example, are sexually receptive every few days in the presence of male—and if not pregnant). Sexual receptivity divorced from time of ovulation is exceedingly rare in most animal species.

self (primarily from the **bulbourethral gland**—see Fig. 8.6), prior to ejaculation, but this fluid is rich in sperm. Hence, even if the penis is withdrawn before ejaculation, some sperm may be deposited in the female tract. Second, there is the notorious "oops!" factor, which results from an error in timing. The frustration involved in withdrawing at the critical point makes this second factor very significant and undoubtedly accounts for the high failure rate of this technique.

During the Renaissance, various types of sheaths were developed which trapped the semen and prevented its deposition in the female tract. These have been refined and are now called **condoms,** or "rubbers." Failure of this method stems not so much from a failure of the device itself as from either a failure to use them regularly or a failure to put them on at the outset of coital activity, because otherwise the sperm-laden lubricating fluid may be left in the female.

The third possible method of prevention used by the male is the **vasectomy,** which is nothing more than a severing of the vas deferens at the point where they pass through the scrotum. Contrary to what is commonly thought, men with vasectomies are still capable of ejaculation, but the semen contains no sperm (review Fig. 8.6 again to understand why this would be so). Moreover, the production of testosterone is unaffected, because by definition hormones require no ducts. The principal disadvantage has been that, until very recently, this was considered an irreversible operation and even today, there is no guarantee of reversibility if one who has had the operation changes his mind.

**Female-initiated techniques**   There are a number of devices or techniques open to women. Least successful of these is a **douche,** which is simply an attempt either to kill or to remove the sperm, or both, after coitus. As entry of the sperm into the uterus takes place extremely quickly, if it is to have any benefit at all douching must be done immediately, and typically this is not the case.

**Spermicidal agents,** used prior to intercourse, frequently in conjunction with a **diaphragm** (a rubber disk that blocks the entrance to the uterus) are quite effective, largely because the female tends to be more motivated to prevent conception than is the male.

Relatively recently, a variety of types of intrauterine devices (**IUDs**) have been developed (although the effectiveness of this method has been known for decades). The precise manner in which these devices work is still a matter of debate, but it would appear that their presence in the uterus acts as a minor irritant, causing the egg to move through the Fallopian tube and uterus rather quickly, before either fertilization or implantation, or both, can occur. Drawbacks involve the spontaneous expulsion of the device (sometimes without the knowledge of the woman, a fact that can later prove embarrassing, to say the least), occasional hemorrhaging (the device seems to work better in women who have already had a child), and increased rates of infection (arguably because the "string" by which the IUD can be removed provides a route by which bacteria can enter the upper reproductive tract).

By far the most well-known technique, however, is "the pill." This technique involves the introduction of estrogen and progesterone mimics into the body, with the result that the pituitary gland is "fooled" and fails to function properly. There are two types of pills. In the first, estrogen–progesterone pills are taken for 21 days, and a normal menstrual period occurs two or three days later. In the second (many brands of which have recently been withdrawn from the market), pills containing only estrogen are taken for 16 days, followed by five days of estrogen–progesterone pills, and then the same normal menstrual period follows in a couple of days. In both cases, the presence of estrogen inhibits the production of FSH and, hence, no egg is released in women on the pill. Furthermore, because there is no gradual estrogen buildup, there is no LH surge, which is required to induce ovulation. As a consequence, of course, conception cannot occur. Progesterone is included to ensure the normal sequence of endometrial growth and sloughing, because of the fear that elimination of the period of menstrual flow might well prove annoying or even dangerous, owing to spontaneous bleeding.

Nonetheless, there are a number of side effects associated with the pill, most of which are undesirable (a reduction in menstrual distress and in acne are two exceptions). The well-publicized increase in the formation of blood clots is still an incredibly rare event, and certainly the risk to life is much less than the risk of becoming pregnant. However, less dramatic (and less provable) disadvantages include an apparent increase in headaches and, at least in some women, a decreased interest in sex. How extensive these are, or even how valid, is a matter of some debate, but the results of one study are given in Table 8.3.

Women can also undergo a sterilization operation (**tubal ligation**) that simply involves the severing or blocking off of the Fallopian tubes. Eggs continue to be produced, because the ovaries are left intact, but they cannot pass into the uterus, nor can the sperm pass the block in order to fertilize them. Until recently, this was a rather serious operation, because it involved opening the abdominal cavity under general anesthesia. (The fact that many more women chose to undergo this operation than did men the much less hazardous vasectomy is a tribute to the unwillingness of men to take responsibility for contraception.) Now it is possible to cauterize the tubes under local anesthesia by inserting a thin wire into the Fallopian tubes by way of the vagina and uterus, and then heating the wires electrically. In either case, sterilization is generally permanent.

Still in the testing stages are various types of "morning after" pills, which are not, strictly speaking, contraceptives (for conception, i.e., fertilization, does occur) but they act rather as abortants. These pills function by inducing menstruation which, of course, either prevents implantation of the fertilized egg or causes the detachment of any implanted egg. There is still considerable controversy about the overall safety of such pills, partly because one type contains the hormone diethylstilbesterol (**DES**), which has been found to be **carcinogenic** (cancer-inducing) in certain circumstances (*see box*). The advantage of such pills, of course, is that, by virtue of not being contraceptives, they are failure-proof. That is, it does not matter whether conception has or has not occurred; the pill will be effective in either event.

Yet another method is simply to be a nursing mother. Especially in the first few months after birth, FSH and LH production in the nursing mother are sharply di-

**Table 8.3**
**Adverse and beneficial associations of various diseases with pill use (per 100,000 users per year)**

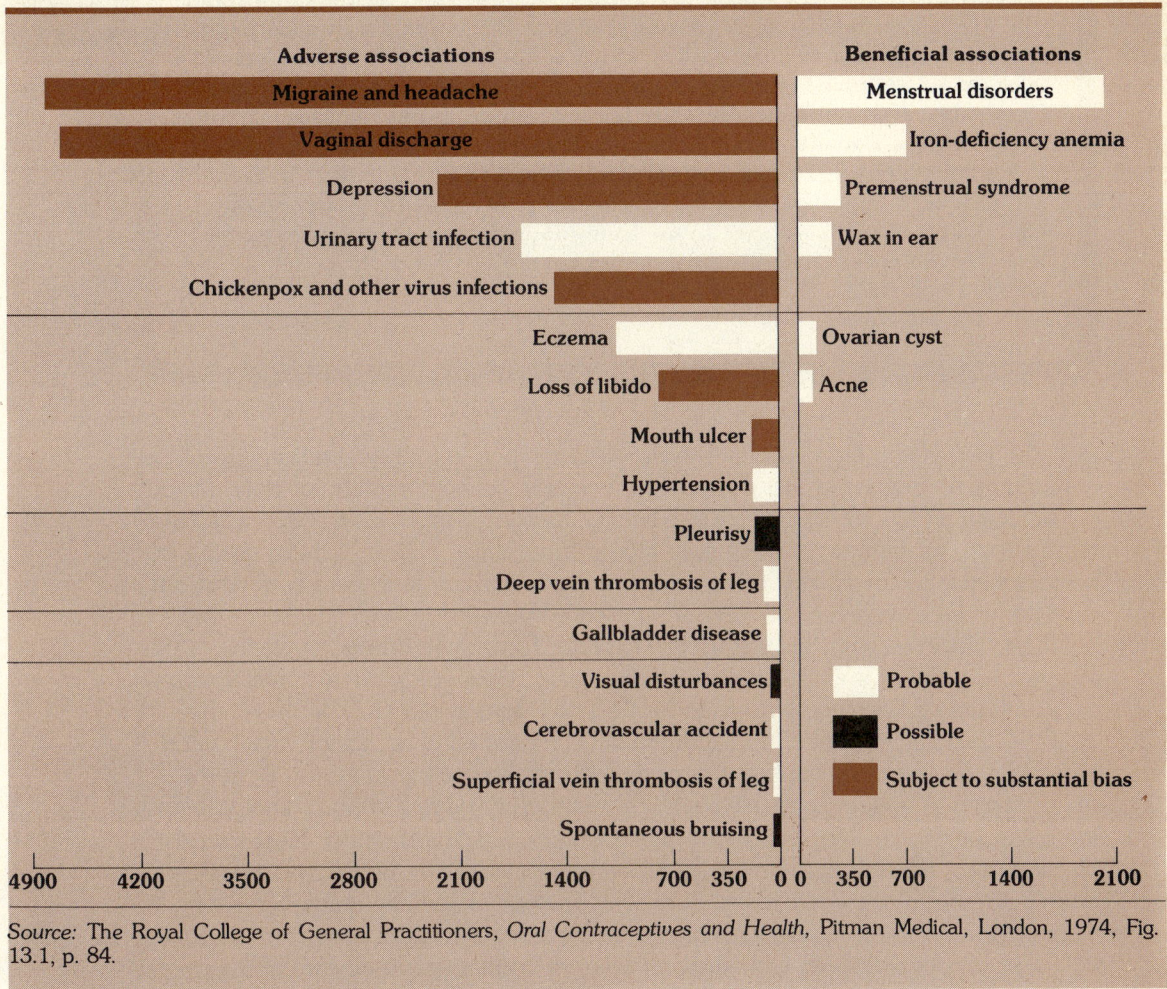

| Adverse associations | | Beneficial associations |
|---|---|---|
| Migraine and headache | | Menstrual disorders |
| Vaginal discharge | | Iron-deficiency anemia |
| Depression | | Premenstrual syndrome |
| Urinary tract infection | | Wax in ear |
| Chickenpox and other virus infections | | |
| Eczema | | Ovarian cyst |
| Loss of libido | | Acne |
| Mouth ulcer | | |
| Hypertension | | |
| Pleurisy | | |
| Deep vein thrombosis of leg | | |
| Gallbladder disease | | |
| Visual disturbances | | ▢ Probable |
| Cerebrovascular accident | | ◼ Possible |
| Superficial vein thrombosis of leg | | ▨ Subject to substantial bias |
| Spontaneous bruising | | |

4900  4200  3500  2800  2100  1400  700  350  0   0  350  700  1400  2100

*Source:* The Royal College of General Practitioners, *Oral Contraceptives and Health,* Pitman Medical, London, 1974, Fig. 13.1, p. 84.

minished. However, in about 50 percent of such women, there is a spontaneous redevelopment of ovulation despite continued nursing. Resulting conceptions have caused this "method" to fall from favor.

One last method of birth control is neither clearly male nor female oriented but rather calls for some mutual involvement. This is the so-called rhythm method, by which sexual activity is limited to those days of the menstrual cycle during which conception cannot occur. One problem associated with this method has been that many couples do not realize that "day 1" is the first day of menstruation, not the first day following menstruation (see Fig. 8.9). This error, of course, has been the source of no end of embarrassment.

**DES and cancer**

In the late 1950s, a recently discovered hormonelike substance called DES was used on several hundred women who had a history of miscarriages. The treatment had limited success, and was discontinued.

During the 1960s it was discovered that DES, by then in wide use in the cattle industry as a substance which would accelerate weight gain, was capable of causing tumors in mice. In the early 1970s the Food and Drug Administration limited its use in farming operations (and some countries, such as Canada, banned its use entirely).

About this same time, a number of cases of vaginal cancer in teenage girls were reported. This is a particularly rare type of cancer, especially in young women, and, in analyzing these cases, it was discovered that in each case, the girl's mother had been given DES during her pregnancy, Regrettably, the accepted treatment of vaginal cancer is a complete hysterectomy.

The underlying theory behind the rhythm method is that fertilization is possible only during an interval of less than 24 hr, because the egg is fertile for only a short period of time (10–15 hr). Sperm can remain viable for several days in the female tract, but are probably motile for only two or three days. Hence, if there has been no sexual activity for at least three days prior to ovulation, and there is none for at least 24 hr after ovulation, conception cannot occur.

The difficulties with this system are twofold. First, even though ovulation occurs on day 14 of the hypothetical 28-day cycle, few women are absolutely regular, and have cycles ranging from about 26 to 31 days in length. Therefore, the "danger" zone either side of ovulation must be expanded to account for such variations. (Indeed, a substantial number of women are so irregular that for them the rhythm method is essentially useless.) This mandatory period of abstinence causes the second difficulty, which is simply perseverance with the method. Most failures of the rhythm method occur because of failure of motivation rather than because of any inherent weakness in the theory. This method is actually quite efficient in women who are both regular and highly motivated.

The irony is that even with such an array of contraceptive devices and techniques, unwanted pregnancies are still very much a problem in this country, even among well-educated people. Many of these pregnancies occur because it was the intention of the couple not to engage in intercourse and therefore neither party made any plans to protect against conception. Moreover, because contraceptive techniques require advance planning, they have been criticized as taking away from the spontaneity of the event. Frequently, a couple engaging in sexual activity rationalizes their failure to use contraception by suggesting that the lack of advance planning demonstrates that they are not sexually promiscuous, but were rather simply momentarily carried away. This rationalization, of course, has proved to be a primary pressure behind the rather recent decision to legalize abortion.

**The future**   Despite recent advances, most notably the pill, birth control remains in its "infancy." A certain amount of sophistication is required to use the pill, and workers in underdeveloped countries relate horror stories as to how the pill has been misused. Perhaps the classic story concerns the worker who visited the village of a woman to whom birth control pills had been given some months previously. The worker found that the woman was visiting relatives in a nearby village, but the woman's husband was quick to offer reassurance: "Don't worry about the pills. I am taking them while my wife is away." The point of this story is simply that although the pill may prove adequate for many Western women, it is not the solution in many underdeveloped countries where the problem of educating the people simply in using the pill is horrendous. A number of other birth control methods are presently under development. Among others, these include:

1 Long-range pills, perhaps in the form of implants under the skin, to thwart the common problem of "forgetting."
2 Some method of controlling the LH surge, which is responsible for ovulation. In essence, this would be a less invasive method of control than the present pill, which affects the levels of several hormones, and thereby increases the probability of undesirable side effects.
3 Prevention of FSH production in men. Because FSH has no function in the production of testosterone, but rather acts only to assist in the maturation of sperm, this would seem an ideal control mechanism. Unfortunately, the control of FSH production in the male is not yet understood.
4 The use of ultrasound to destroy sperm. Still in the experimental stage, this method has promise, because after a single painless treatment, no sperm are produced for several months.
5 The development in the woman of an immune response to sperm. Again highly theoretical, the possibilities of becoming vaccinated against sperm are very intriguing.

**VENEREAL DISEASE**  A simple fact of life in recent years has been that improved contraceptive techniques have fostered more sexual intimacy, since the fear of an unwanted pregnancy has diminished. The consequence of such intimacy has been a sharp increase in the incidence of venereal diseases (Fig. 8.19).

Venereal diseases, by definition, are those which can be transmitted by sexual intercourse. However, the term has unfortunately come to be restricted in many persons' minds to mean only syphilis and gonorrhea, the two most serious venereal diseases, and the only ones that physicians are required by law to report to state health departments. For this reason, many investigators are espousing the term sexually transmitted diseases **(STD)** as one less restrictive in its meaning. At least 14[5] diseases are included under this heading, although not all are transmitted exclusively by sexual contacts. Among the more important are:

1 *Herpes simplex* Type II: Type I causes fever blisters and cold sores, whereas type II causes similar sores on the genitals. Oral sex appears to be responsible for interchange

---

[5] This list includes such serious diseases as hepatitis, which can be transmitted sexually, although other forms of transmission are more common, as well as such minor diseases as "yeast" infections (*Candida albicans*), which are primarily in the nuisance category—and which are more common in women taking birth control pills (Table 8.3). It does not include diseases such as typhus, which is transmitted by body lice, even though sexual contact can be one means whereby the lice themselves are transmitted.

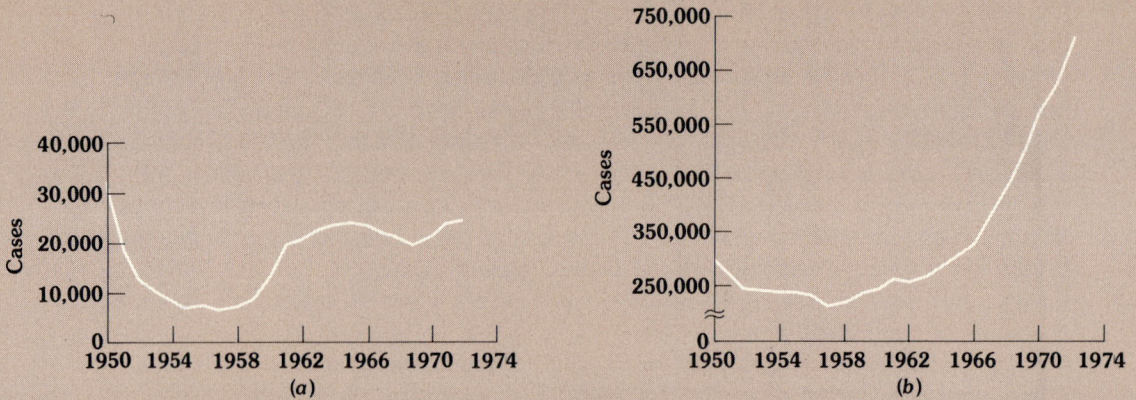

**Fig. 8.19** Incidence of (*a*) primary and secondary syphilis and (*b*) gonorrhea in the United States since 1950. Birth control pills came into extensive use between 1960 and 1965. The correlation between increased sexual freedom and incidence of sexually transmitted diseases (STD) is evident.

between the two. In a recent survey, 10 percent of the women examined had type II virus— and there are indications that it may cause cancer of the **cervix** (the boundary between the uterus and vagina). Babies may pick up the disease from their mothers during delivery, suffering possible brain damage as a consequence. In addition, miscarriages are three times more common in women with *Herpes* type II than in women free of the disease.

2 *Trichomonas vaginalis:* a protozoan infection causing no symptoms in men, but a burning sensation and vaginal discharge in women. It is possibly twice as common as gonorrhea.

3 *Chlamydia:* an organism midway between a virus and a bacterium, which causes extreme pain in men during urination, but is frequently symptomless in women. As in gonorrhea, arthritis, heart trouble, and pelvic disorders can result from this disease.

Despite the recognition of other sexually transmitted diseases, the two principal diseases in this category remain **syphilis** and **gonorrhea.** Syphilis is caused by *Treponema pallidum,* a spirochete bacterium (Fig. 8.20) and was widely reported shortly after Columbus's sailors returned to Europe, which has led to the assumption that the disease had a New World origin (the sailors having presumably contracted the disease while "fraternizing" with the natives). In any case, during the years immediately after Columbus's voyages, syphilis outbreaks were commonplace throughout Europe.

The disease itself passes through three stages. The primary stage occurs anywhere from 10 to 90 days after exposure. This prolonged latency stage results in many difficulties in tracing the route of infection by public health officials, and is a primary reason why syphilis provides such problems in control. In one recent study in

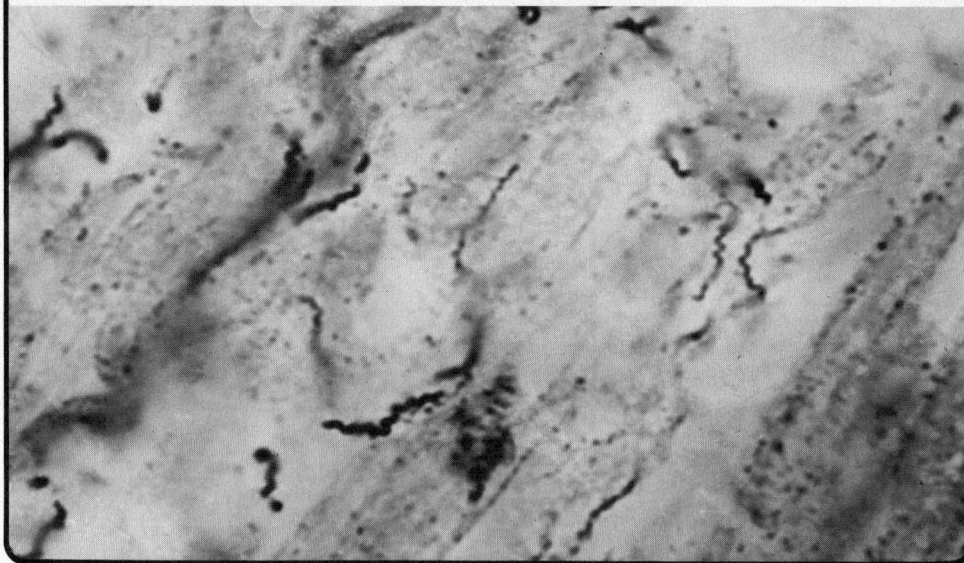

**Fig. 8.20** The spirochete responsible for syphilis. (Lester Bergman and Assoc.)

Britain, it was found that 1639 persons had become infected from a single source in a chain reaction. The principal symptom associated with the primary stage is an open sore, called a **chancre** (pronounced "shanker"—do not confuse this with the white, ulcerous spots in the mouth called **cankers,** which have no connection with venereal diseases). The location of this chancre is variable, but it is always on some mucous membrane (genitals, mouth, or rectum). Because contact of another mucous membrane with the chancre can result in transmission of the disease, actual intercourse is not essential. For example, it is conservatively estimated that at least 10 percent of the infections result from homosexual encounters.

The chancre gradually disappears, and some weeks later the secondary stage occurs. The manifestation here is an extensive rash, not unlike measles or chicken pox, over much or all of the body. Again, these symptoms disappear, but may reappear from time to time.

Most individuals never show further sign of the disease. However, approximately one-third enter the tertiary stage, frequently after an injury or some other illness, at which time many of the organ systems of the body, most notably the central nervous system, may be invaded. Frequently, the severity of this infection proves lethal.

All of the above presupposes no treatment. As it happens, syphilis is among the most easily treated of infectious diseases, and is generally very susceptible to a few doses of penicillin. Therefore, the number of cases that progress to the secondary stage, let alone the tertiary stage, is very few in this country today.

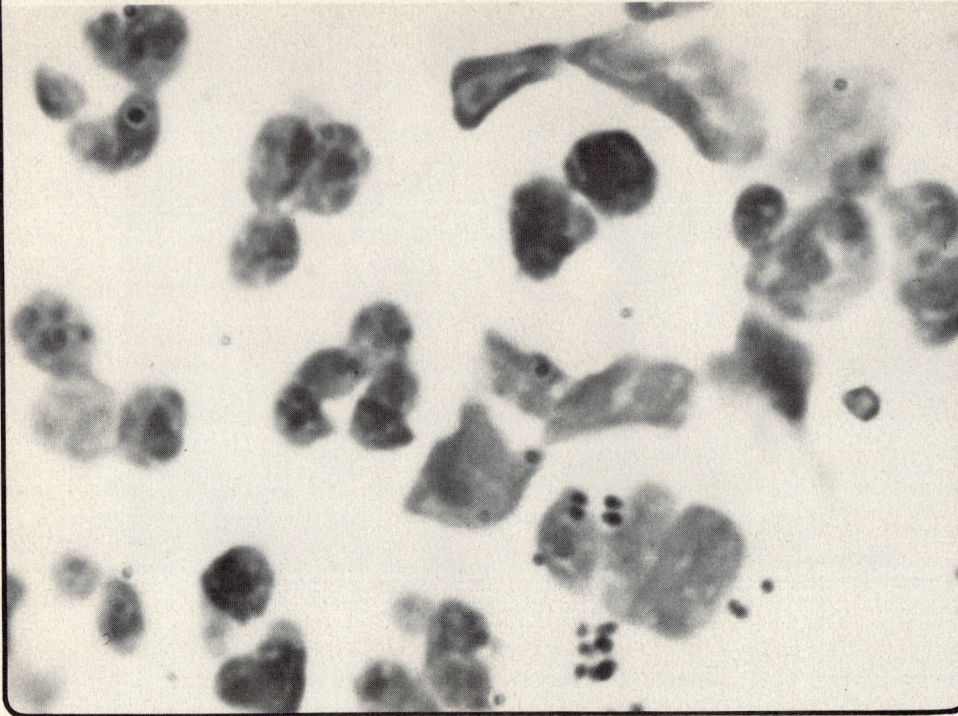

**Fig. 8.21   The gonococcus responsible for gonorrhea. (Lester Bergman and Assoc.)**

Infection of others is possible only during the primary stage, although pregnant women can pass the disease to an unborn child as late as the beginning of the secondary stage. The effects of this disease on an unborn child are devastating, and frequently involve substantial malformations of the bones (especially of the face) as well as destruction of portions of the central nervous system to the point of severe retardation. Because of the effects of the disease on their brains, over 30,000 advanced syphilitics had to be institutionalized in this country as recently as 50 years ago. Many of these were the children of syphilitic mothers. It was this threat to the unborn child which prompted the mandatory blood test (the Wasserman test, named for its discoverer), which is required in most states before a marriage license can be issued, and which is also routinely performed on pregnant women.

By comparison, gonorrhea seems an innocuous disease, although that would be an erroneous conclusion. Gonorrhea is caused by the gonococcal bacterium *Neisseria gonorrhoene* (Fig. 8.21). Unlike syphilis, the disease manifests itself only two to ten days after exposure. However, it has been estimated that symptoms appear

---

[6] The accuracy of these figures is debatable. More probably, mild symptoms are present, but because of their nature (e.g., increased vaginal discharge in women), they are difficult to detect.

counts for the difficulty in eliminating this disease. In males, the most frequent symptom is a burning sensation during urination, which results from infection of the penile urethra. The same symptom may occur in the female, but is less common because the urethra is so much shorter in the female.

Irrespective of the appearance of these symptoms, the disease may spread through the reproductive tract. This poses potential problems, particularly in the female, for an inflammation of the long and very thin Fallopian tubes may result in the formation of scar tissue, which can seal the tubes and render the female sterile. Moreover, because the tubes are open at the top and communicate with the intestinal cavity, extensive involvement of the whole pelvic area (called **PID,** for pelvic inflammatory disease) may result. In the city in which this book is being written, with a college and university enrollment of about 26,000, an average of one hysterectomy per week is performed on a college-age female because of the effects of gonorrhea. Nationwide, some 2,000,000 cases of PID occur each year, although most women do not require a hysterectomy.

Unlike syphilis, invasion of other organs of the body is not common with gonorrhea, although inflammation of the heart sac and infections of joints (gonorrheal arthritis) may sometimes occur. Moreover, gonorrhea is not transmitted by an infected mother to her unborn child. However, earlier in this century, gonorrhea was a leading cause of blindness, because during the passage of the child's head through the infected birth canal, eye infections frequently resulted that ended in blindness. The rather routine use of silver nitrate drops in the eyes of the newborn infant was developed to prevent any such infection.

Gonorrhea is also generally susceptible to standard antibiotics, including penicillin, but resistant strains are increasingly common. Moreover, unlike syphilis, there is as yet no effective blood test for the disease, although by the time this book is in print, such a test may well have been developed.

Neither syphilis nor gonorrhea promotes a lasting immunity (Chapter 5), and thus repeated infections of either can occur upon exposure throughout the individual's lifetime. An interesting question of policy will be presented, however, should a vaccine be developed for either, or both, of these diseases. Given the evident increased sexual activity with improved methods of contraception, would the elimination of the threat of the most serious of the venereal diseases not have the same effect? If so, would parents permit their minor children to be vaccinated? Does the government have the authority, through public health laws, to require vaccinations? These questions will have to be faced by the time you are yourselves parents.

## SUMMARY

Perhaps because of its future-directedness, perhaps because of its obviously different organization between male and female, or perhaps because of the aura of mystery and forbidding in which our society shrouds it, the reproductive system has always had a special fascination for us.

A study of the female reproductive system provides perhaps the most elegant example of hormonal control in the human body. By contrast, the male system is almost mundane. Such a comparison also suggests certain of our societal values such as the fact that the majority of the methods of birth control require active involvement of the female, but not of the male.

Because of the special status we afford the reproductive system, it is not surprising that we afford equal status to diseases of the system. Certainly the primary reason why these diseases are such a problem is that they are so frequently communicated by sexual activity—and such behavior is subject to special restrictions within our society.

Two individuals undergo surgery for enlarged thyroid glands, and equivalent amounts of glandular tissue are removed from each. One individual recovers quickly and is soon leading a normal life once again. The second individual dies of convulsions within a matter of days after the operation. Why the difference?

Why is diabetes a leading cause of blindness? Why does removal of the ovaries not affect a woman's sex drive? Why does a coma result both from too much and too little insulin?

In each case, the proper functioning of one of the endocrine glands—the subjects of this chapter—is impaired.

chapter **IX**

# chapter IX
# Production and control of internal messengers –the endocrine system

Of the ten classic anatomical systems discussed in Chapter 2, each is anatomically continuous, in some fashion—save only the endocrine[1] system. The endocrine system is not really an anatomical system at all, but rather consists simply of a series of tissues and organs which are similar only in their capacity to produce **hormones**—chemical messengers which are secreted into the bloodstream and pass to some other point in the body where they influence the activity of one or more **target organs,** typically by affecting their rate of enzyme production. Many of these secretory tissues are actually a part of organs that have other functions (e.g., the digestive hormones, all of which are produced by various digestive organs—see Chapter 4). However, other hormone-producing tissues are actually distinct and separate glands, having no other function. As such, they are the leftover pieces of the body, once the body has been subdivided into the other nine systems; hence, the necessity of an endocrine "system"—even though its basis is physiological rather than anatomical.

You should note that the endocrine system is generally not responsible for *initiating* biochemical events, but rather merely for *regulating* them. As such, the significance of the endocrine system relates primarily to its role in the homeostatic control of many bodily functions and substances. Thus, a study of the endocrine system provides graphic examples not only of the mechanisms of homeostasis, but also of the fragility of these mechanisms and of the consequences of their breakdown.

**THE NATURE OF HORMONES**   Chemically, hormones are basically of two types. Most hormones are short proteins or amino acid derivatives. A second

---

[1] The term "endocrine" comes from the Greek *endo,* meaning "within," and *krino,* meaning "sift." The two together are meant to refer to the fact that endocrine glands are ductless, and their products are passed directly into the circulatory system.

and numerically somewhat smaller group are **steroids,** a subcategory of lipids (see Chapter 2). Most, if not all, of the first group are incapable of passing through the membrane of the cells of the target organs, primarily because of their size. Rather recently it was discovered that they function by triggering the release of a chemical already present in the cell membrane; this chemical then migrates into the interior of the cell and ultimately causes either an overall increase in the metabolic activity of the cell or an activation of particular cellular enzymes. The membrane chemical is apparently the same in all target organs—why, then, the need for different hormones? Moreover, how is it that different results occur in different target organs if the same chemical intermediary is involved in all cases? The answer seems to be that the target organs are selective in terms of what hormones are capable of binding with their cell membranes to cause the release of the intermediary chemical. Thus, target organ specificity depends on differential membrane-binding capacities for the different hormones—and, of course, each target organ produces its own product. A common middle step is not at odds with these processes.

Conversely, the steroids readily pass through cell membranes and ultimately penetrate the nucleus where they attach to the deoxyribonucleic acid (DNA) of the chromosomes and activate specific genes, thus resulting in the production of specific enzymes. Therefore, the distinction between the two types of hormones is functional as well as chemical.

**THE PITUITARY GLAND**   The pituitary gland which, at 4 g (0.14 oz) is only the size of a lima bean, was for years called the "master gland," not only because it produces so many different hormones, but also because many of its hormones serve to direct the activities of other endocrine glands. Some years ago, however, it was recognized that the more rearward, or **posterior** portion of the pituitary,[2] did not produce hormones at all, but merely stored them. The actual production of hormones occurred in a portion of the brainstem, called the **hypothalamus,** which is located just above the pituitary and which receives input from nerve cells from all areas of the brain and body.

Even more recently, it has been discovered that the more forward, or **anterior,** portion of the pituitary, even though it does itself produce hormones, does not do so autonomously, but rather only under the direction of the hypothalamus. Control is exerted for most of the pituitary hormones by the production of "releasing factors" from the hypothalamus; each anterior pituitary hormone evidently has its own releasing factor. The exception is the hormone **prolactin,** which is automatically produced except when held in check by a hypothalamic inhibiting factor. Normally, this factor is always being produced, but is itself held in check by nervous stimuli from the breasts, produced by the sucking actions of a nursing infant (see Chapter 8). Thus, these stimuli act essentially as positive feedback for the continued production of prolactin. In sum, the brain proves to be the master switch

---

[2] The two portions of the pituitary differ in terms of hormone production because each portion has a different embryonic origin. The anterior portion is an outgrowth from the mouth (the connection to which is subsequently lost); the posterior portion is nothing more than an outpocketing of the hypothalamus itself.

A number of compounds have recently been discovered which have many hormonelike properties. They are all chemically similar and are called **prostaglandins.** Rather than being the products of a discrete block of tissue, they seem to be produced rather generally by many different kinds of tissues.

Their precise role in the normal functioning of the body is unclear, although they are common in the cells that comprise the reproductive tracts—and they cause the contraction of visceral muscle. Indeed, one of the causes of the low back pain which affects so many women just before, or during, their menstrual periods may be the inadvertent release of prostaglandins from the endometrial cells as they are broken down.

Prostaglandins are now being used in place of oxytocin to initiate parturition in women who are late in delivering, and they are also being used as an abortant—visceral muscle contraction is initiated in both cases.

in most of the activities of the endocrine system, just as it is in so many other functions of the body.

This finding is intuitively satisfying for two reasons. First, it establishes a close tie-in between the two coordinating systems of the body (nervous and endocrine). Hitherto, the concept of two coordinating systems operating autonomously was vaguely disquieting. Second, if the hormones of the anterior pituitary were themselves regulated directly by the activities of the target organs in negative feedback loops, there would be no conceivable mechanism whereby the level of either the pituitary hormones or the products of the target organs could be altered. To refer back to an earlier analogy, it is as if you could never change the setting on your furnace thermostat, once it was set. As we shall see, altered output of certain of the pituitary hormones, depending on changed circumstances, is absolutely critical for normal body functioning. Hence, the interposition of the hypothalamus in the negative feedback loop amounts to a finger on the thermostat switch.

**Posterior pituitary**  The posterior pituitary is responsible for the storage and release of two hormones. These are **oxytocin,** which stimulates contraction of the uterus of the pregnant mother at the time of parturition and also functions in the letdown of milk in the nursing mother, and **antidiuretic hormone** (ADH, which is also called vasopressin), which functions to decrease water loss in the urine. Oxytocin production is initiated and maintained by the sucking activities of the baby on its mother's breasts (Chapter 8); ADH operates in a rather complicated negative feedback loop involving blood volume and pressure (Chapter 7).

The failure to produce ADH results in the condition known as **diabetes insipidus.** This is a totally different disease from **diabetes mellitus,** which results from a lack of insulin. In both cases, the volume of urine increases, but in diabetes mellitus the presence of sugar in the urine gives it a sweet taste (*mel* is the Latin word for honey), whereas in diabetes insipidus the urine is not sweet (hence, insipid). (The name of the brave soul who first detected the difference in taste has, unfortunately, been lost to science.)

There are a number of causes of extremely short adult stature. Pituitary dwarfs retain normal body/limb proportions and are sometimes called midgets.

Relatively more common are **achondroplastic dwarfs,** who have an essentially normal-sized head and body, but very short limbs. This condition is caused by a recessive gene.

Pygmies are also of normal body/limb proportions, and are considerably larger than the pituitary dwarfs. Pygmies produce normal amounts of GH, but have developed a genetic nonresponsiveness to it.

The volume of urine produced as a consequence of the absence of ADH is incredible, sometimes reaching 25 l/day (6.6 gal/day). Not only does this require an equivalent ingestion of water, but an enormous amount of energy is required to warm this water up to body temperature. Fortunately, the disease is easily treated by the administration of ADH.

**Anterior pituitary**   The anterior pituitary produces six hormones, three of which are involved in the functioning of the reproductive system. These include follicle-stimulating hormone (**FSH**) and luteinizing hormone (**LH**), which function in the development of the egg follicle and release of the egg in the woman's monthly cycle, and **prolactin,** which functions primarily to cause the mammary glands to produce milk in the nursing mother. FSH is in a negative feedback loop with the ovarian hormone **estrogen;** LH is in a negative feedback loop with the ovarian hormone **progesterone;** and prolactin production is maintained by the sucking movements of the nursing infant on its mother's breasts.

The other three hormones have diverse functions. **Growth hormone (GH),** unlike most hormones, has no single target organ, but rather exerts a body-wide effect to assist in growth. It is essential to point out that a great many factors are responsible for the ultimate size of an individual, of which GH is but one. Specifically, GH aids growth by accelerating the rate of cell division, by increasing the metabolic rate, and by increasing the rate of protein formation.

Underproduction of this hormone results in the failure of the long bones to grow, and the individual is a **pituitary dwarf** (*see box*); overproduction results in *gigantism* (Fig. 9.1). GH has not yet been synthesized, although this seems imminent. At present, children deficient in GH must be given injections from pituitaries obtained from cadavers. The supply of donated pituitaries is regrettably insufficient to treat all of the children suffering from GH underproduction. Underproduction in adults may result in a condition called **myxedema** (Fig. 9.2).

Sometimes GH is overproduced in adults, resulting in the continued growth of the internal organs and the cartilaginous areas of the body, most notably the nose, chin, and joints. This condition is called **acromegaly** (Fig. 9.3).

**Thyroid stimulating hormone (TSH)**   TSH controls the output of hormones from the thyroid gland, which is located just below and on either side of your larynx (Adam's apple). It is in a negative feedback loop with the thyroid hormones.

**Adrenocorticotrophic hormone (ACTH)**   ACTH stimulates the cortex of

Fig. 9.1 Gigantism. (Rotker, Taurus)

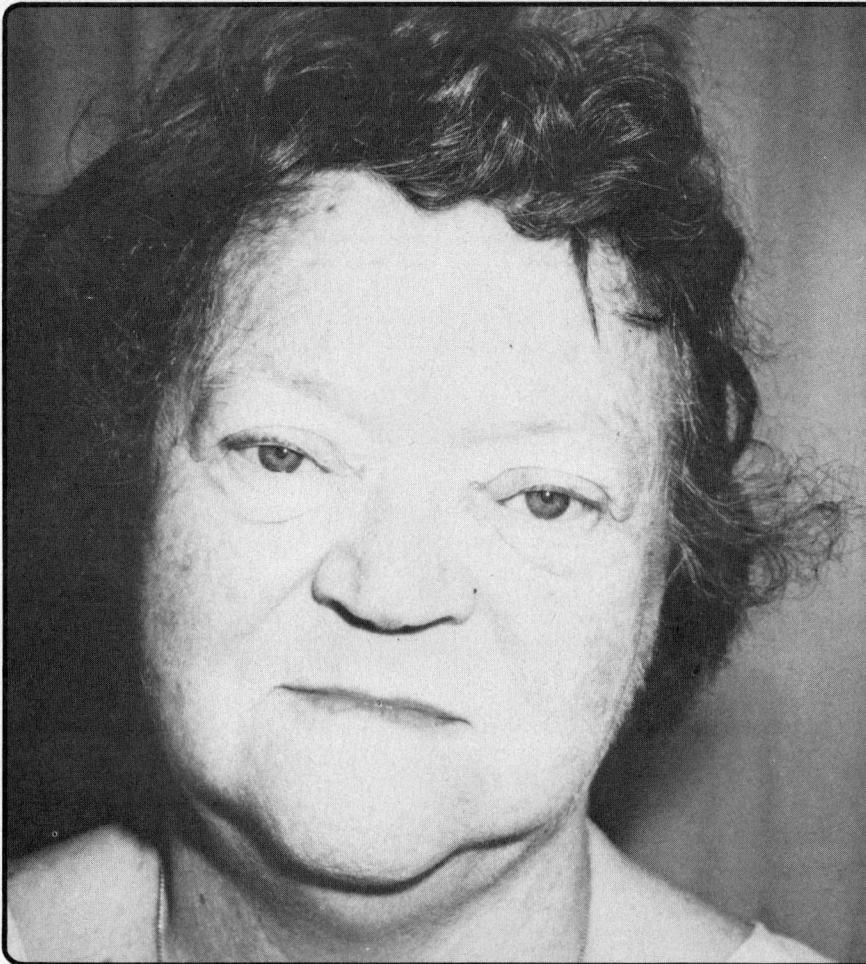

Fig. 9.2
Typical facial
appearance
of an individual
with myxedema.
(Rotker, Taurus)

the adrenal glands, a pair of small glands [5 g (0.17 oz) each] located just above the kidney. The term "adrenal" means "at the kidney." A number of anatomists have pointed out that this is a misnomer in humans, because the gland is actually located above the kidney, and the term "suprarenal" has been proposed as a substitute. This proposed change has met with limited enthusiasm.

The adrenal gland itself consists of two distinct portions, an outer **cortex** and an inner **medulla,** each of which has a very different embryonic origin. In fact, both areas are completely intermingled as clumps of cells scattered over a broad area in the lower vertebrates; it is only in the mammals that they coalesce into separate layers of a distinct gland. ACTH has no effect on the medullary portion of this gland but is in a negative feedback loop with one of the hormones produced by the adrenal cortex.

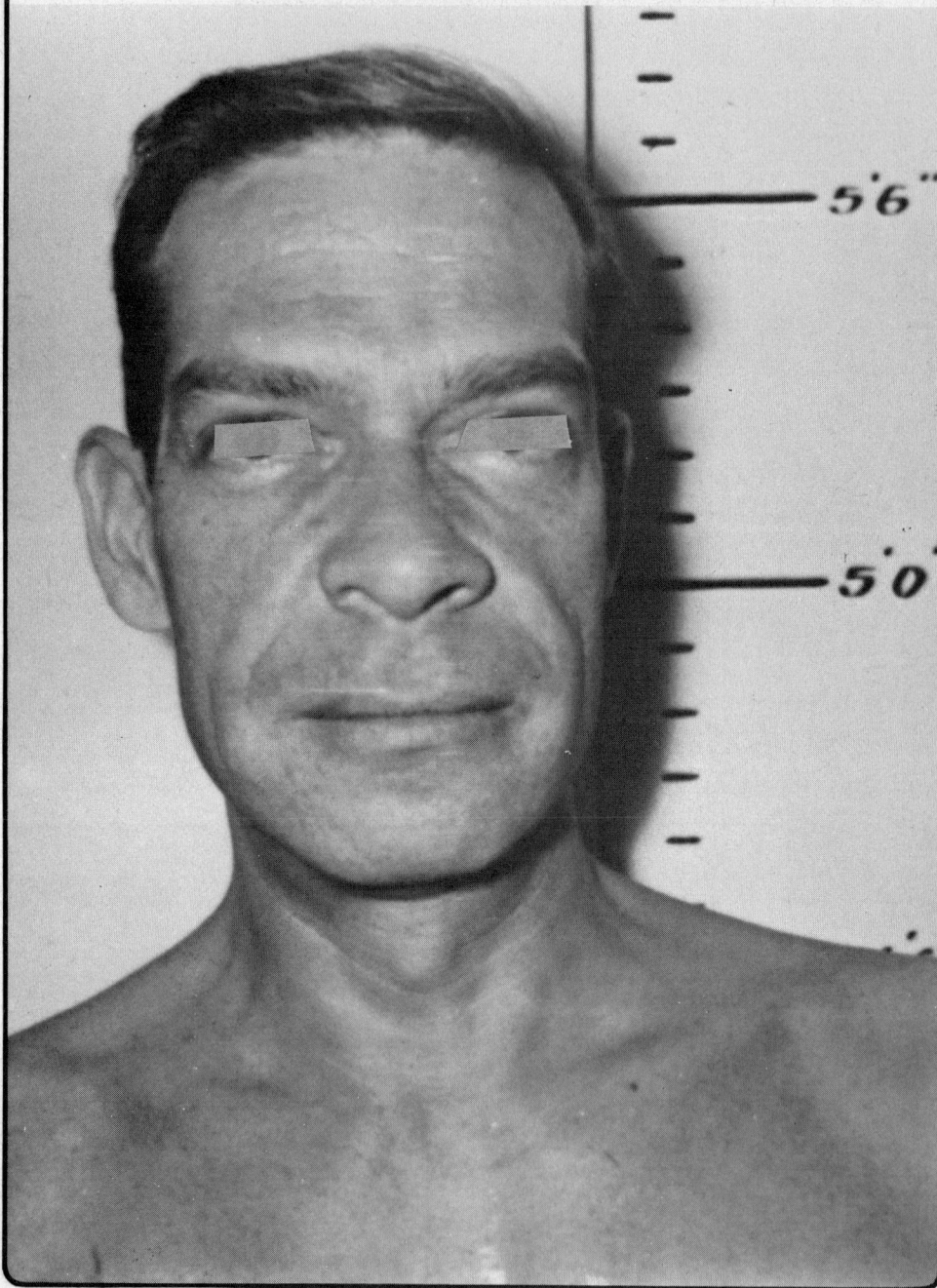

Fig. 9.3 Typical facial appearance of an individual with acromegaly. Note coarsened and thickened nose and chin. (Lester Bergman and Assoc.)

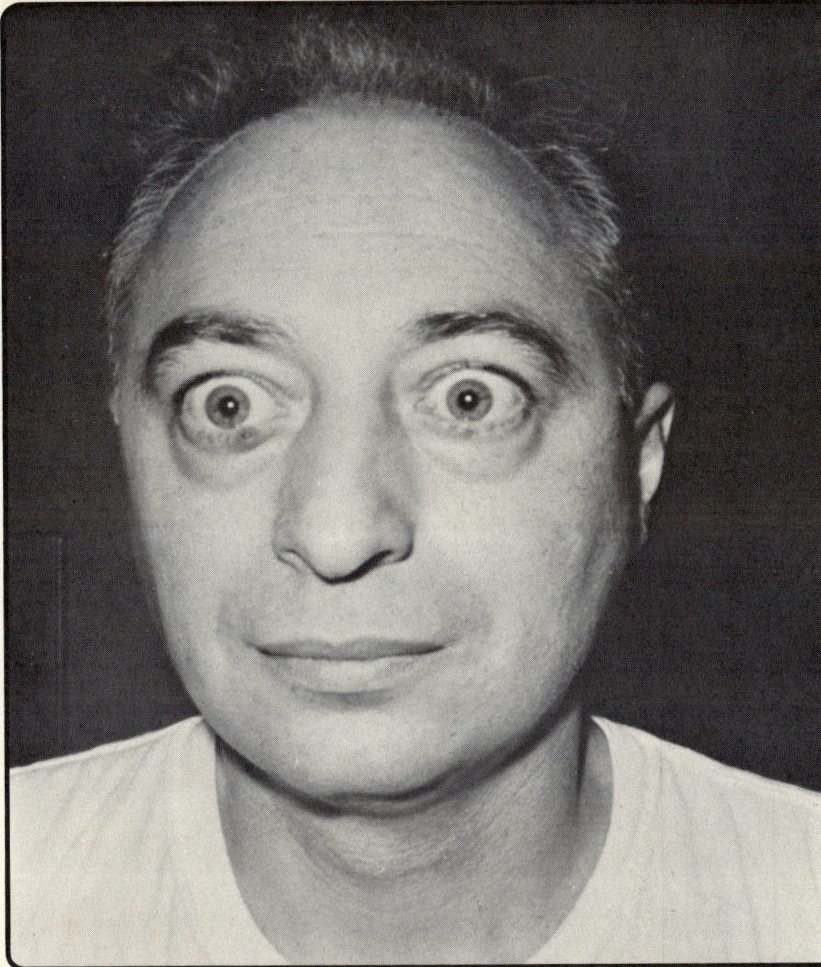

Fig. 9.4
Diffuse toxic goiter
(Graves' disease).
Note moderate
goiter and stare.
(Lester Bergman
and Assoc.)

**THYROID GLAND**  The thyroid gland, at 25 g (0.87 oz) the largest of the endocrine glands, produces two hormones,[3] **thyroxine** and **calcitonin,** which have very different effects in the body.

Thyroxine is a regulator of metabolic rate, and as such affects growth in children. As might be expected, overproduction causes irritability, weight loss, and hyperactivity. Although these symptoms are not in themselves unusual, only very rarely is hyperproduction of thyroxine the cause, as it is an uncommon condition. Typically, it occurs only in cases in which there is a tumor in the thyroid gland, or as a symptom of **Graves' disease** (Fig. 9.4), which is believed to be an inherited condition.

---

[3] Actually, there are three hormones, as thyroxine is produced in two forms. However, because chemically they are very similar and functionally appear to serve the same ends, they are treated as a single hormone here.

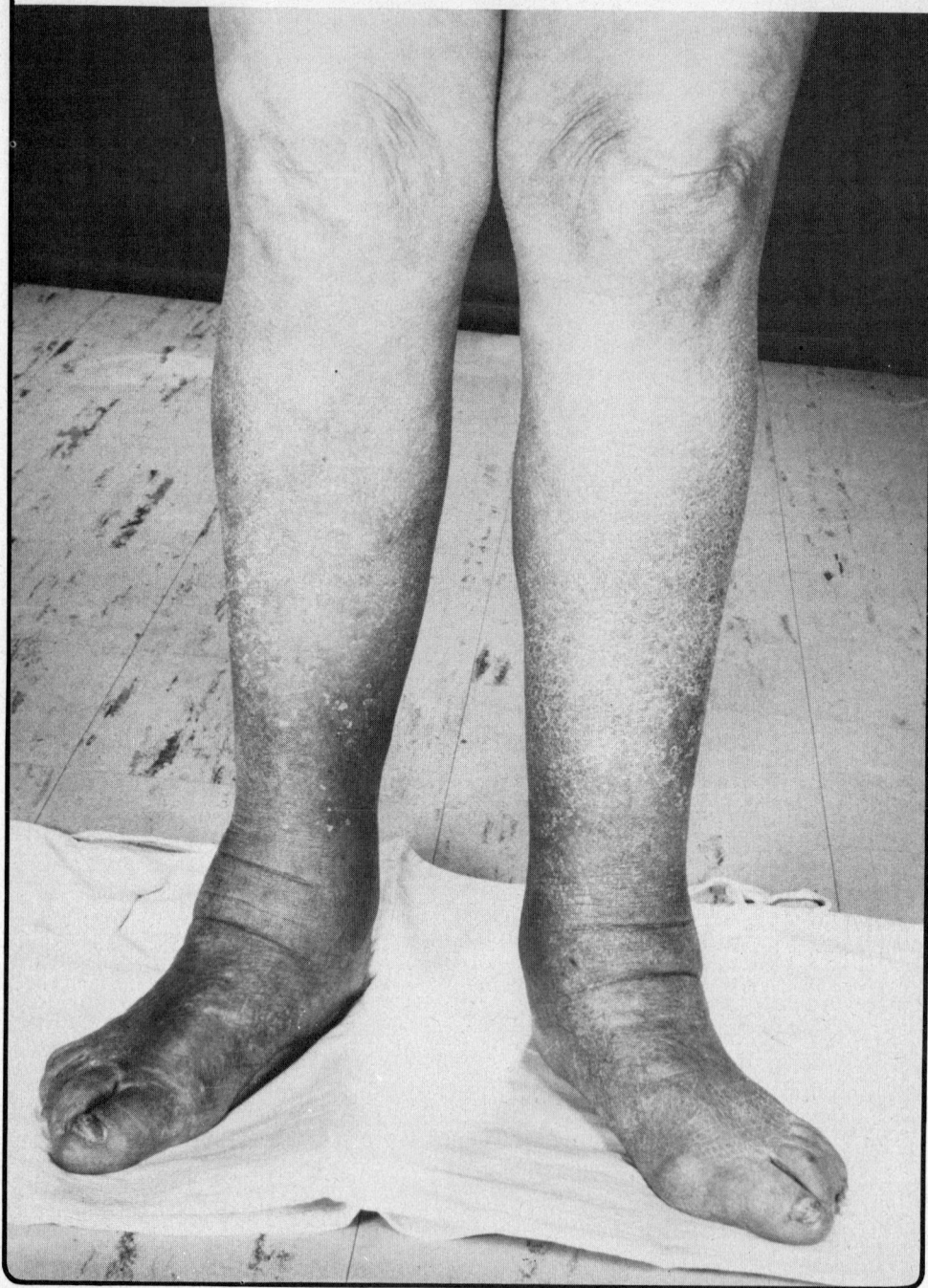

Fig. 9.5 The effects of hypothyroidism on the legs. (Rotker, Taurus)

Underproduction of thyroxine is somewhat more common for reasons which will become apparent shortly. Such individuals are rather sluggish and prone to gain weight. (Women may cease to have menstrual periods.) Again, these symptoms are not uncommon but generally are not caused by underproduction of thyroxine. Unless severe, underproduction of this hormone in adults is merely a major nuisance (Fig. 9.5), but it is of critical significance in infants and in young children. A subnormal metabolic rate in children retards brain growth, and beyond a certain point, subsequent administration of thyroxine will not negate the damage already done to the brain. Such individuals will be permanently (and generally profoundly) mentally retarded, and will also be very short in stature (Figs. 9.6 and 9.7), a condition known as **cretinism.**[4]

Fortunately, the appearance and behavior of infants suffering from underproduction of thyroxine is readily apparent to the trained eye, and cretinism is no longer a serious problem in this country.

The thyroid gland requires about 4 g (0.14 oz) of the element iodine each year for the synthesis of thyroxine (evidently the only requirement for iodine in the entire body). Iodine is relatively abundant in the ocean (hence in marine organisms as well), and also in soils that were at one time part of the sea bed. However, in a

---

[4] A possible origin of the term cretin is the French word *chrétian,* meaning Christian. The rationale was that a mentally retarded individual was thought incapable of sin and therefore a child of God. An alternative origin is from the Latin use of the word *christianus,* which was used derogatorily to mean a barely human creature.

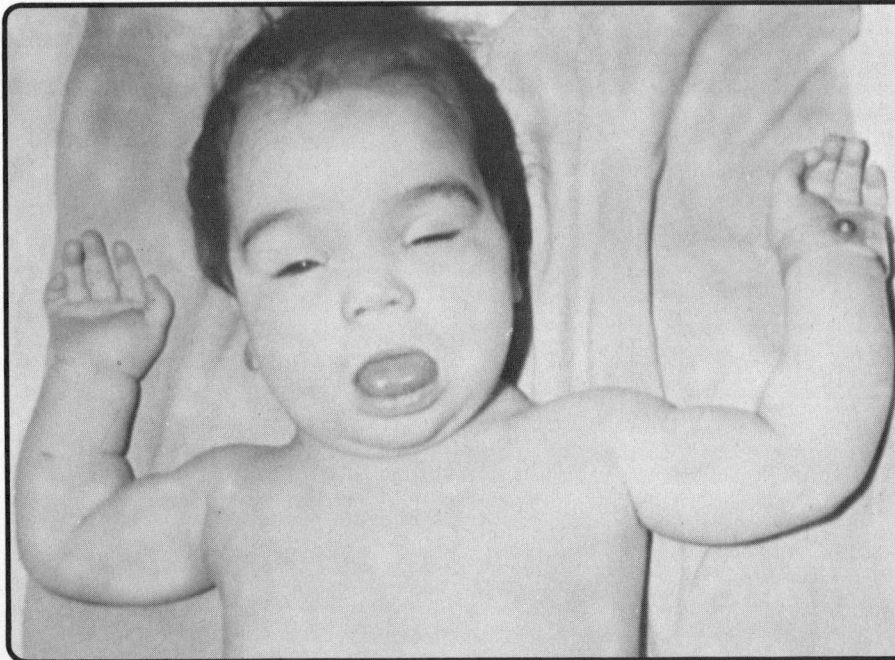

Fig. 9.6
Cretinism.
(Lester Bergman and Assoc.)

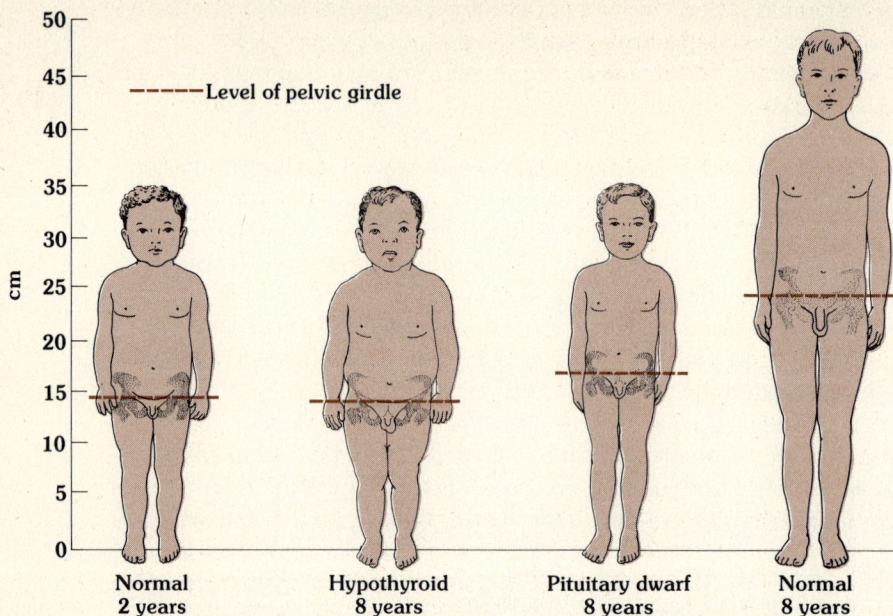

**Fig. 9.7  Body forms of a hypothyroid dwarf, a pituitary dwarf, and two normal children. Note that the body form of the hypothyroid dwarf remains infantile, whereas that of the pituitary dwarf follows chronological age.**

number of parts of the world (including the Great Lakes region of North America and central Europe), iodine is essentially absent from the diet.

The consequences of an insufficiency of iodine in the diet initially is simply an insufficiency of thyroxine. However, an insufficiency of thyroxine also causes a failure of the negative feedback loop with the hypothalamus and pituitary. The effects are just what you would expect if your thermostat were stuck while the furnace was going. TSH continues to be produced, to the point where the thyroid gland begins to grow markedly in size (the effects of TSH, like all the "trophic" hormones, is to stimulate growth and division of the target organ cells). Normally, of course, the more cells in the thyroid gland, the more thyroxine would be produced, but in this instance, thyroxine cannot be produced as there is insufficient iodine. The net result is a greatly enlarged thyroid gland, called a **goiter.** When it was discovered that goiters resulted from an insufficiency of iodine in the diet, this essential element was added to salt[5] (something virtually everyone uses) and, to this day, much of the salt used in this country is "iodized."

---

[5] But only after a pitched battle, much like the more recent controversy involving the fluoridation of drinking water.

The other hormone produced by the thyroid gland is **calcitonin,** which is of importance in the maintenance of blood calcium levels within normal limits. Calcium is important not only because it gives strength and rigidity to bone, but also because it is essential in muscle contraction and in blood clotting.

Calcitonin is formed from different cells in the thyroid gland than is thyroxine, and is not under the control of TSH. Rather, it is in a negative feedback loop with blood calcium. An increase in blood calcium triggers an increase in calcitonin production. Calcitonin stimulates bone cells to remove calcium from the blood and deposit it in the bones.

**PARATHYROID GLANDS**  A common procedure in use for many years in the treatment of goiter is the surgical removal of excess thyroid tissue. When this operation was first performed, occasionally, the patient would die a few days later in a rather horrible, twitching, fashion. Assays of the removed tissues ultimately showed the presence of glandular tissue embedded in the thyroid gland, but of a distinct type. These glands (there are four of them) came to be known as the **parathyroid glands;** each of them is about 5 mm (0.2 in.) in diameter.

The parathyroid glands function to produce the hormone **parathormone,** which operates in concert with calcitonin to aid in the control of blood calcium (*see box*). As would be expected, the action of parathormone is just the reverse of that of calcitonin (although much slower), in that it is produced in response to a drop in blood calcium levels and acts to stimulate bone cells to release calcium from the bones into the bloodstream and to retard calcium excretion by the kidneys. Death results when the parathyroid glands are removed because of the drop in blood calcium caused by the gradual excretion of this ion by the kidneys. The nerves begin firing spontaneously when blood calcium is low, causing the muscles to contract in an uncoordinated fashion. This condition is fatal because of the failure of the chest muscles to operate properly in breathing, and the individual suffocates.

Because only about 1 percent of the total calcium in the body is present in the blood and interstitial fluid (most of the remainder being locked up in the bones), the movement of calcium from bone to blood is nowhere near sufficient to cause bone weakness, except under extraordinary circumstances.

**ADRENAL GLAND**  The adrenal cortex is an extremely important chunk of tissue—without it, we would die in a matter of hours. The exact number of hormones produced by the adrenal cortex is not known, although there are at least a dozen and perhaps as many as four times that number. (Part of the reason for the confusion is in determining whether a given compound is better classified as an intermediate—that is, on the pathway toward making a hormone—or whether it is a true hormone. Not only are the cortex hormones chemically very similar, for all are steroids, but intermediates may sometimes have the capacity to stimulate tissue, just like a hormone.)

There are three major groups of hormones produced by the adrenal

Vitamin D is halfway between a hormone and a vitamin. Our skin produces vitamin D in response to being struck by ultraviolet light from the sun. However, in cool climates, not enough sunlight reaches the skin, and vitamin D must be ingested in a fully formed fashion, just like the other vitamins. Vitamin D was synthesized several years ago and is now frequently added to food, most notably milk. However, early in this century until well into the 1950s, American children were regularly dosed with cod liver oil, because the livers of several types of marine fish, including cod, are high in this vitamin.

The activity of vitamin D is confined to the small intestine. When this compound is synthesized by the skin and transported to its target organ by the blood, it is acting in a hormonelike manner—although its actual activity is not like the typical hormone. Vitamin D acts as a carrier molecule for calcium, and without this carrier, little calcium is absorbed by the intestine, even if it is abundant in the food. A shortage of absorbed calcium is particularly serious in children, who require large amounts of this mineral to strengthen growing bone. However, maintenance of stable blood calcium levels is even more important, and blood calcium levels are maintained at the expense of the bones. As such, the bones, especially those of the legs, may be so soft as to bend under the weight of the child, giving rise to a condition called **rickets**. Note that even though the condition results most immediately from a deficiency of calcium, the actual cause is generally an insufficiency of vitamin D.

**Calcium and Vitamin D**

cortex—**mineralocorticoids,** the **glucocorticoids,** and the **adrenal sex hormones.** Collectively, these affect "salt, sugar, stress, and sex."

The mineralocorticoids are the most significant in terms of maintaining life processes. **Aldosterone** (discussed in detail in Chapter 7) is perhaps the most important of the mineralocorticoids. Collectively, the mineralocorticoids control (as the name would imply) the level of such mineral ions as sodium and potassium in the blood and interstitial fluid. As a proper balance of these ions is of fundamental importance in the functioning of nerves (see Chapter 10), it is the absence of the mineralocorticoids which proves fatal should the adrenal glands be destroyed. The primary feedback is with salt levels in the blood, and not with ACTH.

The glucocorticoids include such hormones as **cortisol** and **cortisone.** These hormones have a number of functions in the body, including promoting glucose synthesis from lipids and proteins (hence, their name) and also aiding healing by reducing inflammation (which is why they are frequently used in treating athletic injuries). Cortisol and cortisone are also used to treat such chronic sources of inflammation as arthritis. This is not without some risk, however, as the reduction of inflammation also decreases the ability of the body to fight infection. For example, glucocorticoids are normally not given to individuals with ulcers.

The adrenal sex hormones include both **androgens** (male sex hormones) and **estrogens** (female sex hormones), although the quantity of each is not great compared to the amounts produced by the mature gonads. Nonetheless, it is significant that both androgens and estrogens are produced in all individuals, regardless of sex. Moreover, androgens (which evidently are more numerous or at least more potent than are adrenal-produced estrogens) have an important role in the female in that they maintain the sex drive, as illogical as that may seem. Women who have had their ovaries removed show no diminishment of sex drive, but should the output of their adrenal sex hormones be reduced, so, too, is their interest in sex. Correspondingly, the administration of testosterone (the most important of the androgens) to women in the treatment of breast cancer[6] results in a marked increase in sex drive.

**ACTH and the formation of cortical hormones**  All the hormones produced by the adrenal cortex are steroids. Not only are they very similar chemically (despite the enormous difference in their effects), but they are all derived from the same *precursor*[7]—the steroid **cholesterol.** ACTH acts only to initiate the first step of the transformation of cholesterol into the cortical hormones; once that step occurs, synthesis of all the hormones in all three categories follows as a matter of course (although glucocorticoids are produced in greatest abundance). Feedback to the hypothalamus and the pituitary, however, evidently is not based on a survey of all the cortical hormones circulating in the blood, but rather just of one—**cortisol,** the most important and biologically active of the glucocorticoids.

As it happens, a genetic malfunction is known wherein the enzyme responsible for synthesizing cortisol is absent. As a consequence, very little cortisol is produced, and thus the negative feedback switch fails to operate, meaning that ACTH continues to be produced. Moreover, because of the nature of the pathways for cortical hormone synthesis, the precursor molecules are instead converted to androgens. Problems arise in young children suffering from this condition, because the presence of androgens, produced in large amounts by these abnormal adrenals, initiates the development of the male secondary sexual characteristics, regardless of the sex of the child (Fig. 9.8). In addition, the presence of the androgens has the effect of inhibiting the production of FSH and LH, both of which are needed for fertility in either sex (see Chapter 8). Fortunately, this is a very rare condition and the effects herein described may be reversed by the administration of large doses of cortisol. The consequences of this condition are less dramatic in adults, although women may develop beards. This is certainly the condition possessed by those bearded ladies of circus fame.

Occasionally, adults suffer from a deficiency of the cortical hormones, usually as a result of the effects of a disease on the cortex (e.g., tuberculosis). The severity of this condition depends on the degree of deficiency, but as mentioned earlier, these hormones are necessary for life, so the condition is potentially fatal. This condition is known as **Addison's disease,** named after its discoverer, and such individu-

---

[6] Cancer tumors result from very rapid cell divisions that are out of control. Testosterone has a retardant effect on the growth (i.e., cell division) of the mammary glands, and is therefore sometimes used in the chemotherapy (chemical treatment) of breast tumors.

[7] A precursor is simply a molecule from which another molecule is synthesized.

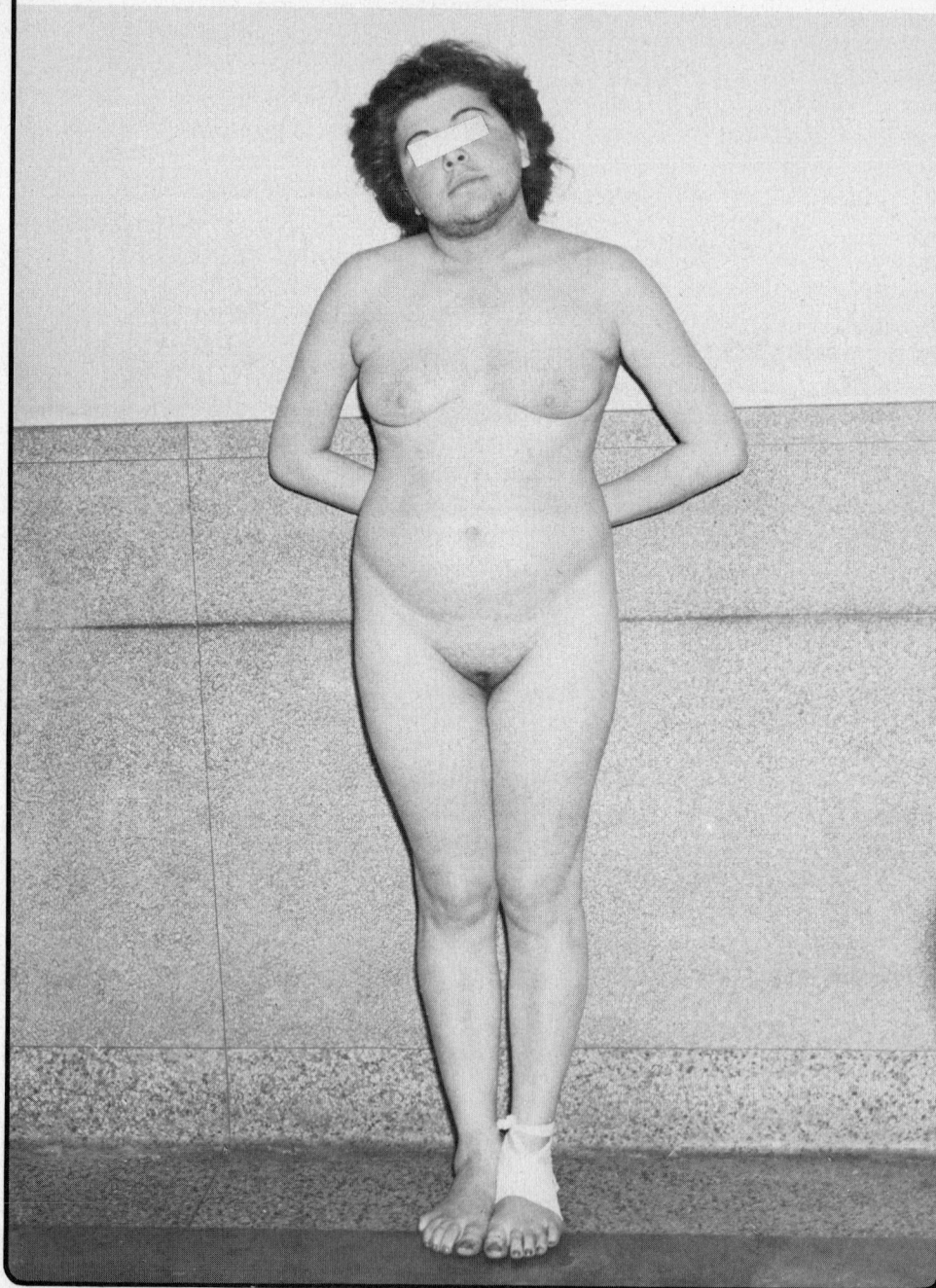

Fig. 9.8 Adrenogenital syndrome. Note the masculine appearance of this young woman. (Lester Bergman and Assoc.)

als have limited resistance to stress, suffer from muscle weakness, lowered blood pressure, and an increase in skin pigment giving a peculiar bronze color to the skin. This is not a rare condition, and is known to be the cause of at least four deaths per thousand in the United States. Probably many more deaths should be attributed to it, but instead are ascribed directly to the effects of some stress condition which a person with fully functional adrenals would have been able to resist. President Kennedy suffered from a mild case of this disease, and received treatment for it (administration of cortical hormones) over a period of years. Interestingly, even individuals who are under treatment are recommended to avoid stressful situations!

Overproduction of the adrenocorticoids (**Cushing's disease**) is even more serious and is inevitably fatal unless the overproduction (usually from a tumor either in the pituitary or in the cortex itself) can be rectified (Fig. 9.9). Diabetes frequently results from overproduction of the glucocorticoids, and these hormones are also responsible for muscle weakness (because of the conversion of muscle protein to glucose). The individual bruises easily and sores are slow to heal. High levels of the mineralocorticoids lead to weakening of the bones and compression fractures of the vertebrae, as well as to spontaneous hemorrhages and high blood pressure.

**Adrenal medulla**  The adrenal medulla, which constitutes only about 15 percent of the weight of the entire gland, is not under the control of ACTH, nor does it produce steroid hormones. It does produce two hormones, however, **epinephrine** (adrenalin), and much smaller levels of **norepinephrine** (noradrenalin), primarily in response to nervous stimulation (as opposed to hormonal control) in times of stress. The adrenal medulla is considered more fully in conjunction with the sympathetic nervous system in Chapter 10.

**PANCREAS**  In addition to its major role in digestion, the pancreas also acts as an endocrine gland in that it produces two hormones, **glucagon** and **insulin,** although less than 1 g (0.04 oz) of pancreatic tissue is devoted to their synthesis. Glucagon acts to convert the glycogen stored in the liver back to glucose, thereby increasing the levels of blood glucose. Insulin acts in just the opposite manner, to lower blood glucose both by increasing the permeability of cell membranes to glucose, thereby effecting a lowering of blood glucose, and also by stimulating the liver to convert blood glucose to glycogen. The primary source of feedback for both hormones is the level of glucose in the blood flowing through the pancreas.

Insulin production seems to be a weak link in the endocrine system as a whole, for reasons to be elucidated shortly. Most of the other endocrine diseases are surprisingly rare, given the sophisticated interrelationships of some of them and given that the amount of hormone involved is so small. However, unfortunately, insufficiency of insulin production is rather common, and if the amount produced is sufficiently low, the blood glucose level rises markedly, a condition known as **hyperglycemia.**[8]

---

[8] From the Greek *hyper,* meaning "above" or "excessive," as opposed to *hypo,* meaning "below" or "under," as in "hypodermic" (under the skin), and *glukus,* meaning "sweet."

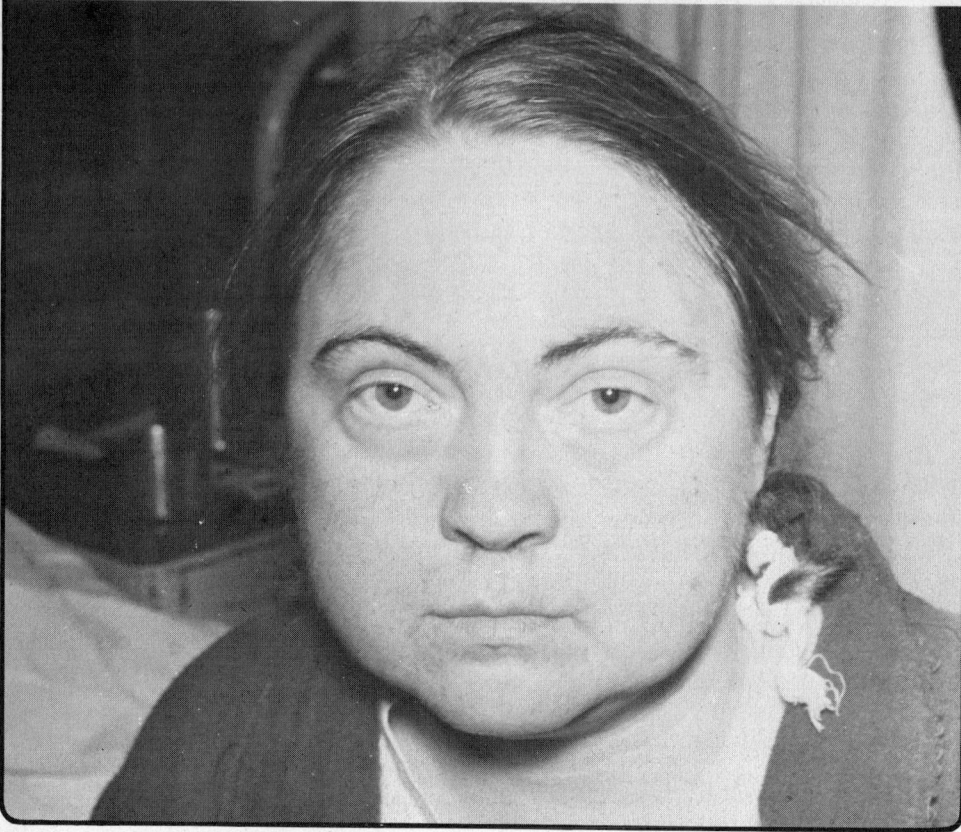

Fig. 9.9    Individual with Cushing's disease. Note beard and puffy face. (Lester Bergman and Assoc.)

By far the most common cause of hyperglycemia is **diabetes mellitus** (the modifying word is necessary to distinguish this disease from the much rarer **diabetes insipidus,** discussed earlier). This disease exists in two forms. One form (**juvenile-onset** diabetes) affects children, and seems to have a strong genetic component. Frequently, there is almost no insulin produced by the pancreas, and in most cases, insulin injections are required. The second form (**adult-onset** diabetes) affects adults, and although again there is evidently a genetic component (as the incidence of the disease tends to run in families), it is simply the **tendency** for the disease, rather than the disease itself, which is inherited. That is, given a sufficient stress on the body (overweight, severe diseases or injuries, pregnancies in rapid succession, advanced age), the disease may be manifested. However, it is usually less severe and may be manageable by dietary restrictions alone (reduced carbohydrate intake), or by diet and oral medication, which act to stimulate pancreatic production of insulin. Insulin injections may be necessary in severe cases. Recent evidence suggests that juvenile-onset and adult-onset diabetes are two different diseases. The juvenile form is evidently a true insulin deficiency, which explains why some form of insulin therapy is so commonly necessary. The adult form is quite different, and results not so much from a deficiency of insulin as from an insufficient amount of a chemical present in the membranes of all cells which normally allows insulin to attach to the cell membranes. Insulin therapy operates to swamp the system, ensuring that a sufficient amount of insulin is attached to the membrane, even though the level of membrane chemical is low. This explanation also accounts for the fact that dietary restrictions may suffice to control the diabetes in adults, a finding which would be difficult to account for, if diabetes were caused directly by insulin insufficiency. Even more important, a deficiency of the membrane chemical can be detected biochemically long before diabetes itself develops. Such potentially susceptible individuals are well advised to adhere to dietary restrictions to avoid full-blown development of diabetes.

Untreated diabetes results in a marked rise in blood glucose because of the absence of insulin to a point where the kidneys are no longer capable of reabsorbing all the glucose (see Chapter 7), and some glucose is lost in the urine. In order to keep the glucose in solution in the urine, more water must also be passed, with the result that undiagnosed diabetics produce larger volumes of urine and show signs of increased thirst. Moreover, in the absence of insulin, fats and proteins are broken down and converted to glucose (see next section), with the result that the individual loses weight, even though appetite is sharply increased. Much more serious is the fact that sodium reabsorption by the kidneys is interfered with, which ultimately has disastrous consequences on brain function.

In addition, the cells of the body do not get a sufficient supply of glucose, despite the extremely high blood glucose levels (recall that insulin acts to facilitate diffusion of glucose across the cell membrane). Muscle cells in particular require large amounts of glucose, and fatigue and muscle weakness may occur as a consequence. (The nervous system is not directly affected because of all the body's systems, its uptake of glucose does not require insulin.)

Much more rarely, there may be an overproduction of insulin (typically resulting from a pancreatic tumor), which produces the condition **hypoglycemia.** Ironically, the symptoms are similar to hyperglycemia, because once again the cells are not obtaining a sufficient supply of glucose—the oversupply of insulin acts to convert too much glucose to the storage form, glycogen. The muscle and nerve cells are the first to suffer, and fainting spells and blackouts may occur. Treatment consists of glucose injections in emergency cases (i.e., if the individual is comatose[9]), followed by surgical removal of the tumor.

Problems may also arise from the insulin injections themselves. Too little insulin results in a mild form of the symptoms just mentioned for diabetes; too much insulin **(insulin shock)** results in hypoglycemia with possible blackouts.

Prior to the 1920s, when insulin was discovered, severe cases of diabetes were in-

---

[9] A hypoglycemic individual given an injection of glucose will regain consciousness in less than 1 min, and the return to normalcy is essentially absolute (although temporary). It is one of the most dramatic events in medicine.

evitably fatal, although death was a slow and painful process. With the availability of insulin, diabetics have been able to lead near-normal lives, with the result that the percentage of diabetics in the population has risen sharply. Unfortunately, as more has become known about insulin, we have come to realize that not only does insulin affect more than just blood glucose levels, but that insulin injections are only a crude approximation of what occurs naturally. For example, diabetics requiring insulin therapy have problems with fat metabolism (which is partially under the control of insulin), and tend to suffer from fatty deposits being laid down in the smaller blood vessels, including those of the retina of the eye and kidneys. Diabetes remains a leading cause of blindness, kidney failure, and heart disease, and the present system of insulin injections cannot be regarded as a cure, but merely a stopgap method of treatment.

**BLOOD GLUCOSE LEVELS—A REPRISE**  Blood glucose levels tend to rise under the influence of (1) glucagon, (2) the glucocorticoids, (3) epinephrine, and (4) the hunger drive, which results in the ingestion of food.[10] Blood glucose levels tend to fall only under the influence of insulin. Put in that way, it is not difficult to understand why insulin failure is both common and severe—there are at least four forces opposing it, and there is no backup system.

Normally, blood glucose levels are maintained on a day-to-day basis by the combined effects of eating and glucagon on the one hand, and insulin on the other. The glucocorticoids come into play primarily in the case of stress—whether emotional or physical—because stress causes the release of ACTH directly, independent of the normal feedback with cortisol. The blood glucose level rises because the glucocorticoids act to convert protein to glucose, which is adaptive if the stressful situation involves having to go without eating, as it must for many animals. However, this reaction in humans explains why it is that postsurgical patients require additional protein in their diets and why diabetics with infections require more insulin (to compensate for the increase in blood glucose brought about by the extra glucocorticoids produced in response to the infection). Similarly, prolonged treatment with glucocorticoids, as in individuals with arthritis, also acts to place an additional burden on the pancreas to produce more insulin, and occasionally the price paid for the relief of arthritic pain is the development of diabetes.

In a related fashion, the release of epinephrine from the adrenal medulla is largely under the influence of the **sympathetic nervous system** (see Chapter 10), although glucagon also stimulates epinephrine release. This is the system that is operational in times of stress, fear, or anger, and among other things, epinephrine operates to convert liver glycogen to glucose. Hence, there is an overlap in functioning of epinephrine and the glucocorticoids, but it seems likely that epinephrine is more important in times of stress of short duration, for instance, when you were in grade school and the class bully told you that he was going to "get you" during recess. Conversely, the glucocorticoids are more significant when the stress is of longer duration, as, for instance, when you are in the hospital recovering from your recess meeting with the class bully.

---

[10] There is increasing evidence that GH also promotes increased blood glucose levels.

A study of the endocrine system has value perhaps as much for the elegance of the negative feedback loops as for the knowledge that comes from understanding the effects of each individual hormone. Certainly no other system exemplifies the workings of homeostatic control mechanisms—and, by inference, the importance of homeostasis itself—as well as the endocrine system.

Recent research has demonstrated that the endocrine system is subject to control by the nervous system, and in that respect it is no more autonomous than any other system.

Finally, one of the most interesting findings to come from a study of the endocrine system is an appreciation of the priorities of the body. For example, blood calcium is directly affected by two distinct hormones, blood glucose by at least four hormones, and so forth. The greater the number of hormones, the more important is the control of the substance in question.

Deep in the Amazon forests, the Jivaro hunter raises his blowgun and propels a feathered dart into the side of a wild pig. The pig squeals and crashes off into the undergrowth, only to collapse and die a few minutes later in a state of total paralysis. How is the paralysis induced?

At the same moment, near some pea fields in New Jersey, the pilot of a spray plane is rolling a drum of the insecticide parathion into his plane, prior to spraying the fields for pea aphids. Suddenly, the top of the drum works loose, and the pilot is drenched in the insecticide. Panicked, he turns and runs toward his car, parked a few hundred meters away, but, before he reaches it, he drops to the ground and dies of convulsions. How are the convulsions produced?

"Two-Ton Tony" McGurk, a local enforcer for the mob, is attacked one night by a rival. In the course of their scuffle, Tony slips and hits the back of his head on the curb. Rushed to the hospital with a skull fracture, he nonetheless quickly recovers—but is permanently blind. How could a blow to the back of the head induce blindness?

In each case, the result is caused by an injury to some portion of the nervous system, which is the subject of this chapter.

chapter **X**

# chapter X
# Coordinating response to environmental change—the nervous system

Although both the nervous system and the endocrine system are involved with coordinating bodily activities, they complement each other in that the endocrine system is adapted for slow, sustained change, whereas the nervous system is adapted for speed.

The nervous system presumably arose as a mechanism for assuring synchrony in muscle cell contractions. That is, the presence of nerve cells allowed muscle cells to work together in a coordinated fashion. Only as organisms grew more sophisticated and developed sense organs, did relay centers evolve for the reception of **sensory** (input) messages and the dispatch of **motor** (output) responses. Ultimately, these relay centers became (at least in the vertebrates) the **spinal cord** and **brain.** This sensory role was secondary, however—the nervous system originated as the controller of the muscles, a conclusion borne out in part by a comparison of the organization of plants with that of animals. Plants have no nervous system, although in terms of evolutionary time they have had quite as long as animals to be able to develop one. Of course, they have no equivalent of a muscle system either. One can conclude that without a muscle system, there was no need for a nervous system.

**ORGANIZATION OF THE NERVOUS SYSTEM**   There are a number of ways in which the nervous system may be compartmentalized. **Central nervous system (CNS)** refers to the brain and spinal cord; this division is in contrast with the **peripheral nervous system,** which comprises the nerves that run from the CNS to the muscles (motor nerves), and from the various sense organs of the body back to the CNS (sensory nerves). The peripheral nervous system is also divided into the **somatic nervous system,** which serves the skeletal muscles and sense organs, and the **autonomic nervous system,** which serves the visceral organs, heart, and so forth. In every case, however, the cellular organization of the system is similar.

**THE NEURON**    The basic unit of the nervous system is the nerve cell, or **neuron.** Although it takes a variety of forms (Fig. 10.1), a neuron is not likely to be mistaken for any other type of cell in the body. In every case, a neuron consists of a small cell body with one or more elongated processes projecting from it. A typical motor neuron is depicted in Fig. 1c. This type of neuron contains a number of short processes called **dendrites** (from the Latin *dendron,* meaning "root") which project from the cell body, and a single very elongated axon, which functions in transmitting messages to other neurons or to receptor organs such as muscles.

**The resting potential**    How does the nervous system work? The basis of its operation lies in changes that can occur within *every* neuron. As is the case with all the cells of the body, the concentration of ions on the inside of the neuron is different from the concentration of ions outside the cell, in the interstitial fluid. Specifically, the concentrations of $Na^+$ (sodium ions) and $Cl^-$ (chloride ions) are much higher outside the cell; the concentration of $K^+$ (potassium ions) is much higher inside the cell (Fig. 10.2). $Na^+$ and $K^+$ are maintained in their respective positions by active transport; $Cl^-$ is free to move across the cell membrane, but remains more concentrated outside the cell because of the presence within the cell of large organic ions (mostly proteins bearing a negative charge), which are much too large to cross

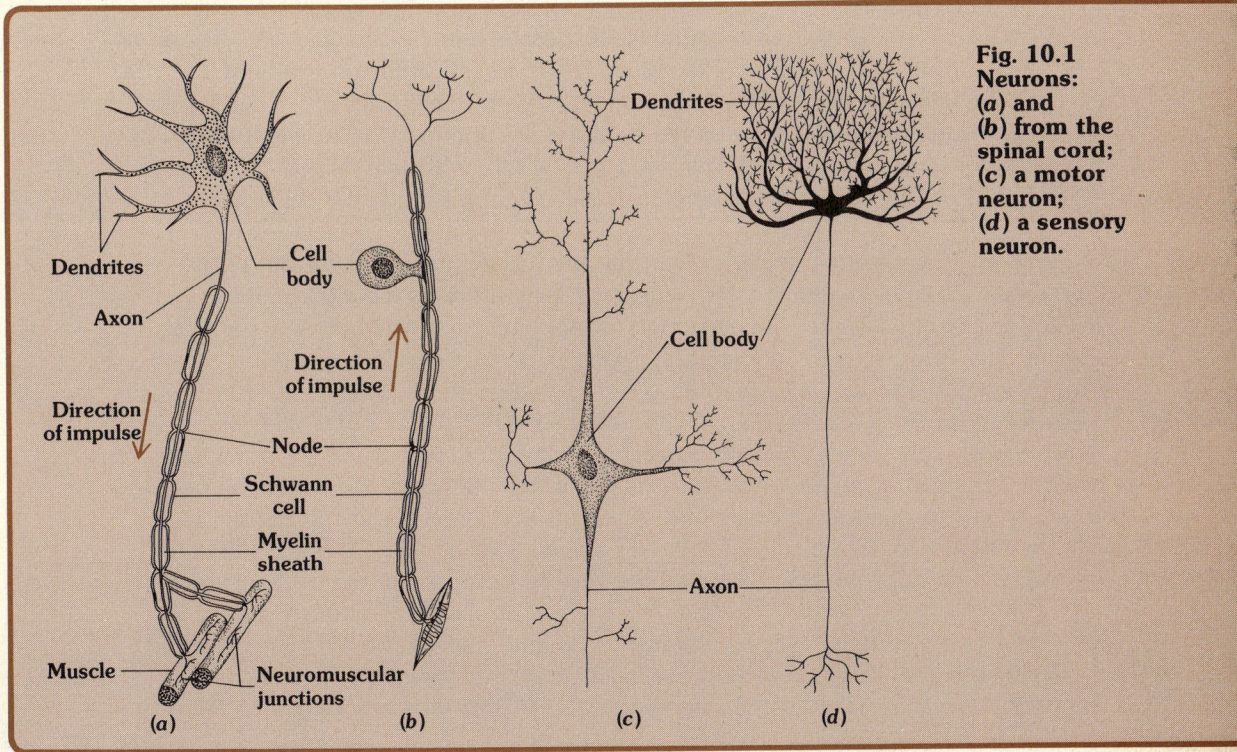

Fig. 10.1
Neurons:
(*a*) and
(*b*) from the
spinal cord;
(*c*) a motor
neuron;
(*d*) a sensory
neuron.

Dendrites

Axon

Direction
of impulse

Cell
body

Direction
of impulse

Node

Schwann
cell

Myelin
sheath

Muscle

Neuromuscular
junctions

Dendrites

Cell body

Axon

(*a*)          (*b*)          (*c*)          (*d*)

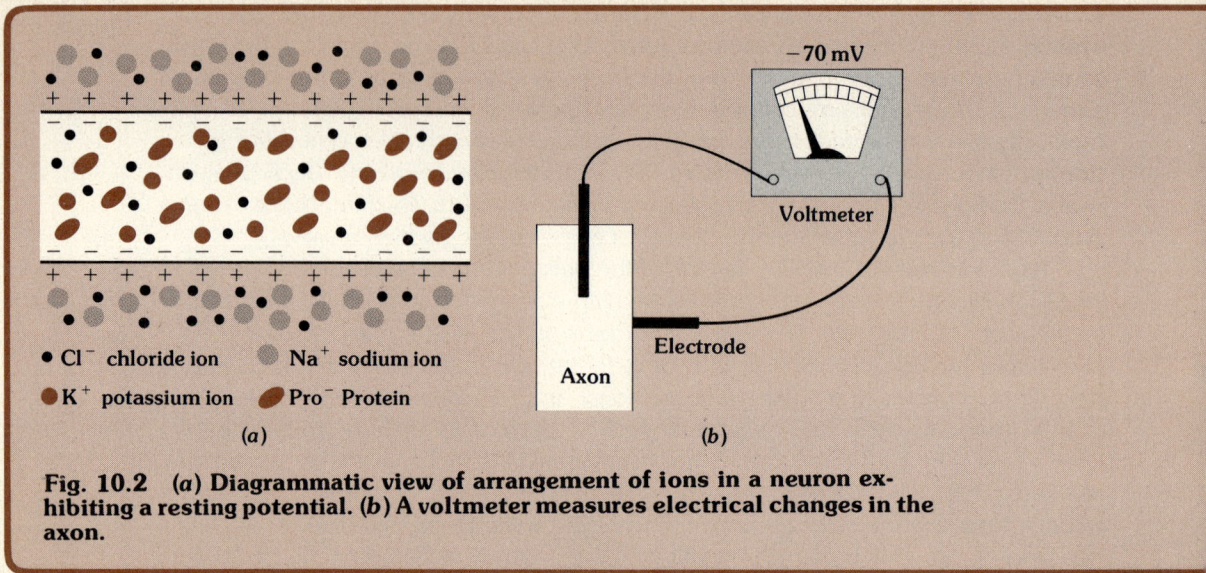

**Fig. 10.2** (*a*) Diagrammatic view of arrangement of ions in a neuron exhibiting a resting potential. (*b*) A voltmeter measures electrical changes in the axon.

the cell membrane. These organic ions largely nullify the diffusion gradient which would predict a movement of $Cl^-$ into the cell, because similar charges repel. Therefore, in this instance, an **electrical gradient** is opposing a **diffusion gradient**.

Are the negative and positive charges in balance on both sides of the cell membrane? It might appear that such would be the case, but, in actuality, the interior of the neuron is some $-70$ mV (millivolts, or 1/1000th of a volt) with respect to the interstitial fluid, indicating more negative than positive charges inside the cell. How is this achieved?

The answer is not yet entirely clear, but it is known that $K^+$ leaks out of the cell some 200 times faster than $Na^+$ leaks into the cell. $K^+$ is a larger ion than is $Na^+$, and might therefore be presumed to move more slowly, but, as it happens, these ions are surrounded by a cloud of water molecules that move with the ions as a unit—and, as $Na^+$ has more water molecules than does $K^+$, it represents a larger, hence slower, unit. This tendency for $K^+$ to leak out of the cell, which is only very partially compensated for by the tendency of $Na^+$ to leak into the cell, results in a surfeit of negative charges in the interior of the cell. (The negative charges represent the protein ions, which are too large to move.) The net result is a cell interior which is $-70$ mV with respect to the exterior. This value can be maintained indefinitely and is called the **resting potential** of the cell.

**The action potential**  The fact that there is an electrical gradient set up across the cell membrane tells us nothing about how messages are conveyed in the system, and it is this ability which is the hallmark of the nervous system. As it happens, despite its stability (in the sense that the $-70$-mV difference can be maintained indefinitely), the resting potential is fragile, because it is subject to disruption. The nature of this disruption may be the action of another neuron, or it may be some environmental change; regardless, the consequences (assuming a strong enough disruption) are a dramatic change in the electrical gradient across the cell membrane of the neuron. Specifically, this change will be the generation of an **action potential,** which is manifested by a brief jump in potential from $-70$ mV to $+30$ mV, followed by an immediate return to $-70$ mV, the whole process taking only 1/1000th of a second (Figs. 10.3–10.5).

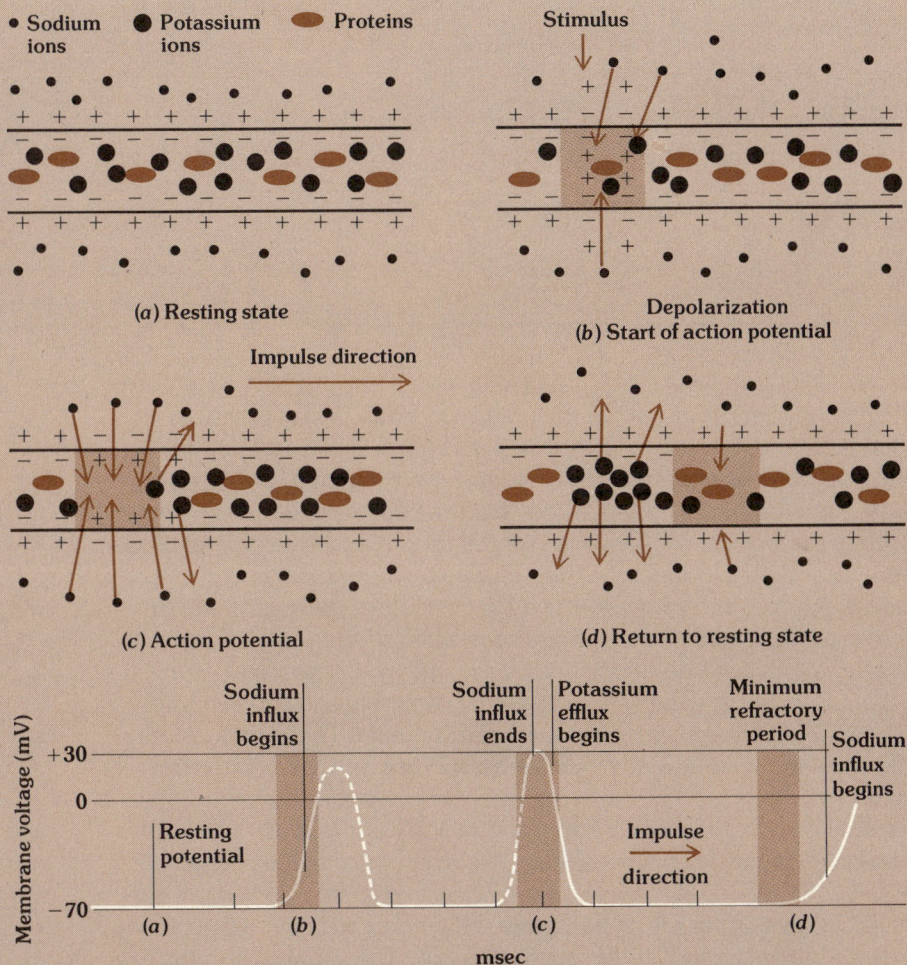

Fig. 10.3 **Movement of ions during the generation of an action potential, followed by a return to normal.**

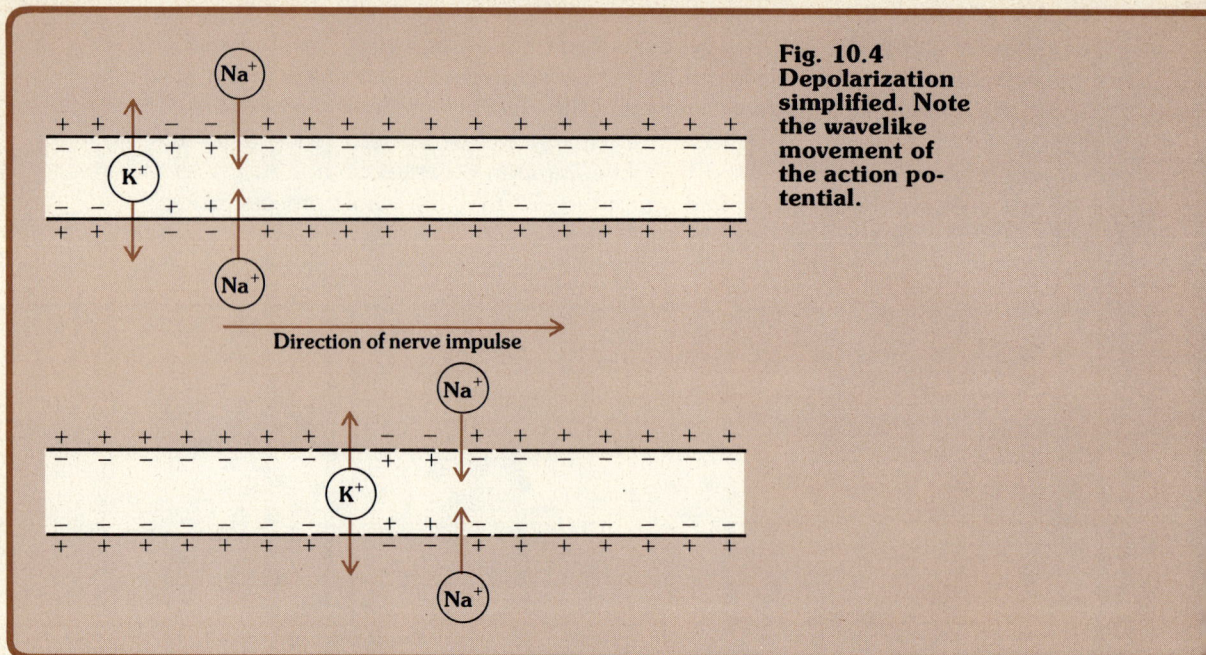

**Fig. 10.4 Depolarization simplified. Note the wavelike movement of the action potential.**

Exactly what changes occur in the cell membrane to allow this alteration in electrical potential are not clear, but it is known that this change is initiated by a sudden influx of $Na^+$, resulting in a momentary $+30$ mV value inside the cell, followed by an immediate efflux of $K^+$ ions, which restores the electrical potential to approximately the normal value ($-70$ mV). The cell membrane is said to be **depolarized** during this period. Active transport then quickly restores the actual ionic balance by transferring the $Na^+$ to the outside and the $K^+$ to the inside of the membrane. As long as the permeability of the membrane remains high to $Na^+$ (about 1/1000 sec), the neuron cannot be depolarized a second time, regardless of the strength of the stimulus; during a subsequent interval of perhaps 1/100 sec, the neuron can only be depolarized by an above-normal stimulus. These periods of time are referred to as the **absolute** and **relative refractory** periods, respectively (Figs. 10.6–10.8).

It is of interest to note that if a given stimulus is of sufficient intensity as to generate an action potential (i.e., to reach **threshold**), the strength of the action potential is always the same ($+30$ mV). This property of the neuron either to depolarize maximally or not at all is known as the **all-or-none rule.** How, then, can the neuron communicate differences in the strength of the stimulus itself? That is, how is your brain informed that you just drove a nail through your foot, as opposed to having pricked the skin with a pin? In part, relative strength of the stimulus is communicated simply by the number of neurons that become depolarized, but the

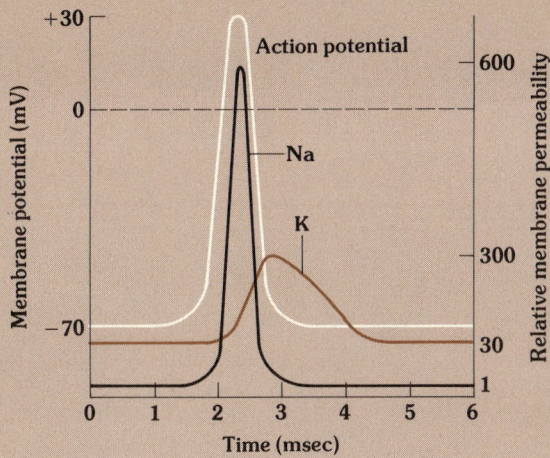

**Fig. 10.5**
Changes in membrane permeability to Na+ (sodium) and K+ (potassium) ions during an action potential.

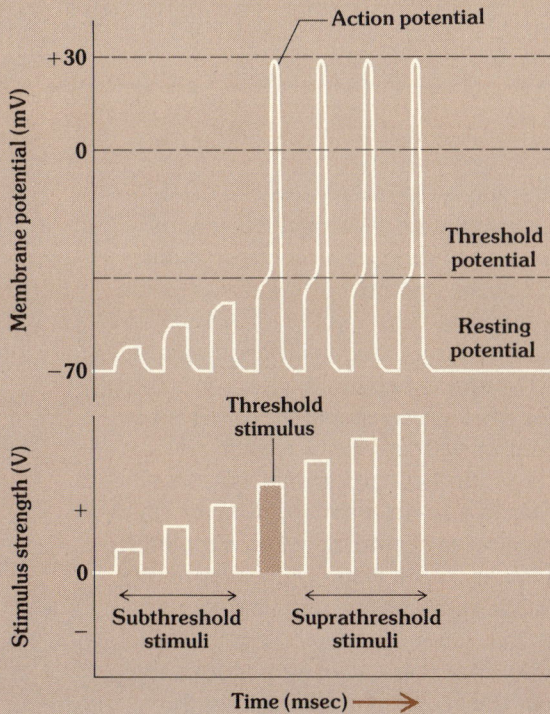

**Fig. 10.6**
Threshold and the "all-or-none" rule, in diagrammatic form. Note that once threshold is reached, increasing the strength of the stimulus has no corresponding effect on the magnitude of the action potential.

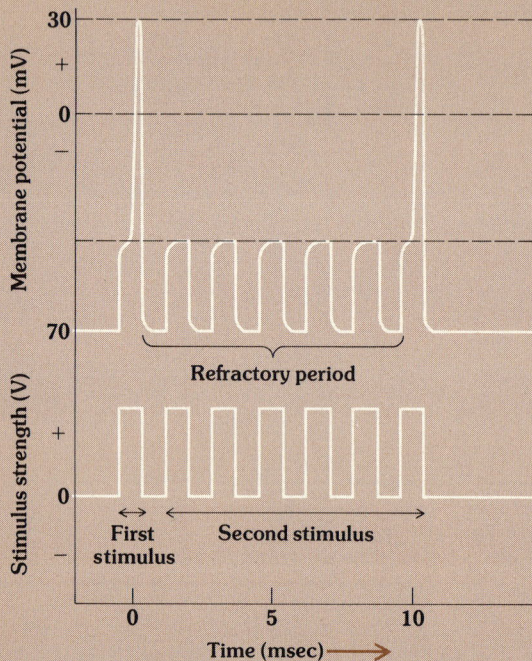

**Fig. 10.7**
**Length of refractory period, as determined by the interval required before depolarization will occur again to a stimulus of the same magnitude.**

single neuron can also communicate stimulus strength by repeated waves of depolarization. That is, the CNS interprets a wave of action potentials arriving 10/sec differently from a single action potential—or from 2/sec. In short, differences in stimulus intensity are communicated by the *frequency,* not the *strength,* of the action potentials.

**The neuron and the action potential**  So far we have been discussing this whole complex as if it were a theoretical system with no basis in the neuron itself. Experimentally, action potentials can be created in neurons simply by giving them a burst of electricity through an electrode in contact with any portion of the neuron. If the electrode is in contact with an axon, an action potential will move both ways from the point of contact with the electrode—in a kind of domino effect, depolarization of one section of the cell membrane serves as the trigger for depolarization of the adjacent areas of the membrane. However, the wave of depolarization moving back toward the cell body is ultimately extinguished, whereas the wave of depolarization moving down the axon away from the cell body may be continued in other neurons in contact with the tip of the axon. Moreover, propagation of an action potential is most readily achieved by stimulation applied at the base of the axon; stimulation at another point will require greater strength to generate an action potential. Both of these facts give an inkling of the ways in which the individual neuron functions as a part of the unified whole, a subject we shall discuss shortly.

The speed of the wave of depolarization as it moves down the axon may exceed 100 m/sec, depending on the thickness of the axon (speed is greatest in thick axons, as it is in thick copper wires, where electrical resistance is reduced), and on the presence or absence of **myelin** (Fig. 10.9). Myelin is typically found around the axons of motor neurons, but is absent from many neurons (most of which are short) in the CNS. Myelin is produced by the spiraling of a specialized type of cell called **Schwann cells** about the axon during its initial development. The Schwann cells are not elongated as are many neurons and, consequently, there are many such cells, arranged like beads on a string, along the longest neurons. Small gaps, or **nodes,** occur between adjacent Schwann cells, and because of the mechanics of action potential generation, myelinated fibers actually depolarize only at these nodes. Thus, transmission of an impulse down a myelinated fiber takes the form of a series of hops, from node to node, which accounts for the fact that conduction is much more swift in myelinated than in nonmyelinated neurons.

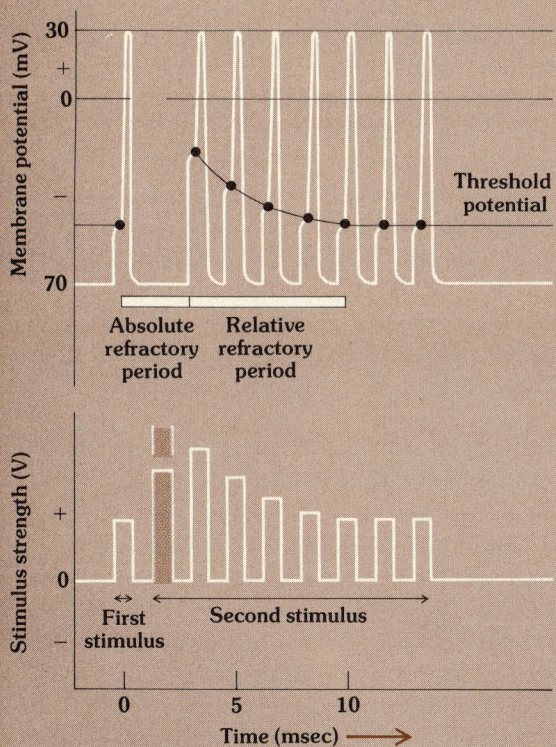

Fig. 10.8
**Diagrammatic view of the absolute and relative refractory periods.**

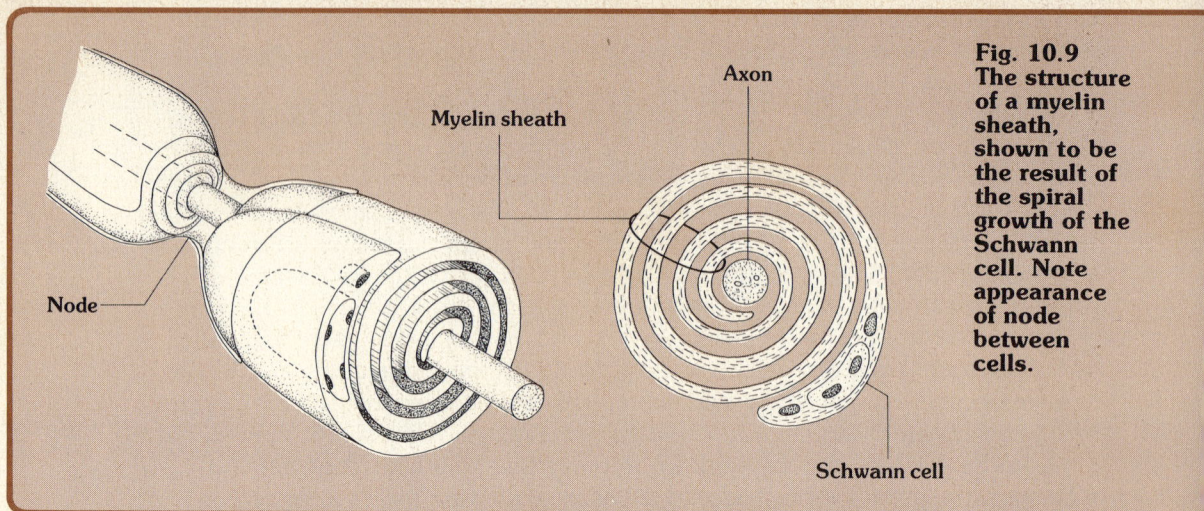

**Fig. 10.9 The structure of a myelin sheath, shown to be the result of the spiral growth of the Schwann cell. Note appearance of node between cells.**

Myelin sheath

Axon

Node

Schwann cell

**THE SYNAPSE**   Strictly speaking, a **synapse** (Fig. 10.10) is a junction between neurons. (The term **neuromuscular junction** is used for junctions between neurons and muscle cells.) The organization of the synapse is rather well known, at least in portions of the peripheral nervous system, and the structure of a typical synapse is depicted in Fig. 10.11. An action potential moves down the axon to the tip and causes the rupture of tiny sacs containing one of several known **neurotransmitters,** of which **acetylcholine** is by far the most common in the peripheral nervous system. Acetylcholine moves across the **synaptic cleft** (a distance of about 1/10,000 mm (4/1,000,000 in.), and lodges in receptor sites on the membrane of the following neuron. If enough of these receptor sites are filled, the second neuron will initiate its own wave of depolarization down its axon. Thus, chemical transmission is substituted for electrical transmission at the synapses.

Studies using radioactive tracers have indicated that with every action potential, about 5,000,000 acetylcholine molecules are released at the synapse, and that only one-third of the 2,500,000 binding sites on the following neuron need to be filled by acetylcholine to effect the generation of an action potential. What prevents repeated depolarizations of the second neuron once acetylcholine has been released into the cleft? Why are more molecules produced than are needed? The answer to both questions is that there is another molecule present in the synaptic cleft called **cholinesterase,** which, as its name would imply, has the function of splitting the acetylcholine molecule enzymatically (Fig. 10.12). Because one molecule of cholinesterase can split 20,000,000 molecules of acetylcholine per minute, not many enzyme molecules are required to nullify the acetylcholine release at the synapse. In essence, the acetylcholine molecules swamp the cholinesterase destructors temporarily, running the gauntlet as it were, and enough of them survive the trip intact to depolarize the membrane of the second neuron. Even so, once they occupy the re-

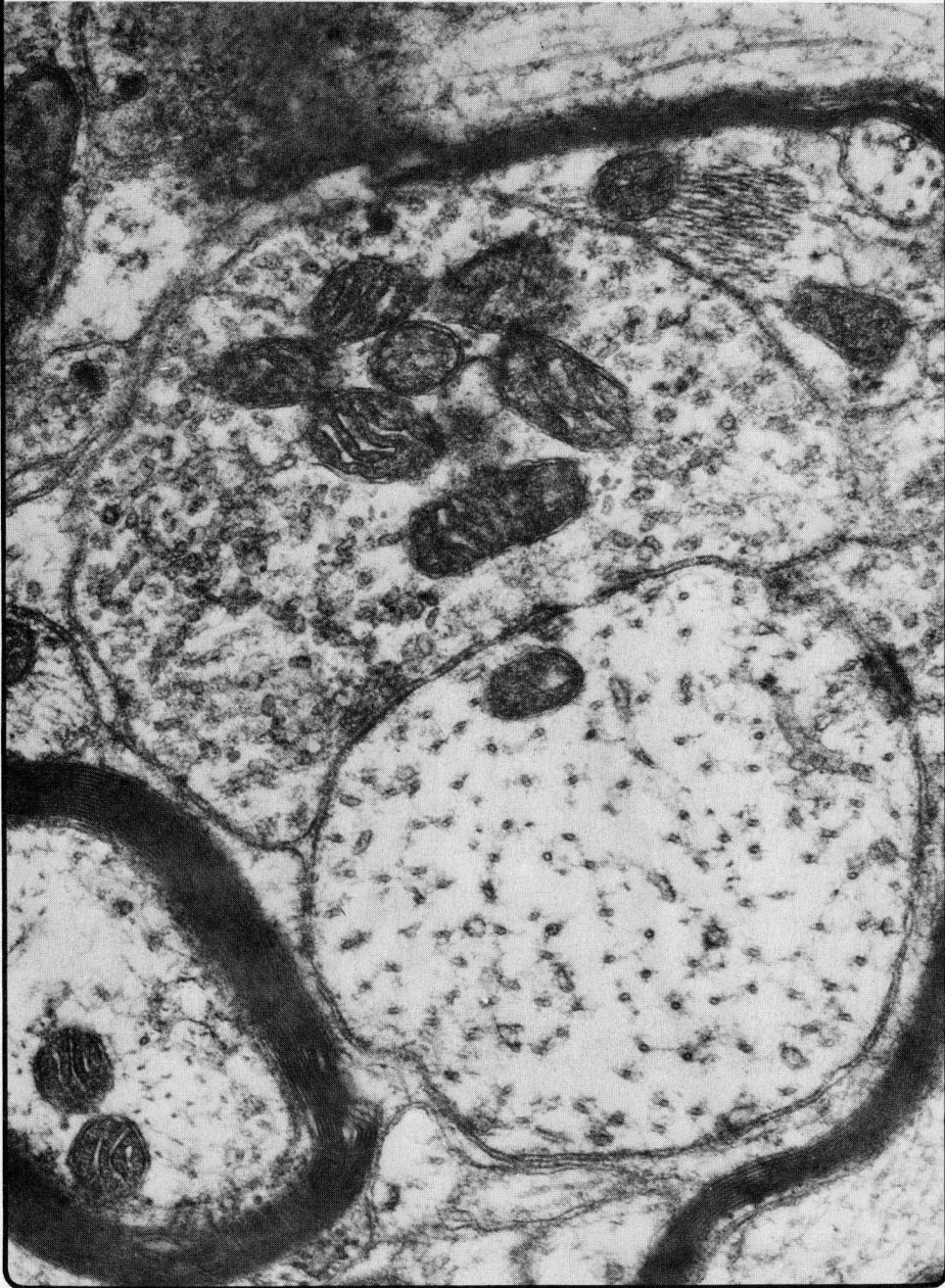

Fig. 10.10   The appearance of synapses on the cell body of a neuron as seen through a scanning electron microscope. (Rotker, Taurus)

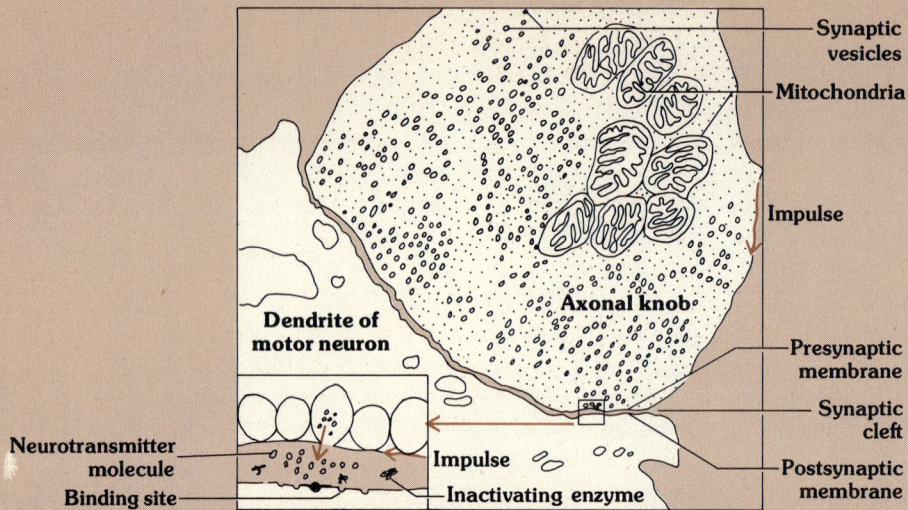

**Fig. 10.11  Parts of synapse shown in relation to each other.**

Labels: Synaptic vesicles · Mitochondria · Impulse · Axonal knob · Presynaptic membrane · Synaptic cleft · Postsynaptic membrane · Dendrite of motor neuron · Neurotransmitter molecule · Binding site · Impulse · Inactivating enzyme

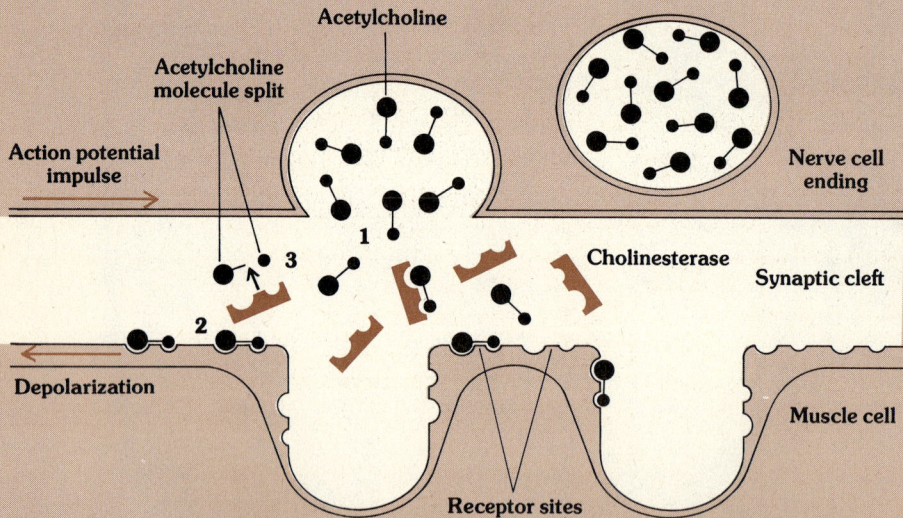

**Fig. 10.12  The steps in transmission of an impulse at a neuromuscular junction: (1) acetylcholine released by the action potential; (2) acetylcholine binds with receptor sites; (3) cholinesterase splits acetylcholine molecule.**

Labels: Acetylcholine · Acetylcholine molecule split · Action potential impulse · Nerve cell ending · Cholinesterase · Synaptic cleft · Muscle cell · Depolarization · Receptor sites

ceptor sites of the postsynaptic membrane, the acetylcholine molecules are still split by cholinesterase, but obviously not before a new action potential has been generated in the second neuron. The two halves of the split acetylcholine molecules are reabsorbed by the first neuron and are reconverted to acetylcholine for subsequent reuse.

But the whole mechanism is a bit more complex than has just been described. For example, neurons need not always act to stimulate—some act to inhibit. Thus, whereas the typical action potential has the effect of causing depolarization in the second neuron, an inhibitory stimulus has the effect of **hyperpolarizing** the second neuron (i.e., increasing its resting potential to $-90$ or $-100$ mV), making it that much more difficult for that neuron to produce an action potential except in response to the strongest excitation. The location of the synapse with respect to the second neuron is also important. If the synapse is at the base of the axon, where the lowest order of stimulation is required to generate an action potential, propagation of an action potential is far more likely than if the synapse is on a distant dendrite. Moreover, because a neuron may have hundreds of synapses on its dendrites and cell body, some of which are stimulatory and some of which are inhibitory, a great deal of flexibility is built into the system, and the action or inaction of a given neuron will depend on the number and strength of the inhibitory and excitatory stimuli at any given time. Despite the obvious advantages of flexibility of response, synapses remain fewer than would be expected (i.e., neurons tend to be very long), because the rate of transmission of an impulse along a neuron is very much faster than is the rate of transmission across a synapse.

As we have just seen, the net effect of a number of simultaneous inhibitory and excitatory action potentials on a given neuron is not strictly additive, as the location of the synapse has a bearing on the strength of the stimulus necessary to generate an action potential in the recipient neuron. Moreover, **summation** may also occur. Summation may involve the additive effects of successive waves of depolarization down the same neuron (temporal summation) or the additive effects of adjacent synapses receiving action potentials at the same time (spatial summation). Without going into any more detail, it is obvious that the functional organization of every nerve cell is extremely complex.

**The neuromuscular junction**   The neuromuscular junction is the area of interaction between a neuron and a muscle fiber (Fig. 10.13). Although it is structurally similar to a synapse, the events that occur at the neuromuscular junction are somewhat different, and the neuromuscular junction is therefore generally distinguished from a true synapse.

Specifically, as the wave of depolarization reaches the end of the axon, acetylcholine is released, just as in a synapse. Similarly, this acetylcholine traverses the cleft and depolarizes the muscle fiber membrane, causing a wave of depolarization to sweep across the membrane (only neurons and muscle cells have the capacity to be depolarized). At this point, the similarity ends, because rather than an influx of $Na^+$, there is an influx of $Ca^{2+}$ (calcium ions) into the cytoplasm of the muscle cell. The $Ca^{2+}$ is normally stored in special sacs along the endoplasmic reticulum (ER)

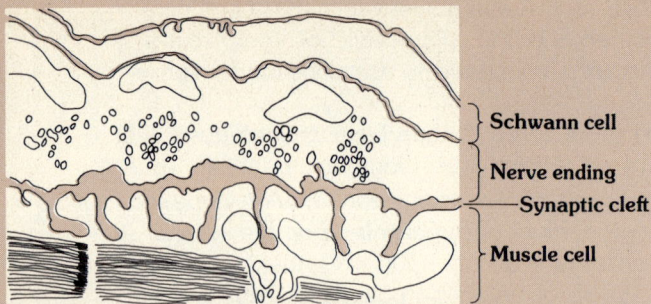

**Fig. 10.13  Detail of the neuromuscular junction, as drawn from an electron micrograph.**

of the muscle cell until they are released by the action potential. The free $Ca^{2+}$ has the effect of removing a physical block on the contractile proteins **actin** and **myosin,** which are now able to utilize ATP to slide across each other, leading to a shortening of the muscle fiber in the form of a contraction (Fig. 10.14). Almost immediately, the $Ca^{2+}$ becomes bound again to the ER, and the muscle relaxes. Because every neuron innervates several muscle fibers, a single action potential can cause contraction in a number of muscle fibers simultaneously. The exact number is a reflection of the degree of fineness of control. Small muscles, such as those of the eye or eyelid, may have as few as six muscle fibers innervated by a single neuron, whereas in large muscles such as the thigh and calf muscle, a single neuron may innervate hundreds or even thousands of muscle fibers. Obviously, the number of neurons activated at any one time determines the strength of the muscle contraction as a whole. Thus, we are able to cradle an egg using the same muscles we would use to crush an empty can.

**Shortcomings of the neuromuscular junction**   It is unfortunately the case that the neuromuscular junction is somewhat more susceptible to interference by outside agents than is the typical synapse. Interference with proper transmission can occur in three ways. These are:

*1. Acetylcholine release may be blocked.* This would result in a failure of muscles to contract, hence paralysis, as occurs in **botulism.** Botulinum toxin, which is produced by the bacterium *Clostridium botulinum* (infrequently found in improperly preserved canned goods) interferes with the release of acetylcholine. It is extremely potent, and less than 1/1,000,000 g may be fatal. Recovery is generally good if an antitoxin is administered quickly, although it may be weeks or even months before full use of the muscles is regained, as the toxin is notoriously long-lasting.

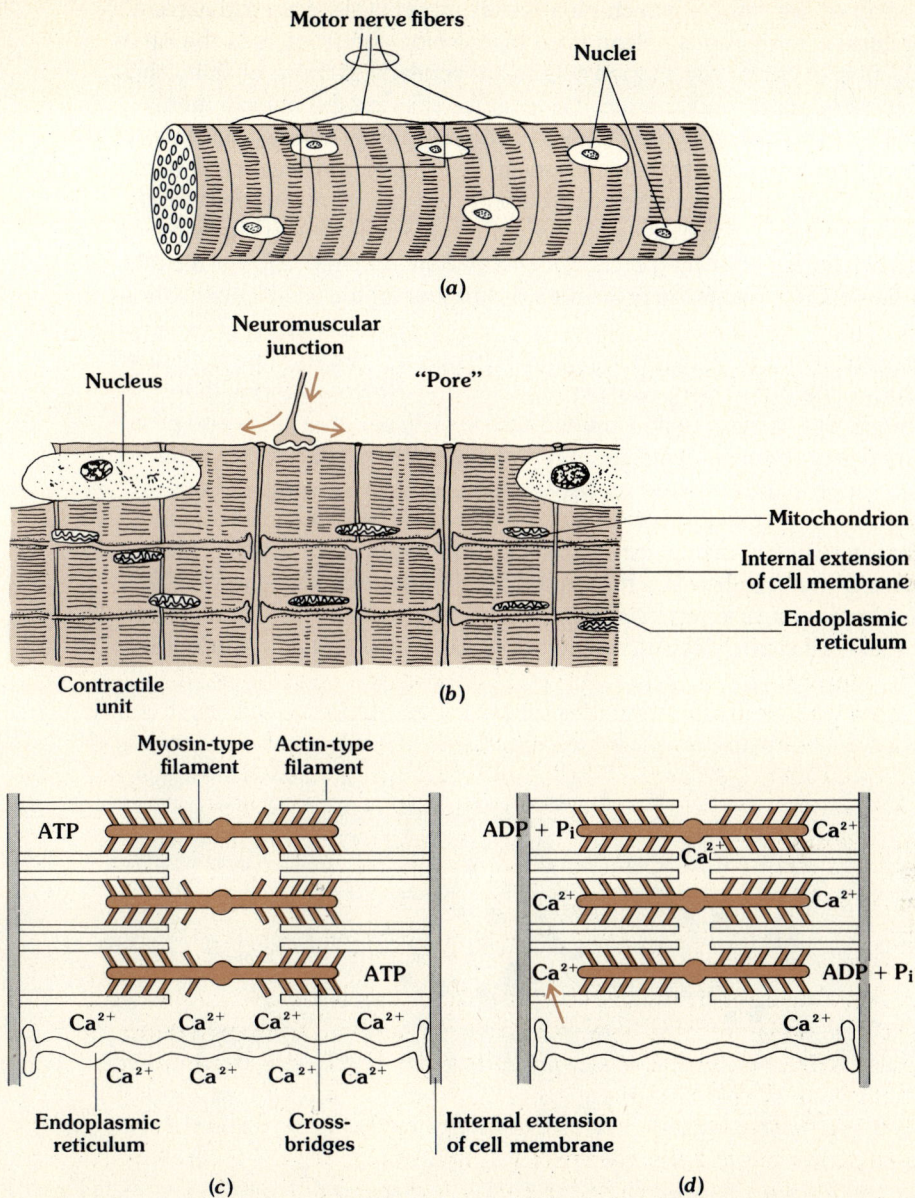

Fig. 10.14  Diagrammatic view of contraction of a muscle cell: (*a*) part of a single muscle cell; (*b*) enlarged region from (*a*)—in thin section; (*c*) contractile unit, relaxed and stretched; (*d*) contractile unit—contracted.

2. *The receptor sites on the muscles membrane may be blocked.* This would prevent stimulation of the muscle, which also results in paralysis. This mechanism is exemplified by **curare,** the famous poison used by the blowgun hunters of the Amazon basin. Molecules of curare are shaped somewhat similarly to those of acetylcholine, and as a consequence, readily enter the receptor sites on the muscle membrane (Fig. 10.15). However, they do not act to trigger muscle cell depolarization, nor are they readily broken down by cholinesterase. In contrast, the acetylcholine molecules released by the neuron are destroyed before they can stimulate the muscle fiber, as the receptor sites on the muscle fiber membrane are filled with curare molecules. Treatment for curare poisoning consists of immediate administration of a cholinesterase inhibitor, which prevents the premature destruction of the acetylcholine molecules.

3. *Cholinesterase activity may be impaired.* This would result in the continued activity of acetylcholine in the cleft, causing repeated muscle contractions that lead to convulsions. This is the situation created by a variety of nerve gases and pesticides. (It is a matter of historical fact that many of our present insecticides, such as **parathion,** were originally developed as nerve gases and, on a dose-per-body-weight basis, are about as toxic to humans as to insects.) These molecules bind with cholinesterase in an essentially irreversible fashion, thereby preventing cholinesterase from splitting the acetylcholine molecules. Continued depolarization of the muscle cell membrane results in repeated contractions of the muscle cell, and the individual quite literally twitches to death. Treatment consists of the immediate administra-

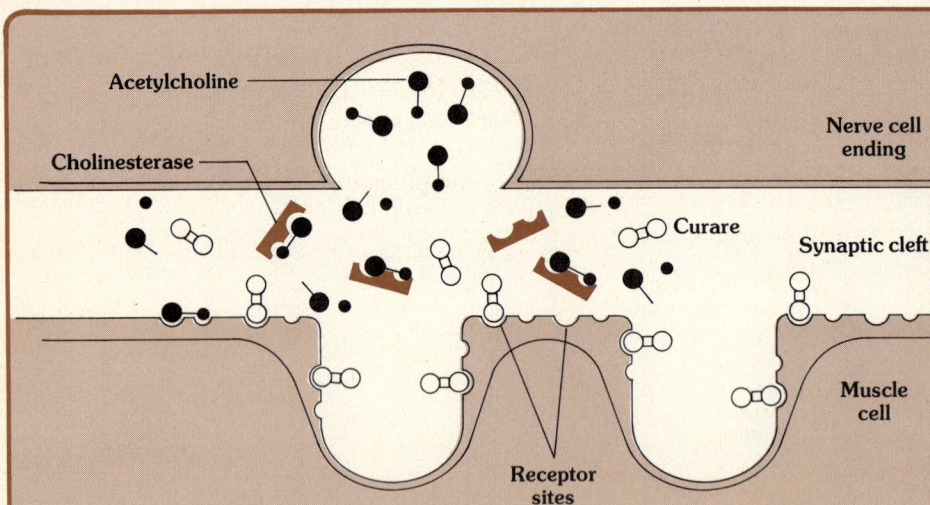

**Fig. 10.15   A schematic view illustrating how curare affects depolarization by competing with acetylcholine for receptor sites.**

A number of the paralytic drugs are extensively used in medicine. **Atropine** has had a long use, although not always as an aid to health. Atropine is the active compound of the plant **deadly nightshade,** so-called because it was a favorite poison of the Middle Ages and early Renaissance. This same plant is also called **belladonna,** because extracts were used at the same time in history by women wishing to be more beautiful ("bella donna"). Drops of the extract were placed in the eyes, causing interference with the acetylcholine receptors of the iris of the eyes, and therefore, a dilation of the pupil (the desired sign of beauty). Modern ophthalmologists use a related compound in examining the eye—increasing pupillary size facilitates examination of the retina.

Atropine is also widely used as an antidote to such poisons as the organophosphate insecticides to counter an immune reaction to bee stings, and in instances in which heart function is threatened, as in snake bite.

**Physostigmine** and **neostigmine,** which are related to the organophosphates, are used in treating such diseases as **myasthenia gravis,** in which the amount of acetylcholine produced by the neurons mysteriously declines. This was the disease that plagued Aristotle Onassis, and that presently affects perhaps 500,000 individuals in the United States. These drugs augment the lowered acetylcholine levels by interfering to a small degree with cholinesterase function, thereby allowing the acetylcholine sufficient time to depolarize the muscle fiber.

Perhaps the most interesting medicinal use of the paralytic drugs was the use of curare as an anesthetic. Only rather recently have the traditional anesthetics chloroform and ether been replaced, despite the fact that both frequently have significant aftereffects, most notably nausea. A number of years ago, curare was tested as a potentially useful anesthetic for certain operations, both because it was considered relatively safe (although some type of artificial respiration had to be used) and because its effects could be almost instantly terminated by the administration of a cholinesterase inhibitor. However, patients experienced an unexpected reaction—they claimed to have been fully conscious during the operation, able to hear everything that was said, and able to feel the scalpel as it cut into their skin, but unable to communicate because of the paralysis produced by the drug. Physicians at first attributed these sensations to hallucinations induced by the drug, but after it was tested by a physician who had the same complaints, the effects of the drug were investigated further. Ultimately, it was realized that the drug interfered only with neuromuscular junctions—the typical neuron–neuron synapses of the sensory system were essentially unaffected, and the patients had indeed been fully conscious and unimpaired in their senses during their operations. Curare is still used in surgery as a muscle relaxant, but only in conjunction with a true anesthetic.

**Paralytic drugs and medicine**

tion of the drug **atropine,** which blocks acetylcholine and artificially recreates the normal situation at the neuromuscular junction. There is a certain irony here in that atropine, a curare mimic, is itself potentially fatal. Therefore, in this instance, one poison is being used to counter another (*see box*).

**The reflex arc**  The basic design of the neurons, including both excitatory and inhibitory impulses, temporal and spatial summation, and hundreds or even thousands of synapses possible with every neuron, allows for enormous complexity. As we shall see shortly, the CNS is even more complex, because **integration** of information from various sense organs typically occurs before a response is sent to a muscle or other effector organ.

The fact remains, however, that we still retain a number of simple **reflexes,** a heritage perhaps of our distant ancestors' less complex nervous systems. Such reflexes may involve as few as two neurons. Why do we retain such a simple arrangement, given the overall complexity of our nervous systems? The answer seems to be that most reflexes must be performed very quickly if they are to have value, and, as we have seen, synapses take time to traverse. Minimizing the number of synapses means reducing the number of neurons involved in a given reflex to the smallest possible number, namely two.

There are a number of reflexes that you can think of for which speed is of the essence. We have eyelashes to serve as detectors of objects coming near the eye. A sliver of glass flying up from a dropped bottle will trigger the blink reflex as soon as the glass touches an eyelash. Obviously, that is not the kind of reaction you want to spend much time thinking about. Similarly, if you put your hand on a hot stove, you pull it away automatically, in a withdraw reflex. There is simply no reason to monitor this response by other senses—theoretically, you could wait until your sense of smell confirmed that your hand was, indeed, on fire, but you are clearly much better off relying exclusively on the pain receptors in your fingertips, as the damage to your hand will be minimized by doing so.

One of the best-known reflexes is the famous knee-jerk reflex (Fig. 10.16), which is generally tested by your physician when you have a physical examination. Typically, this test involves your sitting in a relaxed fashion with your legs dangling over the edge of the examination table. The physician then strikes your leg just below the knee with a little rubber hammer, and your leg straightens to some degree. What is the mechanism whereby the leg is straightened?

You may accept it on faith that the knee-jerk reflex has an adaptive role and that it does not exist simply for the physician's convenience. The action of the hammer is to stretch the tendon running from the large muscles on the front of the thigh, through the kneecap, to the front of the shin. If you think about it a moment, you will realize that this stretching would also occur if your leg were to buckle suddenly and your knee were to bend. The reflex reaction in both instances involves the response of a sensory neuron inside the tendon to this sudden stretch, by generating an action potential. A synapse occurs within a motor neuron in the lower regions of the spinal cord, and a message is returned to the thigh muscles, causing them to contract. The effect of this contraction is to cause the leg to

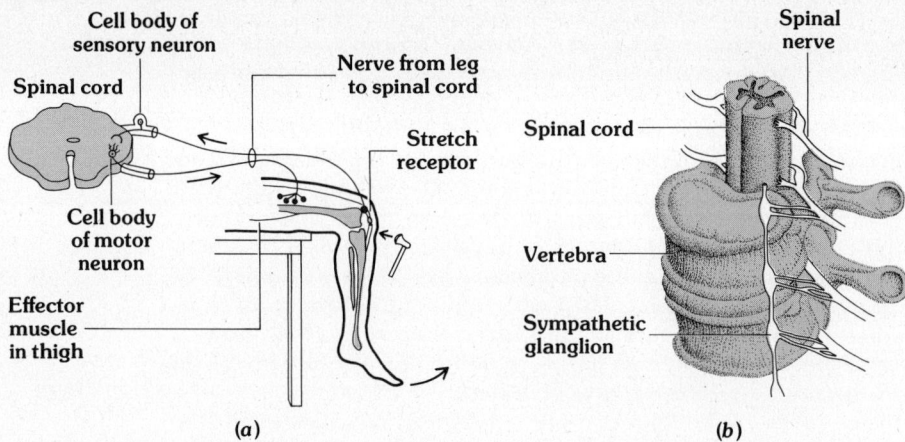

**Fig. 10.16** (*a*) Diagrammatic view of the knee-jerk reflex; (*b*) a segment of the human spinal cord showing the arrangement of the spinal nerves and some of the ganglia associated with the sympathetic nervous system.

straighten at the knee. This is manifested by a weak kick in the rather artificial circumstances of the doctor's office, but this same reaction would prevent your collapsing if it occurred should your leg buckle suddenly as you were standing. More frequently, this reflex is at work as you are running, although it takes awhile to master, as any 10-month-old infant practicing how to walk can attest.

But, you may protest, I can willfully prevent my leg from straightening; I can keep my hand on the stove; I can stop myself from blinking when something touches my eyelash. How does all that fit in with the idea of a simple two-neuron reflex? Two other processes are also at work. First, the brain is ultimately "informed" of the factors surrounding the reflex and of the fact that the reflex occurred. We are certainly aware that we have just placed our hand on a hot stove, for example, because the pain receptors in our hand continue to fire long after the withdrawal reflex has acted to pull the hand away. Moreover, we are conscious of having reacted by pulling our hand away—other neurons are activated in the spinal cord at the synapse between the sensory and motor fibers, but the point is that these other fibers are not necessary to the action of the reflex itself. Second, we are capable of preventing a reflex only when we know what is coming next. You can resist the urge to blink (perhaps!) if someone tells you he is going to brush your eyelashes, but you cannot do so if you are surprised. You can resist pulling your hand away from the hot stove only if you know before you touch it that it is hot. You can prevent your leg from straightening as the doctor hits it with his hammer only if you know he is about to do so. In each instance, the reflex is prevented by the contraction of muscles antagonistic to those which are stimulated by the reflex itself. None of this

The very complexity of the nervous system that allows us to communicate abstract thoughts using arbitrary symbols (letters and words) sets up a situation in which errors in the "wiring," so to speak, are often beyond our capabilities to comprehend fully and even less to remedy. It is not surprising, therefore, that many of the diseases that continue to plague us are, in fact, diseases of the CNS. These include:

*Polio* (poliomyelitis)—a disease of the gray area of the spinal cord and brain, caused by a virus that destroys nerve cells. Depending on the location and the number of cells destroyed, the effects of the disease range from essentially none, to varying degrees of paralysis, and even to death. Because nerve cells cannot divide, the paralysis is generally irreversible. Effective vaccines have existed since 1954 which give total immunity to the disease, and it is now very rare in the United States. In 1952, however, there were almost 60,000 new cases, which gives you an idea of the magnitude of the problem in the prevaccine days.

*Multiple sclerosis*—actually a complex of related diseases in which hard nodules develop in the white area of the spinal cord and disrupt the transmission of nerve impulses up and down the cord. The disease is generally progressive and ultimately fatal, the rate of worsening being a function of the type of sclerosis involved. Some types are fatal within a few years; others allow more normal functioning for many years. The cause of the disease is unknown, although theories abound.

*Cerebral palsy*—involves the destruction of portions of the motor centers of the cerebrum, typically as the result of an injury. Frequently, the condition is present at birth, although head injuries at any age may lead to cerebral palsy. Because it is not caused by a virus or other pathogen, cerebral palsy cannot be totally prevented (as can polio for instance). Moreover, little can be done to improve the condition, once it is present. Cerebral palsy is manifested by varying levels of impairment of motor functions, including speech. There is no interference with intelligence, however, contrary to what many uninformed people may think.

**Diseases of the central nervous system**

takes away from the nature of the reflex, but rather indicates the level of control which our brains can achieve over such reflexes. Such control is a peculiarily mammalian trait—we train our children and our dogs and cats to resist the urinary reflex until they are in an appropriate situation, but try to train your pet turtle. Reflexes are much more a part of the behavioral repertoire of lower vertebrates than they are in ourselves, but it should be obvious that even in ourselves, these simplest of neuronal intergrations are of vital importance.

**THE SPINAL CORD AND THE PERIPHERAL NERVOUS SYSTEM**  In cross section, the spinal cord has the appearance of a gray butterfly on a field of white (Fig. 10.16*b*). The gray matter represents unmyelinated cell bodies; the white matter, the myelinated axons running along the spinal cord.

Between every pair of vertebrae, a pair of **spinal nerves** emerges from the spinal cord, one nerve on each side. Another 12 nerves called **cranial nerves,** emerge from the brain itself. Each nerve consists of hundreds of individual neurons, which are functionally of four types. These are:

1 Somatic sensory—carrying impulses from the peripheral sense organs (e.g., touch receptors in the fingertips) back to the central nervous system.
2 Somatic motor—carrying impulses from the central nervous system to skeletal muscles, and other effector organs, such as glands.
3 Visceral sensory—carrying impulses from the various internal organs of the body to the CNS.
4 Visceral motor—carrying impulses from the CNS back to the internal organs.

It must be reiterated that these are *functional* categories; the neurons themselves appear structurally very similar. Every spinal nerve (but not every cranial nerve) carries all four types.

**THE AUTONOMIC NERVOUS SYSTEM** The **autonomic nervous system** (ANS) consists of the visceral motor and visceral sensory fibers. At first glance, it would seem that this system would be nothing more than the complement to the somatic nervous system, differing only in that the ANS fibers are responsible for regulating visceral events over which we have no conscious control. However, this conclusion would be incorrect for two reasons. First, as we shall see shortly, we do have some degree of conscious control over the ANS. Second, the visceral organs are not designed in the typical antagonistic pair arrangement of skeletal muscles. It is easy to see how an arm would bend if the biceps muscle received a stimulus from its nerve or how the arm would straighten if the triceps muscle were stimulated. However, what happens when the intestine is stimulated by its nerve? Does peristaltic activity increase or decrease? Does glandular activity increase or decrease? Does absorption increase or decrease? The situation is clearly not analogous with the contraction of a skeletal muscle.

As it happens, each visceral organ is innervated by two sets of motor neurons (and two sets of sensory neurons as well), one of which increases the activity of the organ, and the other of which decreases organ activity (Fig. 10.17 and Table 10.1). This dual system allows more fine-tuning of organ activity than would be possible if activity were purely a function of the amount of intensity of stimulation from a single nerve.

Each of these sets of neurons is given a separate name. The set that is primarily responsible for activating the visceral organs is the **parasympathetic nervous system.** The set primarily responsible for inhibiting the visceral organs is the **sympathetic nervous system.** Each system contains both sensory and motor fibers; thus, each visceral organ actually has four types of fibers innervating it, two of which are sensory and two of which are motor. The sensory fibers are very small and not of great interest to us here, but the motor fibers are of particular interest because they are responsible for turning a given visceral organ on or off.

How does a given visceral organ "know" whether it is being stimulated by a sympathetic or a parasympathetic fiber? All it receives is an action potential—how can

**Table 10.1**

Functions of the autonomic nervous system

| Tissue | Location | Parasympathetic functions | Sympathetic function |
|--------|----------|---------------------------|----------------------|
| Smooth muscle | Iris of eye | Constricts pupil | Dilates pupil |
| | Stomach wall | Increased motility | Decreased motility |
| | Intestinal wall | Increased motility | Decreased motility |
| | Anal sphincter | Closes | Opens |
| | Bladder sphincter | Closes | Opens |
| | Bronchioles of lungs | Constriction | Dilatation |
| | Hair follicle | None known | Contraction |
| Gland | Eye (lacrimal) | Secretion | None known |
| | Mouth (salivary) | Abundant watery saliva secreted | A small amount of viscous saliva secreted |
| | Stomach (gastric) | Secretion | Inhibition of secretion |
| | Liver | Diminished glycogen breakdown | Increased glycogen breakdown |
| | Pancreas | Increase in enzyme secretion | No effect |
| | Adrenal medulla | None known | Secretion |
| | Skin (sweat) | None known | Secretion |
| Blood vessel | Body (arteries) | Dilatation | Constriction |
| | Heart (arteries) | Constriction | Dilatation |
| | External genitalia | Dilatation (erection) | Constriction (ejaculation) |
| | Skin | None known | Constriction |
| Heart muscle | Heart | Decelerates rate | Accelerates rate |

these systems be distinguished? The answer lies in the neurotransmitters that are used at the point where the axon contacts the visceral organ. All the motor fibers of the autonomic nervous system, whether sympathetic or parasympathetic, have a synapse in a **ganglion** (a cluster of nerve cell bodies) at some distance from the spinal cord. (The one exception to this statement are the fibers innervating the adrenal medulla, of which we shall say more in a moment.) The neurotransmitter at these ganglionic synapses is, without exception, acetylcholine. However, the fibers leaving this synapse (the so-called **postganglionic fibers**) of the sympathetic system do *not* use acetylcholine at the point of contact with the visceral organ, but instead use **norepinephrine,** whereas the postganglionic parasympathetic fibers again use acetylcholine. Therefore, in answer to our original question, visceral organs can distinguish between stimulation from the parasympathetic system on the basis of the neurotransmitter being used and are thereby capable of responding differently to each.

The exception to all this are the sympathetic fibers that innervate the adrenal medulla, which do not synapse in a peripheral ganglion, but rather pass directly from the spinal cord to the adrenal medulla without interruption. It is interesting to note that the cells of the adrenal medulla, along with postganglionic sympathetic fibers, also produce norepinephrine (as well as larger quantities of the very closely related **epinephrine,** also known as **adrenaline**). Thus, it is assumed that the cells of the adrenal medulla are, in fact, nothing more than modified postganglionic sympathetic fibers. Certainly the functioning of the sympathetic nervous system and the adrenal medulla broadly overlap as we shall see shortly.

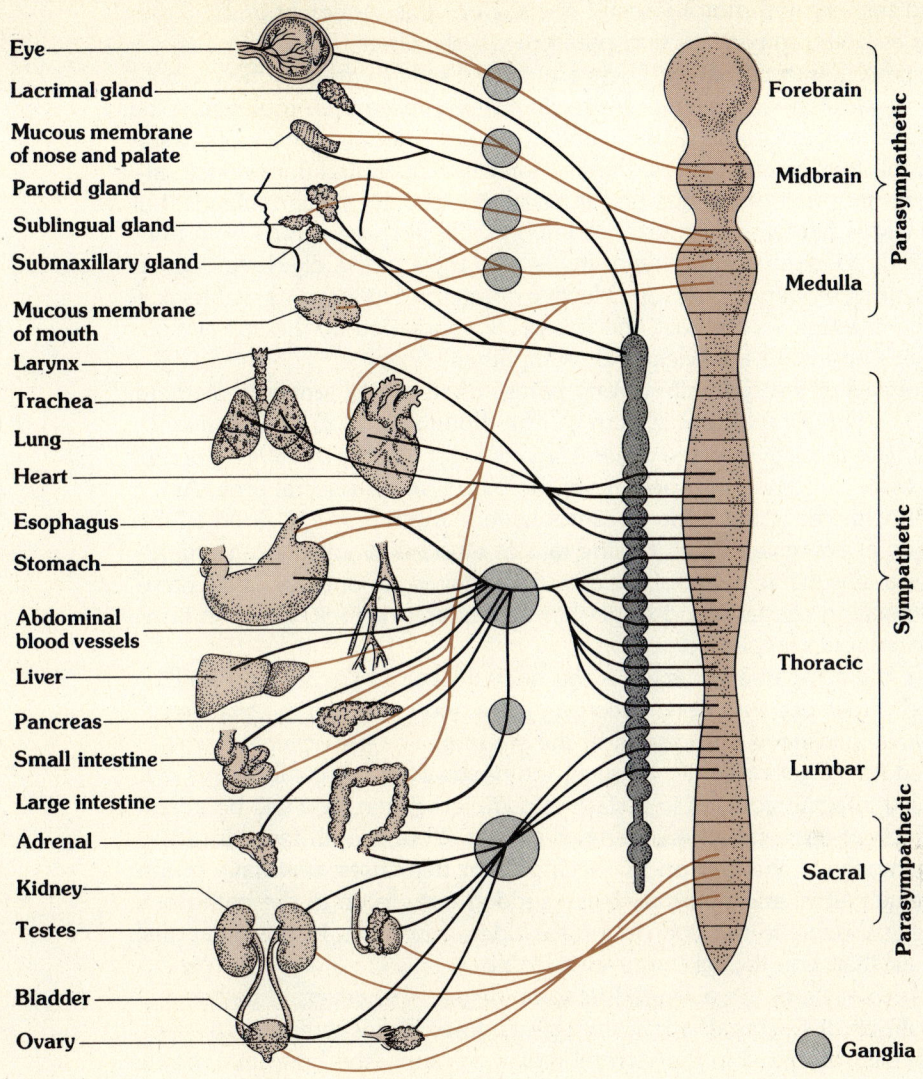

**Fig. 10.17   The autonomic nervous system.**

Eye
Lacrimal gland
Mucous membrane of nose and palate
Parotid gland
Sublingual gland
Submaxillary gland
Mucous membrane of mouth
Larynx
Trachea
Lung
Heart
Esophagus
Stomach
Abdominal blood vessels
Liver
Pancreas
Small intestine
Large intestine
Adrenal
Kidney
Testes
Bladder
Ovary

Forebrain
Midbrain
Medulla
Parasympathetic

Thoracic
Lumbar
Sympathetic

Sacral
Parasympathetic

Ganglia

It was mentioned earlier that every spinal nerve carries sensory and motor somatic fibers and sensory and motor visceral fibers. Does this statement imply that each nerve carries both sympathetic *and* parasympathetic fibers? As it happens, such is not the case. Parasympathetic fibers are carried only by four of the cranial nerves (which, as we discussed earlier, are really nothing more than modified spinal nerves that happen to come from the brain) and by three nerves at the base of the spinal column. Of the four cranial nerves, the first three are of limited importance, affecting the size of the pupil of the eye and the secretions of the salivary glands, but the fourth, the **vagus nerve,** is of enormous importance, because it runs from the brain into the body to innervate the heart, lungs, and most of the digestive organs. The three spinal nerves innervate the reproductive organs and kidneys, as well as portions of the large intestine. In contrast, the visceral fibers carried by the spinal nerves of the neck and chest are exclusively sympathetic fibers.

The functionings of the sympathetic and parasympathetic systems are diametrically opposed. The parasympathetic system can be thought of as the vegetative system—it stimulates the activity of the visceral organs, most notably by causing expansion of blood vessels, resulting in a pooling of the blood in the visceral area. As was discussed in Chapter 5, the body does not have enough blood to flood all the areas of the body at any given time. Yet the rate of blood flow through an organ or area provides a reasonable index of the activity of that organ or area. Thus, parasympathetic stimulation causes blood to pool in the visceral regions, which in turn permits more rapid and complete digestion.

By contrast, the sympathetic system is the body activator. The visceral and peripheral blood vessels are shut down, allowing the blood to pool in the muscles instead. Heartbeat is quickened and certain of the sweat glands are activated. The body is prepared for "flight or fight"; as such, the needs of the visceral organs are overridden. If you imagine yourself in a state of complete panic, you will be able to visualize the effects of the sympathetic nervous system. Your mouth feels dry (saliva production impaired), the hair on the back of your neck rises, you have goosebumps and sweaty palms, and you may feel a sudden urge to go to the bathroom. If someone were suddenly to tap you on the shoulder, you might leap a great deal higher into the air than you thought yourself capable.

Normally, both systems are activated to some degree. However, in certain cases, such as **shock,** the sympathetic system shuts down almost entirely, causing an enormous pooling of blood in the visceral region. As such, the brain may be deprived of blood, and the individual may become unconscious or even die. For this reason, in the treatment of accident victims in whom shock is a possibility, their legs should be raised to prevent blood from pooling in the legs at the expense of the head.

**The ANS and the adrenal medulla**   The adrenal medulla produces large quantities of the hormone **epinephrine** (adrenaline) which is chemically very similar to the **norepinephrine** produced by the postganglionic sympathetic fibers, although the two do function somewhat differently. Why this apparent overlap? Moreover, there is no organ equivalent to the adrenal medulla which acts as a counterpart for the parasympathetic system—why not?

The answer seems to be that the sympathetic system functions on an immediate basis, whereas the adrenal gland functions over a period of hours or even days. Suppose you had no sympathetic system and had to rely exclusively on the adrenal medulla in order to maximize blood flow to the muscles. Suppose, too, that you were suddenly accosted by a knife-wielding maniac. It would require a minimum of 60 sec before the first of the epinephrine reached all parts of the body, by which time you might be somewhat the worse for wear. In actuality, your sense organs convey information regarding this dangerous intruder to your brain and spinal cord, which causes activation of the motor fibers of the sympathetic nervous system within 1 sec. Thus, the sympathetic nervous system and the adrenal medulla operate in concert to ensure the capacity not only for a virtually immediate response, but also for a prolonged response.

**Conscious control of the ANS**  It was stated earlier that the ANS was that portion of the nervous system that is not subject to conscious control. This statement is not entirely true. In recent years, it has been found that many people (perhaps all) have the capacity to affect ANS activity through conscious control. This finding has been put to good use in such disorders as **hypertension** (high blood pressure). Individuals with hypertension can be trained to lower their blood pressure or rate of heart beat consciously, or to perform a host of other activities long thought beyond the reach of the conscious brain. This finding would seem to demonstrate, once again, how much our brains have evolved over those of our vertebrate ancestors, in which autonomic activity is totally autonomous. Conscious control of the ANS is, of course, well known to Indian yogi who practiced such methods for generations before Western medicine was willing to admit that such things were even possible.

**THE BRAIN**  What we know as the brain is actually just the highly modified front portion of the spinal cord. Indeed, the point at which the spinal cord ends and the brain begins is designated, in a totally arbitrary fashion, as the point of entry into the skull—that is, the skeletal system is used as a reference point. Within the nerve tracts that make up the spinal cord and brain, however, there is no dramatic change in appearance at the point of transition.

The human brain can be thought of as consisting of three principal portions (Fig. 10.18). These are:

**1. The brainstem**  This is the region in the floor of the brain with which the spinal cord connects. The most rearward portion of the brainstem is the **medulla oblongata,** from which emanate 8 of the 12 cranial nerves, including the all-important vagus nerve. As the portion of the brain responsible for monitoring such vital functions as breathing and heart rate, it is easy to see that damage to the medulla can be fatal. The reason why hanging is such an efficient mode of execution is that, in hanging, the neck is broken and the medulla or the immediately adjacent portion of the spinal cord is either severed or seriously damaged. The individual actually dies of suffocation, as his breathing is brought to a standstill. For the same reason, an individual with a neck broken in an accident must be handled with extreme care because the chance of recovery is good if the medulla or spinal cord are not damaged in moving the person to the hospital where his head can be immobilized.

Immediately in front of the medulla is the **pons,** which serves not only to initiate certain facial movements and to receive input from the ears, but also to transfer information between the medulla and the parts of the brain lying more anteriorly.

In front of the pons is the **thalamus,** along the sides of the brainstem, and the **hypothalamus,** in the floor of the brainstem. The thalamus in humans has become very important as the site receiving stimuli from the eyes via the optic nerves. In turn, this information is relayed to the higher centers of the brain for further pro-

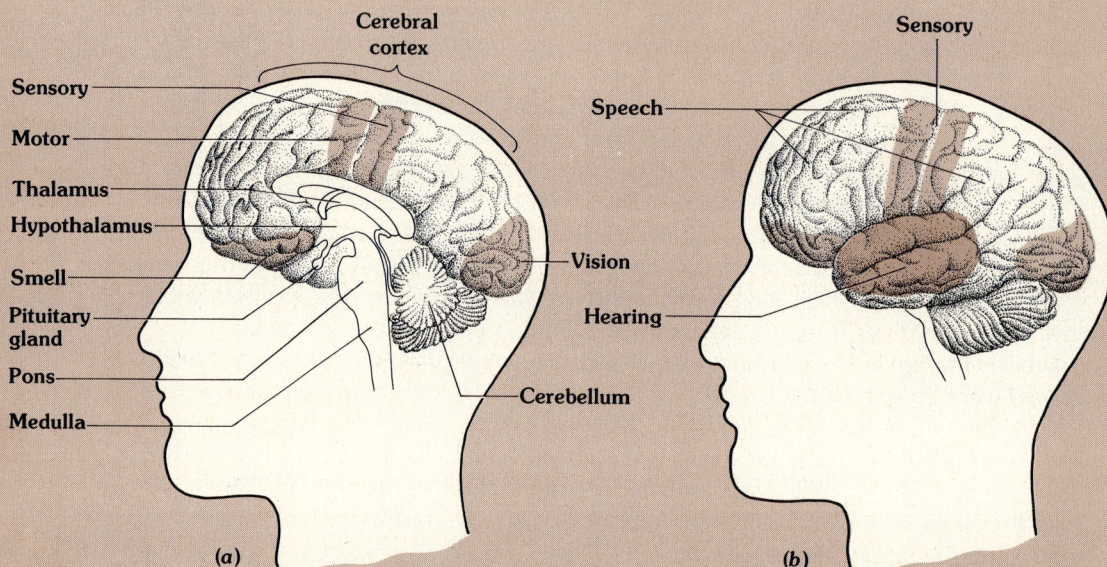

Fig. 10.18  Major regions of the human brain. (a) Median section; (b) outer view.

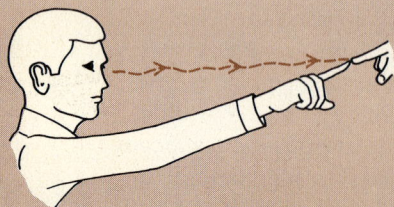

Fig. 10.19
Tremulous movement in an individual with a damaged cerebellum.

The brain and spinal cord are very soft and delicate structures. What prevents them from injury when we fall or bump our head? Certainly there are no muscles to cushion the impact, as we saw at work in Chapter 3. The answer is that the brain and spinal cord are floating in a pool of **cerebrospinal fluid,** and the fluid acts to prevent them from direct contact with the skull and spinal cord. (A **concussion** results from a blow to the head too severe to be completely cushioned by the cerebrospinal fluid.) A set of membranes surrounds the CNS and contains the fluid. These membranes are collectively known as **meninges.** An infection of these membranes (**meningitis**) is a life-threatening event and must be treated rapidly to prevent damage to the CNS itself.

Cerebrospinal fluid

The cerebrospinal fluid is secreted by specialized tissues located at the top of the brainstem. It is derived from blood, and resembles plasma in that it is free from cells, but differs from plasma in its ionic concentration. Injury to the brain or spinal cord frequently may be confirmed by a **spinal tap**—a drawing off of a small amount of cerebrospinal fluid by a needle inserted between the vertebrae. The presence of blood cells in the fluid indicates a rupture at some point in the CNS, a finding that may indicate surgery.

cessing. The hypothalamus is an extremely important area of the brain because it governs such basic behavior as hunger, thirst, and sex, emotions such as fear and anger, and other functions including control of body temperature. Furthermore, the hypothalamus is also the area of the brain that is responsible for the secretion of hormones liberated by the posterior pituitary and for controlling the production of hormones by the anterior pituitary (see Chapter 9).

In primitive vertebrates, the brain is composed almost exclusively of the brainstem. However, in mammals, and especially in ourselves, two other areas of the brain are of extreme importance.

**2. The cerebellum**   This is the baseball-sized portion of the brain immediately above the pons, which is responsible for the "fine tuning" of muscle movement. If you have ever been asked by your physician during the course of a physical examination to close your eyes, stretch your arms out to the side, and then touch your fingertips together, you are being checked to see if your cerebellum is functioning properly. Even with a damaged cerebellum, you would still be able to complete this task, but you would do so hesitatingly, slowly, and very unevenly (Fig. 10.19). Mammals owe their great ability of fast, coordinated movements to the computerlike integrating capacities of the cerebellum.

**3. The cerebrum**   The cerebrum consists of two major lobes lying on the left and right sides. These are so large in humans that they totally obscure the other portions of the brain when the brain is viewed from above. These lobes are, of course, the areas concerned with abstract thought, but as we have seen repeatedly, they have come also to influence, if not control, many other body functions.

Specific portions of the cerebrum are, in some instances, concerned with specific functions—therefore, damage to a given area may result in blindness, deafness, inability to speak, paralysis on one side, and so on (Fig. 10.20). Other areas of the cerebrum seem to have less specific functions. For example, memory is far less impaired by damage to a specific area of the cerebrum than are the other functions just mentioned.

The cerebrum is covered by a **cortex,** which corresponds to the central portions of the spinal cord in that it is gray in color owing to the presence of nerve cell bodies. Interior to it are the white, myelinated axons of these fibers. Hence, in the cerebrum, the relationship of white to gray has been reversed with respect to the spinal cord. The presumptive reason for this shift is that positioning the cell bodies on the periphery allows for the presence of more cell bodies. So, too, does the folding so characteristic of the human brain, another device for expanding surface area while holding volume stable (just as in the lungs, gut, and so forth). It is the total area of the cerebral cortex which allows the behavioral complexity that serves, more than any other single factor, to distinguish us from other vertebrate species.

**Drugs and the human brain** There are a great many drugs which affect the CNS, most of which are pharmaceutical agents of enormous importance in the

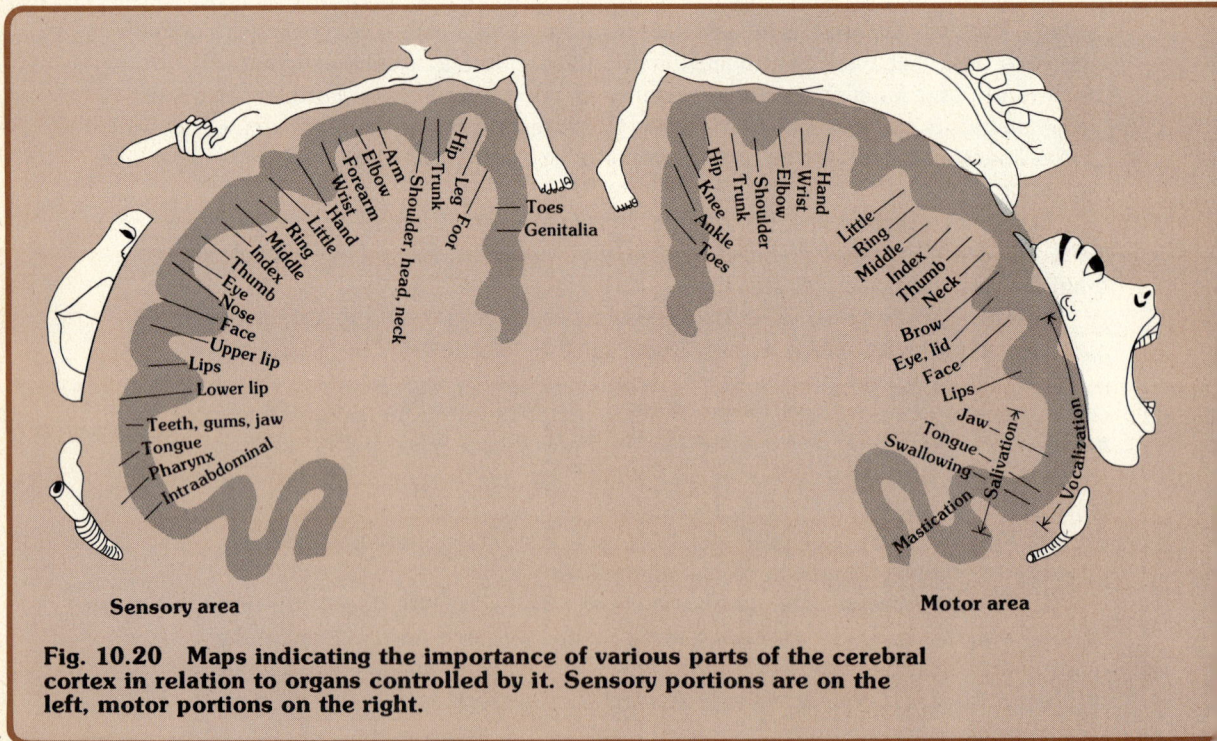

Sensory area

Motor area

**Fig. 10.20  Maps indicating the importance of various parts of the cerebral cortex in relation to organs controlled by it. Sensory portions are on the left, motor portions on the right.**

practice of modern medicine. Ironically, the precise functioning of these drugs is very often incompletely known. Precisely how aspirin functions to relieve pain, for example, is uncertain.

Our concern here, however, is primarily with the drugs of abuse. Such an exposition is not intended as a primer on drug safety, but rather as a set of presumably relevant examples of what we know—and don't know—about how drugs act on our brain and nervous system.

Categorizing drugs is rather difficult, as there are a number of gray areas. Allowing for a certain amount of fuzziness between groupings, however, we can distinguish five major categories relatively easily (Table 10.2). These are: (1) stimulants; (2) depressants; (3) tranquilizers; (4) hallucinogens; and (5) narcotics.

**Stimulants** Stimulants include such common "drugs" as caffeine and nicotine, and such esoteric substances as amphetamines.

*Caffeine* is abundant in coffee, and somewhat less abundant in tea and various cola drinks. It apparently functions by enhancing synaptic transmission in the CNS. Many physicians regard it as a more powerful (hence, potentially a more dangerous) drug than many prescription drugs. Its very ubiquitousness causes most of us to think of it as innocuous, but the fact that people with ulcers, heart disease, or kidney problems are told to eliminate it from their diets suggests its relative danger. Moreover, as is the case with most of the stimulants, the body rather quickly adjusts to its presence, forcing the individual to ingest more to achieve the same effect. Correspondingly, the sudden reduction of caffeine intake has the effect of acting as a depressant, as most inveterate coffee-drinkers can attest, if they have to forego the morning cup.

*Nicotine,* of course, is a primary component of tobacco. Its potential is perhaps best exemplified by noting that nicotine was widely used as an insecticide in the latter part of the last century, before synthetic insecticides became available (and no, you cannot give yourself immunity to mosquitos by smoking so many cigarettes that an insecticide level dose of nicotine is built up in your blood). It operates within ourselves to stimulate the sympathetic nervous system. However, like caffeine, tolerance occurs quickly such that stimulation passes from a "lift" above normal to a lift only from the trough of depression that occurs during periods when the nicotine level in the blood drops.

*Amphetamines* include the so-called "uppers," most notably "speed" (methamphetamine). Chemically, the amphetamines resemble epinephrine, and they evidently act directly on the brain to cause the release of excess neurotransmitters. They are sometimes used in weight reduction, as they act as appetite suppressants. The individual user may feel euphoric at first, but this soon passes, only to lead to a letdown. Continued use leads to hallucinations (perhaps more from lack of sleep than from the drug itself), as well as to the possibility of amphetamine psychosis, which is literally indistinguishable from paranoid schizophrenia. Recent studies also suggest that strokes are a common result of prolonged amphetamine use. Once again, tolerance develops quickly, necessitating larger doses to achieve the desired euphoric feeling.

# Table 10.2
## Drugs and the brain

| Name | Common names and varieties | Sources | Immediate effects | Long term effects | Possibility of addiction Physical* | Psychological** | Tolerance*** |
|---|---|---|---|---|---|---|---|
| **Stimulants** (Drugs that act directly upon the brain, causing increased activity) | | | | | | | |
| Caffeine | Coffee, tea, cola, chocolate | Coffee, tea, cola, chocolate | Mild stimulant | None except restlessness, inability to sleep | No | Yes | Yes |
| Nicotine | Tobacco | Tobacco | Mild stimulant | Lung and heart, circulatory disease, cancer | No | Yes | Yes |
| Cocaine | Snow | From *Erythroxylon coca*, "divine plant" of the Incas | Stimulation, excitability, sense of euphoria | Overdose may depress heart and respiration, causing death | No | Yes | No |
| Amphetamines, including: Methaphetamine Dexidrine Benzedrine | Uppers Speed Dexies Bennies | Synthetic | Dry mouth, heart pounding, stimulation | Restlessness, poor sleeping habits. Prolonged use can result in paranoia and hostility, strokes. | No | Yes | Yes |
| **Depressants** (Drugs acting oppositely to stimulants, reducing activity) | | | | | | | |
| Alcohol Ethyl alcohol | Alcohol | Yeast fermentation, distillation | Relaxation, relief of inhibitions, euphoria | Delirium tremens, nutritional deficiencies | Yes | Yes | Yes |
| Methyl alcohol | Wood alcohol | Synthetic | Similar to ethyl alcohol | *Extremely poisonous.* Even small intake can result in blindness or death. | — | — | No |
| Barbiturates | Downers | Synthetic | Similar to alcohol | Very addicting. Withdrawal can cause emotional instability. | Yes | Yes | Yes |
| Seconal (secobarbital) Amobarbital | Red devils Blue devils | Synthetic | | | | | |

In combination amphetamines and barbiturates can produce euphoria; and in combination alcohol and barbiturates increase the chance of respiratory failure.

## Tranquilizers
(Drugs that reduce anxiety without depressant action)

## Hallucinogens
(Produce distortions of perception)

## Narcotics
(Term applied medically to opium and opium-like depressants all of which produce insensibility. Legally also applies to cocaine, a stimulant.)

| Drug | Common/Slang Names | Source | Effects | Physical addiction* | Psychological addiction** | Tolerance*** |
|---|---|---|---|---|---|---|
| **Tranquilizers** | | | | | | |
| Reserpine | | Synthetic | | No | Slight | No |
| Phenothiazine | Thorazine | Synthetic | | No | Slight | No |
| Meprobamate | Miltown, Equanil | Synthetic | Withdrawal can cause convulsions | Yes | Yes | Yes |
| **Hallucinogens** | | | | | | |
| Marijuana | Pot, grass, Mary Jane | From hemp, Cannibis | Stimulation, euphoria, loosens inhibitions. Effects strongly influenced by environment. | No | Yes | No |
| Hashish | Hash | Concentrated resin from Cannibis | | | | |
| Mescaline | Peyote | From cactus, Anhalohium lewtii; also synthetic | Physical effects uncertain. Tendency to become pre-occupied with drug may reduce useful activities. | | | |
| Psilocybin | | From mushroom Psilocybe mexicana | Both these have long history of religious use by North and Central American Indians. | | | |
| Lysergic Acid Diethylamide | LSD | Synthetic | Markedly greater hallucinogenic effect than marijuana. Strong visual effects especially with LSD. Ordinarily no long term effects although use can worsen an already existing mental disorder. LSD induced suicidal and other dangerous behavior common. Some evidence of chromosomal damage with LSD. | No | Slight | Yes |
| Dimethyltryptamine | DMT | Synthetic, also naturally in S. American plants | | | | |
| **Narcotics** | | | | | | |
| Opium | | From opium poppy, Papaver somniferum | Pupils constrict, reduced sensitivity, reduced tension—euphoria. | Yes | Yes | Yes |
| Morphine / Heroin | Horse | Opium derivatives | Severe withdrawal effects. Anxiety, aches in back and legs, vomiting. Overdose easily lethal and relatively common because of variation in purity of illegally sold narcotics. | | | |

\* Physical addiction—Continued use of drug causes physiological changes such that its discontinuation results in illness.

\** Psychological addiction—Continued use of drug causes psychological discomfort upon discontinuation, but without organic illness.

\*** Tolerance—Progressively stronger doses required to produce initial effect.

**Depressants**   Depressants do not impart a feeling of being depressed, but create a depression of CNS activity, which is quite a different thing. The two principal substances to be considered are alcohol and the barbiturates.

*Alcohol,* of course, is an extremely widely used drug and has been for thousands of years. Ironically, the sought-after feeling of stimulation is brought about by the depressing of central inhibitions. The true depressant effects of the drug are obvious from the rapid impairment in motor skills and, at higher doses, from the impairment of such basic functions as breathing, the usual cause of death in overdoses.

Alcohol is largely broken down by the liver (about 5 percent is also excreted by the lungs, that fact being the basis for breathalyzer tests). Ingesting a larger amount than can be immediately broken down by the liver leads to a temporary rise in blood alcohol levels, which is generally the objective of the drinker. Detoxification by the liver, however, is not accomplished without cost, because liver cells are destroyed in the process. Prolonged destruction of liver cells leads to cirrhosis of the liver (see Chapter 4). Other effects of alcohol may include irreversible brain damage, epilepsy, psychoses, and sexual impotence. Moreover, there is a strong physical addiction, and withdrawal is severe.

*Barbiturates* include many of the prescription sleeping pills. They are also sometimes used to treat tension, but tend to be too short-acting and too sleep-inducing to be of much value for that purpose. The barbiturates act by interfering with CNS neurotransmitters. The effects may be so long-lasting that a person may need amphetamines in order to feel wakeful in the morning—and then require barbiturates again to go to sleep the next night.

Barbiturates have the added handicap of making an individual forgetful, tending to increase the chances for an accidental overdose. In accompaniment with alcohol, the barbiturates may be fatal, for both operate as depressants. Most of the movie and rock stars who have died of drug overdoses (presumably accidentally) reacted to a lethal combination of barbiturates and alcohol, which led to a cessation of breathing.

Barbiturates also lead to a muting of mood control, and individuals who take them may suddenly become violent. Ironically, rapid withdrawal may be more dangerous than addiction, because as the CNS strives to regain normal levels of activity, it may overcompensate, leading to convulsions and sometimes to death.

**Tranquilizers**   The tranquilizers, which are typically described as acting to reduce anxiety and tension without the general depressant activity of the barbiturates, are actually a very diverse group. Generally, they are divided into two groups—the **major** tranquilizers, which act on the brain itself and, in a variety of ways, enhance the action of the parasympathetic nervous system, and the **minor** tranquilizers, which appear to operate largely on the peripheral nerves. The major tranquilizers have been of enormous use since their introduction 20 years ago in treating psychiatric patients, and more than any other single factor, have led to a near-elimination of the padded cells and straightjackets of the asylums of old. No physical dependence or tolerance develops to these drugs, and they tend not to be drugs of abuse.

The minor tranquilizers are quite another story. Despite their name, they can lead to physical dependence, tolerance does develop rather quickly, and they are frequently drugs of abuse (although not in the same context as the narcotics, for example). Moreover, because those addicted to the minor tranquilizers may undergo convulsions upon withdrawal, some of these tranquilizers (such as **meprobamate**) are sometimes classified as barbiturates.

**Hallucinogens**   The hallucinogens include a variety of natural plant products as well as such synthetics as LSD. The least potent is **marijuana,** and the most potent (of those commonly used) is LSD. Not a great deal is known of the physiology of any of the hallucinogens, although it appears that marijuana stimulates epinephrine production in the brain, leading to a mild euphoria. The equally mild hallucinogenic effects have not yet been explained. **Mescaline** (from cactus) and **psilocybin** (from mushrooms), as well as LSD, chemically resemble a brain neurotransmitter called **serotonin,** although, again, their sites of action are unknown. LSD is about 200 times as potent as psilocybin and 5000 times as potent as mescaline, an aspirin-sized sample of LSD being sufficient for about 3700 "trips."

Various side effects have been attributed to the hallucinogens. Spontaneous hallucinating and schizophrenia have occurred in individuals who use LSD repeatedly, and there is some evidence to indicate chromosomal damage in dedicated users. Side effects to marijuana use are less clear, and the issue is controversial, to say the least. One recent report suggests a reduction in testosterone levels and sperm counts as well as the development of certain female secondary sexual characteristics in male users, but this report has not yet been confirmed.

**Narcotics**   From the medical viewpoint, narcotics include the opiumlike compounds, which act as CNS depressants and also relieve pain and induce sleep, although the mechanisms leading to these effects are not well understood. Both natural and synthetic compounds are included in this category, ranging in strength from Demerol (meperidine) to morphine. The euphoria produced by the stronger narcotics, such as heroin, is far more pleasurable and powerful than that produced by the hallucinogenic drugs, but the effects steadily fade with extended use, and periods between taking the drug become more and more violently unpleasant. Ultimately, the drug is taken only to maintain something approximating a normal state. There is a tremendous psychological addiction because withdrawal is so unpleasant, although unlike the case with the barbiturates, withdrawal is seldom fatal because convulsions do not occur.

Legally, the narcotics also include **cocaine,** a stimulant which is chemically far removed from the opiumlike drugs. Cocaine does not lead to a physical addiction, and there is no building up of tolerance, but there can be a psychological dependence, and overdose can be fatal as a result of heart and respiratory failure.

**THE SENSE ORGANS**   Our sense organs are our windows to the world. Changes in any of several environmental parameters are perceived by one or more of our sense organs, and the occurrence of these changes is transmitted to the CNS for analysis and possible action. What types of changes can we perceive?

How is this perception accomplished? Why are some environmental parameters chosen over others?

We have come to think of ourselves as possessing five senses—sight, hearing, smell, taste, and touch—in large measure because of the prominent sense organs associated with all except the last; however, in reality we possess many more than just these five. We can respond to temperature changes, orientation of the body, position of limbs, and so forth. Nor is this by any means a complete list. Some species of fish can sense changes in electrical fields; many insects can detect polarized light and changes in air pressure; some birds are apparently capable of orienting to magnetic fields. Hence, our own sensory capacities are in no sense exhaustive in terms of all the possibilities, but we do possess an impressive variety, the majority of which are relatively acute.

The most satisfactory method of analysis is probably to select the four principal environmental parameters to which we are responsive, and then to analyze the various *methods* of detection within each category. The four parameters are:

1 light rays.
2 pressure (including touch, balance, hearing, etc.).
3 chemicals (taste and smell).
4 temperature.

Before we undertake an analysis of how the various receptor organs function, it is important to note that, at its basis, each such organ consists of sensory neurons which are tied into some change detector. However, these sensory neurons can, in fact, be depolarized by virtually any stimulus of sufficient magnitude—that is to say, the sensory neurons are not in themselves specialized structures. For example, normally the sensory neurons in the eye are activated only when light rays are focused on the **retina** by the **lens** of the eye (the two elements of the change detector, in this instance). However, if someone were to punch you in the eye, the pressure of the blow would be sufficient to depolarize the sensory neurons of the retina, and they would transmit the occurrence of this "change" to the brain. Upon receiving impulses from the eye, the brain would, of course, interpret this information as changes in light intensity. Therefore, if you are punched in the eye you typically "see stars"—the effect of a rather randomized depolarization of the sensory neurons of the retina.

Finally, it should be mentioned that, with respect to any of the sense organs, extreme stimulation may be interpreted as pain (this is most easily demonstrated in the case of the ears, with respect to very loud, high-pitched sounds). We do not consider pain as a sense, as such, as it is actually the subjective interpretation made by the brain of certain types of environmental change. We can also perceive pain directly, through the stimulation of bare sensory neurons located close to the surface of the skin, but it is really a misnomer to call these neurons "pain receptors." More correctly, these are neurons the depolarization of which is interpreted by the brain as pain. The environmental parameter involved may be pressure (thumb hit with hammer), temperature (hand placed on hot stove), or certain types of noxious chemicals (acid spilled on hand). These, of course, are very different environmental parameters, but the interpretation placed on them by our brain is identical—they "hurt."

**Visual receptors**  Although all of our cells show a vague light sense, it is so diffuse as to be of no value to us, especially as compared to the importance of our eyes. The structure of the eye (Fig. 10.21) has been likened to a camera (actually it was the other way round—the structure of the eye inspired the construction of the original camera). In any case, light first passes through an opening (the **pupil**), the size of which is a variable, depending primarily on light intensity. Just as in a regular camera, the **iris** (the "doughnut" surrounding the pupillary "hole") can expand or contract, meaning that the size of the pupil changes accordingly, and so, too, does the amount of light that enters the eye.

Immediately behind the pupil and iris there is a **lens.** It is the function of the lens, in conjunction with the outer **cornea** of the eye, to focus the light rays in such a way that the focal length exactly equals the distance between the lens and the light-receptive areas on the back of the eye. An elongated eyeball may be beyond the capacity of the lens to compensate for—such individuals cannot see distant objects clearly and are said to be **near-sighted.** Other individuals may have just the opposite problem. This condition of **farsightedness** not only occurs in individuals with a relatively short eyeball, but also (and more frequently) in individuals in their 40s or 50s as a function of the loss of capacity of the lens to change shape to the degree necessary to allow clear vision of objects very near at hand. "Reading glasses" become a common appurtenance at this age.

The lens may also become cloudy with age which, of course, dramatically reduces visual acuity. Such a condition is known as a **cataract;** modern surgery

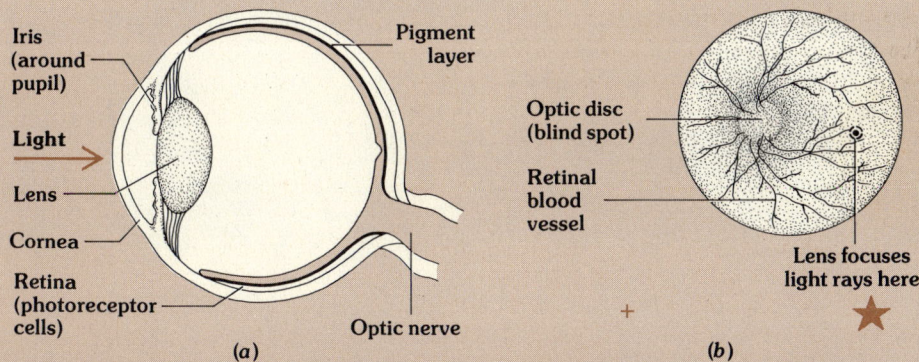

Iris (around pupil)
Pigment layer
Light
Lens
Cornea
Retina (photoreceptor cells)
Optic nerve
(a)

Optic disc (blind spot)
Retinal blood vessel
Lens focuses light rays here
(b)

**Fig. 10.21**  (a) The human eye; (b) the optic disc (blind spot where optic nerve leaves retina). Focus your right eye on the cross and close the left. If you move the page away from your eye, at a distance of 6–8 in. the star disappears.

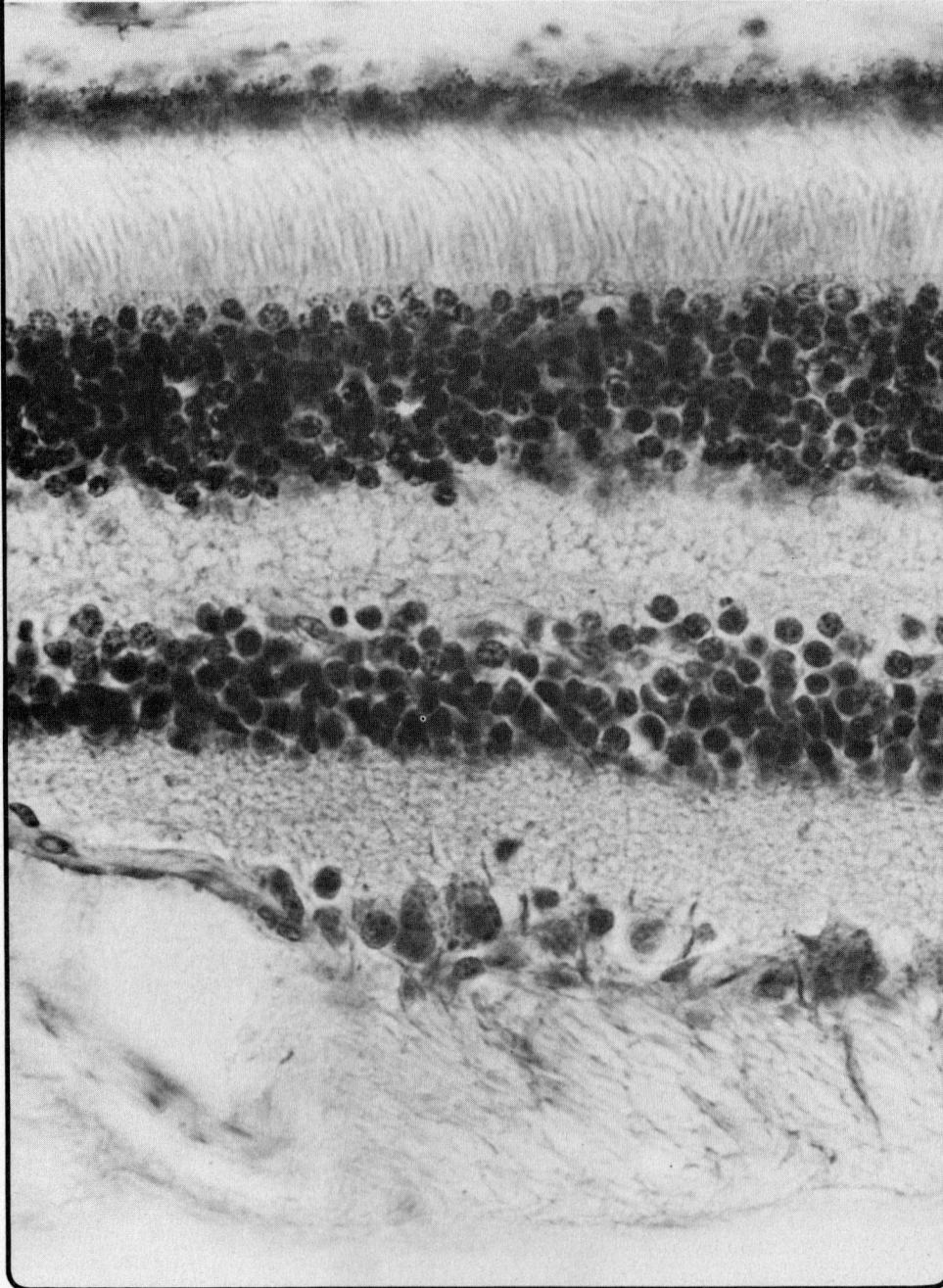

Fig. 10.22   Rods and cones as shown by a scanning electron micrograph. (Rotker, Taurus)

allows removal of these defective lenses and the implantation of lenses from donors who have willed their use. Because the lens has very little connection with the circulatory system, tissue rejection is not a serious problem with this transplant (see Chapter 5). People with cataracts still outnumber available lenses for transplant, however, but even here, some vision can be restored by the removal of the defective lens and the subsequent utilization of specialized glasses.

The focal point of the light is on the **retina,** the layer of sensory cells that covers the interior of the eyeball (the reflection of which gives the black color to our pupils). The cavity of the eyeball is not hollow, of course, but contains a gellike substance. Normally, the amount of this substance is carefully monitored by the body, but in some individuals, for reasons that are far from clear, the gel becomes overly abundant, and the pressure increases in the eye to the point where the delicate cells of the retina are crushed, causing blindness. This condition is called **glaucoma,** and it is a leading cause of blindness in this country. It can be treated quite successfully with medication if detected in time, but, unfortunately, very often there are no symptoms until irreparable damage to the retina has already occurred. If you have ever had a thorough eye examination, you will recall the procedure whereby your eye was temporarily anesthetized and a special type of meter touched quickly to the eyeball. The purpose of this meter is to measure the pressure of the interior of the eye as a check for the possibility of glaucoma.

The retina itself consists of a conglomerate of cells, only the deepest layer of which is directly responsive to light. The specialized neurons which have this capacity are of two types (Figs. 10.22 and 10.23)—**rods,** which are more elongated, have no color-sensing capacity, and are found primarily in the more peripheral regions of the retina, and **cones,** which are shorter, are color-sensitive, and are found at the center of the retina, directly behind the lens. There are a total of four types of visual pigments, all containing a derivative of vitamin A and all very similar chemically. The first of these, **rhodopsin** (Fig. 10.24), is found in the rods. It is very sensitive to low levels of light, but it has no color-detecting capacities. Hence, we can see at night, but only in shades of gray, because only our rod cells are active. There are three different types of cones, each containing a different visual pigment. These respond to red, blue, and green, respectively; stimulation of more than one, at differing intensities, allows us to distinguish a whole series of intermediate colors. Color-blindness (a recessive trait carried on the X chromosome—see Chapter 13) involves the failure of one or more of these color pigments to be produced, and the individual thereby lacks the capacity to distinguish not only the color involved, but also the various intermediate shades normally involving participation by that particular type of cone cell.

The chain of events which we interpret as vision begins when light causes a physical straightening of some of the visual pigment molecules (precise quantification is difficult, because there are some 30 million of these molecules in every rod or cone). This straightening causes the generation of an action potential to two successive types of neurons also in the retina, and in turn the action potential is transmitted through the **optic nerve** to the thalamus of the brain and thence to the

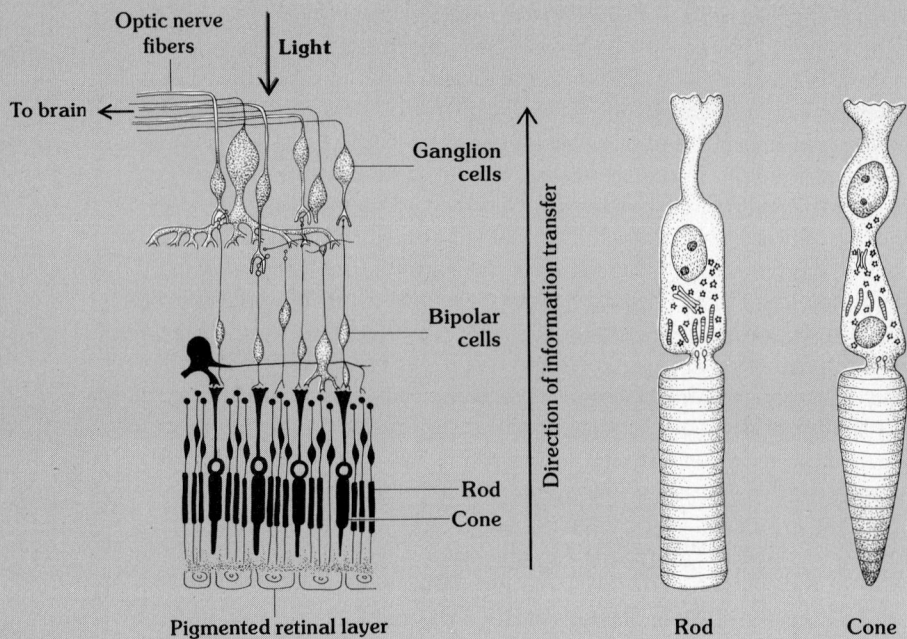

**Fig. 10.23** Structure of the retina, showing diagrammatic views of the rods and cones.

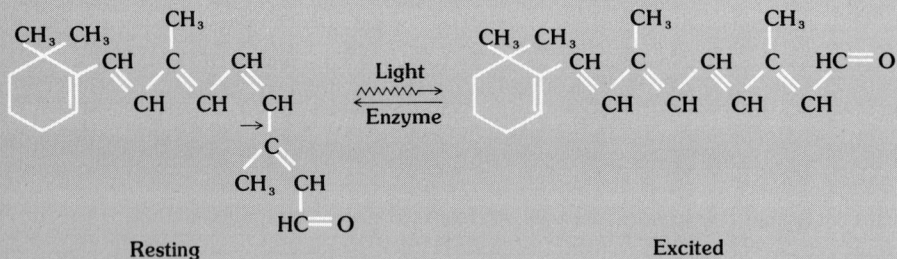

**Fig. 10.24** Change in structure in the rhodopsin molecule when struck by light.

visual cortex at the back of the head (Fig. 10.25). Damage to any portion of this pathway can result in blindness. Thus, it is just as "easy" to go blind from a blow to the back of the head as from a direct injury to the eye itself.

**Pressure receptors** Pressure receptors are of four primary types in the body, involving (1) hearing; (2) balance; (3) touch; and (4) stretch.

**Hearing** Any vibrating object causes bands of compression and rarefaction of molecules in the air (Fig. 10.26), whether the vibrating object is a tuning fork or our vocal cords. If the object is vibrating at a **frequency** (pitch) of between 20 and 20,000 Hz (Hertz, meaning "cycles per second") and if the vibration is of sufficient intensity, we are able to detect it as a **sound**—that is, we are able to **hear** it.

The mechanisms whereby this is accomplished are as follows. Sound waves are intercepted by the eardrum at the base of the ear canal (Fig. 10.27), causing it to vibrate ever so slightly but at the same frequency as the sound in question. Attached to the eardrum are a set of three small **middle ear bones** which, in turn, transmit this sound to the inner ear. The area of the innermost of the three ear bones in contact with the inner ear is much smaller than is the area of the ear drum. This fact, coupled with the ingenious set of articulations of the middle ear bones them-

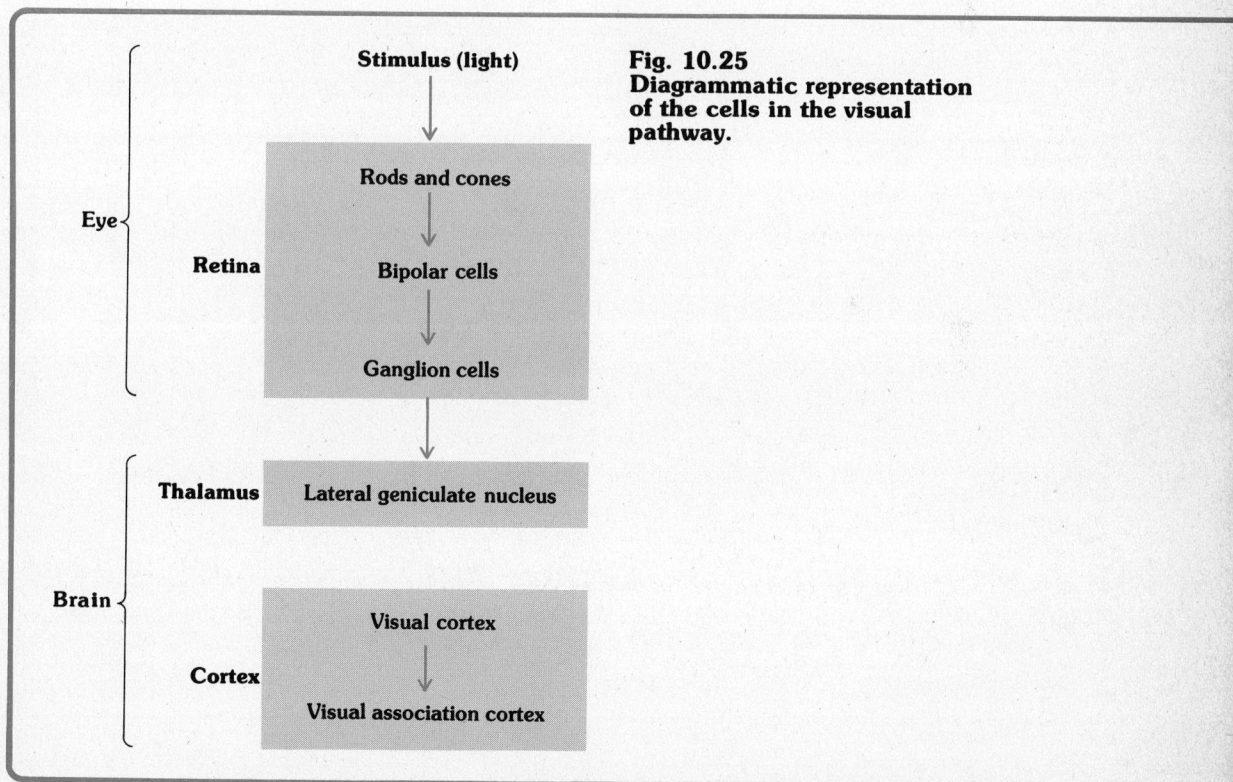

**Stimulus (light)**

Eye {
    **Retina**
    Rods and cones
    Bipolar cells
    Ganglion cells

Brain {
    **Thalamus** Lateral geniculate nucleus
    **Cortex** Visual cortex
    Visual association cortex

**Fig. 10.25
Diagrammatic representation of the cells in the visual pathway.**

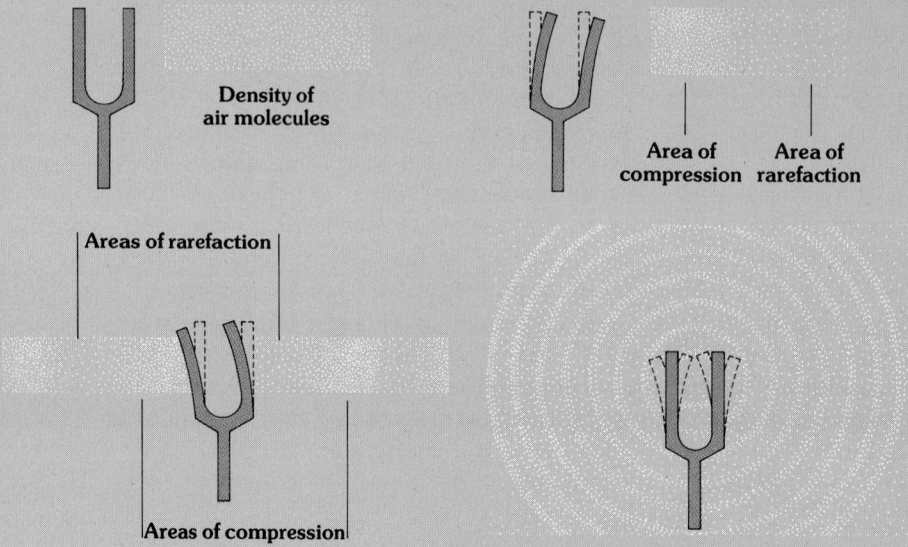

Fig. 10.26 Formation of sound waves.

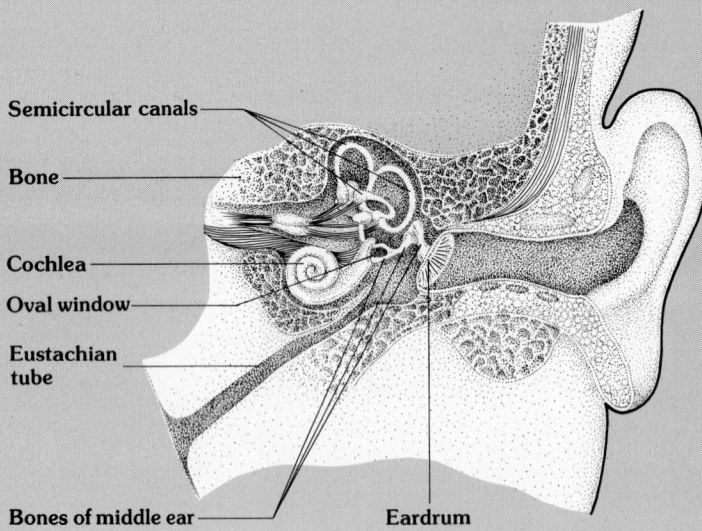

Fig. 10.27
The human ear, and associated semicircular canals.

selves, serves to amplify the intensity of the vibration some 18 times over that received by the eardrum (Fig. 10.28).

The point of contact between the middle ear bones and the inner ear is called the **oval window,** so named because it is thin-walled and capable of being depressed (Fig. 10.29). The inner ear is filled with a fluid that receives the vibrations at the same frequency at which they originally entered the ear. The inner ear is coiled in a snail shape and is divided down the middle by a membrane which is narrow and stiff near the oval window, and broader and more flexible near the tip. This membrane also vibrates at the same frequency as the sound in question, but because of its structure, sounds of different frequencies will cause maximum displacement at different points along the membrane. In other words, a sound of a given frequency will cause maximum displacement at one particular point along the membrane (Fig. 10.30). The membrane itself is a complex structure, consisting of two layers which touch but are not fused. As one layer slides over the other during membrane displacement, it deflects the tiny hairs of the so-called **hair cells,** which are connected to the auditory nerve. Deflection of these hairs causes an action potential to be generated. Thus, sounds of different frequencies will cause different hair cells to be stimulated, which, in turn, will activate different fibers in the auditory nerve, all of which ultimately will allow the brain to make discriminations according to sound frequency.

What about loudness, or sounds of more than one frequency? Loudness is detected simply by the magnitude of the deflection of the eardrum (which is ultimately reflected by the magnitude of deflection of a given set of hair cells). Excessive loudness, however, can destroy the hair cells (Fig. 10.31). Complex sounds are detected simply by the simultaneous deflection of hair cells responsive to the particular frequencies in the sound. This rather simple-sounding system is responsible for allowing us to distinguish an estimated 400,000 different sounds.

**Balance** Anatomically associated with the inner ear, although functionally far removed from it, are the **semicircular canals** (Fig. 10.27). We have three such

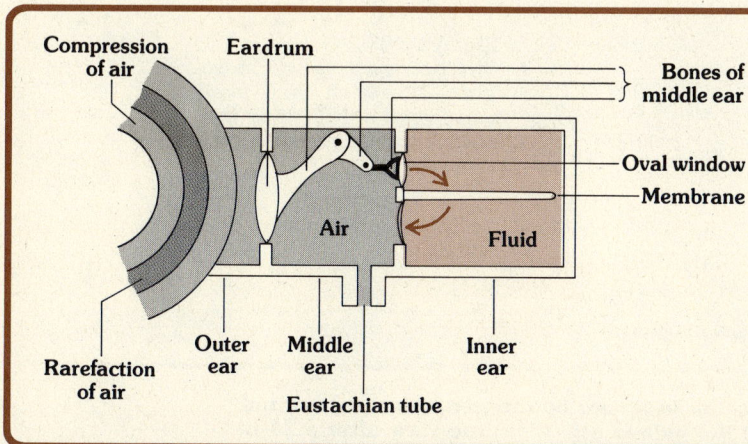

**Fig. 10.28**
**Model of ear function.**

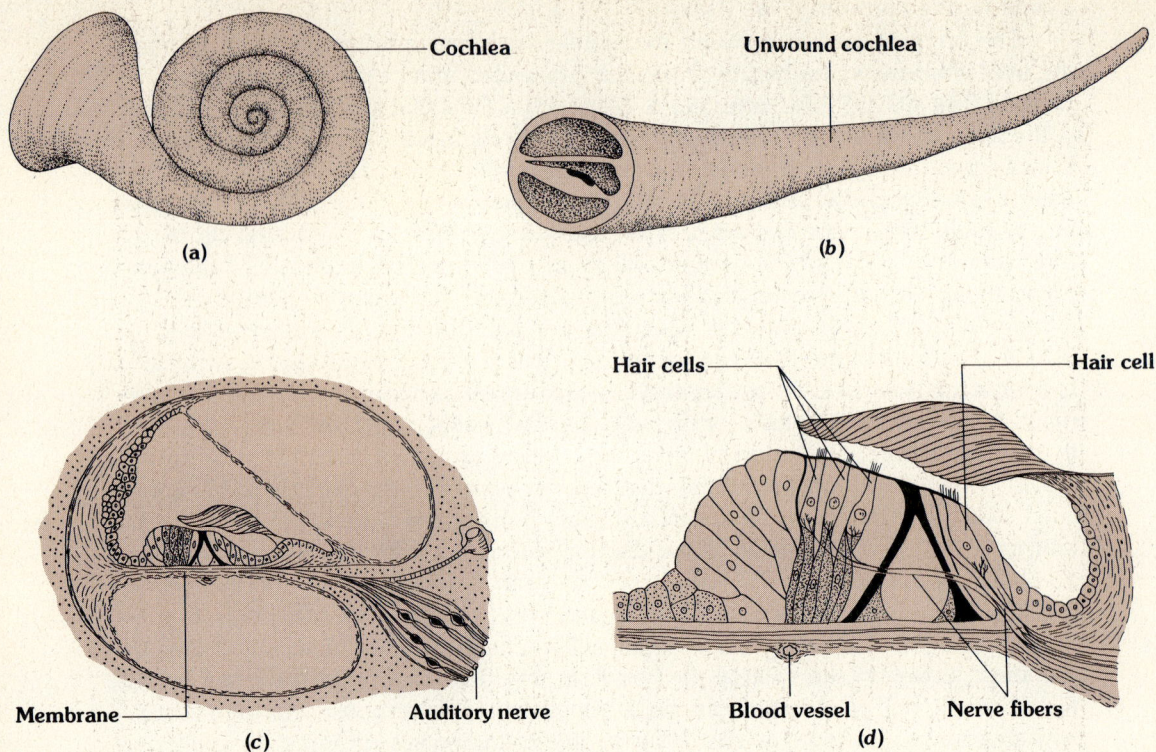

Fig. 10.29 The inner ear or cochlea: (a) as it normally appears; (b) unwound; (c) in cross section; (d) in enlarged cross section.

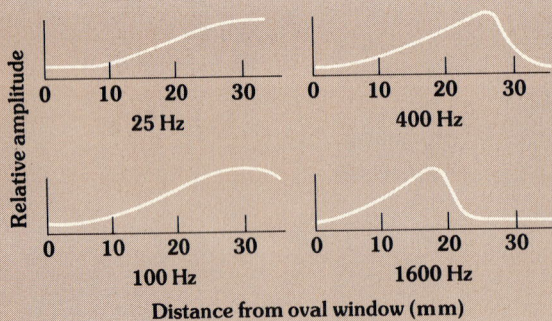

Fig. 10.30
The point along the membrane at which the traveling wave peaks differs as the sound frequencies differ.

Fig. 10.31 (*opposite*) Injury to the inner ear by intense noise: (a) normal hair cell structure in the ear of the guinea pig; (b) same view after a 24-hr exposure to rock music at 120 decibels. (Jos. E. Hawkins, Kresge Hearing Research Institute, University of Michigan Medical School)

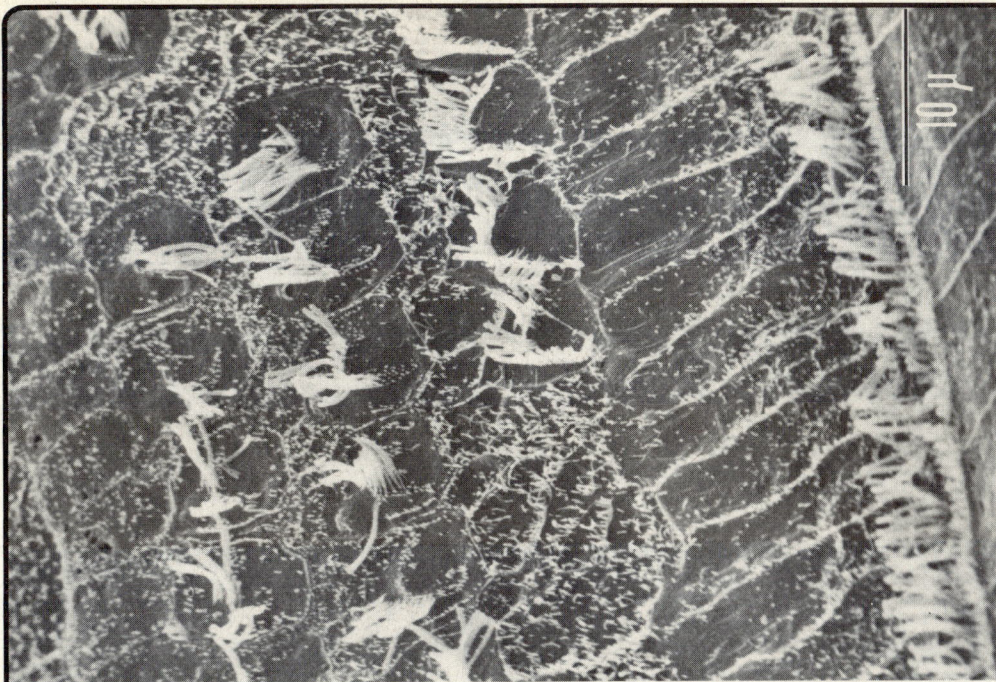

(b)

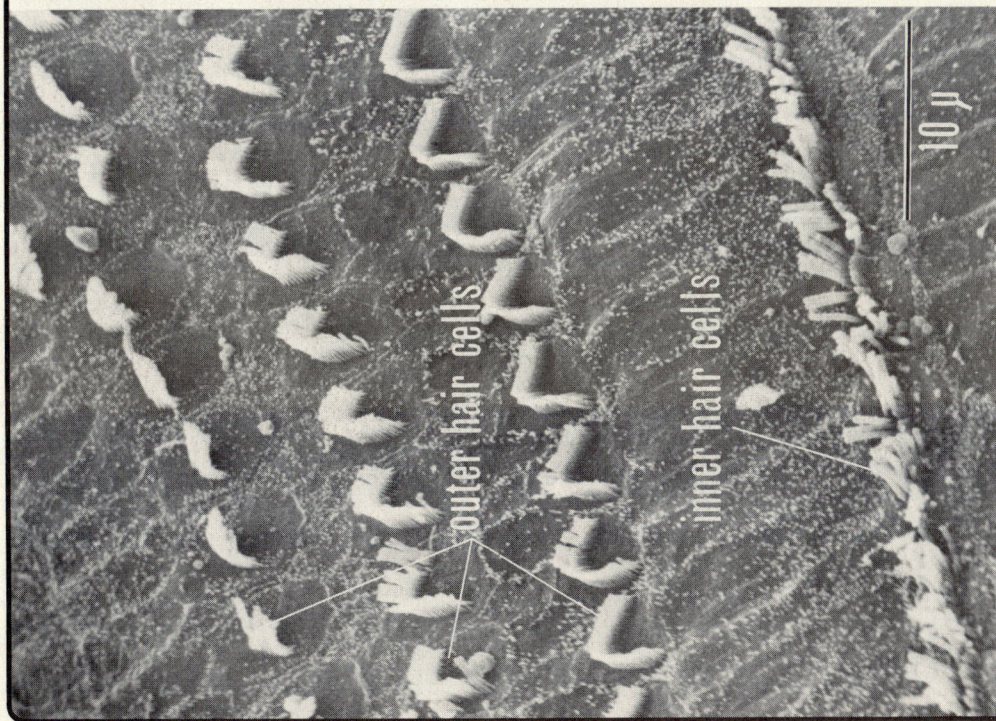

outer hair cells

inner hair cells

(a)

canals associated with each ear, each canal lying in a different plane. The canals are filled with a fluid in which hair cells are embedded. When the head is moved, this fluid tends to stay in one place, and then shifts when the head stops moving. (Imagine a cup filled to the brim with coffee which you rapidly move horizontally through the air before coming to a sudden stop. The coffee, as a liquid, moves less slowly than the cup at first, but continues to move once the cup has stopped, and, of course, you spill the coffee. Had the lip of the cup been covered with hair cells, obviously some of them would have been deflected as the coffee sloshed about.) As the semicircular canals are arranged in three planes, no movement of the head takes place without detection. Moreover, because there are two complete sets of canals, movement of the head will be detected differently by each set if the movement is to the right rather than to the left.

The semicircular canals have two primary purposes. First, they serve to direct the muscles that move the eyes. Thus, we are able to move our head and still focus on the same point. (This may not seem a significant function, but consider the difficulties you have in viewing a movie made with a hand-held camera—all too often the picture emerges as a series of jerks, because the operator was unable to maintain a precise focus as he moved about.) Second, they serve as an aid in posture. We use other information, most notably the eyes and receptors in the muscles to tell us about our posture (i.e., orientation with respect to the horizon), but, if you have ever made yourself dizzy by spinning rapidly about (i.e., if you have caused strong impulses from the semicircular canals), you know how hard it is to overcome the tendency to fall down, such being the magnitude of the stimuli coming from the canals. Indeed, damage to the semicircular canals may give rise to extremes of dizziness so severe that the individual may not be able to stand up.

**Touch**   Our sense of touch is perceived by tiny organs in the skin (Figs. 10.32 and 10.33). Some of these are specialized pressure receptors surrounding a single neuron; these are called **Pacinian corpuscles.** Others are attached to the base of hairs, such that the deflection of the hair causes an impulse to be generated.

The mechanics of these organs is very simple. The magnitude of the stimulus, whether a feather or a lead weight, is conveyed to the brain by the rate of action po-

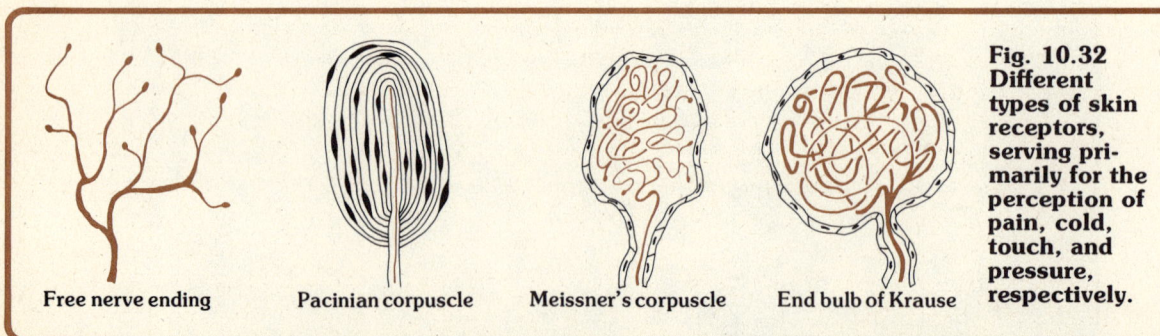

Free nerve ending        Pacinian corpuscle        Meissner's corpuscle        End bulb of Krause

Fig. 10.32 Different types of skin receptors, serving primarily for the perception of pain, cold, touch, and pressure, respectively.

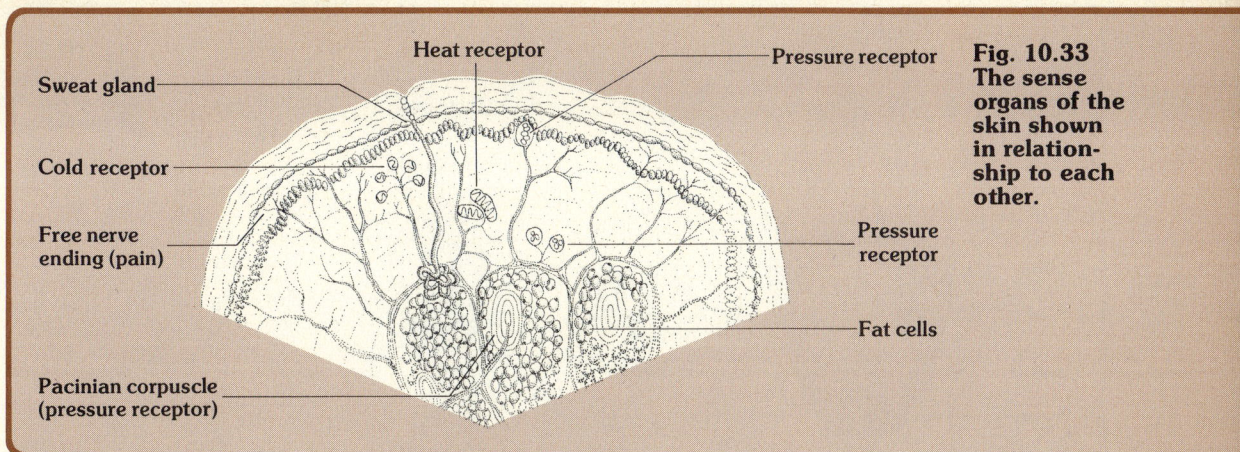

Fig. 10.33 The sense organs of the skin shown in relationship to each other.

Sweat gland

Heat receptor

Pressure receptor

Cold receptor

Free nerve ending (pain)

Pressure receptor

Fat cells

Pacinian corpuscle (pressure receptor)

tentials produced by the detecting neuron or neurons. Fortunately, touch receptors fatigue quickly. Therefore, a constant level of touch rather quickly results in a diminishment of firing. When we pull on our wool socks in the morning, we are conscious of their "feel," but (unless a rash develops) we soon stop feeling them. This is just as well, lest our central nervous systems be overwhelmed by useless information. Suppose every touch receptor in the body were to fire constantly as long as it were being touched by anything, even clothing. The consequences would be disastrous.

**Stretch**   Stretch receptors are also structurally very simple and are found in such connective tissue as tendons and muscles. Quite simply, they are neurons wrapped around a portion of a tendon or muscle fiber, and they are stimulated as the tendon or muscle is stretched. Thus, we are constantly receiving sensory input from our tendons and muscles, information which we use in moving about. It is from these receptors that you "know" what position your arm is in, without having to look at it. Similarly, if you happen to sit or sleep in an odd position, you may cut off circulation to some of the peripheral nerves for a time, causing your arm or leg to "fall asleep"—and you subsequently find it very difficult to control the limb for a few seconds, until the blood circulation begins again. We tend not to think much about these receptors, but we are using them (or at least our cerebellum is using them) whenever we make a movement.

**Chemical receptors**   Chemical receptors include the taste buds of the mouth (Fig. 10.34) and the olfactory epithelium of the nose (Fig. 10.35).

**Taste buds**   It is usually said that we have four kinds of taste buds, responding to the four tastes of sweet, salt, sour, and bitter, but that is the equivalent of saying we can see only three colors. In both instances, there is a wide array of subtle changes and combinations possible. It would appear that each of the taste buds can, in fact, respond to any of the four tastes, but some respond more rapidly to a given taste than do others. However, the mechanics of action are not well under-

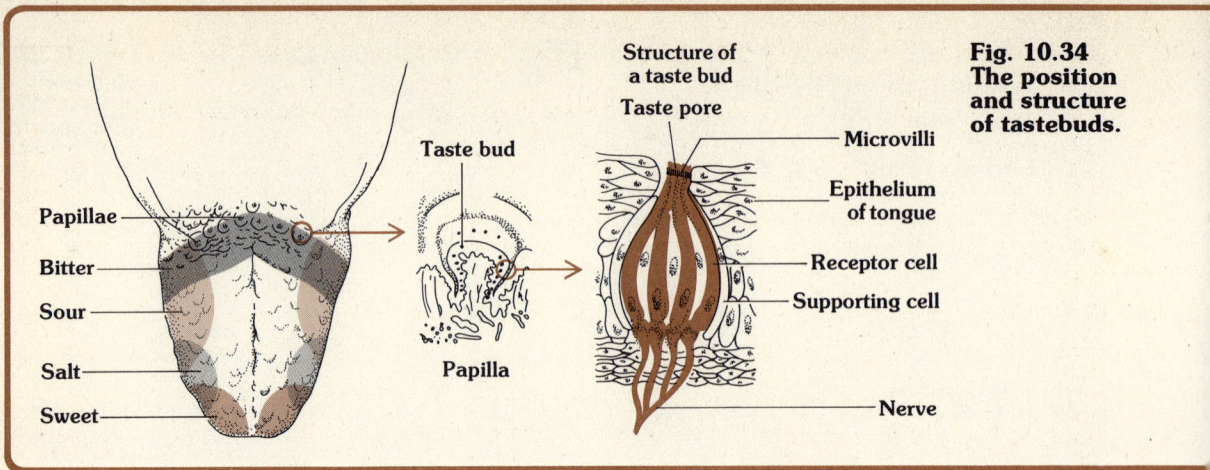

Fig. 10.34
The position
and structure
of tastebuds.

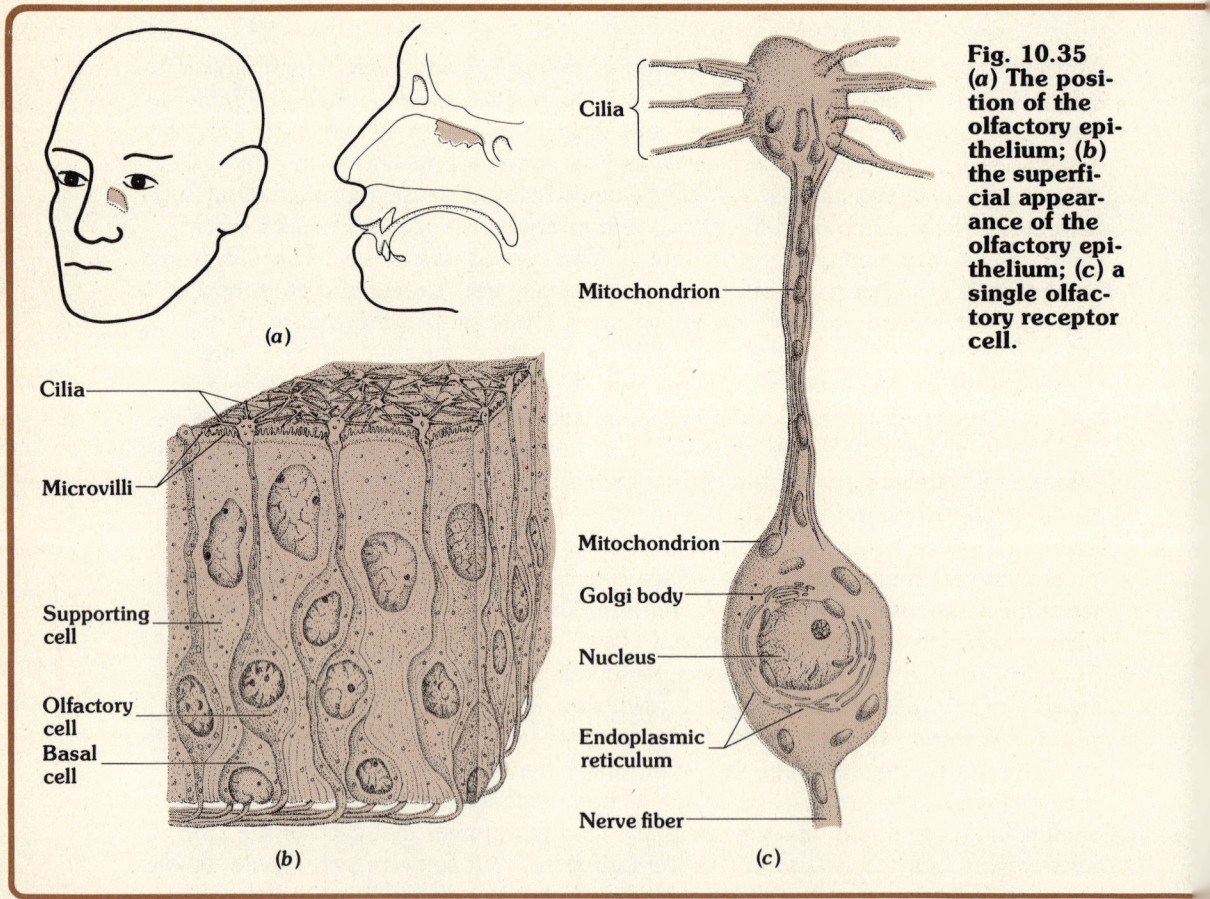

**Structure of
a taste bud**

Taste pore

Papillae

Bitter

Sour

Salt

Sweet

Taste bud

Papilla

Microvilli

Epithelium
of tongue

Receptor cell

Supporting cell

Nerve

Fig. 10.35
(*a*) The posi-
tion of the
olfactory epi-
thelium; (*b*)
the superfi-
cial appear-
ance of the
olfactory epi-
thelium; (*c*) a
single olfac-
tory receptor
cell.

Cilia

Mitochondrion

(*a*)

Cilia

Microvilli

Supporting
cell

Olfactory
cell

Basal
cell

Mitochondrion

Golgi body

Nucleus

Endoplasmic
reticulum

Nerve fiber

(*b*)

(*c*)

stood, beyond the obvious fact that waterborne chemicals of the four types mentioned can initiate action potentials within the neurons of the taste buds, which information is then relayed to the brain.

**Smell**   Lining the upper reaches of our nasal passages are large numbers of cells which are responsive to airborne chemicals of a variety of types. The mechanics of odor detection are perhaps more poorly known than for any of the other senses. It is thought that each cell may be responsive to a number of different types of odors, based on chemical structure of the molecule in question, with different cells possessing different types of receptive sites. However, these events are at the molecular level, and all the olfactory cells are identical when viewed microscopically. Therefore, proving or disproving the model just advanced will be difficult. Moreover, a number of external factors influence the nature of the response. For example, women tend to have a more acute sense of smell than do men, and hunger increases overall acuity. Finally, fatigue of the olfactory cells is rather rapid. Therefore, we are quick to notice a strange odor, but we also cease to note it rather quickly. Obviously, the mechanisms of odor detection are complex, and a complete understanding of the underlying physiology is some time off.

**Temperature**   Heat and cold receptors take the form of small organs located in the skin. Of course, heat and cold are simply extremes on the same continuum, but there are, in fact, two types of receptors, one type which fires at low temperatures and one which fires at high temperatures. Impulses from these receptors are transmitted through the spinal cord and brainstem to the hypothalamus, which is responsible for the homeostatic control of body temperature. It is presumed that the heat receptors function by forming an action potential whenever heat causes changes in the structure of the protein molecules of the nerve ending; the mechanism for cold reception is unknown.

## SUMMARY

The complexity of the human brain is belied by the relative simplicity of the neurons of which the brain is composed. Furthermore, because the generation of an action potential follows the "all-or-none" rule, a given neuron would seem to be limited in the amount of information it can transfer. Like a computer, it would appear that the neuron is limited to a binary system—it is either "on" or "off." However, because of multiple synapses, temporal and spatial summation, and excitatory and inhibitory synapses, enormous complexity can be produced, the culmination of this complexity being the functioning of the human brain.

The synapses, and even more the neuromuscular junctions, are subject to invasion and interference by a host of potentially lethal agents. Ironically, the very synapses that allow for complexity also represent the weak links in the system. Many organisms with far simpler nervous systems are able to survive doses of poison that would kill us almost instantly. Thus, just as a finely tuned sports car is more of a joy to drive than a staid sedan but is more difficult than the sedan to keep running smoothly, there is a trade-off between degree of nervous system complexity and the degree of jeopardy from poisons.

To paraphrase Aldous Huxley, our sense organs are our doors of perception, and relative to most animal species, we are very well endowed. Again, a complex nervous system is a prerequisite to the development of sophisticated sense organs, for complexity is required in order to be able to assimilate and utilize the sensory information. Therefore, the price we pay for being able to see and enjoy a beautiful sunset is our unfortunate ability to smell air pollution the next morning.

Why do blind children smile? Why do humans kill each other with an abandon not shared by any other species? Why don't women show seasonal sexual receptiveness as do virtually all other animals? These are the kinds of questions addressed by the newest branch of biology —ethology, the study of behavior.

chapter **XI**

# chapter XI
# Ethology–our behavioral heritage

Ethology is a relatively new branch of biology (although its roots are very old), which deals with the biological basis of behavior. Even though the foundations of this discipline were established from studies on species simpler than our own, the potential of ethology as an aid in the understanding of human behavior was demonstrated by the awarding of the 1973 Nobel Prize in physiology and medicine to the three men who are generally acknowledged to be the founders of the discipline—Konrad Lorenz, Niko Tinbergen, and Karl von Frisch. Because biologists are now beginning to explore human behavior using models developed on lower animals, it is worthwhile to examine ethology as one of the areas in which future generations of biologists will undoubtedly be spending much time.

**ETHOLOGY AND COMPARATIVE PSYCHOLOGY**   Ethology had its origin in Europe, and it remains an important force there. However, in the United States psychology reigns supreme. How do these disciplines differ? Ethologists are primarily interested in four facets of behavior:

1  The adaptive value of behavior;
2  The evolution of behavior;
3  The ontogeny (development) of behavior;
4  The physiological basis of behavior.

Psychologists are much less interested in the first two than in the last two facets. Because their interests differ, so, too, do their theories, terminologies, and, most notably, their methods. Methodological differences can be summarized by the oft-quoted saying that a psychologist places an animal into a box and watches it, whereas an ethologist climbs into the box himself and looks out at the free-ranging animal. Put another way, the psychologist uses an experimental approach, whereas the ethologist uses naturalistic observations, generally of animals in their natural habitat.

**BASIC ETHOLOGICAL THEORY**   From their study of lower animals, ethologists have developed a rather sophisticated theory of behavior. Many types of behavior are classified as **fixed action patterns (FAPs).** These consist of highly stereotyped behaviors, varying little from performance to performance in different animals of the same species. Such activities are typically performed in response to a **sign stimulus** that triggers the behavior, but in some instances the behavior may be performed in the absence of an appropriate sign stimulus.

For example, consider the behavior of a caged male dove, presented with a female dove in the early spring. The female acts as a sign stimulus and releases the mating behavior of the male, which actually consists of a number of consecutively performed FAPs. It is important that the male's performance be highly stereotyped and typical of the species, for it is through his mating behavior that the female recognizes the male as one of her own species and consequently accepts his advances.

Now suppose we kept the male alone during the mating season. We would find that the standards required by the male for the sign stimulus would gradually relax. After a time, he would court a stuffed female dove placed in his cage, later a pigeon would suffice, then a ball of rags, and ultimately the male might perform the mating dance just to the corner of the cage. We would say that the male is performing a **vacuum activity,** as there is no sign stimulus present. (You might have your own name for such desperate behavior.) The performance of vacuum activities, after the prolonged absence of the appropriate sign stimulus, is a characteristic feature of many FAPs.

Consider some additional examples. You are walking through the rain forests of eastern Zaire (formerly the Congo), when you encounter a group of mountain gorillas feeding on some shrubs. The big silverback male is offended by your presence, rips a small tree out of the ground, and dashes it to bits on the rocks. Then he stands erect and thumps his chest. Even if you had never seen a gorilla before, you would probably assume that he was not performing in this way because of a sudden hatred of trees. On the contrary, this is an example of **redirection activity,** whereby an FAP (attack behavior), released by an appropriate sign stimulus (intruding human), is not directed at the sign stimulus, but rather at another object (i.e., the tree). We do the same thing when we might kick the wall, or throw a pillow when we are frustrated with a low grade.

Watch a pair of bull elk squaring off for control of a harem of females. Between jousts, they frequently stop to browse on some shrubbery. Are they hungry as a result of their exertions? On the contrary—this is an example of **displacement activity,** in which an inappropriate FAP suddenly appears in an out-of-place context. In a totally analogous fashion, a third-grader, asked a question to which he or she does not know the answer, may well begin to scratch his or her head.

Both redirection and displacement activities frequently occur in **conflict situations**—instances in which the tendency to attack is evenly balanced with the tendency to flee. Hence, in the gorilla example, had you taken a step forward while the gorilla was bashing the tree about, you would have forced the issue. You might well have had the pleasure of watching this enormous creature flee from your advance—or the obvious displeasure of having him attack and dismember you. Either way, he would have forgotten about the tree, because either flight or attack would have become dominant, and the redirection activity would have disappeared.

**FAPS and reflexes**   Ethologists talk about FAPs, whereas psychologists discuss reflexes. How do these two terms differ? Actually, there are several similarities, but there are differences which go beyond the fact that the two terms had their origin in different disciplines. Some differences are:

1 Reflexes are always in response to a stimulus; FAPs may occur as a vacuum activity.
2 FAPs may appear in other contexts (i.e., as displacement and redirection activities); reflexes never do.
3 Reflexes tend to be simpler, frequently involving only a portion of the body; FAPs are generally more complex, and frequently involve the entire body.

**Innate versus learned**   A major debate that has only recently begun to settle down has pitted the biologists against the psychologists (with some defections from both camps) on the question of whether all behavior is learned (sometimes phrased as "environmentally determined") or whether some is **innate** (instinctive; genetically programmed). The debate has quieted in large measure because it became generally recognized that no behavior is exclusively genetic or exclusively environmentally determined. Nonetheless, the argument is far from resolved, partly because of semantic problems. Genetic determination may be the hallmark of innateness, but "learning" is a much narrower category than "environmentally determined." For example, some of our ultimate adult height is environmentally determined (a function of nutrition during childhood), but we certainly do not "learn" to grow tall.

Moreover, the outright dismissal of genetic determination as the sole control of any behavior is a bit facile. Basically, the argument for its dismissal is based on the observation that the genes never operate in a vacuum, but rather in an environment that is subject to variations. Therefore, any characteristic, whether structural or behavioral, is always the result of an interaction of environmental and genetic forces, and can never be exclusively genetic. The fact is, however, that such anatomical characteristics as height, weight, and number of limbs differ in the relative importance of genetic and environmental influences, and we should expect nothing less from behavioral traits. Of the three characteristics just mentioned, we accept as a matter of course that adult weight is largely determined by environmental factors, whereas we acknowledge virtually total genetic control over the development of limbs. The fact that such drugs as **thalidomide** (see Chapter 13) can alter the environment so severely as to prevent normal limb development hardly takes away from the fundamental genetic control of this trait.

Hence, the question has now become the **relative** significance of learning versus genetics for given behavioral traits, but there is far from universal agreement on specific examples. This is a most important area for ethologists because it is a fundamental tenet of ethological theory that FAPs have a genetic basis, even though some FAPs may be modifiable to some degree by environmental influences.

**Do humans have FAPs?**   The knee-jerk response to this question is an emphatic "no!"—we do not want to acknowledge anything less than total cerebral, voluntary control over our behavior, any more than our great-great-grandparents wanted to acknowledge evolution—or than our ancestors in the sixteenth and seventeenth centuries wanted to acknowledge that the earth was not the center of the universe. However, the answer is very important to us, for at least two reasons. First, we need a little humility. If we constantly think in anthropocentric terms, we will never be willing to acknowledge the constraints that physical and biological laws place on us, nor will we gain the perspective necessary to determine our place in nature. Second, we must be willing to acknowledge and compensate for our shortcomings, such as our aggressive behavior, which has led to so many wars in the past.

Although the presence of FAPs in humans is increasingly being accepted more widely, a number of anthropologists and psychologists continue to deny their occurrence. The evidence marshaled by ethologists in their attempt to refute this belief comes from three types of study:

**1. Different populations** Even though half the people you know seem just to have returned from Europe, most human populations are remarkably sedentary and have only limited contact with other populations. Therefore, the presence of the same behavioral trait used for the same purpose in widely separated populations may be taken as evidence to support the concept that such traits have a genetic, rather than a learned, basis, that is, that they are FAPs.

Several such traits have been described, of which the **eyebrow flash** may be the most interesting (Fig. 11.1). The chances are you have used this behavior your-

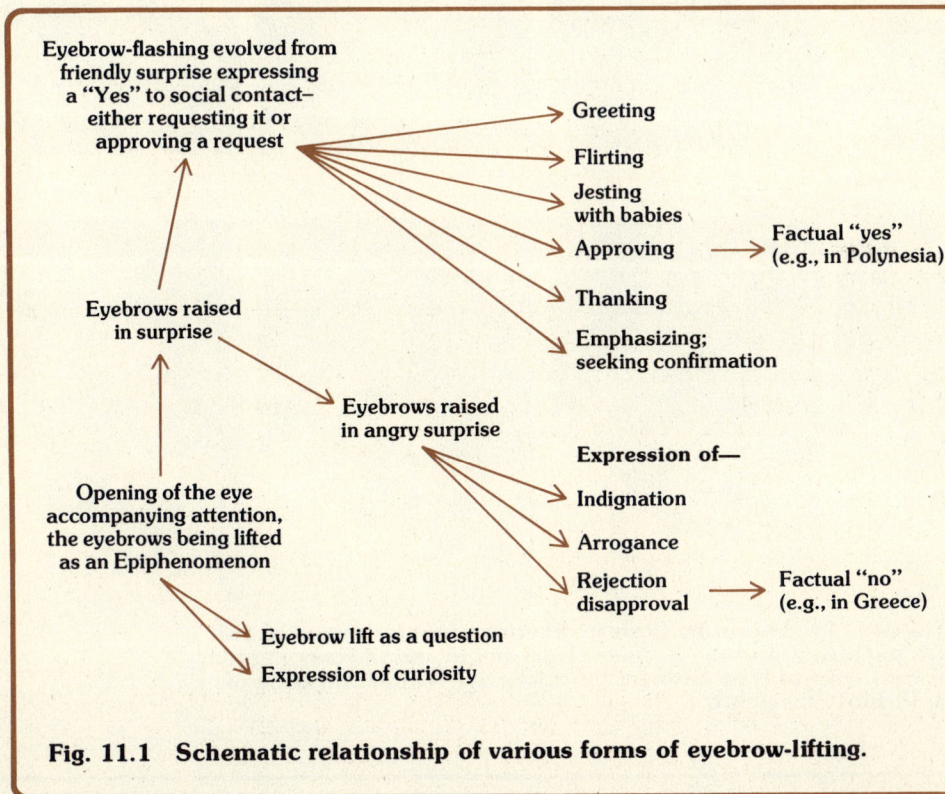

**Fig. 11.1  Schematic relationship of various forms of eyebrow-lifting.**

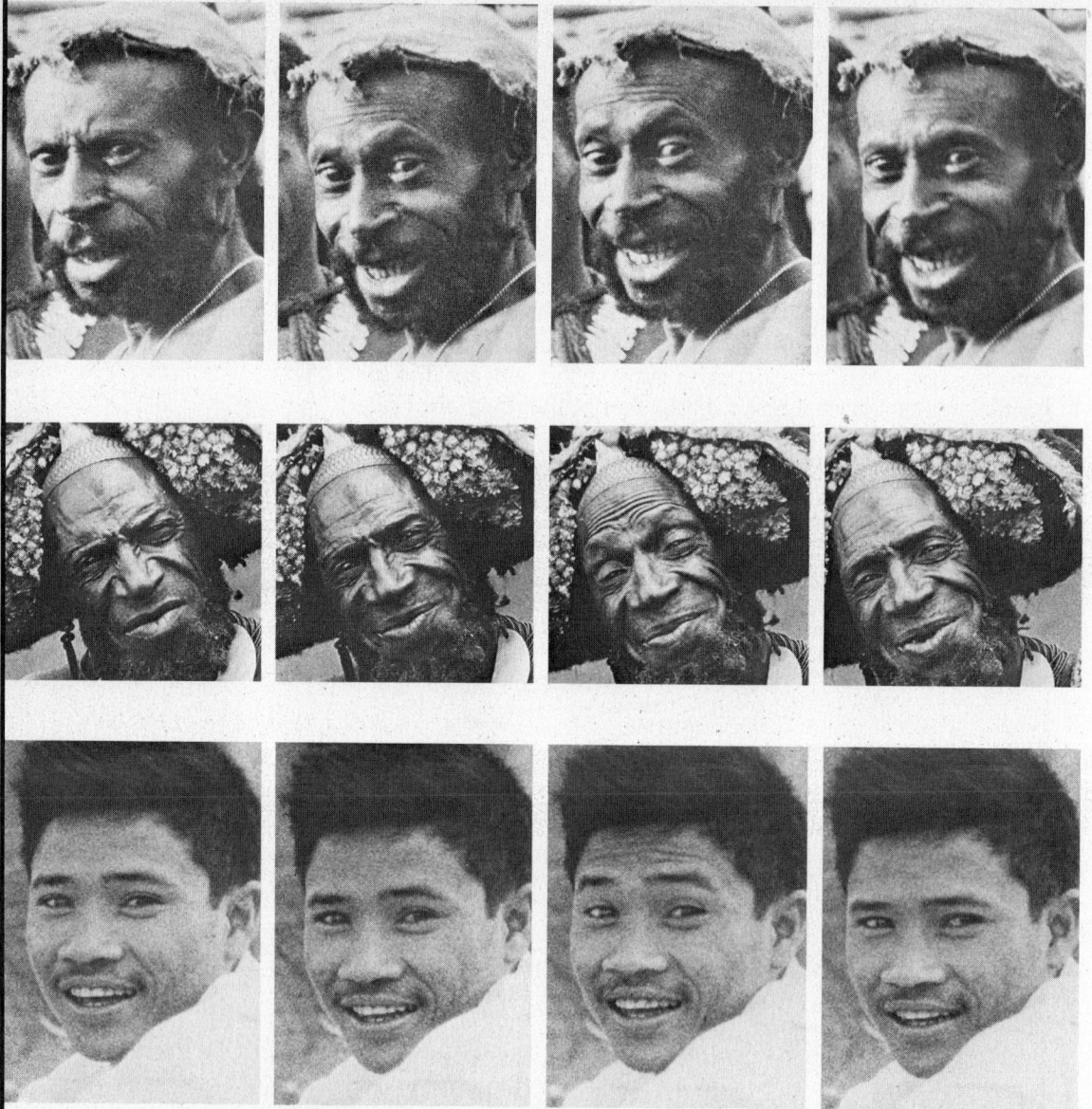

Fig. 11.2 The use of the eyebrow flash in greeting among a series of culturally diverse peoples: (*top row*) Balinese from the island of Nusa Panida; (*middle row*) Huri tribe of New Guinea; (*bottom row*) Woitapmin tribe of New Guinea. (I. Eibl-Eibesfeldt)

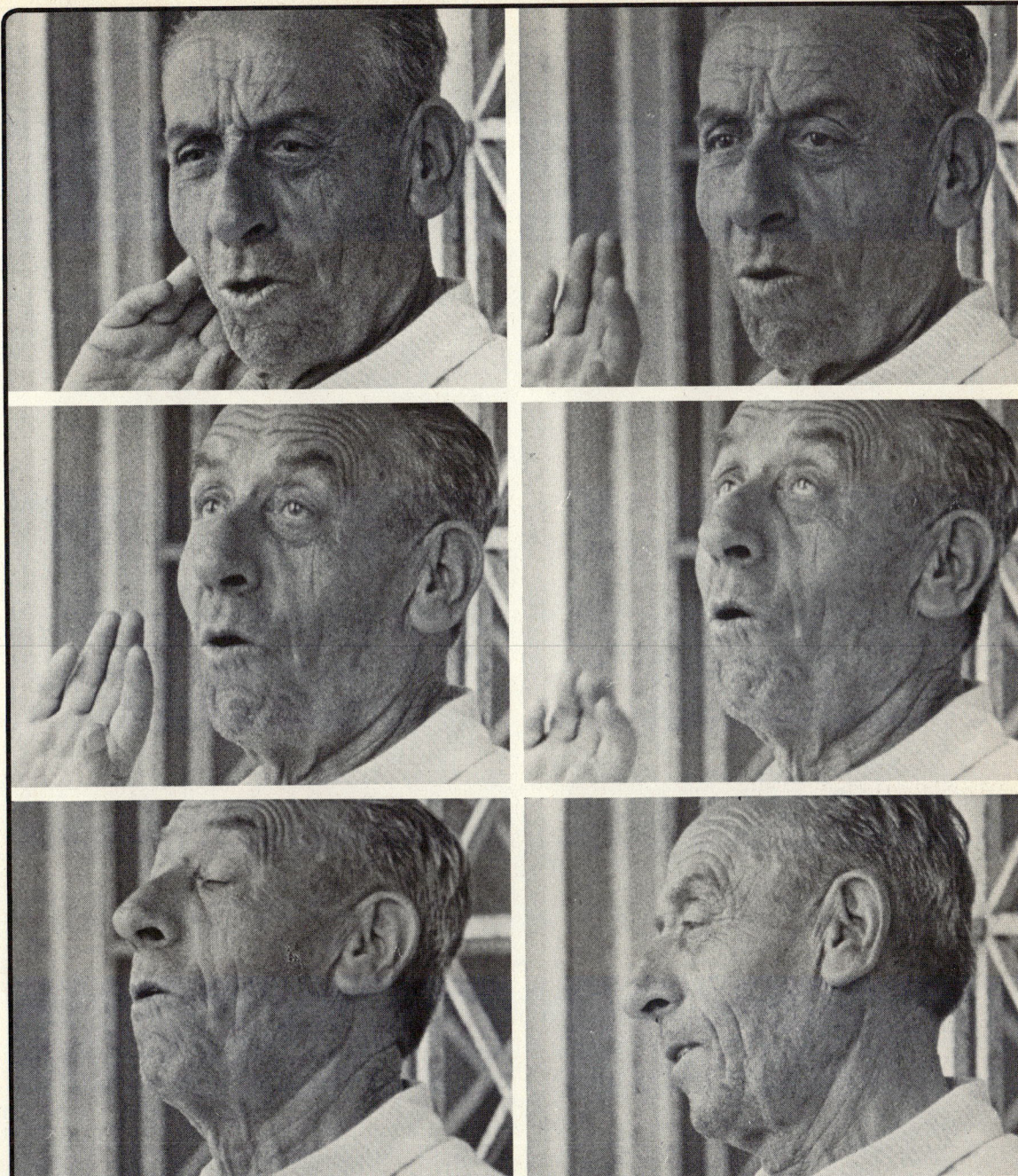

Fig. 11.3  A Greek expressing negation. (I. Eibl-Eibesfeldt)

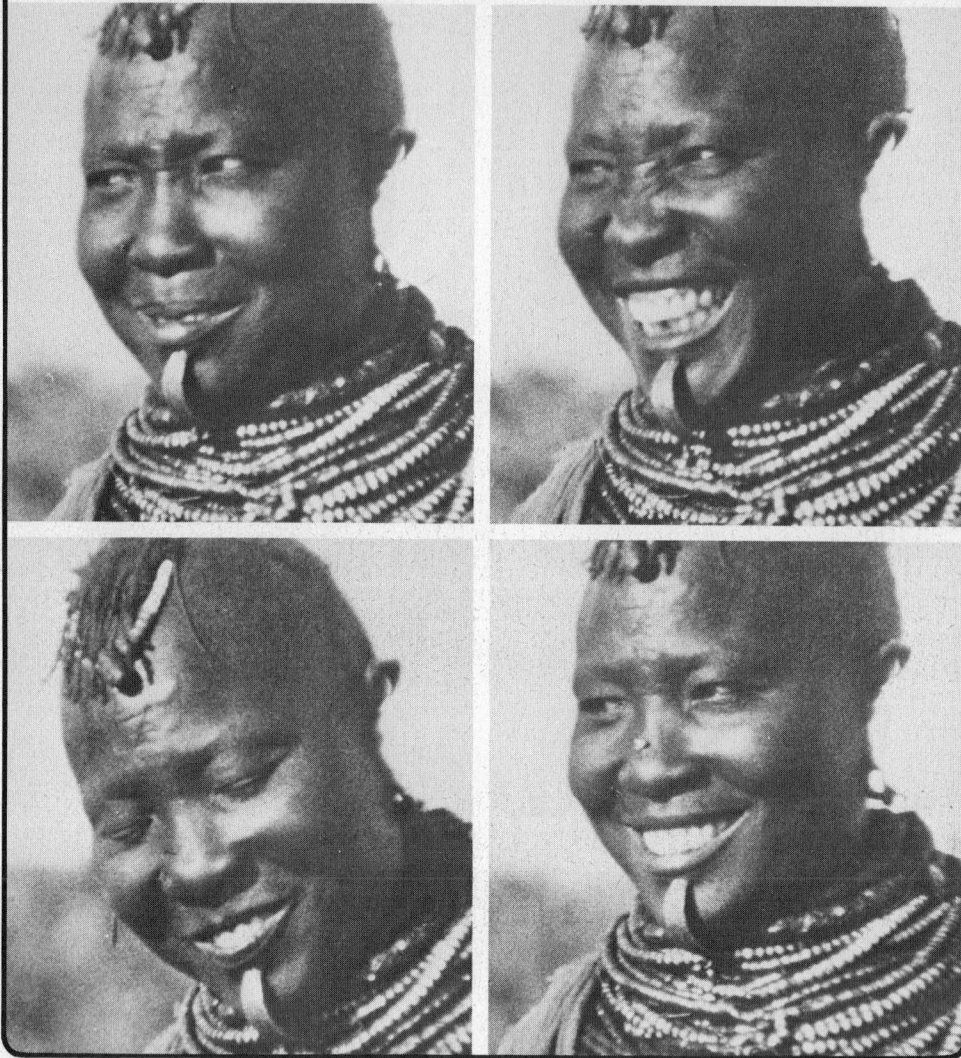

Fig. 11.4 Flirting Turkana woman of Kenya. (Hans Haas)

self, in certain contexts, such as when you are engaged in conversation with a friend, and an acquaintance walks by. Not wanting to interrupt your conversation, but not wishing to snub your acquaintance either, you acknowledge his presence by flipping your eyebrows up and down quickly. The eyebrow flash, as an affirmative greeting, is widely, but not universally, used in many human populations, most of which have little contact with each other and cannot easily be thought of as having learned this trait from outsiders (Figs. 11.2–11.3). A similar argument can be made for the embarrassment response, smiling, frowning, and so on (Figs. 11.4–11.7).

Fig. 11.5 (*Left*) Pouting Bushman girl of the Kalahari; (*right*) pouting Waika Indian of South America. (I. Eibl-Eibesfeldt)

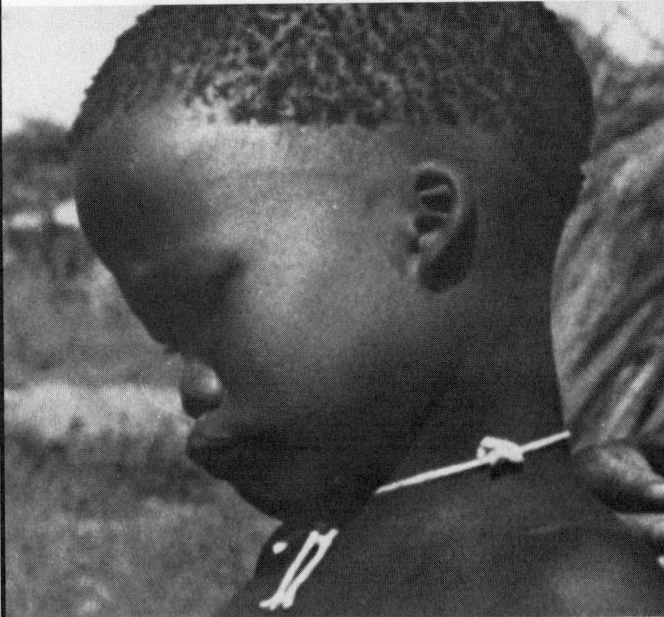

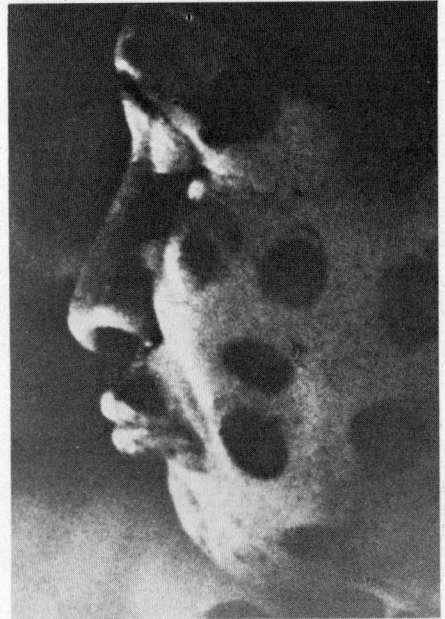

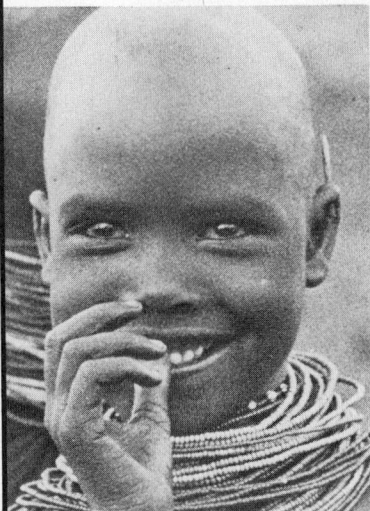

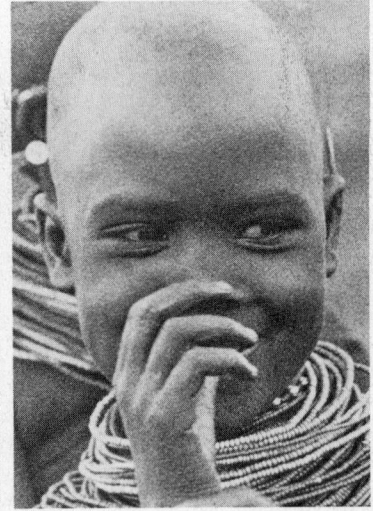

Fig. 11.6 Flirting Samburu girl of Indonesia. (I. Eibl-Eibesfeldt)

Fig. 11.7 (*Left*) Japanese Kabuki actor showing rage; (*right*) unrehearsed rage in a young boy. (I. Eibl-Eibesfeldt and Whightsil, Taurus)

**2. Different species**  FAPs are described as being species-specific, but the term simply means that the behavior in question is performed by virtually all members of the same species. It does *not* mean that the behavior is *limited* to that one species. In looking at other primates—monkeys and apes, for instance—we are struck by the great similarity in the facial expressions of these animals to the expressions of humans (Figs. 11.8–11.14). Indeed, so striking are these similarities that we can frequently quite correctly identify the context of the behavioral interaction from a photograph showing the facial expression of one of the participants. How did this similarity come about? Did we learn the expressions from monkeys, or did they learn them from us? Neither alternative is tenable. A more reasonable explanation is that these behaviors were present in our common ancestors and some have been retained in both monkeys and in humans. This conclusion naturally suggests a strong case for innate behavior, that is the presence of FAPs in humans.

Fig. 11.8 (*Left*) fear scream and (*right*) gesture of appeasement in a female baboon. (Russell Mittermeier, Anthro-Photo and DeVore, Anthro-Photo)

Fig. 11.9 (*Left*) human grimace (golfer Tony Jacklin) and (*right*) smile (his wife Vivien). (Wide World)

Fig. 11.10 (*Left*) Gorilla open-mouth threat and (*right*) baboon tense-mouth face. (Heidrich, Anthro-Photo and DeVore, Anthro-Photo)

Fig. 11.11 Human open-mouth threat (*left*) and tense-mouth face (*right*) Heads of state are dominant males often photographed making confident-threat gestures. (UPI)

Fig. 11.12  Savannah baboon in a bared-teeth scream. (DeVore, Anthro-Photo)

Fig. 11.13   Golden lion marmosets making an open-mouth threat. (Lyster, Zoological Society of London)

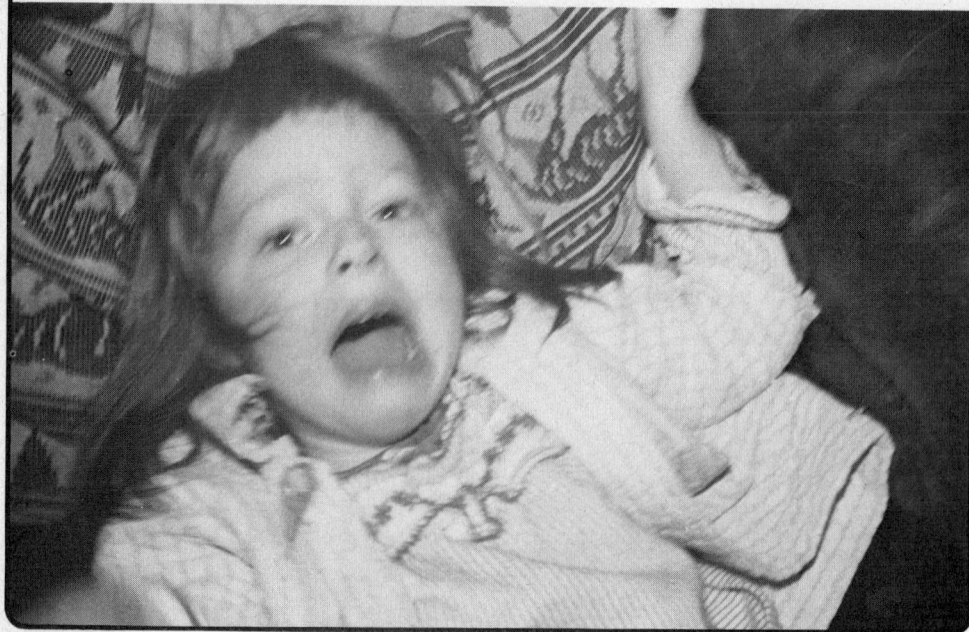

Fig. 11.14   Child with open-mouth scream. (A. Jolly)

**3. Human development**   The newborn infant exhibits a number of highly stereotyped behaviors, including random searching movements for the mother's breast and clinging (Figs. 11.15 and 11.16). Walking and swimming movements can also be demonstrated, although these behaviors soon disappear, as the infant's overall behavior regresses for a time during the first three months after birth. Prematurely born children may be even more adept at such behavior. Regardless of whether these behaviors are sophisticated reflexes or FAPs, they unquestionably are not learned.

Deaf–blind children, isolated from sensory input of two of the primary senses, and who, as a consequence, present many teaching problems, nonetheless develop the normal complement of facial expressions for the major emotions, including smiling, laughing, frowning, pouting, and so on (Fig. 11.17). Many of these expressions subsequently regress, presumably because of the lack of feedback. Given that the regression occurs in the **absence** of environmental stimulation, it is difficult to explain the acquisition of the behaviors as a **product** of environmental stimulation (i.e., learning). On the contrary, the ethologist would say that these are FAPs which become manifest at the appropriate developmental stage in the infant.

A different type of developmental evidence can be gleaned from comparisons of human and nonhuman primate behavioral developmental sequences (Fig. 11.18). It is apparent that many of the same behavioral traits occur among these dif-

Fig. 11.15
Rhythmic
search
for the breast
in neonate
human.

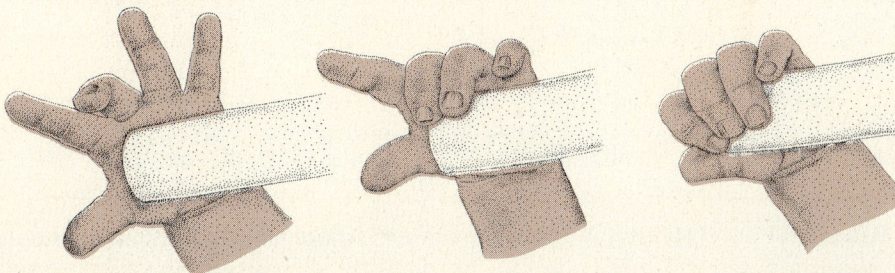

Fig. 11.16
Grasping
reflex
of the human
infant.

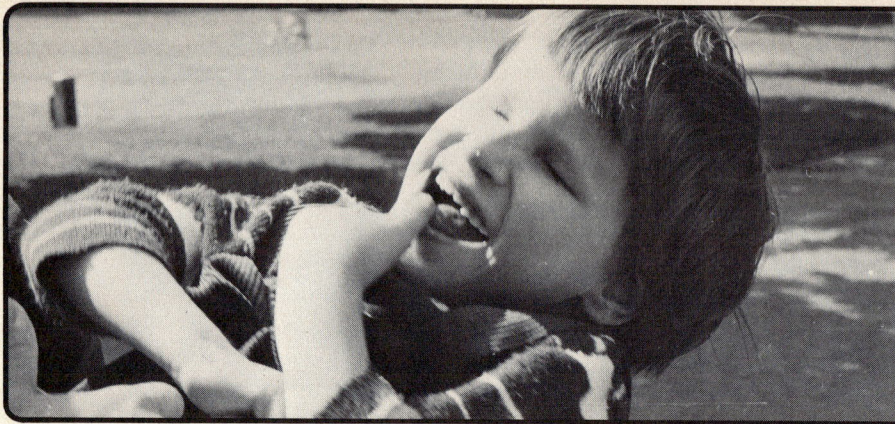

Fig. 11.17
Laughter in a
seven-year-old
deaf-blind child.
(I. Eibl-Eibesfeldt)

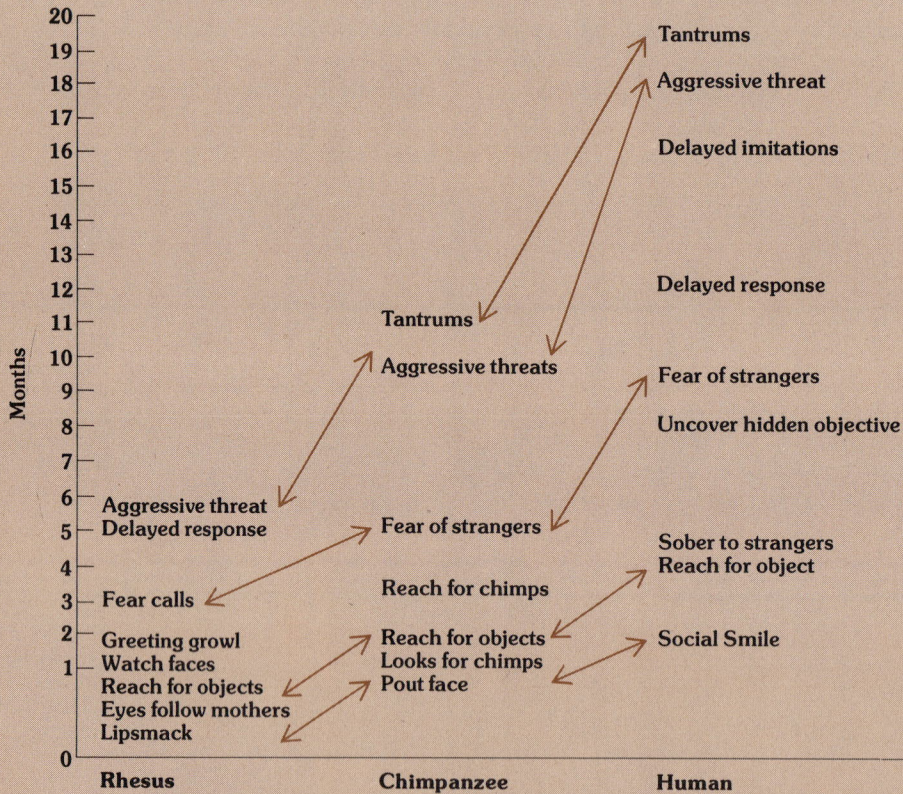

Fig. 11.18   Behavioral development in rhesus monkey, chimpanzee, and human compared.

ferent species, and that they develop in the same sequence, even though the age of the infant at which a given behavior is first seen may be quite different.

In balance, then, three rather independent lines of evidence all support the presumption of genetically determined FAPs in humans. The fact that FAPs have a genetic basis does not mean, of course, that we *must* perform them once the proper sign stimulus is presented. We have conscious control of our emotions, at least to some degree, and we can control such FAPs as facial expressions as well—although some of us are better at it than others. Those really adept individuals who can feign various emotional expressions at the drop of a hat go on to become great actors—or great con men!

**SOCIAL ORGANIZATION**　Social organization in primates ranges from virtually nil in the solitary species to very complex in species living in large groups. Because of inferences which can be drawn regarding our own social structure, it is only logical that we should be interested in the social organization of the other primates. It is not only the nature of the organization itself that is of interest, however, but more importantly, it is the mechanisms that caused the organization to come into being and to be maintained. Foremost among these is communication—the adhesive that holds societies together.

**Communication**　An important aspect of the behavior of virtually all animal species is communication. The significance of this behavior is in large measure a function of the social structure to which the animal belongs. Largely solitary species tend to have relatively poorly developed communicative behavior, whereas highly social species may have very sophisticated forms of communication.

Communication may involve the use of one or more sensory modalities. These include olfaction, audition, and vision.

**1. Olfaction**　Odor is a very common method of communication in many species, as you are no doubt aware if you have ever tried to walk a male dog past more than three consecutive vertical structures. Chemical communication has the advantage of being long-lasting and highly specific, but it is not an *effective* method of conveying a succession of messages quickly. (Imagine trying to carry on a "conversation" by wafting odors back and forth!) Nevertheless, this modality is the one of choice among invertebrates, wherein the number of messages utilized by a given species is rather few.

**2. Audition**　Communication of sound is paramount for ourselves, of course, but is much less important in most lower organisms. Other than in a few species of insects, sound communication is essentially a vertebrate invention, presumably because, to be effective, it requires not only a mechanism with which to perceive sound (an "ear" of some sort), but also requires some apparatus that is capable of producing a variety of sounds. As it happens, most animal species have a very limited "vocabulary"—frogs are limited to as few as three or four sounds, birds generally do not exceed ten, and even the apes have fewer than 30 distinctive sounds.

Sound has the advantage of being instantly perceivable over a broad area, and it is highly effective even in wooded areas, where visual communication is limited. Typically it has a "here I am" function, which serves to intimidate others of the same sex and to lure members of the opposite sex. In primates, it is more broadly used to reinforce visual communication.

**3. Vision** Visual cues disappear instantly, of course, and are effective only over a short range (one has to be seen). Moreover, they are easily disrupted in a forest setting. Nevertheless, they remain an important method of communication for mammals in general and for primates in particular. Typically, visual cues involve distinctive body postures coupled with distinctive facial expressions. As we move from the higher vertebrates, to mammals, primates, and ultimately to humans, we find that the facial musculature is increasingly more refined, allowing increasingly greater numbers of different facial expressions. Over 30 such expressions are

**Fig. 11.19**
**Aggressive displays by chimpanzees.**

Fig. 11.20 A specialized threat gesture, the yawn, in a Savannah baboon. (DeVore, Anthro-Photo)

known in humans, for example, collectively serving to communicate a whole range of feelings and emotions.

**Ritualization and aggression** The more socially complex a species is, the more intraspecific interactions will occur. These interactions frequently take the form of encounters over a limited resource, such as food, resting places, or females. Many mammals are equipped with lethal weapons of offense or defense, such as antlers, elongated canine teeth, claws and hooves, and so on. It is obvious that many encounters could rather easily result in bloodshed and death, results that are undesirable if a species is to survive beyond a single generation.

As it happens, very frequently these interactions have evolved a certain *pro forma* aspect, which limits physical damage. Over time, various displacement activities and **intention movements** (so-called because they anticipate an action, as when a bull paws the ground before charging) have become systematically built into the behavior of a species solely for communicative purposes. Such behavior is said to be **ritualized,** meaning that it is highly exaggerated stereotyped behavior that serves purely communicative purposes, very frequently in aggressive encounters (Figs. 11.19–11.21).

Fig. 11.21  Human aggression. (Karales, Peter Arnold)

The effectiveness of ritualized behavior in communication, both in terms of preventing ambiguity of message and of reducing the probability for actual damage, is impressive. Were you to face that gorilla who was bashing a tree around, you would know enough not to assume that he was using it to shoo mosquitoes. His message is very clear, not only to you, but also to other gorillas. A dog or wolf will thwart an attack by another of its species by exposing its neck or belly (the two most vulnerable areas) to its aggressor. Bull elk always "attack" each other antler to antler, rather than broadside.

What about ritualized behavior in humans? Some authorities maintain that humans have developed very little in the way of ritualized aggressive behavior, but this is only partially true. A clenched fist, a scowl, or a finger wagged under the nose all communicate anger and hostility, and, to some extent, these behaviors are mitagated by such appeasing gestures as an open hand, a smile, or looking away. The fact is, however, that, as a species, we have only recently developed the capacity for lethality. It is rather difficult to beat someone of your own size to death with your fists, but it takes nothing more than a twitch of your finger to put a bullet in his head. Therefore, a more correct statement of ritualized behavior in humans would be that we have never evolved the *level* of ritualized aggression and appeasement postures necessary to deal with our recent technological advances. Moreover, because of our ability to override genetically based behavior by conscious thought, what little ritualized behavior we have seems to be easily subverted by appeals to "the fatherland" or "the flag," or by the swish of new uniforms. We can hardly evolve effective ritualized appeasement gestures overnight, but by comparing our actions with those of lethally equipped animals, we can (and should) recognize our shortcomings and try to compensate for them.

**The evolution of human social organization**  There are many dangers in extrapolating from the behavior shown by our primate cousins to the behavior of humans. However, if we proceed cautiously, we can make some educated guesses as to the events that must have taken place from a period several million years ago until the development of recorded history in terms of the acquisition of social behavior by our immediate ancestors. It is generally accepted that three changes must have occurred, each of which represents a dramatic departure from anything observed in modern nonhuman primate species. These include the construction of **tools,** the development of a sophisticated **language,** and the formation of the **pair bond.**

**Use of tools**  For many years, it was widely accepted that humans differed from other animal species in that only humans used tools. We have subsequently found birds using cactus thorns to dig insects out of trees, sea otters cracking clams open on flat rocks, and chimpanzees impaling termites on twigs in a shish kebab fashion. Moreover, it is not even enough to attempt to distinguish humans as being uniquely able to *construct* tools (as opposed to using naturally occurring formations), as chimpanzees are capable of connecting a series of short sticks to make one long enough to reach a banana that would otherwise be out of reach. The fact remains, however, that the *extent* of our use of tools far outstrips that of any other species.

One component of ritualized behavior deserves a special note, and that is the nature of the interactions between young and adults. It is very unusual for an adult organism to attack the young of the same species, especially in the more social species. How are young differentiated from other species members, other than on the basis of size? The answer is that body shape is markedly different in the young of most animals, as compared with the adults. The juvenile profile (Figs. 11.31 and 11.32) evokes

The cuteness response: a comparison of juvenile profiles and adult profiles.

what has been called the **cuteness response** in ourselves, and it is arguable that some similar response must be operative in other animals as well. Look at Fig. 11.22 and see if your reaction isn't a satisfied "Aaahhh."

In addition, infants and children reinforce the cuteness response by smiling. So pervasive is this reaction to gratifying situations that it is difficult to prevent oneself from smiling in such contexts. Compare Figs. 11.7 (right), 11.14, 11.17; most people will feel warm toward Fig. 11.17, cool toward Fig. 11.14, and outright hostile toward Fig. 11.7 (right). ("I'd like to slap that expression right off his face!") Collectively, these illustrations also demonstrate the significance to our species of nonverbal communication.

Fig. 11.22    Newborn rhesus monkey. (DeVore, Anthro-Photo)

Tool development in prehumans can be traced back more than 1,000,000 years when the long bones of prey animals were split and the sharp edges were used as tools. Stone tools date back several hundred thousand years, and they show increasing sophistication as we move toward the present. Tool development, then, is relatively easy to trace, as it leaves a record, in the same way as do fossils.

Tools were important to social structure because they ultimately allowed a **division of labor** within the society. Such an arrangement is quite unlike any of the nonhuman primate societies, wherein, despite the appearance of sophisticated interactions, in the final analysis it is every ape for himself—except for nursing infants, each individual must be capable of foraging for himself or he faces starvation. In short, there is no Social Security for elderly apes. Presumably, the prehumans underwent this same stage, but the development of tools meant that an individual too old to hunt, for example, could still be useful to the society by making weapons, and so forth. Therefore, others of the group would be willing to provide him with food.

**Communication**  Apes and monkeys, of course, are quite capable of making a large number of sounds for communicative purposes, and obviously they regularly employ sounds in just this way. But it would be erroneous to suggest that they have developed a language, for many of the parameters of language (e.g., factors 10–16 in Table 11.1) do not exist in the communicative system of all species save that of humans.

At what point the development of language among prehumans began is uncertain, but the argument has been made repeatedly that the advantage of language was so great that it provided the selective pressure for the evolution of the large brain by which we characterize modern humans.

The evolution of man is discussed in Chapter 13, but it is interesting to note that a major difference between modern humans and Neanderthal man is the shape of the skull. It has been suggested that the shape of the Neanderthal skull, especially the formation of the mouth region, was such that little verbal refinement would have been possible in their language. As such, they would have been at a considerable disadvantage to Cro-Magnon man (our immediate ancestor) who was presumably much more articulate, and this may have been why the Neanderthalers disappeared so suddenly from the fossil record.

**Pair bond**  Although several species of birds and mammals mate for life, pair bonds within a larger social organization are unknown among nonhuman primates. Moreover, the absence of the pair bond in the other social primates seems to hinge on the relative absence of sexual receptivity on the part of the female. As in most animals, sexual receptivity in nonhuman primates is confined to a brief fraction of the adult female's life. In our primate cousins, this time of receptivity is restricted to the time approaching ovulation, a pattern not unlike that of dogs and cats. The dominant male in a baboon troop or in a group of gorillas associates with any female who is ovulating and who is therefore sexually responsive.

In contrast, the human female is sexually responsive virtually at all times. As such, the possibility of a permanent association between a given male and female

**Table 11.1**
**Design features of language**

DF 1   Vocal-auditory channel
DF 2   Broadcast transmission and directional reception
DF 3   Rapid fading (the sound of speech does not hover in the air)
DF 4   Interchangeability (adult members of any speech community are interchangeably transmitters and receivers of linguistic signals)
DF 5   Complete feedback (the speaker hears everything relevant of what he says)
DF 6   Specialization (the direct-energetic consequences of linguistic signals are biologically unimportant; only the triggering consequences are important)
DF 7   Semanticity (linguistic signals function to correlate and organize the life of a community because there are associative ties between signal elements and features in the world; in short, some linguistic forms have denotations)
DF 8   Arbitrariness (the relation between a meaningful element in a language and its denotation is independent of any physical or geometrical resemblance between the two)
DF 9   Discreteness (the possible messages in any language constitute a discrete repertoire rather than a continuous one)
DF 10   Displacement (we can talk about things that are remote in time, space, or both from the site of the communicative transaction)
DF 11   Openness (new linguistic messages are coined freely and easily, and, in context, are usually understood)
DF 12   Tradition (the conventions of any one human language are passed down by teaching and learning, not through the germ plasm)
DF 13   Duality of patterning (every language has a patterning in terms of arbitrary but stable meaningless signal-elements and also a patterning in terms of minimum meaningful arrangements of those elements)
DF 14   Prevarication (we can say things that are false or meaningless)
DF 15   Reflexiveness (in a language, we can communicate about the very system in which we are communicating)
DF 16   Learnability (a speaker of a language can learn another language)

After Hockett and Altmann (1968) and Marler (1969b).

must have arisen based on their sexual responsiveness. Such a development would have had profound implications on the social structure of the prehumans, for it would have increased the division of labor. Quite possibly, the females stayed in camp, cared for the children, and gathered food found growing nearby, while the males cooperatively hunted prey too large for any one male to hunt alone. Food would then have been shared, depending on whether the male or the female was the more successful.

With the females leading a relatively stationary existence, as opposed to being constantly on the move as in the nonhuman primates, more prolonged child care was made possible, and the infant ultimately became increasingly more dependent on the mother, in the sense that adulthood was not achieved for many years. The pattern would thus have been established whereby extended periods of learning could take place between generations.

In sum, the development of language, of tools, and of pair-bonding were presumably the forces responsible for the evolution of a highly integrated and cooperative society, with a degree of sharing and a division of labor among the members totally unknown among the nonhuman primates. Such a complex society in turn gave rise to **culture**—the transmission of information and beliefs between generations, without which we would be compelled to rediscover fire and the wheel with every generation.

## SUMMARY

As much as any branch of biology, ethology tells us that we differ in degree, but not in kind, from other animals. Such a revelation should not be greeted with dismay, however, but rather with an acceptance of our own shortcomings and a recognition of the necessity to compensate for them. We also need to recognize that pair-bonding is at the heart of our social structure and has been in all likelihood for hundreds of thousands of years. Hence, when we become iconoclastic, we must take care that the beliefs we shatter are not those which make us what we are, but rather, are only those which hold us back from what we might become.

Why isn't it possible to be vaccinated against cancer the way we are vaccinated against other serious diseases? How can a single fertilized egg, simply on the basis of successive divisions, develop into the specialized tissues and organs that characterize the mature adult? Is it possible to treat specialized cells in some way and cause them to begin dividing like a fertilized egg?

All of these questions relate to development, which is the subject of this chapter.

chapter XII

# chapter XII
# From fertilized egg to adult–division, development, and differentiation

Multicellular organisms reveal their heritage from simpler organisms in that, for the most part, new individuals begin as a single undifferentiated cell. (To be sure, in such vegetatively propagating organisms as strawberries, the beginning of a new individual in the form of a single cell is less obvious. Such examples, however, are rare compared to the number of species that follows the general rule.) One of the areas of greatest activity in modern biology is aimed at understanding the mechanisms of **development**—the process whereby a single cell divides many times and ultimately becomes a mature member of its species. It is one of those quirks of nature that the fundamental mechanism underlying **differentiation,** in which cells *change* in form and function, is precisely the same mechanism which, in the mature organism, is responsible for the replacement of worn-out or damaged cells with *identical* counterparts. This phenomenon which, depending on the context, is responsible either for change or for stability, is termed **mitosis**—the division of a single cell into two daughter cells.

**MITOSIS** Mitosis has been known for more than 100 years, because it is both a common[1] event as well as one easily observed through a light microscope. The actual process of cell division is very similar throughout both plant and animal kingdoms. As the early observers noted, mitosis begins with the gradual deterioration of the nuclear membrane, coupled with the appearance of the **chromosomes** as distinct bodies. Their appearance results from their change from an elongated condition into a series of short, tight coils. In the uncoiled state, the chromosomes of each of our cells would total about 1.8 m (6 ft) in length. At the same time, a small cytoplasmic organelle, the **centriole,** splits and the two halves begin

---

[1] In ourselves, mitosis occurs at the rate of 50,000,000 divisions every second.

moving in opposite directions around the outside of the nucleus, ultimately taking positions directly opposite each other. The events just described typify that portion of mitosis known as **prophase** (Fig. 12.1).

The next stage, **metaphase,** occurs with the lining up of the chromosomes along a vertical axis at the center of the cell. At this point, it is apparent that each chromosome is actually present in a doubled form, with each half (called a **chromatid**) linked to the other at a single point, the **centromere.** The centromere may be located at any point along the chromosome, although for a given chromosome its position is constant. Also appearing at this same time are a series of thin strands, which stretch from the now-opposed centrioles to each centromere. These strands are called **spindle fibers.**

Almost immediately, the chromatids separate from each other and begin moving toward opposite sides of the cell. Once separated, the chromatids are termed chromosomes. This stage of division, which involves chromosomal migration, is called **anaphase.**

At the end of the migration, the chromosomes cluster about each other and begin to uncoil, reversing the process described for prophase, gradually fading from sight. At the same time, the nuclear membrane begins to reappear, and a **cleavage furrow** begins in the cytoplasm, which ultimately culminates in the total separation of the daughter cells. This stage is called **telophase,** and completes the stages of mitosis.

All these details were known to the light microscopists of the nineteenth century, but it was the significance of the events, beyond the obvious dividing of the cell, that was not understood. For example, "chromosomes" (meaning "colored bodies") were so called simply because they stained darkly in microscopic slide preparations. Their significance in heredity was suspected but not proved until the 1940s. Moreover, the period between cell divisions (now termed the **interphase**) was originally called the "resting stage," which implied that the cell was relatively quiescent. We now know that all of the processes for growth, including replication of the chromosomes, occur during interphase (hence, the term "resting stage" hardly seems appropriate), and mitosis is simply the culmination of all of these prior events. In addition, it is now appreciated more than ever that the four "stages" of mitosis are nothing more than arbitrary, human-imposed categories for what is actually a continuous process, and the labels are retained only because it is convenient to place the many events that transpire during mitosis in some sequential ordering. Indeed, the nineteenth-century level of understanding of this process is perfectly exemplified by the name "mitosis," which means nothing more than "thread condition," an apparent reference to the sudden appearance of the chromosomes. As has occurred in so many other aspects of biology, description far preceded an understanding of process—and, as we shall see, the level of our understanding of mitosis even today is none too impressive.

**Cancer—mitosis run amok**  It is easy to become smug at the ignorance of the nineteenth century light microscopists, who were descriptive biologists little given to the construction of theories by which biology became transmuted into the qualitative and analytical science we now know. Needless to say, biology only be-

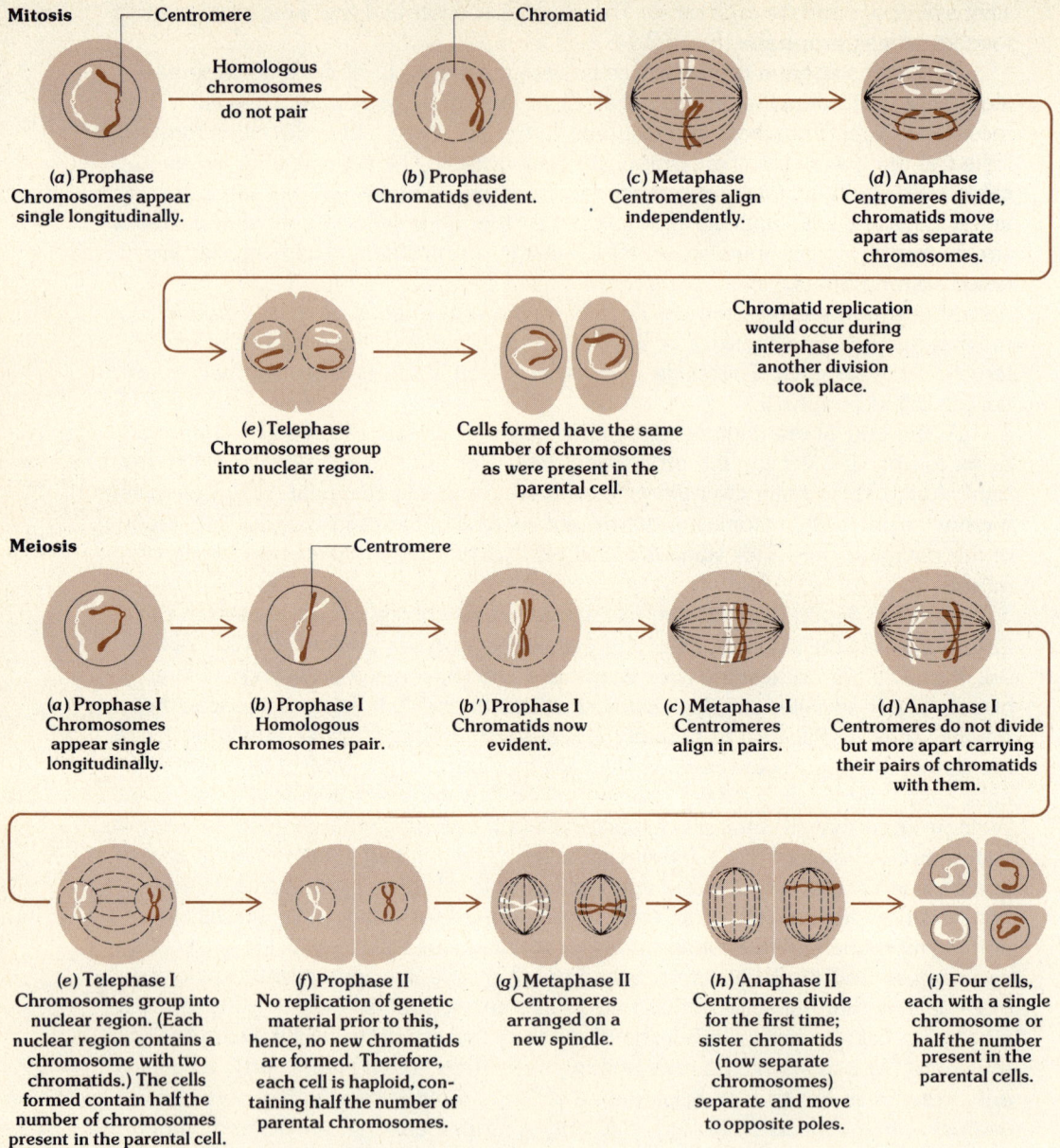

**Mitosis** — Centromere — Chromatid

Homologous
chromosomes
do not pair

**(a) Prophase**
Chromosomes appear
single longitudinally.

**(b) Prophase**
Chromatids evident.

**(c) Metaphase**
Centromeres align
independently.

**(d) Anaphase**
Centromeres divide,
chromatids move
apart as separate
chromosomes.

**(e) Telephase**
Chromosomes group
into nuclear region.

Cells formed have the same
number of chromosomes
as were present in the
parental cell.

Chromatid replication
would occur during
interphase before
another division
took place.

**Meiosis** — Centromere

**(a) Prophase I**
Chromosomes
appear single
longitudinally.

**(b) Prophase I**
Homologous
chromosomes pair.

**(b′) Prophase I**
Chromatids now
evident.

**(c) Metaphase I**
Centromeres
align in pairs.

**(d) Anaphase I**
Centromeres do not divide
but more apart carrying
their pairs of chromatids
with them.

**(e) Telephase I**
Chromosomes group into
nuclear region. (Each
nuclear region contains a
chromosome with two
chromatids.) The cells
formed contain half the
number of chromosomes
present in the parental cell.

**(f) Prophase II**
No replication of genetic
material prior to this,
hence, no new chromatids
are formed. Therefore,
each cell is haploid, containing half the number of
parental chromosomes.

**(g) Metaphase II**
Centromeres
arranged on a
new spindle.

**(h) Anaphase II**
Centromeres divide
for the first time;
sister chromatids
(now separate
chromosomes)
separate and move
to opposite poles.

**(i) Four cells,**
each with a single
chromosome or
half the number
present in the
parental cells.

**Fig. 12.1   Mitosis and meiosis compared and contrasted.**

came theoretical once a large descriptive foundation was laid. However, it is perhaps ironic that we still do not know a great deal about the most important aspect of mitosis—that is, what causes it to happen.

Thus far, we have treated mitosis as an abstraction, unlinked to any particular cell. Yet it is not a universal phenomenon in all cell types, and even when it does occur, the rate of mitotic divisions may vary enormously. Generally speaking, cells of a given tissue divide at a certain rate. The cells lining the gut are replaced completely every 36 hr; those comprising our skin, every few days; and red blood cells, every few weeks. Certain cells, such as muscle and nerve cells, survive a lifetime. Of course, in times of injury, the normal mitotic rate might prove too slow. For example, a cut in the skin initially leads to the formation of a blood clot, which in turn becomes a scab (Fig. 12.2). Subsequently, the injured area is invaded by

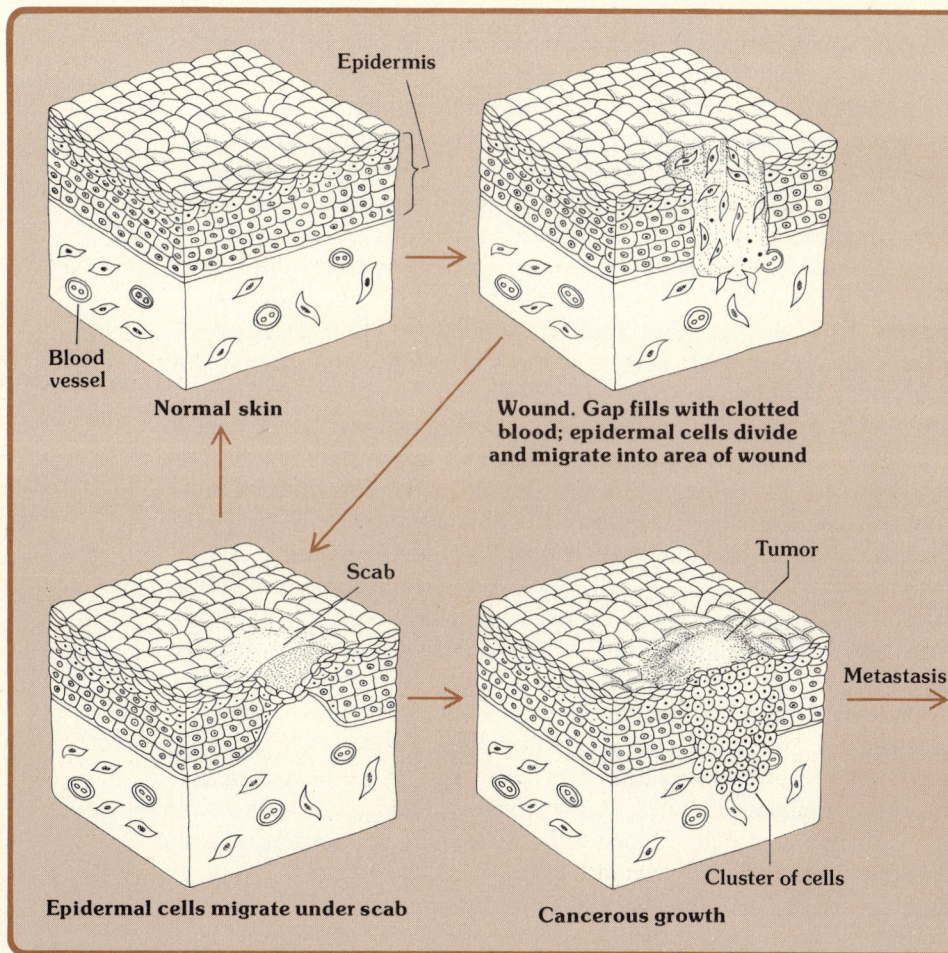

**Fig. 12.2 Wound-healing and cancer formation contrasted.**

Epidermis

Blood vessel

**Normal skin**

**Wound. Gap fills with clotted blood; epidermal cells divide and migrate into area of wound**

Scab

**Epidermal cells migrate under scab**

Tumor

Metastasis

**Cancerous growth**

Cluster of cells

**Specialized cells and immortality**

surrounding skin cells which divide rather quickly and soon fill the gap created by the wound. A return to normal rates of division occurs once the gap is filled, which suggests that the presence of similar cells in close proximity somehow slows the mitotic rate. Just how this **contact inhibition** is accomplished is unknown, but a very similar situation occurs during embryonic development, which is discussed later in this chapter. Such highly specialized cells as nerve, muscle, and red blood cells never divide—multiplication of these cells must come from precursor cells which divide prior to becoming specialized (*see box*). For example, it is the loss of mitotic capacity in nerve cells which accounts for the general failure of severed nerves to regenerate. This failure is most graphic, of course, in instances wherein the spinal cord has been damaged. However, the loss of mitotic capacity by highly specialized cells is not as troubling as is the lack of information about what controls mitotic rates in cells that do divide. This is not a totally esoteric question as it is the increase in mitotic rate which characterizes **cancer**—the cause of more than 300,000 deaths per year in this country, or about one-fifth of deaths from all causes.

Cancer is not really a disease but, like fever, it is only a symptom. This is an important point, as it accounts for the fact that "cancer" can result (as can a fever) from many different causes. The symptom of cancer is the development of clumps of cells, most commonly in epithelial tissue, in which mitosis is occurring at above-normal rates and in an uncontrolled fashion. To be sure, those very traits are true, at least to a limited degree, of such noncancerous diseases as psoriasis or warts. But the critical difference between these conditions and cancer is that in cancer these clumps of tissue are capable of spreading throughout the body and invading other tissues, a process known as **metastasis**. In most cases, these tissues are localized as a **tumor** (or **neoplasm,** meaning "new growth"), although in such cancers as **leukemia** the tissue involved (white blood cells) by its very nature precludes the development of a distinct tumor.

Tumors are characterized as either **benign** or **malignant** (Fig. 12.3). Only the latter type are considered cancerous. A benign tumor is one which, although it may increase in size, is, by definition, incapable of spreading to other parts of the body. In contrast, a malignant tumor can spread, by metastasis. This designation would seem to suggest, however, that benign tumors are not hazardous, which is not necessarily the case. The effects of tumor growth may cause crowding of other organs with resultant pain or, as in the case of a brain tumor, possible death as the brain is squeezed against the unyielding skull. Moreover, benign tumors may sometimes become malignant, and for that reason are generally removed once discovered.

The threat to life from malignant tumors is not so much from the growth of the tumor itself as from the hazards of metastasis, in which bits of tumor tissue break off and are transported by the circulatory system to other parts of the body. (Cancer cells are known to be less "sticky" than are normal cells, which may account for their tendency to break away from the tumor.) Initially, these tissue clumps are destroyed by the lymph nodes, but a point is reached whereupon the lymphatic system is overwhelmed. This point may not be reached for months or even years after the initial onset of cancer, which explains why it is that continued vigilance is necessary after the removal of a malignant tumor, in case the process of metastasis is already underway. Once the lymphatic system becomes overwhelmed, it may itself be the focal point for renewed tumor growth (for this reason, all surrounding lymph nodes are removed along with the malignant tumor), or tumor growth may begin in another tissue. Interestingly, not all cancers show the same propensity for metastasis. Some types of skin and brain cancer rarely metastasize, whereas breast and lung cancers metastasize readily. Similarly, the various types of cancers are capable of invading only certain types of tissues. Lung cancers tend to invade the brain, whereas prostate cancer commonly invades bone. These facts provide further evidence of the earlier statement that "cancer" is a hodgepodge of conditions, sharing only the common trait of metastasis.

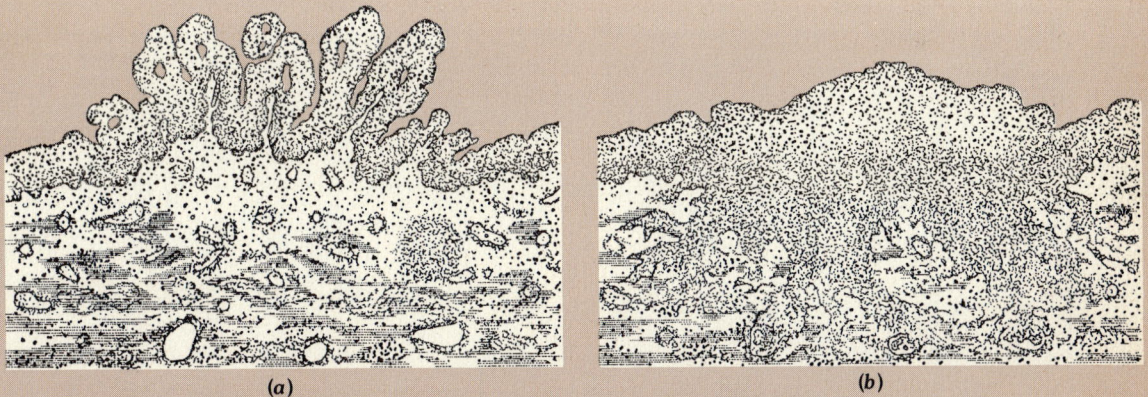

Fig. 12.3  (a) Benign and (b) malignant tumors. Note the clawlike penetration of the tumor in (b)—hence the term "cancer," from the Latin *cancri*, meaning crab.

**Causes of cancer** Because cancer is not a specific disease, but rather is a symptom characterized by uncontrolled cell division, it should not be surprising to learn that the cause of cancer can be anything that affects mitotic rate. At present, there is no definitive evidence that any human cancer is caused by a virus, although the fact that certain animal cancers are known to be virus-induced has prompted continued investigation on this route of causation. On the other hand, it is known quite definitely that a variety of environmental conditions may induce cancer. Lung cancer is highly correlated with pollutants in the air, most notably cigarette smoke (Fig. 12.4). There is increasing evidence that cancer of the large intestine is linked with diets high in animal protein (or low in cereals—because these tend to go together, it is difficult to determine which is the causative agent). Dyeworkers have been found to have a high rate of bladder cancer—and chimney sweeps have a tendency for cancer of the scrotum. Cancer of the penis is linked to the absence of circumcision. Breast cancer seems to have a hormonal relationship—it is more common in women who have never had children or who have not breast-fed than in women who have, and it is far more common in women than in men. A number of chemical agents, including certain pesticides and artificial sweeteners have been banned because in high doses, they can cause cancerous tumors in experimental animals (and, by inference, perhaps in humans as well). Various forms of radiation, ranging from the ultraviolet light present in natural sunlight to radiation released in atomic explosions may also cause cancers.

A common pattern found in most of these cancers is that they tend to occur with increasing frequency as a function of age. For example, the incidence of lung cancer parallels the smoking habits of individuals of 20 years ago. Thus, there is frequently a long lag period between the time of initial environmental insult and the ultimate development of a cancer. The incidence of lung cancer is only now beginning to show a marked increase in women, whereas the increase in men began 30 years

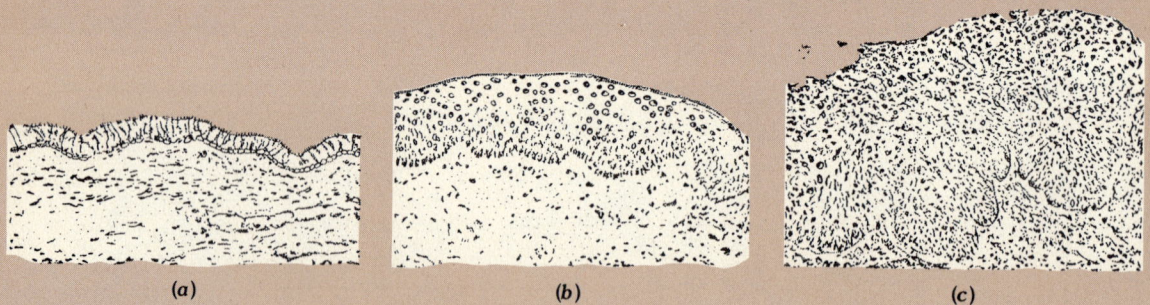

(a)                                      (b)                                      (c)

**Fig. 12.4  Stages in the development of bronchial cancer: (a) normal epithelium; (b) thickened epithelium; (c) malignancy and invasion of underlying connective tissue.**

ago (Fig. 12.5). This sexual difference is attributed to the relatively low cigarette consumption of women until World War II, whereas cigarette smoking in men has been common since the turn of the century. Comparable patterns can be demonstrated for many other cancers as well.

The prevailing theory behind this apparent lag is that before a cell is to become cancerous, several independent genes must mutate. It is argued that the accu-

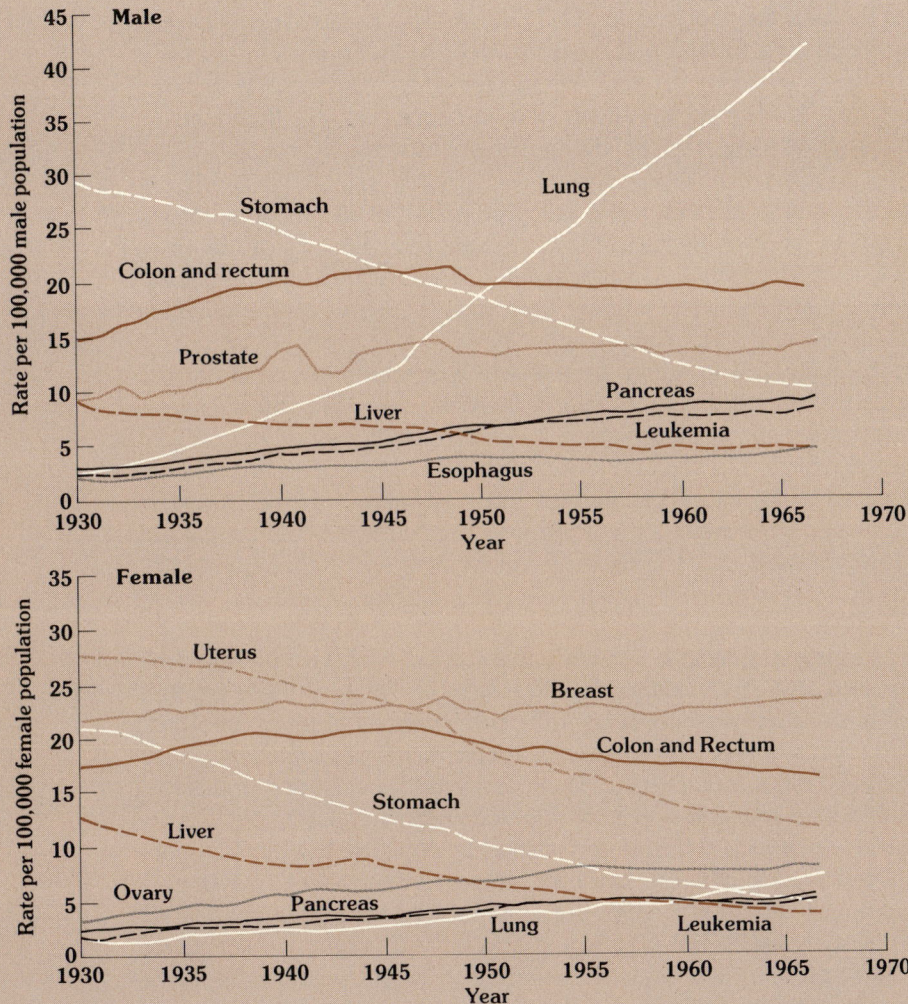

Fig. 12.5  Cancer death rates by site in the United States, 1930–1967.

mulation of such mutations requires the passage of time, as well as continued exposure to environmental insult (although the time required may be relatively short to such profound insults as radiation exposure). Consequently, we can in no sense be guaranteed free from the threat of cancer merely by eliminating an environmental insult at a time prior to the development of a tumor, because tumors may arise 20 years hence, even in the absence of continued insult.

**Treatment of cancer**  The problem in attempting to cure an individual of cancer is that the immediate threat posed is the potential colonization of other organs of the body by the body's own cells. These cells are malignant and therefore aberrant, to be sure, but they are the body's own cells nonetheless. Therefore, unlike diseases caused by a foreign pathogen, it would seem impossible to immunize the body, as there would be no apparent way in which the immune system could be programmed to respond selectively only to cancerous cells (although some work is being done to determine whether or not this can, in fact, be achieved).

The ugly facts of cancer are that fewer than 50 percent of all cancer patients survive more than five years after the cancer is first diagnosed. Although a few cancers do respond well to specific treatments, most cancers are treated by whatever combination of drugs, surgery, and radiation therapy best meet the needs of the individual patient.

Ultimately, of course, it is expected that all of the factors that control normal mitosis will be discovered and that more specific mechanisms will consequently be developed to prevent the spread of malignant cells. However, many knowledgeable scientists propose elimination of such environmental hazards as cigarettes and other **carcinogens** (cancer-causing agents) as the best and most logical method of reducing the incidence of cancer. Preventative medicine was largely responsible for the eradication of infectious diseases, and a similar approach might well be followed with cancer.

**MEIOSIS**  In many organisms, including many species of bacteria and protozoa, mitosis is not merely the mechanism for growth, for because these organisms are unicellular, it is also their method of reproduction. The complexities involved in using mitosis as a means of reproduction increase enormously as organisms grow larger and increasingly more multicellular. Thus, as would be expected, mitosis is not the typical method of reproduction in multicellular plants and animals. You can imagine the problems you would face if you wished to duplicate another of yourself by a kind of giant mitotic division. Not only would all the cells of your body have to divide essentially at the same time (an impossibility for certain specialized cells, as we have already seen), but they would also have to migrate in such a way as to allow a splitting off of a new organism. All in all, it would seem an impractical method, to say nothing of the problems of deciding which of you had rights to your nondividing tennis racket or rock albums.

However, there is another serious drawback in utilizing mitosis as a method of reproduction—mitosis has no mechanism built into it to achieve variability. In fact, the whole "objective" of mitosis is to produce a cell that is genetically identical to the one that divided. So equal is this division that we do not even think of the process in terms of parent and offspring—both cells are identical, so who is to say which is the parent and which the offspring? This is an unfortunate drawback for, as we shall see in Chapter 13, a population of organisms (whether amoebae or elephants) that vary genetically amongst themselves are more likely to survive environmental changes than are organisms that are all genetically identical. Therefore, early in the evolution of living organisms, a variant form of mitosis developed in which changes in the genetic constitution of the next generation could occur. This outgrowth of mitosis is called **meiosis,**[2] and it is the basis of sexual reproduction.

Undoubtedly, you already know that the messengers of inheritance are called **genes,** and these are not found rattling around in the nucleus like a bunch of marbles, but rather are strung together in a consistent way so as to form chromosomes. If you were to design a method of ensuring genetic variability in offspring, relative to their parents, you might conclude that, in creating the genetic constitution of the offspring, the easiest method would be simply to combine the chromosomes of the two parents in some fashion. The immediate difficulty you would face, however, would be that just by cramming the chromosomes of the parents together, you would double the number of chromosomes possessed by the offspring. In a few generations, there would be no room in the cells for anything but chromosomes. Obviously, your design must include some mechanism of ensuring that no automatic doubling of chromosomes occurs with each generation. Preventing such a doubling is, in fact, the primary function of meiosis—and, although it may seem a bit complex at first, if you bear in mind what the "objective" of the process is (i.e., the retention of a constant number of chromosomes in every generation), it makes a certain amount of sense.

Meiosis consists of two successive cell divisions, which result in the formation of four cells, each with half the number of chromosomes that were present in the original cell. The first division is called the **reduction division,** because it is at this stage that the actual reduction in chromosome number occurs. The phases followed are exactly like those in mitosis and, in fact, they are given the same names (Fig. 12.1). However, a major difference occurs in the prophase of the first cell division (called *prophase I*) in that the chromosomes line up not as single chromosomes but rather as pairs of chromosomes. Our chromosomes consist of 22 visually indistinguishable **homologous** pairs, one member of each pair coming from each of our parents. In addition, there are two more chromosomes which are a true pair in females (and are designated as XX), but which are an unequal pair in males (and are designated XY), in that one chromosome (Y) is much smaller than the other. Hence, our body cells contain 46 chromosomes. At the end of prophase, a cell undergoing mitosis would show a lineup of 46 chromosomes, each present as a set of two chromatids, linked to each other at the centromere. In contrast, a cell undergoing meiosis would show 23 **pairs** of chromosomes, again each chromosome being present as a set of two chromatids. Thus, separation of the chromosomes during anaphase I of meiosis does

---

[2] The term *meiosis* comes from the Greek, meaning "to lessen," a reference to the decrease in chromosome number which typifies this process.

not consist of the rupturing of the centromeres, but rather of a separation of each of the 23 pairs of chromosomes. Thus, the chromosome number of the resulting cells has been reduced from 46 to 23. You may find that statement hard to accept, as each of those 23 chromosomes is present as a set of two chromatids, but the number of chromosomes in a cell is determined by convention on the basis of the number of centromeres present—and following meiosis I, there are only 23 centromeres in each of the daughter cells.

The important thing to remember in all of this is that the lining up in pairs occurs totally at random. That is, there is no mechanism whereby, for example, each of the paternal chromosomes lines up on the left side, and each of the maternal chromosomes lines up on the right. Rather, there is no pattern at all, with the result that the cells produced by the first meiotic division contain a unique combination of chromosomes, and not merely the same precise chromosomes possessed by the individual's mother or father. Thus, the effect of meiosis I is to recombine the genetic material (the chromosomes) contributed to the individual by his or her parents in a totally new way. Rather clearly, such a reshuffling has the effect of providing the genetic variability which we earlier argued was a welcome part of sexual reproduction. In fact, in the human cell, there are $2^{23}$, or over 8,000,000 possible ways of combining the maternal and paternal chromosomes. The chance of two individuals being genetically identical (i.e., the chance of two identical sperm fusing with two identical eggs), even assuming common parents, is less than 1 in 64,000,000,000. Thus, despite their frequency on television dramas, unrelated but identical individuals would seem a virtual impossibility.

Years ago, it was noted that at the end of prophase I, homologous chromosomes become entangled with each other prior to being pulled apart during anaphase I. This distinctive component of prophase I, which has no counterpart in mitosis, is called **crossing over.** The significance of this phenomenon was not at first understood, but is now clear that it represents much more than a parting handshake of the separating chromosomes. On the contrary, crossing over is a mechanism for the interchange of genetic material, because during the process equivalent portions of homologous chromosomes may switch positions. Thus, after crossing over, the pairs of chromosomes no longer represent "pure" maternal and paternal sets, but each individual *chromosome* may have both maternal and paternal portions. In theory, this chromosomal interchange allows for a total reshuffling of all the maternal and paternal genes. Thus, the potential for unique gene combinations is vastly increased over what would be possible if maternal and paternal chromosomes continued as inviolate entities.

Crossing over has other ramifications as well. Without the process of crossing over, it would be advantageous to have the largest possible number of chromosomes, because variability in terms of genetic constitution would be exclusively a function of chromosome number. With crossing over, however, variability can be achieved independently of reshuffling whole chromosomes. Thus, the existence of crossing over accounts for the otherwise remarkable variation in chromosome number among different species.

The second meiotic division is an **equational** division, which implies that there is no further reduction in chromosome number. Rather, it is a pure mitotic division, differing from regular mitosis only in that the number of chromosomes lining up at the end of prophase II is just half (i.e., 23 in the case of ourselves) the number normally present in a mitotically dividing cell of the same species. Theoretically, this division would not even be necessary, but meiosis is sufficiently closely related to mitosis that a doubling of the chromosomal material (a seemingly unnecessary step) takes place before prophase I; thus, the second meiotic division is required to separate homologous chromatids.

Although meiosis is exceedingly common in both the plant and animal kingdoms, it is a very restricted process within the individual organism. It is important to remember that meiosis creates the cells used in sexual reproduction. Thus, in ourselves, meiosis results in the formation of sperm and egg cells, and as such, is restricted to the organs responsible for the formation of these cells—the testes and ovaries, respectively. By contrast, mitosis is found in every organ of the body, including the reproductive organs.

**FERTILIZATION**  As discussed in Chapter 8, fertilization occurs in the upper reaches of the Fallopian tube. The actual process is complicated somewhat because one (and only one) sperm nucleus must be allowed to fuse with the egg nucleus in order to restore the chromosome number to the 46 typical of human body cells.

The egg is not released from the ovary as a single naked cell, but rather is enclosed in a protective layer of nurse cells. These cells are held together by a kind of organic glue. One reason why millions of sperm must enter the female's reproductive tract in order to ensure fertilization is that passage through the protective layer of cells surrounding the egg apparently can be achieved only by the collective action of a great many sperm cells. Each sperm cell is capable of producing a small amount of enzyme which dissolves the organic glue, but many thousands of sperm are apparently required before enough of the glue can be dissolved to allow the passage of sperm to the egg.

Once the egg is reached, the first sperm to penetrate the cell membrane evidently initiates the production of a **fertilization membrane** by the egg, which splits away from the surface of the egg and effectively prevents the entry of any additional sperm. Some time later, the sperm that has entered the egg fuses with the egg nucleus thereby restoring the chromosome number to the original 46. Note that fertilization is defined as the combining of the nuclei of two gametes—thus, production of the fertilization membrane occurs before fertilization has actually taken place. This is logical enough, of course, because otherwise many sperm would enter the egg before the first sperm fused with the egg nucleus. Obviously, the hazards of many sperm nuclei fusing with the egg nucleus are at least as severe as a situation in which none fused.

**DEVELOPMENT**  Within a matter of hours after fertilization occurs, the egg begins to divide. If the cells formed from the first division separate, they may develop independently into two complete individuals which are genetically identical—**identical twins.** (**Fraternal twins,** which are much more common, result from the release and fertilization of two eggs.)

Successive divisions (Fig. 12.6) over the next few days produce a ring-shaped ball of cells called a **blastocyst** (Fig. 12.7). A portion of this structure is destined to develop into the embryo proper, whereas the rest (the so-called **extraembryonic membranes**) are destined to surround and protect the embryo and to aid in the formation of a **placenta.**

Subsequently, the cells of the embryo proper split so as to form two layers (Fig. 12.8). The outermost layer is the **ectoderm,** an embryonic tissue that will form the skin and nervous system of the adult organism. The inner layer is the **endo-**

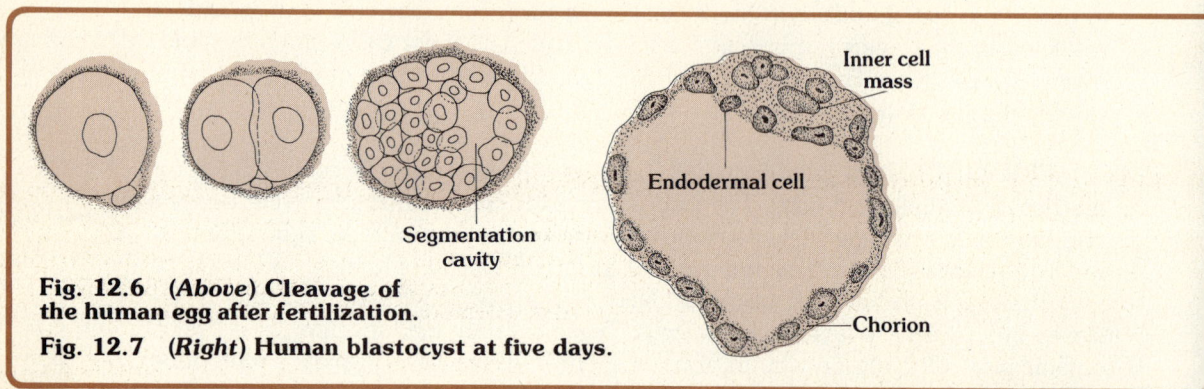

**Fig. 12.6** (*Above*) **Cleavage of the human egg after fertilization.**

**Fig. 12.7** (*Right*) **Human blastocyst at five days.**

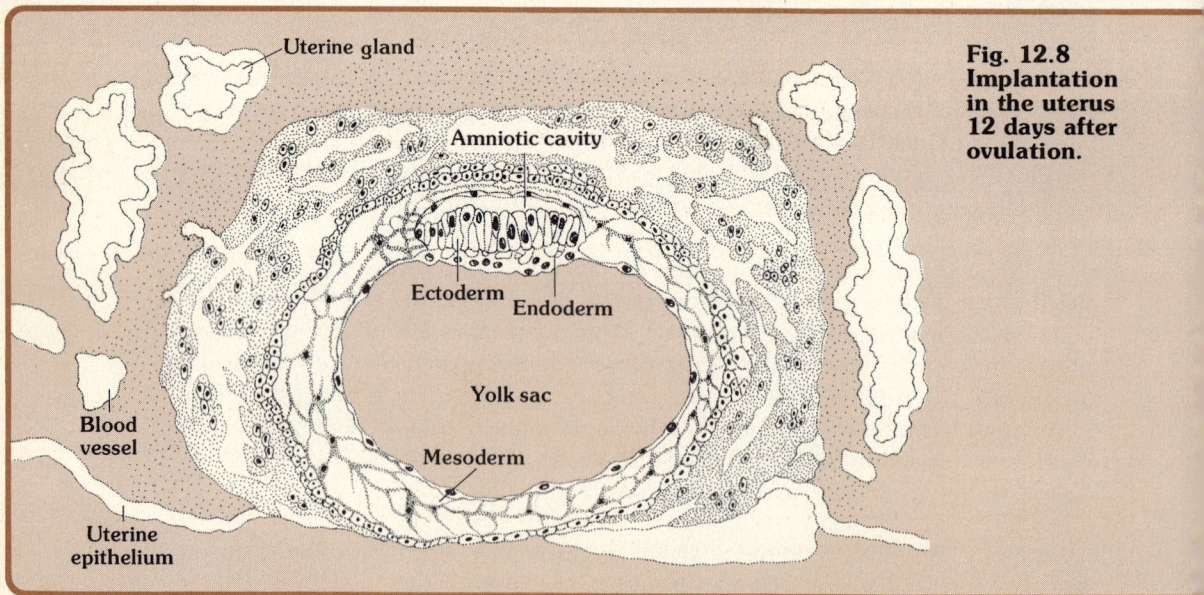

**Fig. 12.8 Implantation in the uterus 12 days after ovulation.**

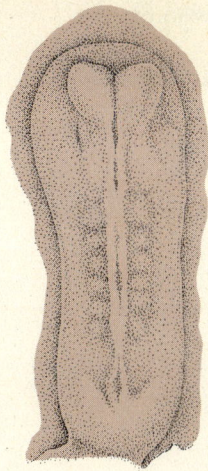
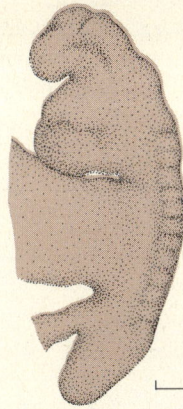

1 mm

1 mm

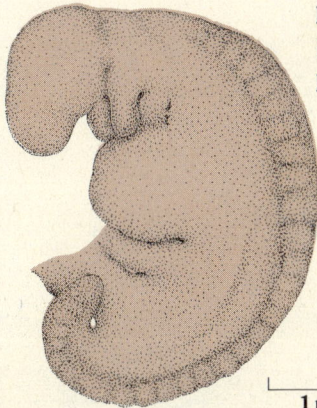

1 mm

**Fig. 12.9** (*Top left*) Human embryo at the end of the third week.
**Fig. 12.10** (*Top right*) Human embryo during the fourth week.
**Fig. 12.11** (*Left*) Human embryo at the end of the fourth week.

**derm,** the embryonic tissue that will become the lining of the gut tube and the lining of the digestive organs in the adult organism. Finally, a third layer called the **mesoderm** is produced between these two layers, forming the bulk of the body tissue in the adult.

The progression from an undifferentiated mass of cells to a recognizable embryo begins rather quickly. The **neural tube** (the primitive spinal cord) begins to form at about three weeks of embryonic age (Fig. 12.9), and a definite head fold can be seen shortly thereafter (Fig. 12.10). Before the beginning of the fourth week, gill slits in the pharyngeal region (a characteristic trait of all vertebrates) begin to form (Fig. 12.11), and at about the same time, the heart starts to develop from the fusion of veins in the chest area. Even so, by the end of this first month of development, the human embryo is still rather nondescript, and far from recognizable as human (as opposed to any other species).

During the fifth week of development, several additional changes occur. The head is more clearly defined, and the heart begins to beat. However, the embryo is still only about 1 cm ($\frac{3}{8}$ in.) in length. The beginnings of the arms and legs appear, initially just as undefined lumps of tissue (Fig. 12.12a–c). Outgrowth of brain tissue reach the ectoderm and induce the formation of the lens of the eye (Fig. 12.12d). **Induction** is a process whereby one group of cells influence the devel-

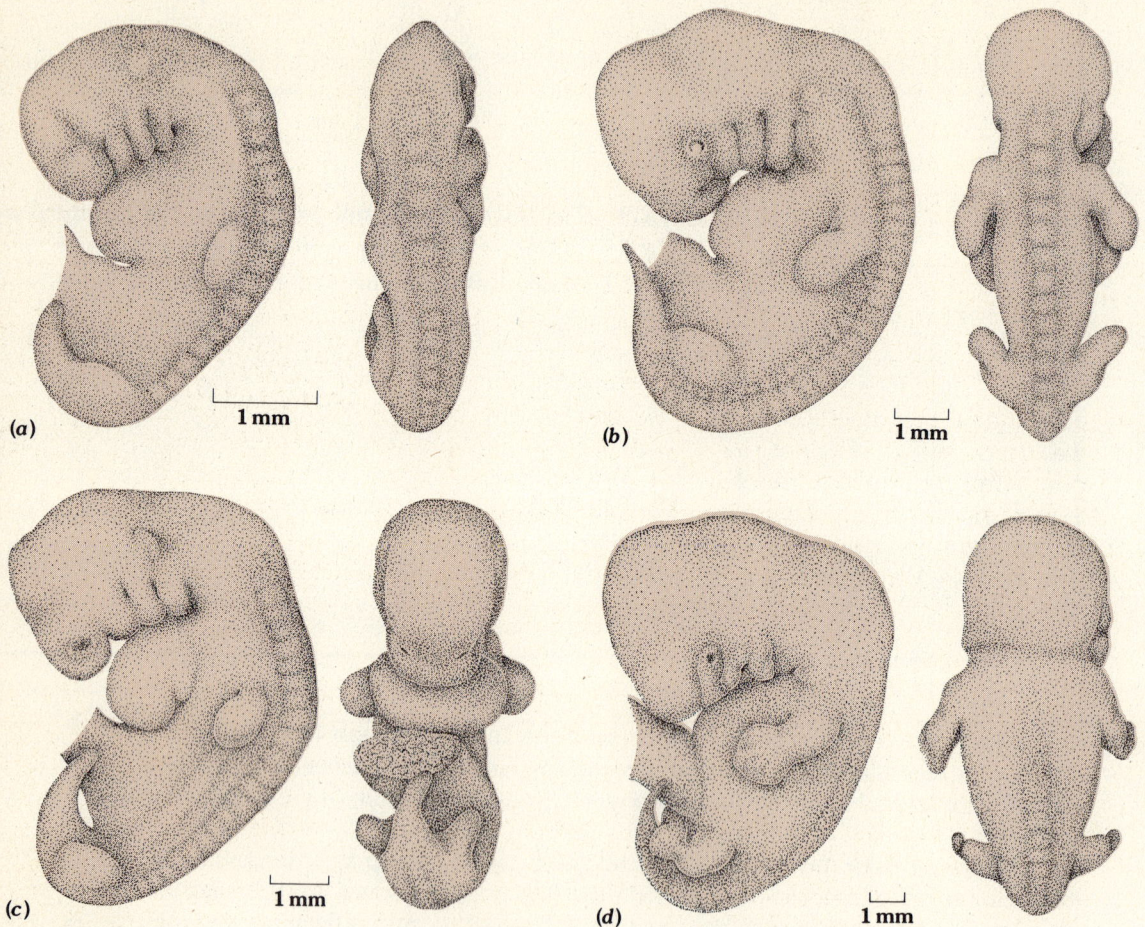

(a)  1 mm

(b)  1 mm

(c)  1 mm

(d)  1 mm

**Fig. 12.12  Human embryo at (a) beginning, (b) and (c) during, and (d) end of fifth week.**

opment of another group of cells in a specific manner. The precise mechanisms involved are far from clear, although it appears that the production and release of chemicals by the inducing cells is involved. Presumably, these chemicals serve to activate specific genes in the recipient cell.

During the sixth week (Fig. 12.13a–c), the arm and leg buds begin to differentiate, the heart is divided into chambers, and the liver is sufficiently well developed that it begins to manufacture red blood cells. (It is the liver, and not the bones, that serves to produce red blood cells during embryonic development.) The gut tube continues to differentiate into distinct regions. The embryo is by now about 1.3 cm ($\frac{1}{2}$ in.).

During the seventh week (Fig. 12.13d–f), the embryo begins to assume a somewhat more human appearance, primarily because the eyes shift from a lateral to a frontal position and, as such, a face begins to emerge. Growth is concentrated in the head region, where the developing brain is responsible for almost half the length of the embryo. External ears begin to develop, as do the fingers and toes. The eyes are pigmented. The embryo is now almost 2.5 cm (1 in.) in length, but still weighs less than 1 g (0.035 oz).

Largely because of its more human appearance, after seven weeks the embryo is termed a **fetus.** During the eighth week, and on through the third month of development, the fetus becomes increasingly more human in appearance. The kidneys begin to function, and the fetus reaches 7.5 cm (3 in.) in length, with a weight of about 20 g (0.7 oz) by the end of the 12th week. The nose is developed, as is the mouth. Deposition of bone begins to occur in the hitherto exclusively cartilagenous skeleton. There is a tendency for the fetus to straighten from its former C shape, The external sex organs are developed, and blood begins to form in the bones, as well as in the liver. The nervous system is matured to the point that simple reflexes occur. This movement of the fetus is termed **quickening,** and was formerly thought to indicate the entry of the soul into the fetus. Such movements are not generally detected by the mother until the fourth or fifth month, but it was arbitrarily decided by the ancients that quickening occurred on the 40th day in males and on the 80th day in females. (Male chauvinism has had a long and pervasive history.)

Development during the second trimester (fourth through sixth months) is less dramatic, the most obvious change being a huge increase in size, from 7.5 cm (3 in.) at the outset, to more than 30 cm (12 in.) by the end of the sixth month [and with a weight of as much as 0.9 kg (2 lb)]. Reflexes such as gripping and thumb-sucking occur, and such details as hair on the scalp and eyebrows develop. However, a separate existence is still generally not possible, because the lungs, digestive system, and capacity to regulate temperature are still insufficiently developed.

During the third trimester, growth continues such that, at birth, the fetus is about 50 cm (20 in.) long and weighs about 3 kg (7 lb). The chance for survival increases markedly throughout this period, in the event the child is delivered prematurely (the principal problem being respiration). There is a marked increase in subcutaneous fat, with the result that the fetus is less wrinkled than previously.

**Extraembryonic membranes**  The membranes that enclose and protect the

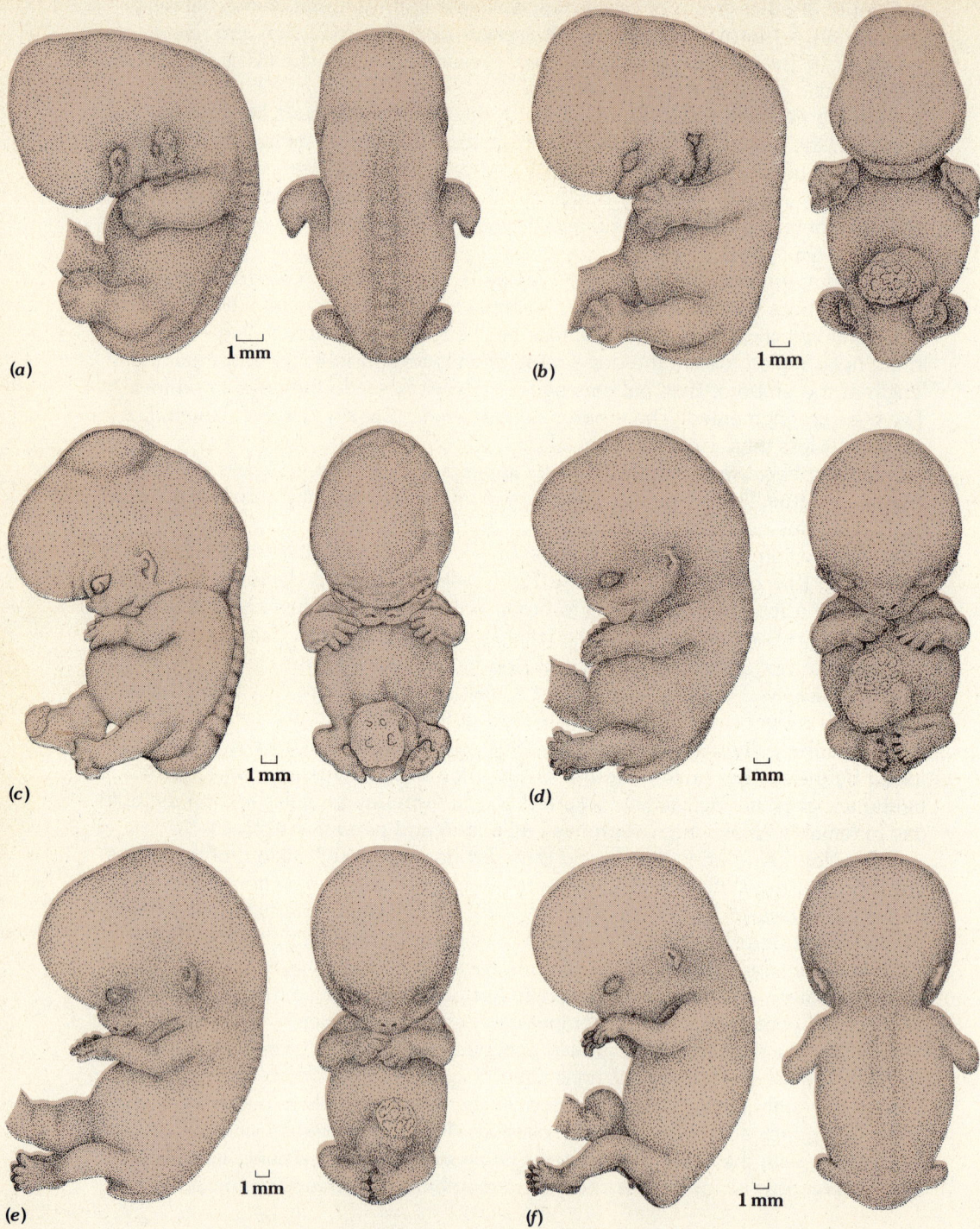

**Fig. 12.13** Human embryo at (a) beginning, (b) during, and (c) end of sixth week; and at (d) beginning, (e) during, and (f) end of seventh week.

developing human fetus are a legacy of the reptiles from which mammals originally evolved. The development of reptile eggs is not unlike that of amphibian eggs, the principal difference being the development of substantial extraembryonic tissue in the former, which permits them to survive on dry land, an impossibility for amphibian eggs, which would quickly dessicate. Four such membranes are involved in the reptilian egg (and the bird egg as well, birds being little more than glorified reptiles). These include an inner **amnion** in which the embryo develops, bathed in fluid; a **yolk sac**, which contains the yolk required during embryonic development and which is tied directly into the primitive gut tube; an **allantois,** which serves initially to store the nitrogenous wastes produced during embryonic development and which also connects to the gut tube; and an outer **chorion,** which surrounds the other membranes. A shell is laid down on top of the chorion as the egg proceeds down the mother's oviduct.

Mammals have retained and modified this system. The amnion remains as the "bag of waters" in which the embryo develops. The chorion also remains, but the ultimate growth of the embryo and fetus is so great that the space between the amnion and the chorion is all but eliminated, and the two appear virtually as a single membrane. The yolk sac is present only during the earliest stages, as mammalian eggs possess relatively little yolk. The allantois, which serves initially as a storage sac for wastes in reptiles and birds and which subsequently fuses with the chorion to provide a kind of "egg lung" whereby gas exchange can occur, is retained in mammals as part of the umbilical cord; the area in which the allantois and chorion grow together is retained as the embryonic contribution to the **placenta** (the rest of the placenta being maternal in origin). These relationships are shown in Fig. 12.14. Note in Fig. 12.15 that identical twins share a common chorion and placenta, although each has its own amnion.

The placenta, of course, takes the place of lungs, kidneys, and food supply. The fetal and maternal circulations are in intimate contact, but never actually mix (Figs. 12.16 and 12.17). Exchange efficiency is maximized between the two circulations by the employment of the **countercurrent principle,** which was explained in detail in connection with the kidney (Chapter 7).

**Recapitulation theory**   Many of the events described in connection with human development are duplicated, or at least paralleled, in the development of other animals, with the more closely related forms undergoing the most similar embryonic development. Moreover, a number of steps that occur in the developing human embryo or fetus have little reference to the adult condition, but rather suggest similarities with other vertebrates. The two most obvious of these are the possession of gill slits in the region of the pharynx (which, of course, are nonfunctional in all land-based vertebrates) and the presence in ourselves of a distinct tail during the fourth to seventh weeks.

Observing these and other facts, a German scientist of the nineteenth century, Ernst Haeckel, proposed his Theory of Recapitulation. Translated from the German, his theory states "Ontogeny recapitulates phylogeny"—a sufficiently cryptic phrase as to warrant the conclusion that perhaps it is still in German. What

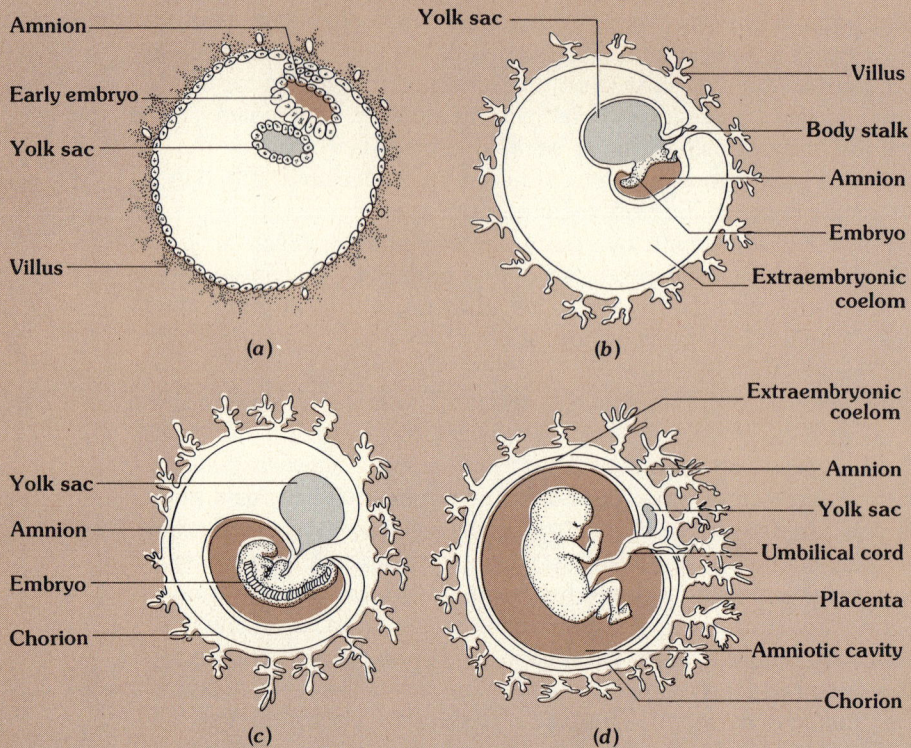

**Fig. 12.14  Development of the extraembryonic membranes:** (a) at 12 days; (b) at 16 days; (c) at 28 days; (d) at 12 weeks.

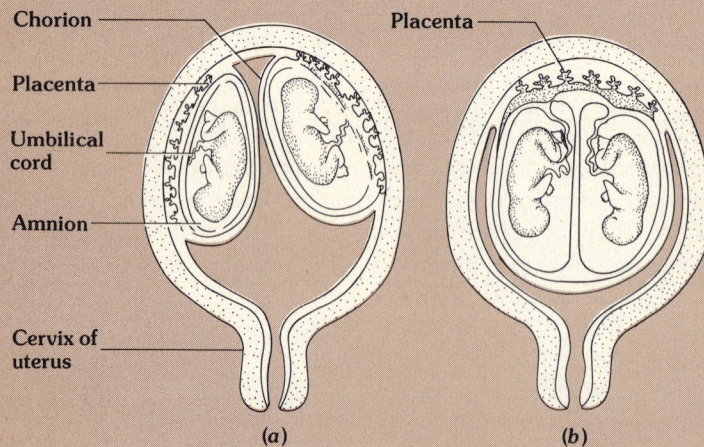

**Fig. 12.15 Fetal membranes in (a) fraternal twins and (b) identical twins. Note the common chorion and placenta in the latter.**

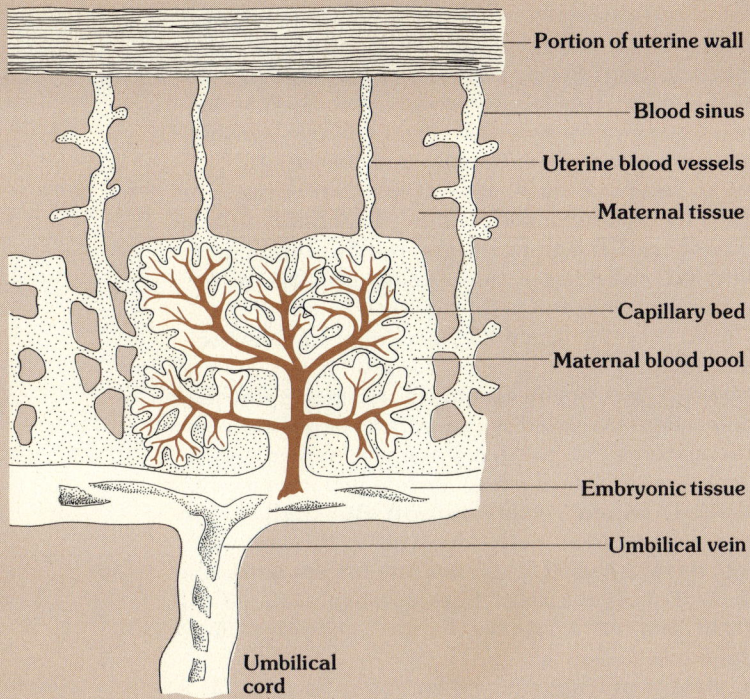

Portion of uterine wall

Blood sinus

Uterine blood vessels

Maternal tissue

Capillary bed

Maternal blood pool

Embryonic tissue

Umbilical vein

Umbilical cord

**Fig. 12.16
Interrelationship
of fetal and
maternal
circulation.**

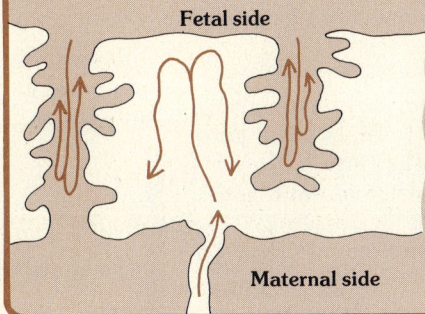

Fetal side

Maternal side

**Fig. 12.17
The principle of countercurrent flow of
blood in the human placenta.**

Perhaps the most fascinating thing about development is that it does not consist merely of a series of cell divisions, but rather results in specialized cells and tissues. Interestingly enough, however, this specialization appears to be accomplished without the occurrence of any fundamental change in the nucleus of the cells. This is only logical, because, as we have seen, mitosis is designed to reproduce the chromosomes entirely faithfully—but if the nuclei are unchanged, how does differentiation occur? The answer seems to be that only a small proportion of the total genetic message possessed by the nucleus of any cell is actually utilized by that cell—and the more specialized the cell, the smaller the number of genes used. Exactly how this masking of the chromosomes is accomplished is far from clear, but certainly the cells themselves affect the differentiation of each other, no doubt through the formation of chemical messages passing between cells.

That each cell possesses a genetically complete nucleus is not mere idle speculation. It has been demonstrated repeatedly in frogs and some other organisms that if the nucleus of a fertilized egg is removed and replaced with the nucleus from a skin cell of a frog normal development frequently ensues, and a perfectly normal adult from will ultimately be produced—but one, of course, that is genetically identical to the individual that donated the nucleus of its skin cell. (These experiments are more successful when the transplanted nucleus comes from an unspecialized cell, which implies that a certain amount of irreversible nuclear specialization occurs in the formation of adult tissues.)

This capacity of the nucleus of a specialized cell to substitute for the nucleus of a fertilized egg in differentiation is called **totipotency**. The term suggests that all nuclei are totipotent—that is, that they possess a complete genetic message—and it is only the inhibitory influences of the cytoplasm of the specialized cell which prevents each cell from acting as a fertilized egg and forming a new individual.

This technique of producing multiple genetically identical individuals is called **cloning**. An even more dramatic example of cloning occurs in such plants as lilies and carrots, in which single cells from an adult plant have the capacity, in the proper fluid medium, to develop into mature plants—but ones that are genetically identical to the original adult plant. Obviously, there are serious considerations to be made in deciding whether or not such techniques should ever be used on humans. How do we select individuals for cloning? What criteria will be used in making the decision? Although human cloning is not yet a possibility, it will almost certainly become so during the next decade or two—and these esoteric questions will become highly pragmatic very quickly.

**Differentiation, totipotency, and clones**

is meant by this is that the development of an individual (ontogeny) duplicates the ancestral history of the species (phylogeny). As originally stated, this theory held that, during its embryonic development, an organism proceeds through the adult stages of its ancestors. Therefore, because fossil evidence suggests that mammals are descended from a group of primitive reptiles, and reptiles in turn are descended from amphibians, and amphibians from fish, then the embryo of a mammal should pass through a series of developmental stages in which it resembles each of these presumed ancestral types in turn.

This was a very insightful observation and conclusion on the part of Haeckel, but unfortunately, somewhat overly ambitious. As it became increasingly clear that mammalian embryos do not, in fact, resemble fish, or amphibians, or reptiles, less and less attention was paid to Haeckel's theory, except for an occasional sneer. However, in relatively recent years, the theory has been resurrected, albeit in a modified form, so as to read, "During embryonic development, the embryo passes through a series of steps wherein it resembles the **embryos** of its ancestors." This is a fair approximation of the truth, for, as we have seen, the embryos of fish and humans both possess gill slits at one point in development, although the human embryo at no point resembles an **adult** fish. Although this theory in no way *explains* embryonic development, it does allow certain conclusions to be made. These include: (1) embryonic development is evolutionarily very conservative, for certain developmental stages are retained despite the absence of any apparent necessity for their retention; and (2) evidence for the phylogenetic relationships of a given species can be obtained by studying the embryonic development of that species.

**Events at birth**  All species must undergo certain rather dramatic physiological changes at the moment of birth, regardless of whether they are hatched from an egg or are born live. These changes are especially profound in the human, because of the sudden cessation of utter dependence on the placenta as the organ supplying all the needs of the growing fetus. Most critical, perhaps, is the proper functioning of the lungs, which, *in utero,* are nonfunctional. Improper functioning can lead to **hyaline membrane disease,** wherein an insufficient amount of fluid is produced by the lungs, rendering breathing very difficult. Typically, respiratory difficulties are the most serious problems faced by prematurely born infants, because the lungs are the last organs to mature; hyaline membrane disease is the most common of these problems. Patrick Bouvier Kennedy, President Kennedy's third child, died of this disease.

However, anatomically, the most dramatic changes occur in the circulatory system (Fig. 12.18). Although the heart begins to beat after only a few weeks of embryonic development, the fetal circulatory pattern is quite unlike that of the infant's, because of the substitution of the placenta for the lungs. Two events occur virtually instantaneously at the moment of birth. First, the placental circulation is shut down by virtue of the collapse of the umbilical artery and vein. Second, the momentary delay in blood flow resulting from the collapse causes a flap of tissue to swing closed on an opening existing through the membrane separating the left and right atria. In the event this membrane does not close properly, there is leakage between

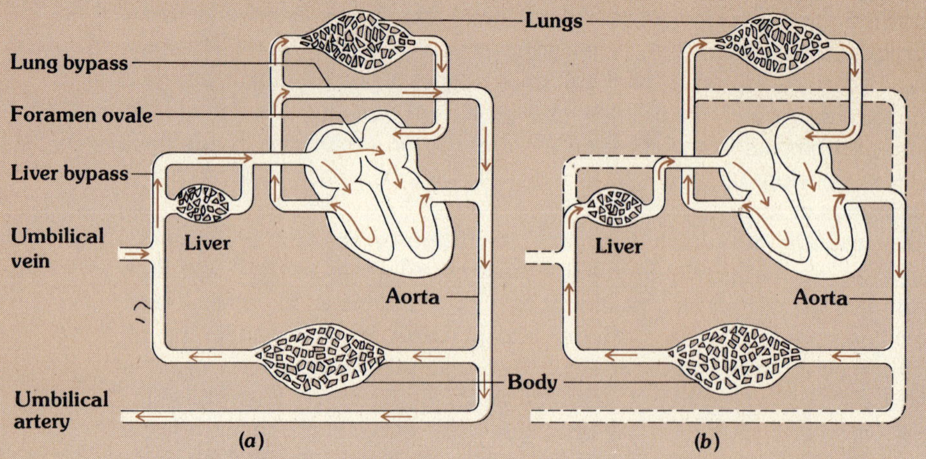
Lungs
Lung bypass
Foramen ovale
Liver bypass
Umbilical vein
Liver
Aorta
Body
Umbilical artery
(a)
Liver
Aorta
(b)

**Fig. 12.18** **(*a*) Fetal circulation and (*b*) infant circulation.**

the two atria which, of course, results in the mixture of oxygenated blood (returning from the lungs) with deoxygenated blood (returning from the body). Such infants have a blueish cast,[3] owing to large amounts of the darker, deoxygenated blood flowing through the body; logically enough, they are called "blue babies." A generation ago, these children had little hope of surviving more than a few years, and in any event were forced to take a vicarious view of life, as their condition disallowed any physical exertion. With the development of new surgical techniques in recent years, these children now have an excellent chance for a totally normal life.

[3] Any condition involving incomplete oxygenation of the blood is termed "cyanosis," from the Greek word for a blue mineral. Cyanide has the same etymological root.

## SUMMARY

**Cell division, or mitosis, was the first cellular event to be observed by the light microscopists of the last century. We tend to take this event for granted, as virtually everyone is at least vaguely aware that cells have the capacity to divide; yet this is an enormously important phenomenon, which underlies the potential for growth, reproduction, specialization—and, where performed erroneously, for cancer.**

**Mitosis is incapable of serving in a reproductive capacity in multicellular organisms, and a specialized form of cell division, called meiosis, is used in the formation of gametes. Although developed out of necessity, meiosis has the added benefit of providing a mechanism for maximizing genetic variability, such variability being highly desirable to any population.**

Queen Victoria ruled England for more than 60 years. She had eight children and outlived her husband by 40 years. In short, she was the picture of health—but one of her sons died of hemophilia, and the thrones of Russia and Spain fell in part because this disease was transmitted to Victoria's descendents. Why did Victoria show no signs of the disease?

Anne and Mary, identical twins, both recently married, contract German measles. Five months later, Anne gives birth to a healthy child. Two months after that, Mary gives birth to a child who is blind, deaf, and mentally retarded. What accounts for the difference?

Humans of modern appearance are first known from fossils that date back 25,000 years. What were we like before then? How closely are we related to apes and monkeys?

These questions involve genetics and evolution, which are the subjects of this chapter.

chapter **XIII**

# chapter XIII
# Retaining stability and generating change–genetics and evolution

The focus of biology from its earliest beginnings has, in large measure, been directed toward generating responses to two questions: (1) How are parental traits inherited by the offspring? (2) In what ways, if any, are different species of organisms related? Answering these and related questions has been the task of the disciplines of **genetics** and **evolution.** As we shall see, the two disciplines are interrelated, and there is a touch of irony in the additional fact that the basics of each were formulated at about the same time (the middle of the last century). Mendel, who later became known as the father of genetics, worked in such obscurity that for several years his "child" might have been considered stillborn. Darwin, who became inextricably tied to the theory of evolution, labored largely in the limelight; but it is also ironic that his ignorance of Mendel's work posed his greatest problem in explaining the mechanics of the evolutionary process. Indeed, much of the criticism of Darwin's ideas on evolution was directed toward his erroneous views of inheritance, which was unfortunate because it shifted the spotlight away from his much more accurate statements on natural selection.

**MENDELIAN GENETICS**   In honor of Gregor Mendel, the monk who first developed an accurate theory of genetics, that branch of genetics which deals with the inheritance of traits as units, has been termed **Mendelian genetics.** This is in contrast to **molecular genetics,** which involves the biochemical interrelationships of the molecules involved in inheritance, a subject beyond the scope of this book.

Mendel, who lived a quiet, monastic life during the middle of the last century, was by no means the first individual to investigate the nature of inheritance. However, unlike his predecessors, he had a strong interest in mathematics and a powerful scientific mind. Equally important, as we shall see shortly, he had the good fortune to select a relatively simple genetic system—seed and pod characteristics of the ordinary garden pea.

Mendel grew several varities of peas in his garden at the monastery, including some peas with wrinkled seeds (as opposed to the more common round shape) and some with green seeds (as opposed to the more common yellow). When he pollinated flowers of the wrinkled seed variety with pollen from flowers from the round seed variety (or vice versa), invariably all the seeds produced were round. It appeared as if the characteristic of wrinkling had totally disappeared. But when Mendel then pollinated flowers produced from this $F_1$ generation[1] of seed with pollen from other $F_1$ flowers, a minority of the seeds produced were wrinkled, although most were round. Because of his mathematical interests, Mendel counted the numbers of each type, and found that there were 5474 round seeds and 1850 wrinkled seeds, or approximately three round seeds for every wrinkled seed. This ratio held true for successive experiments. Hence, the wrinkled trait not only did not disappear—it reappeared in the second generation with predictable frequency. Mendel then repeated the experiment using six other traits of the pea plant and invariably achieved the same results, with an $F_2$ (second generation) ratio of between 2.82 and 3.15 to 1. From these data, Mendel concluded that such traits were not lost when a plant possessing them was crossed with a plant possessing different traits, but rather were merely masked for a generation. This conclusion became known as Mendel's First Law—the **Law of Segregation**—which states that traits are not blended and lost during fertilization, but rather are segregated only to reappear in subsequent generations.

Mendel then extended his work to crossing plants that differed in two traits, and the results were, in part, predictable. Plants with round yellow seeds were crossed with plants having wrinkled green seeds, and the $F_1$ generation was found to have exclusively round yellow seeds. The $F_2$ generation, however, consisted not only of the two expected parental types, but also of plants with round green seeds and with wrinkled yellow seeds. These results were the basis of what is now known as Mendel's Second Law—the **Law of Independent Assortment**—which states that multiple characteristics are not inherited as a single parental unit, but rather are inherited in an independent assortment. Moreover, Mendel found the ratio of types in the $F_2$ to be nine round yellow seeds, three round green seeds, three wrinkled yellow seeds, and one wrinkled green seed. This $9:3:3:1$ ratio is the equivalent of $(3:1)^2$, and Mendel calculated that the ratio of any multiple cross of $n$ characters would produce an $F_2$ ratio of $(3:1)^n$.

It is important to recognize that Mendel's work was done at a time when the role played by chromosomes in inheritance was completely unknown. Therefore, at the time, his mathematical ratios had no known anatomical explanation. That fact, coupled with the fact that Mendel had published his findings in a rather obscure journal, led to his principles becoming ignored and forgotten for 35 years, before being simultaneously rediscovered in 1900, some 16 years after Mendel's death, by three geneticists who were working independently.

---

[1] $F_1$ stands for "first filial generation" (from the Latin *filius*, meaning "son").

In the years since the rediscovery of Mendel's findings, we have learned a great deal about the mechanics of inheritance. We know, for example, that the reason why there were no wrinkled seeds in the $F_1$ generation is because the gene that codes for the wrinkled trait is masked in its expression by the gene that codes for round. For that reason, the masked trait (wrinkled) is referred to by convention as the **recessive** trait, whereas the masking trait (round) is referred to as the **dominant** trait. Generally speaking, genes possessing a dominant–recessive relationship to one another are found at the same point (or locus) on each member of a pair of homologous chromosomes; such genes are called **alleles.** If both genes are represented by the same allele (e.g., for wrinkling), the organism is said to be **homozygous** with respect to that trait; if each is a different allele (e.g., one for wrinkling and one for roundness), the organism is said to be **heterozygous** (or **hybrid**) for that trait. A cross between two organisms differing in only one trait is a **monohybrid** cross and in two traits, a **dihybrid** cross, and so on.

Although Mendel deduced the necessity of discrete heritable particles in order to explain his results, the term "gene" was not introduced until 1909. Moreover, Mendel was primarily interested in the appearance, or **phenotype,** of the character, rather than in its genetic constitution **(genotype).** As such, Mendelian ratios are phenotypic, not genotypic, ratios. For example, the typical monohybrid ratio of $3:1$ is a phenotypic ratio; as can be seen from the Punnett square, the genotypic ratio of this same cross is $1:2:1$. The significance of this difference is considerable. For instance, even recessive alleles which, if present in the homozygous condition might be dangerous or even lethal, can be carried with impunity in the heterozygous state because there is no difference between the homozygous dominant and the heterozygous condition in terms of phenotype.

**Post-Mendelian findings**   In many respects, Mendel's choice of experimental organisms and traits was fortuitous, for he "chose" the simplest genetic situations. There are many instances in flower colors, for example, for which alleles are not fully dominant and recessive to one another—thus, a red flower crossed with a white may yield a pink, a result which, in Mendel's day, would have been interpreted as evidence for **blending,** the popular theory of inheritance of his time. In such crosses, the parental types do emerge again in the $F_2$, but, without dominance, the genotypic and phenotypic ratios are the same (i.e., 1 red: 2 pink: 1 white), and these results would have forced a reconsideration of Mendel's First Law had they been obtained in Mendel's time.

As it happens, Mendel himself ultimately became aware of similar problems. As if to rub salt in the wounds of the obscurity in which he labored, Mendel had the misfortune of being advised by a prominent botanist, who doubted Mendel's findings, to repeat his experiments on another plant species. The botanist suggested a specific plant and Mendel labored unsuccessfully for five years attempting to verify his findings on this second species. Failing eyesight and a total lack of success forced him to give up, never knowing whether the genetic system he had devised was valid only for peas or if it was valid for all organisms save only the species of plant recommended him by the famous botanist. Needless to say, the second alternative was correct, but this was not proved for years after Mendel's death. The excep-

Shortly after the turn of the century, an English geneticist named Punnett came up with a simple method of mapping crosses in a grid, thereby providing a picture of Mendel's mathematical ratios. Beginning students have been in his debt every since.

Suppose one is crossing plants with round seeds with plants with wrinkled seeds. It is known that the round trait is dominant, so we will represent it (as Mendel did) as *R*; correspondingly, the recessive wrinkled trait will be represented as *r*. If two homozygous individuals, one for each trait, are crossed, all of the gametes of the round strain will contain the dominant allele *R*, and all of the gametes of the wrinkled strain will possess the recessive allele *r*. Thus, a grid will look like this:

| Gametes | *r* |
|---------|-----|
| *R*     | *Rr* |

All of the progeny will possess the heterozygous configuration *Rr*. If these are now crossed amongst themselves, on the average, 50 percent of the gametes will contain *R* and 50 percent half *r*. Hence the grid:

| Gametes | *R*  | *r*  |
|---------|------|------|
| *R*     | *RR* | *Rr* |
| *r*     | *Rr* | *rr* |

**The Punnett square**

Because *RR* and *Rr* look the same (the presence of a single dominant *R* allele is sufficient for the round condition), it is obvious that there will be three round seeds for every wrinkled seed—a ratio of 3:1, as Mendel had found.

tional species, as it happens, even though it produces seeds, reproduces asexually most of the time, a fact unknown both to Mendel and to the famous botanist. It seems a pathetic finale for one of the great minds in nineteenth century biology.

Mendel was fortunate, however, in avoiding the knotty problem of **linkage.** We now know that genes are actually portions of chromosomes and, although the number of genes in an organism may reach the tens of thousands, the number of chromosomes rarely exceeds 100. Had Mendel chosen two traits that were genetically linked on the same chromosome, he never would have found a 9:3:3:1 dihybrid ratio, nor would he have formulated the law of independent assortment.

As we discussed in Chapter 12, actual breakage and reformation of homologous chromosome pairs may occur during the crossing-over stage of prophase I of meiosis. As a consequence, traits carried on the same chromosome may sometimes be separated. This separation of linked traits was a very troublesome finding to the geneticists of the early part of this century who were still groping their way toward an understanding of crossing over. Ultimately, it was realized that the chance of two traits carried on the same chromosome becoming separated was a function of the distance between the genes responsible for these traits. That is, the farther apart the two genes, the more likely they are to be separated from each other during the period of crossing over. This was a finding of enormous importance, not only because it accounted for some otherwise inexplicable ratios, but also because it permitted an actual mapping of the chromosomes. For the first time, it became possible to position specific genes on specific chromosomes.

Still other complications have been added to the basic Mendelian plan. A given gene may affect more than one trait, a phenomenon termed **pleiotropy.** A gene may fail to be expressed in the presence of another nonallelic gene, a phenomenon called **epistasis.** Depending on the genetic or environmental framework in which the gene is found, it may or may not be expressed—such a gene is said to have **limited penetrance.** An example is the gene responsible for adult-onset diabetes (see Chapter 9), which may not be expressed in individuals who avoid environmental situations that stress the insulin-producing capacity of the pancreas. In totality, the picture is a good deal more complex than Mendel described, but we owe to him that most important first step.

**DARWINIAN EVOLUTION**   In 1858, seven years prior to the publication of Mendel's paper, a middle-aged specialist in barnacles named Charles Darwin published a summary of his forthcoming *Origin of Species* as a paper in the *Journal of the Linnean Society.* This publication, like Mendel's, was received with something less than earthshaking applause. In fact, at year's end, the secretary of the society was moved to note that little of significance had been published in the journal that year.

Darwin persisted, however, and his 1859 book became a best-seller. The impact, although delayed, was enormous, and it triggered a wave of response which has not totally died even today.

The underlying premise of Darwin's work was that species were not immutable, as most people of his time believed, but rather that they changed in accordance with the pressures placed on them by changing environments (a process Darwin termed **natural selection**). In a second book, Darwin committed the ultimate heresy of including *Homo sapiens* as a species subjected to the whims of natural selection, and this concept was immediately transformed into something quite unlike what Darwin had proposed. Specifically, Darwin's beliefs were perverted into statements to the effect that humans evolved from monkeys, which was an error of the first magnitude. Darwin had in fact proposed that humans and present-day apes shared a common ancestor, but never suggested that a species of ape living today was our ancestor. Nonetheless, this erroneous transmutation of Darwin's beliefs made great copy for the popular press, and cartoon satires of Darwin and his monkey "relatives" abounded (Fig. 13.1). The idea of such a relationship seemed so farfetched that many people immediately rejected the whole concept of evolution. The list of disbelievers included not only many prominent church figures, but also many scientists. Foremost among the latter was Louis Agassiz, the most renowned natural scientist of his day. It is ironic that despite the fact that Agassiz decried the theory of evolution until the day he died, he founded the Museum of Comparative Zoology at Harvard University, which quickly became (as it remains today) a center for the advancement of evolutionary theory.

Darwin was by no means the first to propose a theory of evolution, and he was not even the first to hinge evolution on the process of natural selection. However, no one had ever presented an evolutionary theory so forcefully and so well documented. In a way, Darwin was a biological Columbus—not the first to discover new land, but the one primarily responsible for the revolutionary consequences that followed.

In addition to spotlighting natural selection as the creative force of evolution, almost by necessity Darwin came to regard any given species as being definable only as a population, with each individual member slightly different from every other member. This concept was in sharp contrast with the prevailing view of scientists of the time, who espoused the "type" concept—every species was represented by a "type" specimen, which possessed all the attributes of the species, and variations on this type, although they might occur, were of no significance. This may strike you as a sort of "the glass is half full—the glass is half empty" dichotomy—that is, a purely semantic difference—but such is not the case. Accepting a species as a population of differing individuals makes it easy to see evolution as a force for change over time. In contrast, the type concept does not allow room for the possibility of evolution because evolution implies change, and the type concept implies maintenance of the status quo. Thus, natural selection under the type concept would at best be a purely static selection, with any deviation from the type being rigorously weeded out.

This type concept, emphasizing the ideal individual rather than the variable population, was a sufficiently esoteric concept as to be well understood (and accepted) only by biologists. Thus it was that Darwin's views were actually more readily accepted by laymen, who were largely uninitiated into the intricacies of the type concept, than by Darwin's fellow biologists, who were better aware of the magnitude of his heresy.

**Neo-Darwinian evolution**  The one glaring weakness in all of Darwin's writings was his failure to elucidate a satisfactory explanation for inheritance. Darwin was obviously discomforted by the prevailing views of **inheritance of acquired characters** (e.g., giraffes' necks were long because their parents had stretched their own necks reaching for leaves and passed this trait on to their offspring) and **blending** (e.g., a white flower crossed with a red flower will produce either red- or pink-flowered progeny, depending on the "strength" of the white color, but in any event, the white color is forever lost). However, Darwin had no better alternative to offer, and evidently remained ignorant of Mendel's work (as did all other biol-

A DARWINIAN.

SCIENTIFIC MONKEY. "Cut it off short, Tim; I can't afford to await developments before I can take my proper position in Society."

**Fig. 13.1 Contemporary cartoons expressing typical beliefs regarding Darwin's views on natural selection. (Culver)**

ogists). Darwin was most tellingly criticized on his theories of inheritance, and although he squirmed his way through various theories in later editions of his book, he was never able totally to rebut the arguments made against them, and, by extrapolation, against his whole concept of evolution by natural selection.

Darwin actually suggested three quite distinct methods whereby variability could be introduced into a population. He ranked these in importance as follows:

1 A use and disuse theory (characters not used are lost);
2 The direct influence of the environment on the introduction of variability, with the varied characters being passed on to the offspring;
3 The chance emergence of new characters in a spontaneous manner.

**MR. BERGH TO THE RESCUE.**

The Defrauded Gorilla. "That *Man* wants to claim my Pedigree. He says he is one of my Descendants."

Mr. Bergh. "Now, Mr. Darwin, how could you insult him so?"

The first two methods are, of course, variations on the now-defunct theory of the inheritance of acquired characters. The last, however, is the equivalent of mutational theory, the only method now accepted by biologists whereby variability can be introduced into a population.

In large measure, Darwin was forced to subscribe to some sort of inheritance of acquired characters theory, because the prevailing view of the day on inheritance involved variations on the "blending" theory, whereby different characters were averaged when crossed. It was seemingly impossible that a single variation would ever be perpetuated in a population, because it would be blended out of existence after the first generation. Thus, Darwin found himself on the horns of a dilemma. To avoid the blending argument, he would have to suggest that many identical variations must arise simultaneously, in order to be perpetuated—yet this explanation would amount to a sort of special creationism, and not natural selection. Alternatively, he could accept inheritance of acquired characters, as this theory would allow for a number of identical variations occurring together without necessitating an invocation of special creationism.

Mendel, of course, had the answer. Not only did blending not occur, but much variation was not new, in the sense of just having arisen; rather it represented the recombining of existing variation, as occurs in the expression of a hitherto masked recessive trait in a homozygous individual. However, even with the rediscovery of Mendel's work in 1900 (18 years after Darwin's death), the critics of evolution were not immediately silenced, as the new breed of geneticists was unable to resolve all of the problems overnight. In fact, some added to the confusion by interpreting natural selection as a wholly negative force, which lopped out individuals possessing traits different from the majority. Darwin, of course, had recognized that natural selection was fundamentally a positive force and one which could result in change over time, if a changed environment favored individuals that were different from the majority. However, as geneticists added new data on the mechanisms of inheritance, Darwin's prophetic views were enhanced, and the magnitude of his discovery increasingly became more widely recognized.

### MUTATIONS—THE BASIS OF VARIABILITY

The actual genetic material found in chromosomes consists of linear chains of the nucleic acid DNA (see Chapter 2), which is present in the form of two complementary chains twirled about one another, like the strands of a rope. Duplication of chromosomes, which, of course, occurs in interphase, consists of a sequential unraveling of these two strands, with each of them then acting as a kind of mold, or template, for the formation of a new chain of DNA. These new chains are constructed out of the sugar, nitrogenous bases, and phosphate groups found floating freely in the cell. Because the sequence of these nitrogenous bases (several of which collectively constitute a gene) ultimately determines the way in which amino acids are assembled to form a given protein, it is imperative that each DNA chain form its complementary chain without error. An introduction of error into a DNA chain also introduces error in the assembling of amino acids in protein formation. No process can be absolutely error-free, of course, and errors occur periodically in the assembling of complementary strands of DNA. Similarly, from an evolutionary standpoint, it is important that this duplicating process not be completely error-free, as it is only through errors (mutations) that variability can be introduced into a population—and without variability, natural selection and evolution could not occur.

Even though the error in DNA formation may be as minor as the substitution of one nitrogenous base for another, the consequences may be far-reaching. For example, the difference between normal and sickle-cell hemoglobin is limited to the substitution of a single amino acid (out of the hundreds that form a hemoglobin molecule), and this substitution itself can be traced back to a single out-of-place nitrogenous base in the DNA of the chromosome. The effects on the body of this type of hemoglobin are, of course, devastating (see Chapter 5).

Substitutions of nitrogenous bases in the formation of DNA strands are called **gene** (or **point**) **mutations.** These mutations are in contrast with mutations involving larger portions of the chromosome, which are called, appropriately enough, **chromosomal mutations** (Fig. 13.2). Chromosomal mutations include:

**1. Inversions**  The rotation through 180 degrees of a segment of a chromosome during the crossing over stage of prophase I of meiosis, and the ultimate reattachment of the segment to the original chromosome. You may wonder why this would be called a mutation, as there has been no change in the nitrogenous bases themselves, but rather only a change in the juxtaposition of these bases at one particular point. As it happens, however, it is not only the absence or presence of bases that has an effect on inheritance, but also the physical relationship of these bases to each other. This effect may take the form of one gene negatively interfering with the performance of another gene (epistasis), or it may be that one gene enhances the activity of another. In any case, a change is detectable.

**2. Translocations**  Like inversions, a piece of a chromosome breaks free during crossing over, but rather than reattaching to the original chromosome, the detached segment attaches to another chromosome in a translocation. This means that consequently some gametes will possess an extra bit of chromosomal material and other gametes will lack an equivalent amount. The consequences of such mutations are related to the amount and function of the chromosomal segment involved, but it is not unusual for either an excess or a deficiency to be lethal.

**3. Duplications**  A portion of a DNA strand, during the formation of a comple-

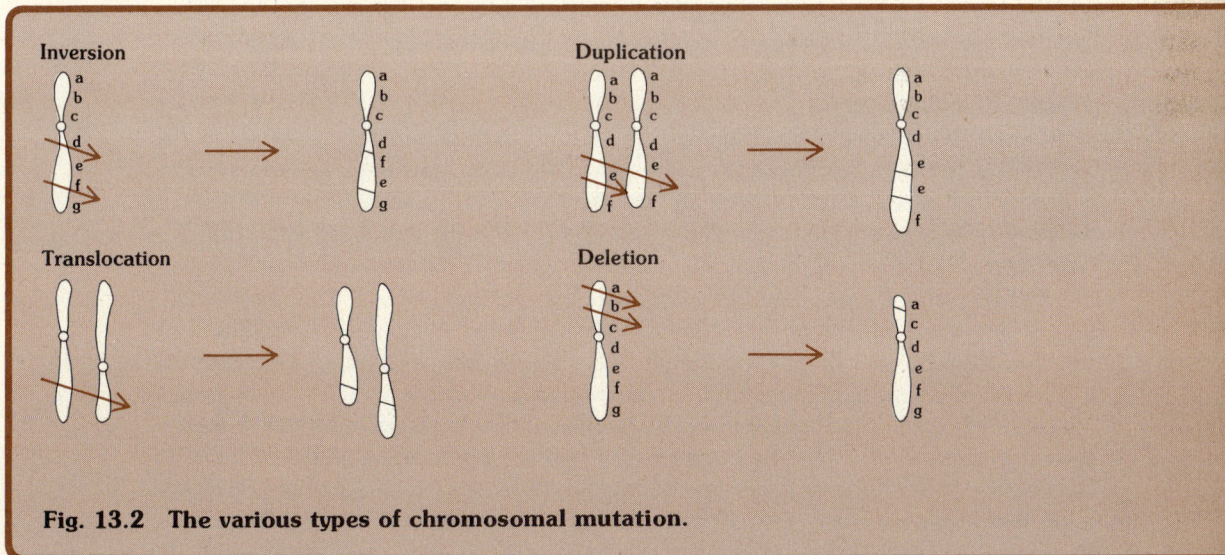

**Fig. 13.2   The various types of chromosomal mutation.**

mentary strand, repeats a series of nitrogenous bases successively, meaning that one or more genes may be represented twice on the same chromosome, in a side-by-side fashion. Again, the effect of such duplications is a function of the number of genes involved and their relative importance to the organism.

**4. Deletions**  Mirror-opposites of duplications, involving the omission of a series of nitrogenous bases during the formation of a complementary DNA strand. As a consequence, certain genes are not present at all on the affected chromosome. The effects of such an error range from virtually nil to lethal conditions.

**5. Nondisjunction**  Probably the most serious of all chromosomal mutations. It arises from the failure of homologous chromosomes (meiosis I) or chromatids (meiosis II) to separate. As a consequence, one group of gametes has an extra chromosome, and another group lacks a chromosome. Either condition is typically lethal if it involves the nondisjunction of a large chromosome, and even where the smallest chromosomes are involved, the effects may be devastating. Specific mutations are discussed later in this chapter.

**What is a "good" mutation?**  Once a mutation occurs, what determines whether it will be maintained in the population? First, it is not the mutation upon which natural selection acts, but rather the *possessor* of the mutation. It should be obvious that natural selection must operate on whole organisms and not on individual genes, but this point is frequently missed. Second, it is not the genotype which is the unit of selection, but rather the phenotype. Again, this should be obvious, but understanding it explains why even deleterious recessive genes are protected from the ravages of selection, as long as they are present in a heterozygous condition.

A mutation is said to be **adaptive** or **nonadaptive,** depending on whether it is beneficial or deleterious to the organism, but these slippery terms must be carefully defined so as to avoid generating a circular argument ("Mutations are adaptive if they remain in the population. Mutations remain in the population if they are adaptive.") Specifically, we speak of adaptive mutations as increasing **fitness,** which itself is defined as anything that increases **reproductive success**—the probability of producing fertile offspring.

Darwin understood that all species, whether houseflies or elephants, have a sufficiently high reproductive potential to cover the face of the earth, were that reproductive potential ever realized. Obviously, the numbers of all species (except *Homo sapiens?*) are held in check, and the mechanisms whereby this is achieved are collectively termed "natural selection." However, Darwin erred in placing emphasis on the elimination and death of unfit individuals, for we now know that natural selection is generally not so overt or bloody. Rather, natural selection is now defined in terms of reproductive success. This is a critical and eminently logical point, because it serves to allow measurement of fitness not by some abstraction, but rather by the increase in the rate of perpetuation of the mutation. For example, a mutation that served to increase the size and strength of a bull elk might be assumed, *a priori,* to be adaptive, based on our subjective impression of what is "desirable" in elk. However, the mutation clearly would not be adaptive if it were also to cause the elk to be sterile, because such an animal would, by definition, leave no offspring, and the mutated gene would immediately be lost from the population, once the bull elk possessing it were to die.

In summary, natural selection is generally a matter of statistical probabilities, and not a series of bloody battles. The individual who makes the greatest numerical contribution to the population of the next generation (by virtue of producing the largest number of fertile offspring) is, by definition, selected for, regardless of how scruffy and bedraggled he might appear to an outside observer. Indeed, frequently the victor of a bloody struggle is so weakened by repeated battles that he is unable to reproduce at all, in which case a completely erroneous conclusion would be drawn in an attempt to predict fitness from the outcome of battles.

Three corollaries can be drawn from this statement on fitness and adaptation. First, by expressing evolution in terms of reproductive success, we recognize that the measure of evolution is the speed with which gene frequencies change in time. As such, we also realize that individuals cannot evolve, for an individual cannot change his own gene frequency. Therefore, evolution is an event that affects populations only.

Second, mutations are never neutral; in the long run, they are either adaptive or nonadaptive, as it is inconceivable that a mutation would not have some effect on reproductive success, regardless of how infinitesimal, over successive generations. Thus, we always expect to see the frequency of a given mutation either increasing or decreasing with the passing of generations, unless there is some countering selective pressure to maintain a balance (see **polymorphism,** next section).

Third, only mutations that are adaptive will increase in frequency within the population. For many years, it was widely accepted that the saber-toothed tiger became extinct because its teeth grew so large that the animal had trouble eating. This erroneous conclusion was long ago laid to rest, but it continues to pop up occasionally in introductory texts and in the popular press. Such a situation would clearly be impossible, for any increase in tooth length that operated in a deleterious manner in feeding would certainly be selected against, given that starving animals can hardly be expected to reproduce in large numbers. In fact, the saber-toothed tiger was capable of opening its jaw to an angle of almost 120 degrees, and had no trouble feeding. It evidently preyed on thick-skinned, rhinoceroslike animals, which it killed by slashing through the thick skin of the back with its teeth, riding about on the back of the prey animal until the prey bled to death. With climatic changes, most of these thick-skinned animals became extinct, and the saber-toothed tiger followed suit.

A similar and equally preposterous argument was made about the Irish elk, which supposedly became extinct because its rack of antlers grew too large. It may well have been that improved hunting capabilities of humans were responsible for its extinction (the Irish elk died out only about 1000 years ago), and that the antlers of the elk were of little use in combatting hunters. But to imply that extinction came about because the antlers grew too large is to create visions of an animal suddenly becoming unable to lift its head off the ground, or periodically wedging itself between two trees, and this was certainly not the case.

**What maintains genetic variability?**   It is certainly advantageous to a population to possess a method of maintaining variability; this would prove beneficial should an environmental change occur of sufficient magnitude to outstrip the capacity of the species to generate new variability through mutations before extinction could occur. Yet if mutations are either adaptive or nonadaptive, why does the population not quickly become homozygous for all of the most adaptive genes? The answer involves several interacting forces. These include: (1) recurrent mutation; (2) random assortment: (3) dominant–recessive relationships; (4) changed environmental circumstances; (5) polymorphism.

**1. Recurrent mutation**   To begin with, mutations occur spontaneously at a very low, but predictable, rate. Thus, deleterious mutations which have previously been eliminated from the population frequently occur spontaneously again, and once again are "measured" as to their adaptive value, if any. At any one time, most of the individuals of a population are close to the adaptive peak, to be sure; most mutations are therefore nonadaptive and are lost. However, because environments do change, the adaptive value of the mutation may change as well, and even the strongly deleterious mutations of today may one day prove adaptive. In any event, recurrent mutation is not a function of how well or poorly a given population is adapted to its environment, but is rather an inherent property of DNA replication.

**2. Random assortment**   This force ensures that mutations are spread about in different gene combinations. Random assortment, of course, is the Mendelian restatement of the cellular events that occur during meiosis (separation of homologous chromosomes and crossing over). Even though mutation is the sole route for the introduction of genetic variation into the population, the recombinant effect of crossing over is of much greater importance in the formation of unique genotypes. Because virtually all phenotypes (the unit on which selection acts) possess a unique combination of genes, the particular genetic setting in which a mutated gene finds itself may be of critical importance in determining whether the organism that possesses it will be favored or disfavored.

**3. Dominant-recessive relationships**   Most mutations are recessive and are therefore not exposed to natural selection when they first arise. As such, these mutations may spread throughout the population before finally showing up in a homozygous recessive condition, when finally their possessor will be subjected to selection (Fig. 13.3). In the meantime, however, they may become rather numerous in the population.

**4. Changed environmental circumstances**   Of course, the principal reason why organisms do not achieve homozygous perfection and remain in that state is that the environment is constantly changing, however subtly. As such, there is always a gradual shift in what represents genetic perfection, which means that natural selection is an ongoing process and remains as such as long as the environment is subject to change.

**5. Polymorphism**   Finally, a very important method of retaining population variability is polymorphism, a phenomenon whereby multiple alleles are retained in a population in a balanced fashion. Perhaps the classic example is human blood

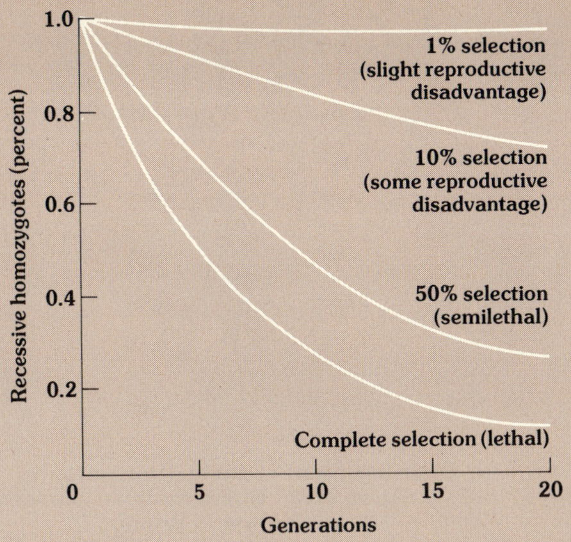
1.0
0.8
0.6
0.4
0.2
Recessive homozygotes (percent)
1% selection (slight reproductive disadvantage)
10% selection (some reproductive disadvantage)
50% selection (semilethal)
Complete selection (lethal)
0
5
10
15
20
Generations

**Fig. 13.3**
**Although the intensity of selection will affect the frequency of individuals possessing a given homozygous recessive trait, such individuals will never be totally eliminated from the population.**

type. All of the major types of human blood (A, B, AB, and O) are created by alleles of a single gene, and it might logically be expected that one type would carry greater adaptive value than the others. However, the frequency of each allele is relatively stable in most populations, even though these levels may differ markedly between populations. Evidently, some type of balanced selection must be at work. The nature of this balance is not known for blood types, but it is known for sickle-cell anemia.

Sickle-cell anemia is a condition found widely in tropical Africa and in American blacks as well, most of whose ancestors came from those regions of Africa (Fig. 13.4). The disease is manifested genetically only in the homozygous recessive state, and it is characterized by abnormally shaped red blood cells (see Chapter 5). There is incomplete dominance, such that the heterozygote shows some sickling of the red blood cells, but not to a degree sufficient to create disease symptoms. How could such a condition become established, given the relative seriousness of the disease in the homozygous recessive state? The answer is that the tropical regions of Africa are heavily infested with malaria, a parasite of red blood cells, some forms of which are virulent to the point of causing death. The changed shape of the heterozygous red blood cell is not sufficient to cause the condition of sickle-cell anemia. However, once the heterozygous cell is invaded by a malarial parasite, it changes into a sickled shape, apparently because the parasite uses up oxygen in the cell. This aberrantly shaped cell is then destroyed, along with the parasite it contains, by white blood cells, thereby preventing a spread of the parasite through the blood. In short, homozygous normal individuals are selected against by the parasite, and homo-

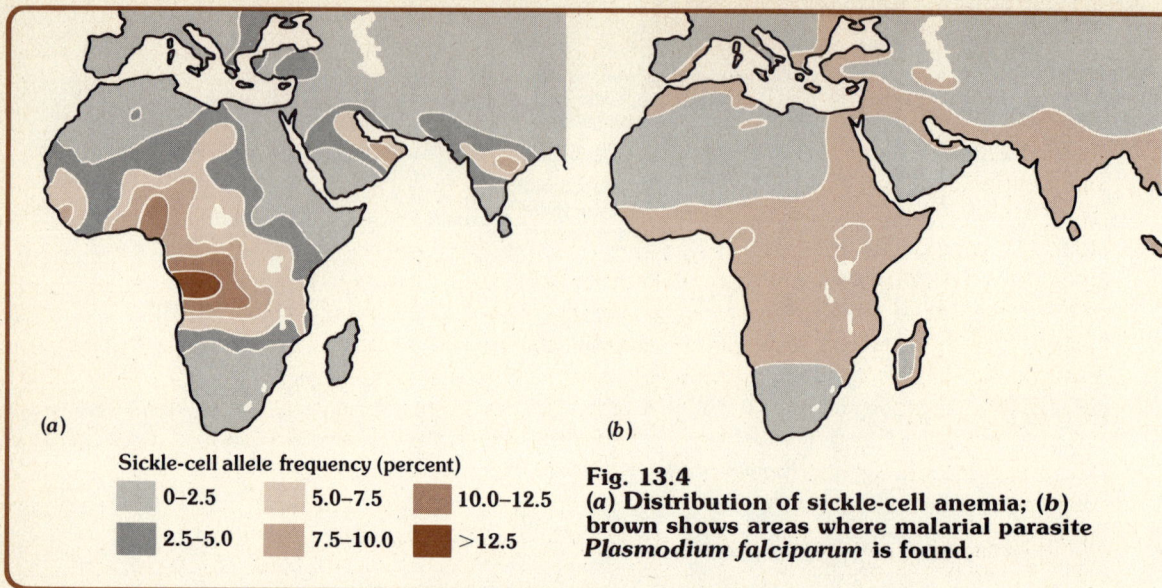

Sickle-cell allele frequency (percent)

| | |
|---|---|
| 0–2.5 | 5.0–7.5 | 10.0–12.5 |
| 2.5–5.0 | 7.5–10.0 | >12.5 |

(a)

(b)

**Fig. 13.4**
**(a) Distribution of sickle-cell anemia; (b) brown shows areas where malarial parasite *Plasmodium falciparum* is found.**

zygous recessive individuals are selected against by the sickle-cell condition, whereas the heterozygotes are protected from both. This heterozygote advantage is at the heart of many polymorphisms and, of course, will guarantee the continued production of homozygous dominant and homozygous recessive individuals, which means that the polymorphism will be perpetuated. Polymorphism is an exceedingly common situation, and one that is coming to gain increasing respect by geneticists as a method of maintaining genetic variability in the population.

**EVIDENCE FOR EVOLUTION**   The bulk of Darwin's *Origin of Species* is devoted to a massive compilation of data that indicate that evolution has occurred in the past and, by extrapolation, continues to occur today. Since Darwin's time, even more evidence has been compiled. This evidence can be relegated to four reasonably distinct categories:

1 fossil evidence;
2 phylogenetic evidence;
3 comparative embryological and anatomical evidence;
4 experimentally obtained evidence.

**Evidence from fossils**   Fossils may be formed in a variety of ways, but they occur typically when an organism dies in a body of water in which sedimentation (silt, etc.) is occurring at a relatively rapid rate. The soft parts of the organism generally decay rather quickly and are seldom fossilized, but the skeleton may be preserved either in the form of petrified bone or as a mold, the latter condition resulting when the skeleton is initially surrounded by sediment which then hardens.

Subsequently, the bony tissue is leached out by acidic waters, leaving a mold of the organism in the hardened sediment.

Given that most fossils are of marine organisms, which possess hard shells or skeletons, it is amazing that the fossil record is as complete as it is. Although there are many gaps in the record, there are also many instances in which it is startlingly clear. The fossil history of the vertebrates, in terms of their relationships to invertebrate organisms, is not very good, in large measure because the ancestral forms seem to have been small, soft-bodied organisms, with little potential for fossilization. The gaps between the major vertebrate groups, however, are generally well bridged by fossils. The earliest amphibians, for example, are rather well represented in the fossil record, and they appear very much like the fossil lungfish of the same period. Similarly, the gap between the amphibians and the reptiles is replete with fossil evidence. The early history of birds is not well represented in the fossil record, in large measure because birds have fragile skeletons, they seldom die in sedimenting areas, and their evolution as a group appears to have been very rapid.

The gap between the reptiles and the mammals is particularly well bridged. Mammals, of course, are distinguished from reptiles in that they produce milk to nurse their young. Mammals are also characterized by the possession of hair, by being endothermic, by giving birth to live young, and by a number of features of the circulatory and nervous system, none of which fossilizes, of course. Therefore, a determination of the taxonomic position of a fossil, whether reptile or mammal, must generally be based on extrapolation from skeletal features. Modern mammals possess only one bone in their lower jaw, whereas the lower jaw of reptiles may consist of as many as ten bones. The fossil record indicates a continuum between reptiles having ten bones and mammals having only one. At what point along that continuum did the development of mammary glands (the primary mammalian characteristic) occur? We have no way of knowing. The point, then, is simply that the fossil lineage between reptiles and mammals is so complete that many of the intermediate fossils must be classified arbitrarily, as they possess both mammalian and reptilian characteristics.

Not only are there numerous instances in which rather complete lineages of fossils linking two groups occur, but these fossils are also found in a very precise sequence in sedimentary rock, as opposed to being scattered randomly throughout various rock strata. That is, the ages of these fossils are precisely as expected, given the relative degree of development of earlier and later fossils. The age of fossils is determined in one of two ways. First, the rate of sediment deposition can be determined with reasonable accuracy, and the thickness of the overlying strata of rocks thus provides a good indication of the fossil's antiquity. Second, many elements occur in more than one atomic structure, some of which are unstable. These **radioactive isotopes,** as they are called, **decay** (i.e., give off subatomic particles in the form of radiation) at an extraordinarily regular rate. (In fact, so regular is the rate of decay that world time standards are maintained by "atomic" clocks.) Thus, the amount of these radioactive isotopes still left in the fossil or in the surrounding rock can furnish a highly accurate time measure as to the age of the fossil.

In summary, fossil evidence, although by no means complete, supports the concept of evolution in three ways:

1 Many major groups (e.g., reptiles and mammals) are very thoroughly linked by fossils.
2 The fossils themselves frequently are of species not present on earth today, but are similar to modern species.
3 The ages of these fossils are highly consistent, as reckoned by alternative methods, and the fossils are sufficiently old as to provide enough time for the very slow process of evolution to have occurred.

**Phylogenetic evidence**   Since at least the time of Aristotle, there has been an interest in grouping species in terms of the degree of similarity or dissimilarity among them. The formulation of "family trees," or **phylogenies,** of related species provides evidence for evolution simply by virtue of the fact that they can be constructed at all. That is, it would be reasonable to expect a totally random pattern of similarities and dissimilarities among different species if evolution did not occur, and it would therefore be impossible to construct phylogenies because relationships, as such, would not exist. The fact of the matter, however, is that relationships between species do exist, and in most animal groups phylogenies have been constructed which are not only in accord with the degree of similarity and dissimilarity among the various species, but which also strongly suggest the actual evolutionary pattern of relationships that interconnects these species.

Similarly, the pattern of geographic distribution of species is far from random, a fact strongly emphasized by Darwin. Indeed, in his initial examination of the fauna of South America, long before he had formulated his theory of evolution, Darwin noted three facts that were later to provide the basis of much of his evidence for the existence of evolution. These were:

1 The fact that South American fossils were obviously more similar to living South American animals than to animals of other parts of the globe.
2 The fact that there was greater similarity between South American animals in different climatic regions than between South American and African animals of the same climatic region.
3 The fact that island fauna were most closely related to the fauna of the nearest portion of the mainland.

Again, there is no reason to assume such a patterning unless evolution has occurred.

**Comparative embryological and anatomical evidence**   If there were no such thing as evolution, and species were to remain essentially immutable, the embryonic development of a given species would be expected to show only random similarity to the development of any other species. However, as has been known at least since the time of Haeckel (see Chapter 12), not only do species show definite similarities in their embryonic development but, as a general rule, the degree of this similarity is directly correlated with their overall similarity as adult organisms. That is, just as apes are far more similar to humans than to opossums as adults, so, too, are they more similar to humans during embryonic development. That, of course, is hardly surprising, but our complacency in learning this fact is hinged directly on the fact that evolution is now very widely accepted. The same finding created a sensation in the nineteenth century, when Haeckel first elucidated it as his Recapitulation Theory.

Perhaps even more dramatic is the existence of developmental similarities among species that are very different as adults. Early in human development (as in all vertebrates), there is developed in the neck region a series of openings between the throat and the external surface of the body. In fish, these are to become the gills of the adult. In humans, virtually all regress (although the first remains as the **Eustachian tube,** the canal connecting the middle ear with the throat). Why are they formed, as they have no function in the adult human? Why do human fetuses develop a tail, which normally regresses later (although occasionally babies are born with small tails, which must be surgically removed)? Why does the fetus develop a coat of hair all over its body, which is lost prior to birth? All of these features, as well as many others, make a certain amount of sense in the light of evolution and phylogenetic relationships, but are totally inexplicable to one who accepts species as being immutable.

An even stronger case can be made for evolution based on adult anatomical structures. The whole discipline of comparative anatomy is based on recognizing and describing these relationships, which exist in every organ system. Indeed, the idea of evolution, although coupled with an incorrectly designated cause, first arose among anatomists of the eighteenth century, who were struck by the magnitude of the similarities in all organ systems among different species of animals.

To take just a single example, consider the development of the aorta and anterior arteries in vertebrates. Primitive fish have six pairs of anterior arteries, each passing through its own gill, although this number may be reduced to as few as four in a more advanced fish. Most amphibians have just three pairs. Of these, the first serves to carry blood to the head, the second (the aorta) carries blood to the body proper, and the third carries blood to the lungs to be oxygenated. Reptiles have a similar arrangement, except the two aortas are of different sizes, the left being smaller than the right. In birds, which are presumed to be closely related to reptiles, the left aorta is lost, leaving just a single vessel. Mammals are similar, although presumed to be more distantly related to reptiles—and have lost the right aorta, leaving only the left. In all vertebrates, during embryonic development, all six pairs are formed. It is very difficult to explain away this remarkable progression of events other than by accepting an evolutionary sequence.

As a final bit of anatomical evidence supporting evolutionary theory, consider vestigial characters. These are anatomical structures which apparently serve no function, and are difficult to explain away except by accepting them as remnants of structures that once had a distinct purpose in our more distant ancestors. The muscles of the midabdominal wall are segmented, as can be seen from the washboard effect so assiduously cultivated by body-builders—but why segmented? Segmentation serves no purpose. Why do we still possess muscles to move our outer ear? We may use these muscles to help break the ice at parties, but certainly never to direct the ear toward the source of a sound. And why do we possess an appendix, a structure of which we are only aware when it becomes inflamed and must be removed? In each case (and in many others), it is easy to see the role these structures played in ancestral species, but difficult to explain away their presence in humans purely by chance.

**Experimental evidence**   There are literally hundreds of experimental studies in which a species with a short generation time (such as the fruit fly) has been subjected to a high selection pressure by the investigator (e.g., selecting only those flies for breeding with the highest and lowest number of bristles on the thorax) for 20 or 30 generations, by which time two populations have emerged which are distinctly different in number of thoracic bristles. Such experiments prove only that evolution can occur in controlled environments, but not necessarily that it can occur under natural conditions. The problem is that, by definition, only populations evolve, the measure of evolutionary rate being the change in gene frequency over time. Moreover, only rarely do environments change quickly. Taken together, these facts explain why thousands of years are required for the development of a new species—a time frame simply beyond the capacity of direct human observation. Hence, most of the evidence for evolution is inferential.

However, there are a number of instances involving natural environments, but environments so profoundly affected by human influence as to provide experimental evidence for evolution. One example is the peppered moth of England. In collections of 100 years ago, this moth was known almost exclusively in a single color mode, a kind of mottled gray, which seemed to serve a protective function for the moth which spends the daylight hours resting on lichen-covered tree trunks (Fig. 13.5). A gradual shift followed such that, by 1950, virtually all the moths in the English midlands were the black phase, the gray being very rare. This situation was recognized as being a balanced polymorphism, and the evidence indicated that the shift in color modes was caused by the effects of industrial soot in killing off the lichens and blackening the tree trunks. Under these conditions, no camouflage was afforded the grays, and they were eaten by birds, whereas the blacks were benefited by the soot on the trees. In recent years, diminished industrial pollution has allowed the return of the lichens, and the frequency of the gray form of moth is once again on the increase. This factual situation illustrates the mechanisms of natural selection in a nutshell.

A similar picture can be painted for resistance. Many pathogenic bacteria have developed strains that are resistant to such bactericides as penicillin, and many insect pests have become resistant to such insecticides as DDT. In each case, a powerful selective force (the lethal agent) has been met by genetic mutations which, in various ways, allow the possessing organisms to survive the formerly lethal agent. As we have now redefined evolution to state "a change in gene frequencies over time," both these examples qualify as evolutionary events.

**HUMAN EVOLUTION**   Most of the evidence for evolution just discussed involved the human organism peripherally at best. What is the evidence for assuming

Fig. 13.5 Dark and light forms of the polymorphic peppered moth. (American Museum of Natural History)

that evolution is also a process that affects humans? This thorny question was implied by Darwin in his *Origin of Species* (1859), but was not fully discussed until his *Descent of Man* (1871). Many individuals were quick to assume from the earlier work that Darwin had intended to include humans as being subject to the processes of evolution, but few were prepared for the strength of Darwin's arguments in his later book. That humans were related to a bunch of smelly, dirty, old monkeys! The very idea! As one refined lady of the time said, "My dear, let us hope that it is not true, but if it is let us pray that it will not become generally known."

It seems odd to us, retrospectively, that these ideas created quite the furor they did. After all, a primary reason why the monkey house has always been among the most popular exhibits in any zoo is that the anatomy, behavior, and expressions of these animals remind us of ourselves. In a tacit way, we·are acknowledging a relationship. However, this has been a relatively recent development. In the middle of the last century, few people were willing to acknowledge anything more than vague similarity. Humans, after all, were  . . .  well . . .  human, and animals were  . . .  animals.

A troublesome feature in postulating human evolution in Darwin's time was that virtually no fossils had then been discovered which would substantiate such a theory. The earliest discovery was made in 1856, just three years prior to the publication of his *Origin of Species*. This specimen was found during quarrying activity in the Neander valley (*tal* is "valley" in German) in Germany. This discovery created a minor sensation, not because the bones were immediately recognized as being a near relative of modern man, for they were not, but rather because the bones were promptly identified as belonging to something else. The quarry owner thought them to be the bones of a bear; more knowledgeable individuals were sure they belonged to some hapless individual who drowned in Noah's flood; another scientist proclaimed them to be the bones of a Mongolian, a member of the Russian cavalry, who had deserted during the push against Napoleon in 1814; yet another anatomist thought the bones to be those of "an old Dutchman."

A few years later, in 1868, another ancient find, this time near Cro-Magnon hill in southern France, revealed the presence of much more modern-appearing skulls. This discovery was greeted with a wave of relief, for it demonstrated that humans had "always" looked the way we do now. However, when more Neanderthal skulls were discovered, this time in Belgium a few years later, concern was generated once again. The first find could not now be dismissed as an aberrancy—the skeleton of a diseased individual, for example—because here were several more, a few hundred miles from the first find. Then, in 1891, a much more primitive, but still humanlike skull was found in Java, and there was renewed interest in the status of the Neanderthal types. In 1908, some very complete skeletons were found and a series of measurements and reconstructions ensued which were to have a profound effect on the status of the Neanderthalers in the public eye. Retrospectively, the magnitude of the errors was surprising. The anatomist in charge simply ignored the bones in front of him and, focussing on the heavy brow and flattened skull, pronounced Neanderthal a shuffling, brutish, clumsy, stupid individual with the foot of

**Fig. 13.6  Dioramic figures of Neanderthalers. (American Museum of Natural History)**

**Fig. 13.7  Reconstruction of Neanderthal head and torso. (American Museum of Natural History)**

**Fig. 13.8  Reconstruction of Cro-Magnon head and torso. (American Museum of Natural History)**

an ape (a divergent big toe). This image has etched itself into the public mind, indelibly it would seem. Neanderthal man is usually portrayed as having spent most of his time shuffling and grunting, except when he was otherwise occupied bonking nubile females on the head with his club (Figs. 13.6 and 13.7). In contrast, Cro-Magnon man is usually depicted as clear of mind and steady of hand; with ramrod straight posture and blond hair and blue eyes; a primitive Viking, as it were, whose task it was to rid the earth of the smelly Neanderthal and make the world safe for humanity (Fig. 13.8). Needless to say, with that type of introduction, Neanderthalers were to have a protracted wait before their role in human evolution was to become clear.

**Primates, Hominoidea, and Hominidae**   Humans are classified as primates, along with the apes, monkeys, and such primitive species as lemurs and tarsiers. Primates as a group are rather distinct among mammals, with a flattened face, eyes in front, considerable lateral movement of the arms (coupled with a strong collarbone), some opposability in the first digit of the hand and feet, and so on.

The superfamily Hominoidea includes the gorilla, chimpanzee, orangutan, gibbon—and humans. These are distinguished from the monkeys because of size, shape of chest, tooth structure, and other anatomical features. Only humans (*Homo sapiens*) occupy the family Hominidae. Humans differ from the apes (e.g., the gorilla) in having a lighter jaw, with a more rounded tooth row, and with teeth all about the same size (Figs. 13.9–13.11), among other features.

The classification of living organisms is a relatively easy task. What about the fossils? Where do they fit in?

**The human heritage**   Before we begin to look at any type of human lineage, as determined from fossils, it is necessary to state that, in most cases, only partial skeletons of any one type are available (Table 13.1). In many cases, there may be only one or two bone fragments to work from. Although the trained scientist can extrapolate from a small number of bones to a complete skeleton with considerable accuracy, how do we know that the individual was characteristic of the population of which he was a member? If you were on a collecting trip for the intergalactic zoo, and were told to bring back a representative earthling, which would you choose—a pygmy or a Norwegian basketball star? It might be difficult, at first glance, to accept that all modern humans are interfertile and members of a single species, but such is the case.

**Table 13.1**
**The meager remains of Australopithecus**

| Skull | | Shoulder girdle | |
|---|---|---|---|
| Crania | | Shoulder blades | 1 |
| (Skulls without jaws but with some facial | | Collarbones | 4 |
| bones) | 11 | **Arms** | |
| Casts of skull interior | 13 | Unae (one of two lower arm bones) | 3 |
| Face fragments | 22 | Wrist bones | 4 |
| Skullcaps | 45 | Radii (lower arm bones) | 6 |
| Upper jaws | 77 | Finger bones (palm) | 9 |
| Lower jaws | 79 | Humeri (upper arm bones) | 10 |
| Baby teeth | 110 | Finger bones (digits) | 22 |
| Permanent teeth | 933 | **Legs** | |
| **Skeleton** | | Fibulae (outer leg bone) | 1 |
| Sacra (base of spine) | 1 | Ankle bones | 1 |
| Ribs (mostly fragments) | 13 | Shin bones | 2 |
| Vertebrae | 20 | Toe bones | 5 |
| **Hip girdle** | | Arch bones | 12 |
| Ischia (bottom of pelvis) | 1 | Thigh bones | 14 |
| Ilia (hipbone of pelvis) | 3 | | |
| Fossa coxae (side bone of pelvis) | 9 | | |

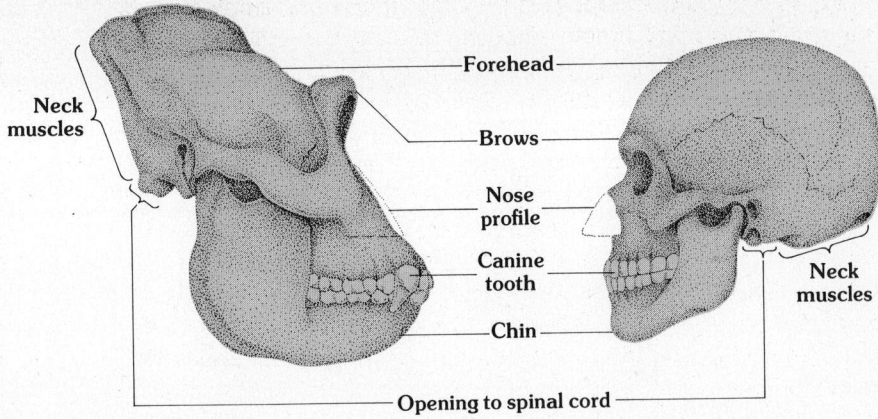

Fig. 13.9
Skulls
of gorilla
and human
compared.

Neck muscles

Forehead

Brows

Nose profile

Canine tooth

Chin

Neck muscles

Opening to spinal cord

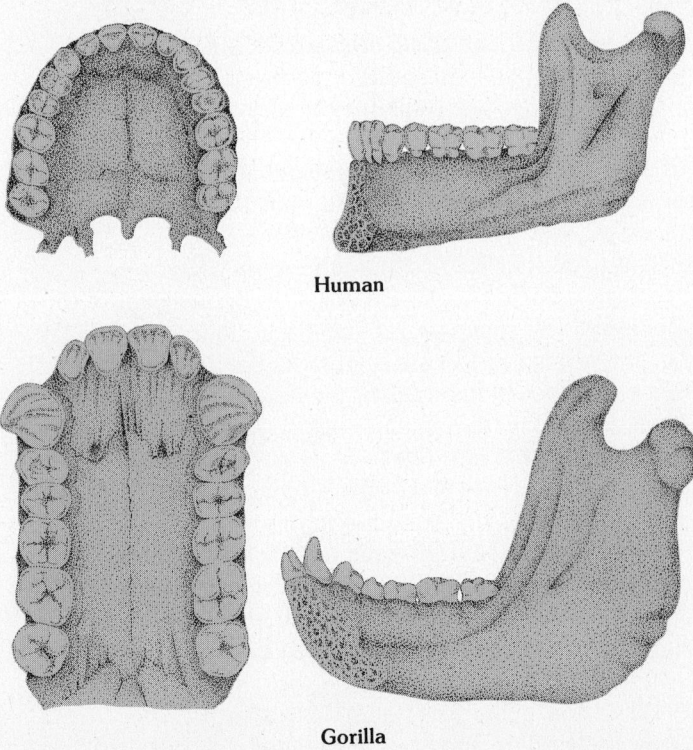

Fig. 13.10
Jaws of human and
gorilla compared.

Human

Gorilla

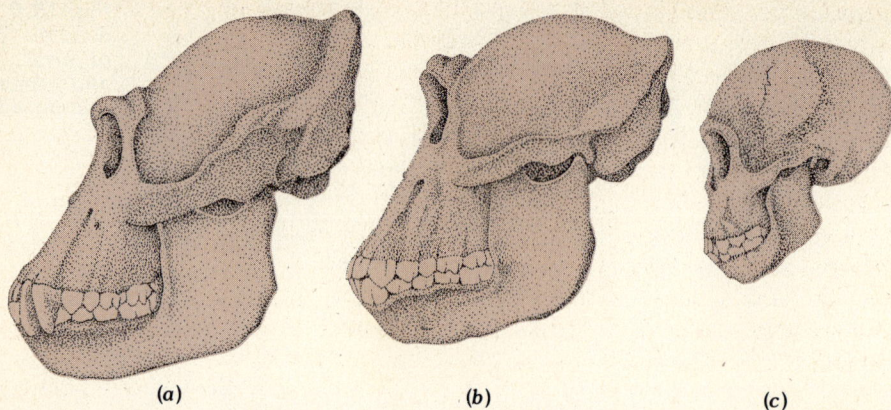

**Fig. 13.11  Gorilla as (a) adult, (b) juvenile, and (c) infant.**

In the age of fossils, populations were few and small, and we should expect a great deal of variation within a given fossil type—and, indeed, such appears to be the case.

The first fossils found—Neanderthal, Cro-Magnon, Java—were all more or less humanlike. It was possible to view them as leading back toward an apelike ancestor, but there was obviously still a very sizable gap. Where and what was this "missing link"?

The first answer came from South Africa in 1924, although the significance of this first discovery was not widely recognized until after World War II, as the discoverers were not well-known anthropologists. The skull was no larger than that of a baboon, but it had a flat face rather than a muzzle, and an even tooth row, without the long canine tooth characteristic of a baboon. In addition, the opening to the skull from the spinal cord was situated such that the animal must have stood erect. Human? No, the brain was too small—only 400 cm³ (24 in.³), as compared with the 1400 cm³ (85 in.³) average of modern humans. It was small, only about 1.2 m (4 ft) tall and it weighed only 23–40 kg (50–90 lb). Nonetheless, it was a hominid, and obviously the most primitive then known. It was named *Australopithecus africanus* (southern African ape) (Fig. 13.12*a*). Later excavations in South Africa revealed the presence of a second, heavier-jawed type, which was called *Australopithecus robustus* (Fig. 13.12*b*). This was a type that stood about 1.5 m (5 ft), and weighed perhaps as much as 70 kg (150 lb). Unfortunately, the rock in which these fossils were found was not stratified and, at the time, there was no way of determining their age. Their discoverer estimated their age at 2 million years, a figure so staggering as to render to shreds any credibility he had left.

There the matter stood until 1959 when Louis and Mary Leakey discovered similar fossils in Olduvai Gorge, Tanzania. These fossils were of an individual still more robust than either of the South African types and it was named *Australopithecus boisei* (Fig. 13.12c). This individual was estimated to be 1.7 m (5.5 ft) tall and weighed perhaps as much as 90 kg (200 lb). Even more important, the discovery was made in some well-known strata, allowing a date to be assigned—1.75 million years. Suddenly, there was renewed interest in the South African fossils, because evidently the 2-million-year figure was not too far off the mark.

Still, problems remained. Further discoveries in South Africa suggested that some *Australopithecus robustus* were as much as 1 million years younger than were the more human-appearing *A. africanus*. Moreover, *A. robustus* appeared to be exclusively a plant-eater, rather like the modern gorilla—and the teeth and skull crest seemed to confirm this. The discovery of a still more robust form (*A. boisei*)

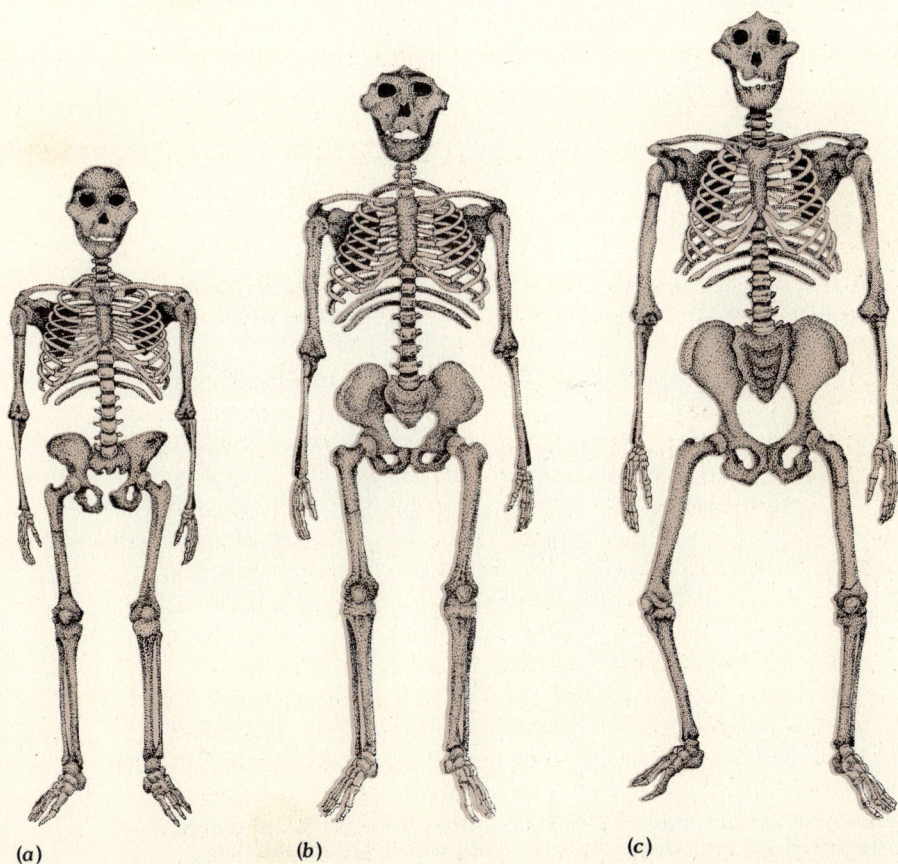

Fig. 13.12 The three dominant skeletal types of Australopithecus: (a) A. africanus/habilis; (b) A. robustus; (c) A. boisei.

(a)    (b)    (c)

added to the problems—it began to appear as if modern man had evolved rather suddenly from a somewhat primitive form. However, in 1960, the Leakeys found another skull, also about 1.75 million years old, which was even more refined than that of *A. africanus*. The correct designation of this form has been in some dispute, some authorities regarding it simply as a more refined, and later, type of *A. africanus,* but with Leakey himself assigning it to the genus *Homo,* as *Homo habilis.* Certainly his brain was larger—at 650 cm³ (40 in.³), some 200 cm³ (12 in.³) larger than typical South African *A. africanus* (Fig. 13.13). At the very least, *Homo habilis* seems to be poised on the edge of full human status—the argument is simply to which side of the line he belongs.

More recent discoveries confirm that at least two quite different types of prehumans existed contemporaneously. Fossils found at Lake Rudolph in northern Kenya and at Omo, just across the border in Ethiopia (both of which are about 800 km (500 mi) north of Olduvai Gorge) indicate the presence of *A. boisei* from as long ago as 3.7 million years and as recently as 1 million years ago. Hence, it is clear that at least one, and perhaps two, robust vegetarian species of hominids lived throughout much of eastern and southern Africa for several million years. A recent discovery at Lake Rudolph of a leg and skull of a *Homo habilis* type of individ-

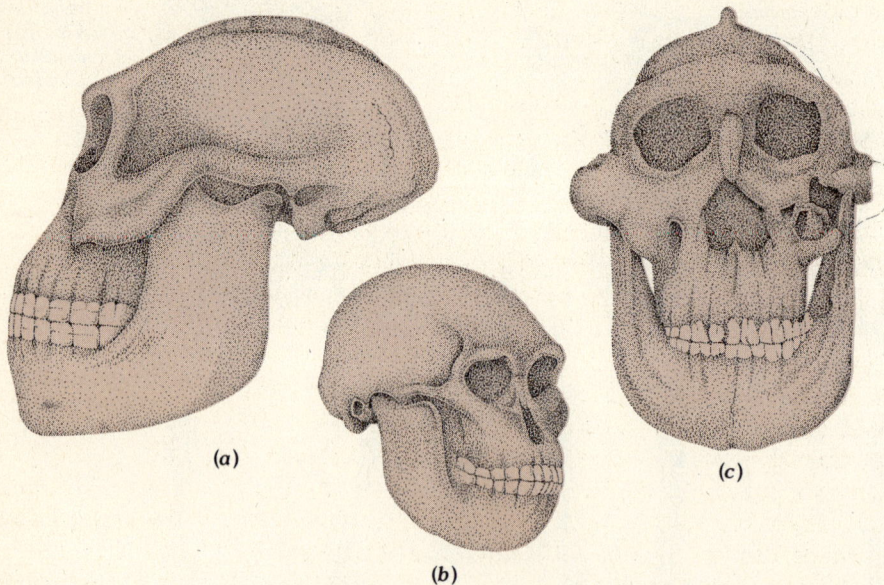

**Fig. 13.13  Skulls of various types of Australopithecines:** (a) *A. robustus;* (b) *A. africanus;* (c) *A. boisei.* **These are all one-fourth their actual size.**

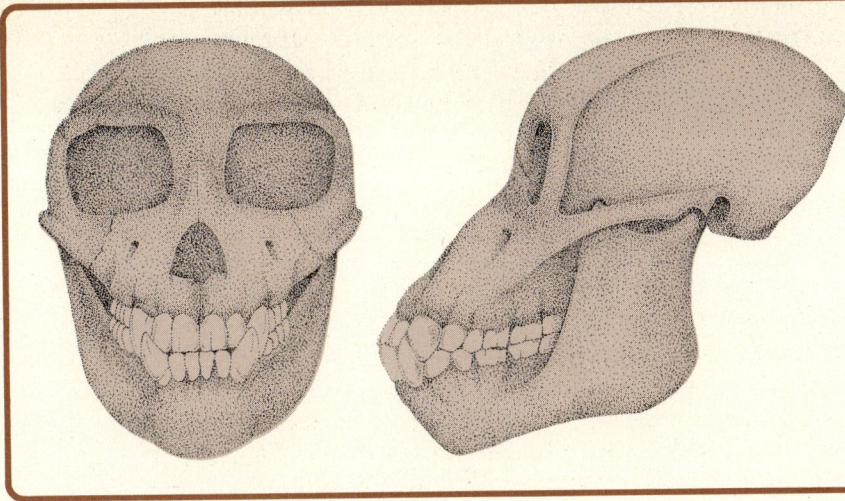

**Fig. 13.14**
**The skull of**
***Dryopithecus,***
**one-half**
**actual size.**

ual, but dated at 2.8 million years, clouds the theory that *A. africanus* evolved into *H. habilis,* insofar as this newly discovered specimen would have existed at the same time as the oldest of the *A. africanus* fossils. Regardless, whether there were one or two species, this more slender meat-eater was clearly separate from the *A. robustus/boisei* type.

About 1 million years ago, *A robustus/boisei* disappears from the fossil record, indicating an extinction at about that time. Were they killed off by *A. africanus/habilis*? We may never know for sure, but since that time, evidently only one species of hominid has been on the earth at any given time.

In recent years, both the picture between *Australopithecus* and the apes, and that between *Australopithecus* and modern humans have become clearer. The earliest known fossil that possesses definite ape and human traits—that is, the earliest known hominoid—is *Dryopithecus* (Fig. 13.14), discovered on an island in Lake Victoria in Africa. Similar remains have also been discovered in India and in Europe. The age of these remains has been estimated to be as much as 20 million years. There are disputes, of course, as to whether the lineage leading to humans diverged before, or after, the time of *Dryopithecus,* but this animal is our best candidate for the ancestor of both modern apes and man we have at present.

A somewhat more recent fossil is that of *Ramapithecus,* first found in India, with an age of perhaps 12 million years, and more recently also found in Africa, dating to 14 million years. This fossil is generally thought of as being the earliest known hominid, as its teeth are decidedly more humanlike than apelike. However, nothing is known of its body, as only portions of the skull have been found. Hence, it is not known if *Ramapithecus* walked upright or even how large it was.

As we move into the more recent past, since the time of *Australopithecus,*

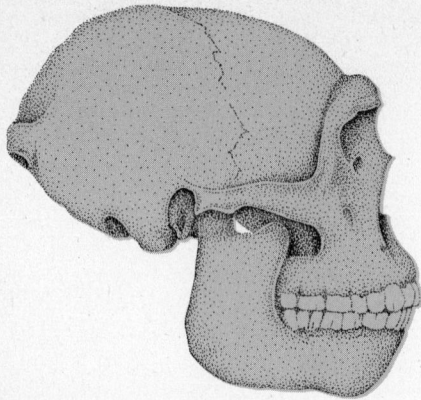

Fig. 13.15
The skull of Java man,
one-fourth actual size.

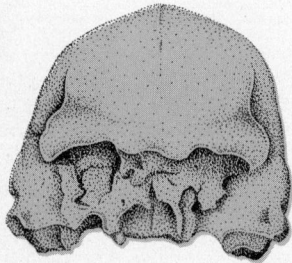

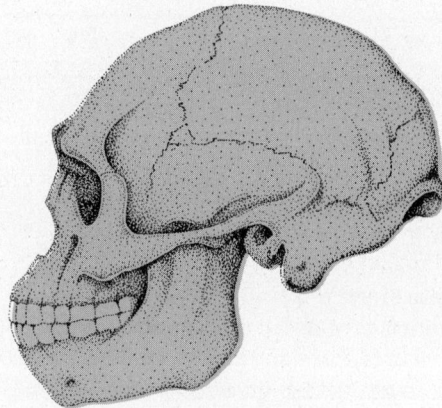

Fig. 13.16   The skull of Solo man from Java, a late version of *H. erectus*.
The drawing is one-fourth actual size.

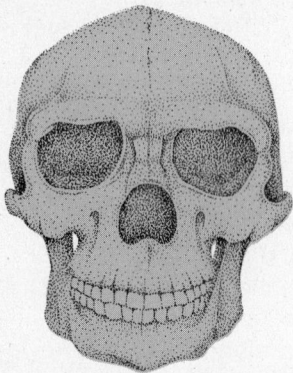

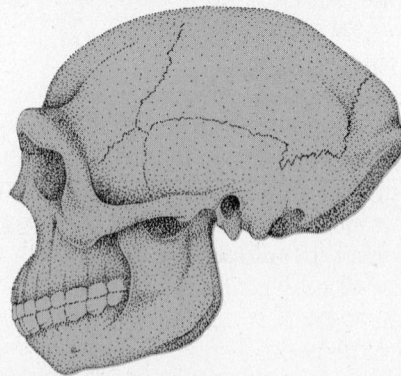

Fig. 13.17   The skull of Peking man, one-fourth actual size.

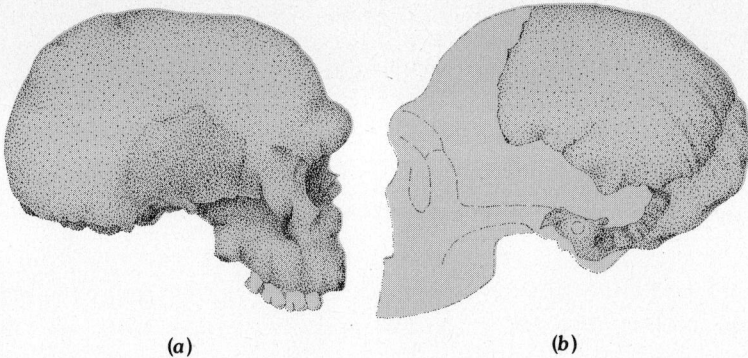

**Fig. 13.18**
**(a) Steinheim and**
**(b) Swanscombe**
**skulls.**

(a)  (b)

there are increased numbers and completeness of skull and skeletal fragments. There is considerable diversity among these, often indicating the geographic separation between the finds, which stretch from China, through Southeast Asia, Africa, and finally into Europe. Collectively, these skeletons and skulls have been lumped under the heading *Homo erectus,* and include such well-known names as Java man, Peking man, Steinheim and Swanscombe man, and so on (Figs. 13.15–13.18). Fossils of this type have been dated at between 250,000 to 1,000,000 years ago. Most have a skull capacity of 750–1100 cm$^3$ (45–70 in.$^3$), and they are larger and heavier than *A. africanus.*

The actual line into our own species, *H. sapiens,* was crossed somewhere between 250,000 and 100,000 years ago. (Swanscombe man, a skull dating back to 250,000 years ago, is transitional, some authorities regarding it as *H. erectus,* others as *H. sapiens.*)

All of this leads us back to a reconsideration of the Neanderthalers. What was their role in human evolution? The Neanderthalers had a brain capacity (Fig. 13.19) at least the equal of modern humans—what happened to them? The most recent Neanderthal fossil is dated at 40,000 years, but the oldest Cro-Magnon fossil is only 26,000 years old. What happened during those intervening 14,000 years?

During much of the time between 200,000 and 10,000 years ago, glaciers covered as much as 30 percent of the earth's surface. Obviously, throughout this period, hominid populations must have been constantly migrating, as the glaciers alternately expanded and contracted. It is very likely that small populations were isolated by surrounding ice for many generations. Such isolated populations could be expected to diverge anatomically from each other, and there is some evidence that the Neanderthalers did so. Thus, even though Neanderthal fossils are most common from Western Europe, the Neanderthalers ranged widely across Europe and Asia, and even extended into Africa (Fig. 13.20). Moreover, the skulls are the most different from modern humans in Western Europe—in other parts of the world, such as the Near East, the skulls appear to be halfway between the Western

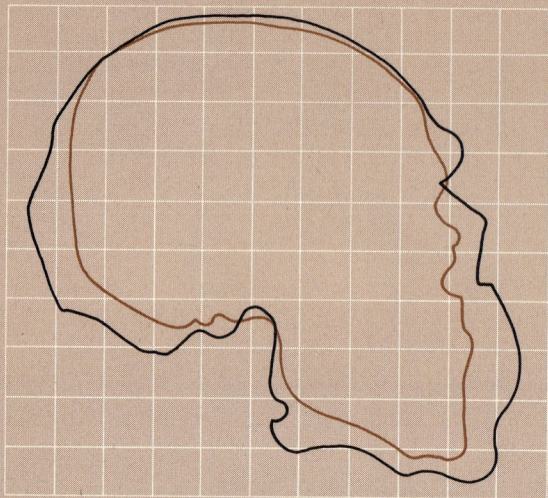

Fig. 13.19
Neanderthal (black line) and modern human (color line) skulls compared, drawn to the same scale.

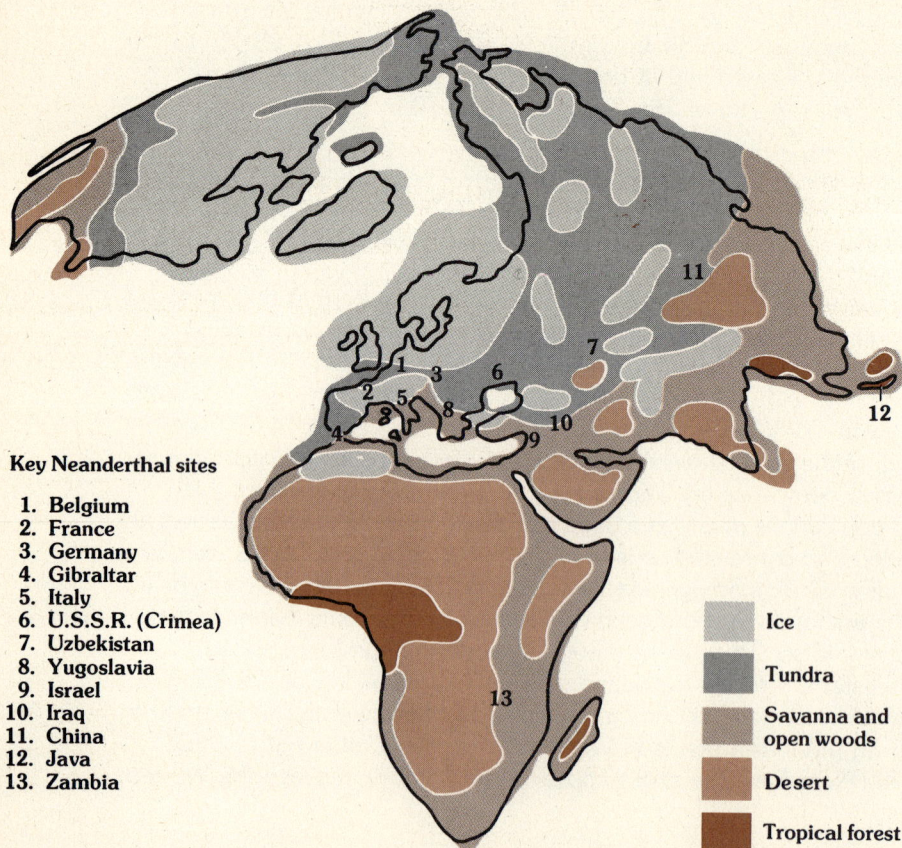

**Key Neanderthal sites**

1. Belgium
2. France
3. Germany
4. Gibraltar
5. Italy
6. U.S.S.R. (Crimea)
7. Uzbekistan
8. Yugoslavia
9. Israel
10. Iraq
11. China
12. Java
13. Zambia

Ice

Tundra

Savanna and open woods

Desert

Tropical forest

**Fig. 13.20    Range of the Neanderthalers 60,000 years ago.**

European Neanderthal and Cro-Magnon man (Figs. 13.21–13.23). The prevailing theory has it that these more generalized Neanderthalers evolved into Cro-Magnon man, perhaps because of improvement in language skills or in tool-making, and the Western European Neanderthalers either assimilated or died out, as the Cro-Magnon types moved in from the east. Thus, even though the more extreme forms of Neanderthalers may have died out, there is little question that the basic stock gave rise to ourselves—and, as such, we owe the Neanderthalers a bit more respect than they have received in the past.

**Races**  The term ''race'' is not widely used in biology, primarily because of the variety of interpretations which this term has suffered in other disciplines. Biologists, along with social scientists, are interested in cohesive groups which are largely, but not completely, reproductively isolated from other such groups. This interest exists for reasons ranging from the esoteric (measuring evolutionary rates) to the

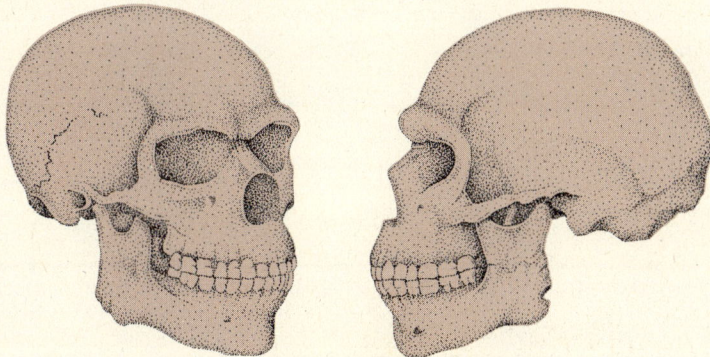

**Fig. 13.21** The most modern-appearing of the Neanderthal skulls, from the Near East (Skhul Cave on Mount Carmel), drawn to one-fourth actual size.

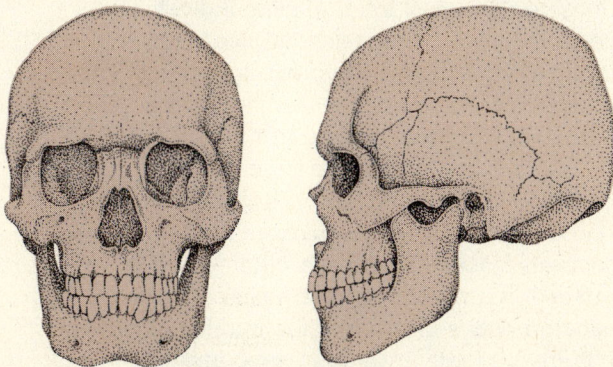

**Fig. 13.22** A generalized modern human skull, one-fourth average size.

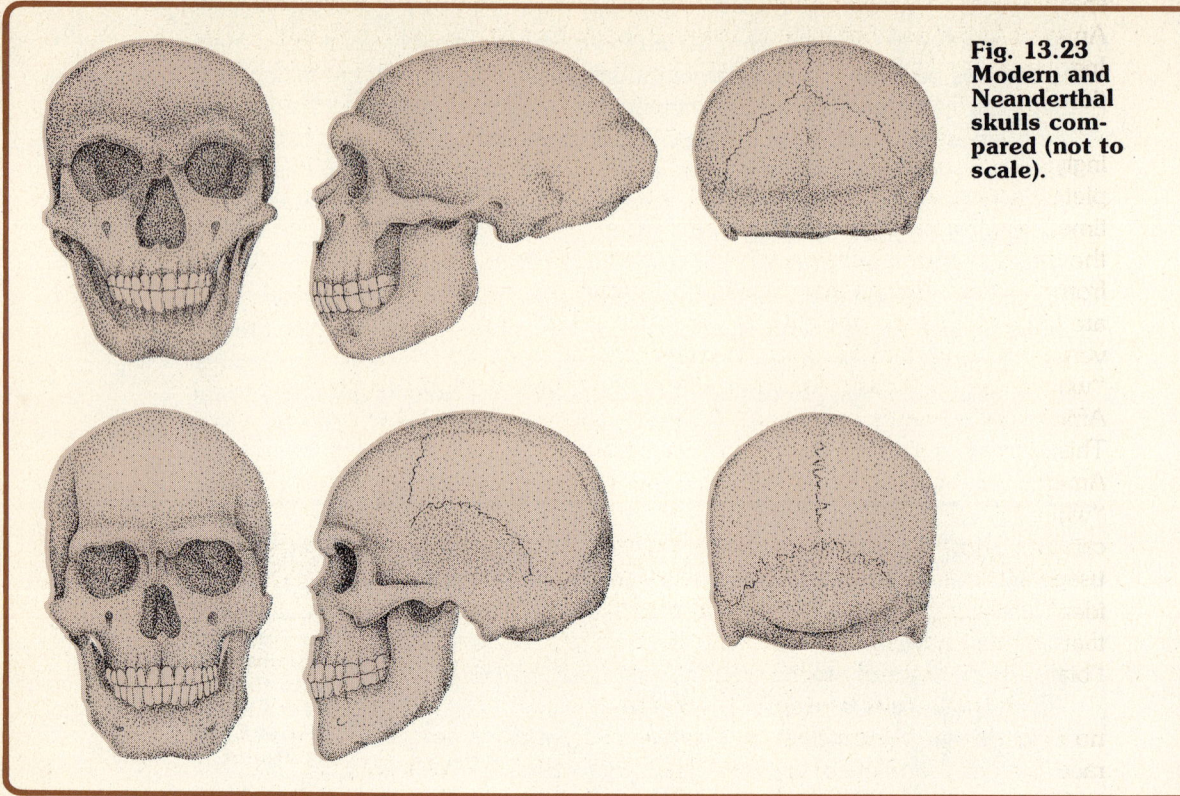

**Fig. 13.23 Modern and Neanderthal skulls compared (not to scale).**

highly pragmatic (genetic counseling). However, such groupings are generally not termed "races" by biologists for the following reasons.

**1. Inefficient terminology**   There are other less controversial terms available. The term "population" refers to a localized group of interbreeding individuals of a given species, be they people or sparrows; the term "subspecies" is used in reference to populations which are geographically separated and anatomically distinct.

**2. Insufficient consensus**   With respect to ourselves, there is little agreement as to how many races there might be, with estimates ranging from three to more than 30. However, even the latter number is not sufficiently large to recognize all of the reproductively isolated populations. "South American Indians" is a fairly restrictive category, but it is too broad to be of much use, because many of the individual Indian tribes are reproductively isolated from each other.

**3. Insufficient geographic separation**   It is impossible to set limits on human "races." Though there is geographic separation between some of the traditionally accepted racial groups in certain regions, there are large areas of intermixture elsewhere. For example, the Sahara Desert separates "whites" and "blacks" relatively sharply—but to the east, there is a broad zone of mixture in the Sudan and Ethiopia. An even larger zone of mixture exists in India, and things get totally out of hand in Indonesia, the Philippines, and the other Pacific Islands. Whatever utility the traditionally accepted racial groups may have had, it is lost in these areas.

**4. Increase of genetic mixture**  Zones of mixture are becoming increasingly broader. It is likely that the major racial groups were, at one time, rather completely separated for long periods of time. However, certainly during historical times, and increasingly in more recent years, the groups have been mixing. To take the most obvious example—American blacks are now genetically quite different from the West African populations from which they are descended. Several genes are known in which one allele is very common in West African populations, and very rare in Western European populations, the homeland of most American "whites." In each case, American blacks show heterozygosity in the two alleles, the African alleles being from two to three times as common as the European alleles. This heterozygosity is a measure of interbreeding between blacks and whites in America, and indicates that American blacks are, on the average, about 30 percent "white." In recognition of this fact, some anthropologists have proposed that American blacks are a separate "race" or an "emergent race," but of course such a usage has little practical application. The only valid purpose of racial designation is to identify populations that do not interbreed with other populations, and it is obvious that very considerable interbreeding has occurred and continues to occur between "blacks" and "whites" in this country.

**5. Insufficient biological criteria**  Perhaps most importantly, there is in no case a single characteristic that serves to separate all members of one human race from all members of another. The characteristics most commonly used are such obvious external features as hair and skin color, both of which are affected by many genes. Race, then, is a statistical concept, a type of averaging whereby one population may (sometimes) be separable from another population, but without implying that any given individual can be distinguished from any other individual based on racial designation. (Obviously, where the term has been most abused is in precisely this area—in the blind insistence that everyone belongs to one race or another, and in the continued search for ways of making the identification.)

The plain fact of the matter is that even though the reproductive isolation of human populations from one another may have once been the norm, and even though some populations still exist which are reproductively isolated, increasing numbers of individuals, especially in the Western Hemisphere, are in no way reproductively isolated from members of another of the traditional racial groups. Thus, although there may still be some justification in screening only "blacks" for sickle-cell anemia, for purposes of genetic counseling, the day is not far off when, because of "mixed racial" marriages, everyone may have to be tested. By the same token, interbreeding between previously isolated populations decreases the probability of potentially dangerous homozygous recessive conditions. Thus, we can expect to see a reduction in the frequency of a number of genetic diseases as these "racial" barriers are broken down.

**CONGENITAL DEFECTS**    Congenital defects are anatomical or physiological anomalies present at or before birth. The frequency of such defects is difficult to measure, not only because many are very minor and frequently escape being recorded, but also because serious defects may cause miscarriage and spontaneous abortion and, in such instances, it is uncommon that a cause for the termination of the pregnancy is ever determined. The best estimate, however, is that at least 20 percent of all conceptions terminate in miscarriage or spontaneous abortion and, of that number, between 20 and 40 percent involve congenital defects. Including these failed pregnancies, it is estimated that 10 percent of all conceptions possess a congenital defect.

The causes of congenital defects are varied. About 20 percent are known to be caused by a single gene; an additional 10 percent involve major chromosomal aberrations; another 10 percent are caused by viruses or other environmental agents; and the remaining 60 percent involve some combination of the above, or the cause is unknown.

**Single gene effects**    Relatively few congenital defects are attributable to a dominant gene for the simple reason that, if serious, such defects would quickly be eliminated from the population. An example of a minor congenital defect caused by a dominant gene, however, is **brachydactyly** (abnormally short fingers).

Far more common are the recessive conditions. It is known, for example, that over 100 human disorders are the result of the failure of a given individual to produce a particular enzyme. In **phenylketonuria (PKU),** for example, the absence of an enzyme that converts a particular amino acid to another amino acid leads to interference in the synthesis of a vital brain chemical. The PKU child appears normal at birth, but within six months is profoundly and irreversibly mentally retarded. Fortunately, this disease is readily detected at birth by the presence of abnormal chemicals in the urine. In all states, this condition is now routinely tested for in all newborn infants. If such children are placed on a special diet low in the amino acid in question, they can avoid the retardation and lead essentially normal lives. An equally dramatic but much less serious enzymatic failure results in albinism. Sickle-cell anemia is yet another example of a congenital defect caused by a single gene.

**Sex determination and sex linkage**    One of the most important causes of congenital defects stems from the mechanisms involved in sex determination. In all animal species which reproduce sexually, males and females are chromosomally distinguishable by the composition of one or more pairs of **sex chromosomes.** In humans, for instance, males and females are indistinguishable on the basis of 22 of the pairs of chromosomes; however, the 23rd pair is a true pair in females (and therefore is designated XX) but is an odd pair in males (designated XY) (Fig. 13.24). Males as a sex need not be in possession of an odd chromosome pair—in birds, butterflies, salamanders, and many fish, the males are XX and the females XY—but, in all cases, the sexes differ in at least one chromosome.

In humans, the X chromosome is large and carries many genes not directly associated with sexual development. In contrast, the Y chromosome appears to be solely concerned with sex determination. As such, genes on the X chromosome in males cannot be considered as either dominant or recessive, because every gene will be expressed, as there is no mechanism of suppression with only one X chromosome. Therefore, females who are heterozygous for traits carried on the X chromosome will act as carriers for recessive conditions which will be expressed in 50 percent of the male offspring (i.e., those inheriting the X chromosome bearing the recessive allele). For example, such conditions as hemophilia, in which the blood fails to clot (Fig. 13.25), and color-blindness, among more than 75 hereditary disorders, are caused by recessive alleles on the X chromosome. Females are almost never hemophiliacs or color-blind, because a homozygous recessive condition rarely occurs for such defects, but both conditions are relatively common in males.

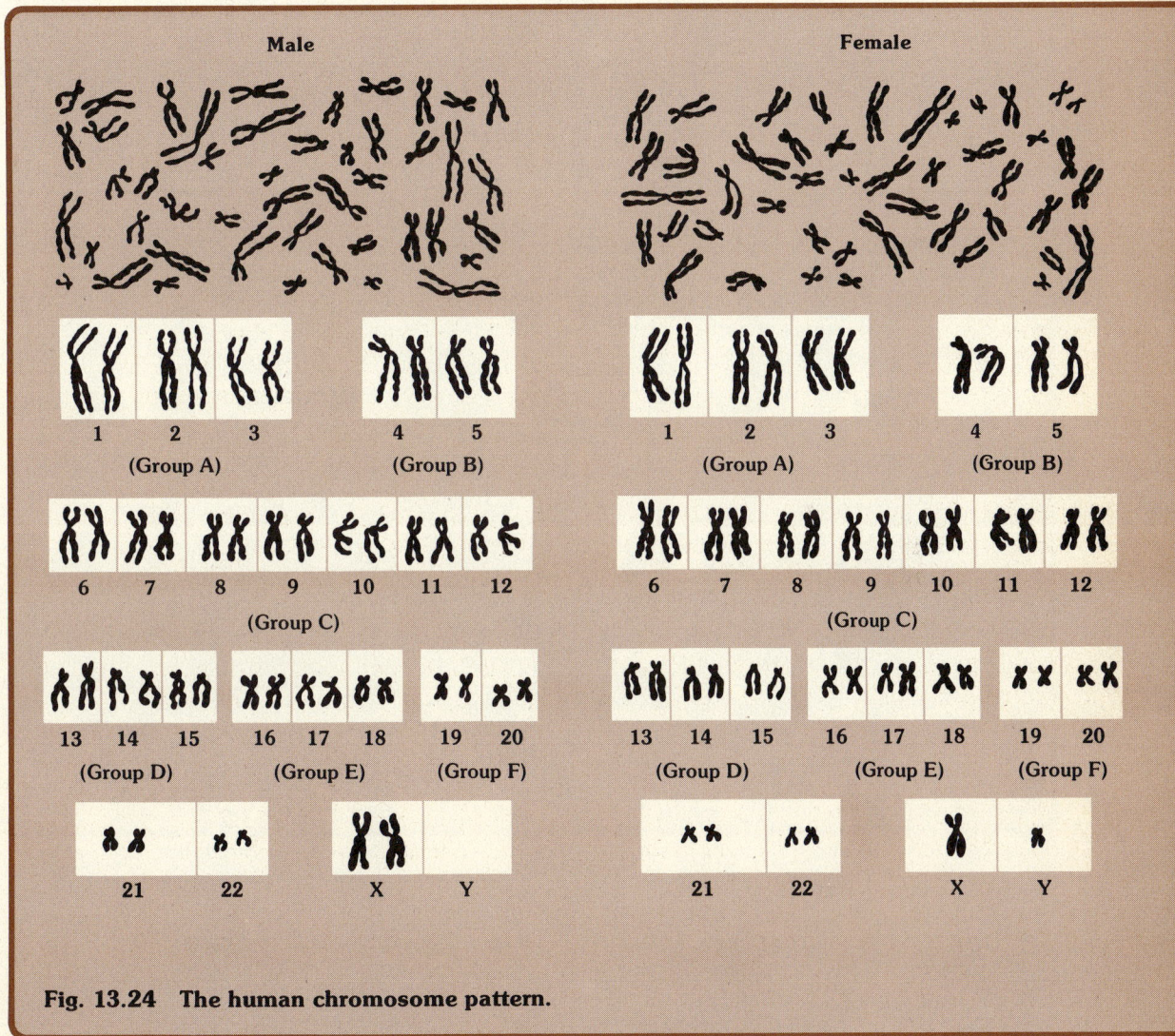

Fig. 13.24 **The human chromosome pattern.**

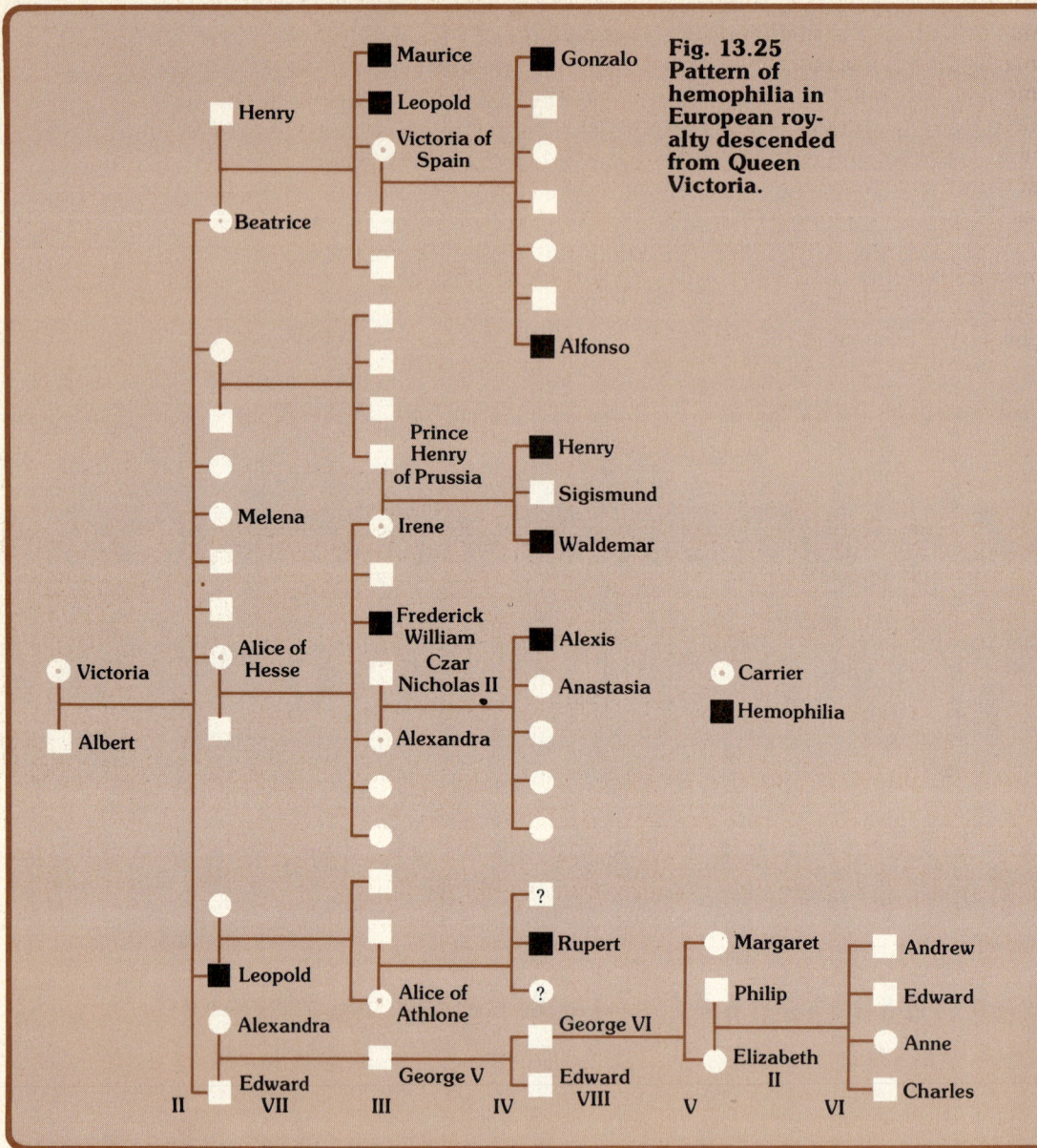

Fig. 13.25 Pattern of hemophilia in European royalty descended from Queen Victoria.

Maurice
Leopold
Victoria of Spain
Gonzalo
Alfonso
Henry
Beatrice

Prince Henry of Prussia
Irene
Henry
Sigismund
Waldemar

Melena
Alice of Hesse

Frederick William
Czar Nicholas II
Alexandra
Alexis
Anastasia

Carrier
Hemophilia

Victoria
Albert

Leopold
Alexandra
Edward VII

Alice of Athlone
George V
?
Rupert
?

George VI
Edward VIII

Margaret
Philip
Elizabeth II

Andrew
Edward
Anne
Charles

II          III          IV          V          VI

**Chromosomal abnormalities**   As indicated earlier, the most serious chromosomal abnormality is nondisjunction—the absence or presence of an extra chromosome. Typically, this condition is lethal. However, an extra one of the small chromosomes, such as number 21, produces mongolism, or **Down's syndrome** (Fig. 13.26a,b) as the condition is more correctly known. Despite the relative frequency of this syndrome and all the time spent in studying it, the only direct correlation that can be shown is with the age of the mother—the incidence of this syndrome rises sharply after a maternal age of about 37 (Fig. 13.26c).

There are also a number of conditions that result from extra, or missing, sex chromosomes (Fig. 13.27). If the Y chromosome is missing, the individual is designated XO and the consequences are **Turner's syndrome** (Fig. 13.28). The individual in question is female, but she suffers from a number of anatomical and

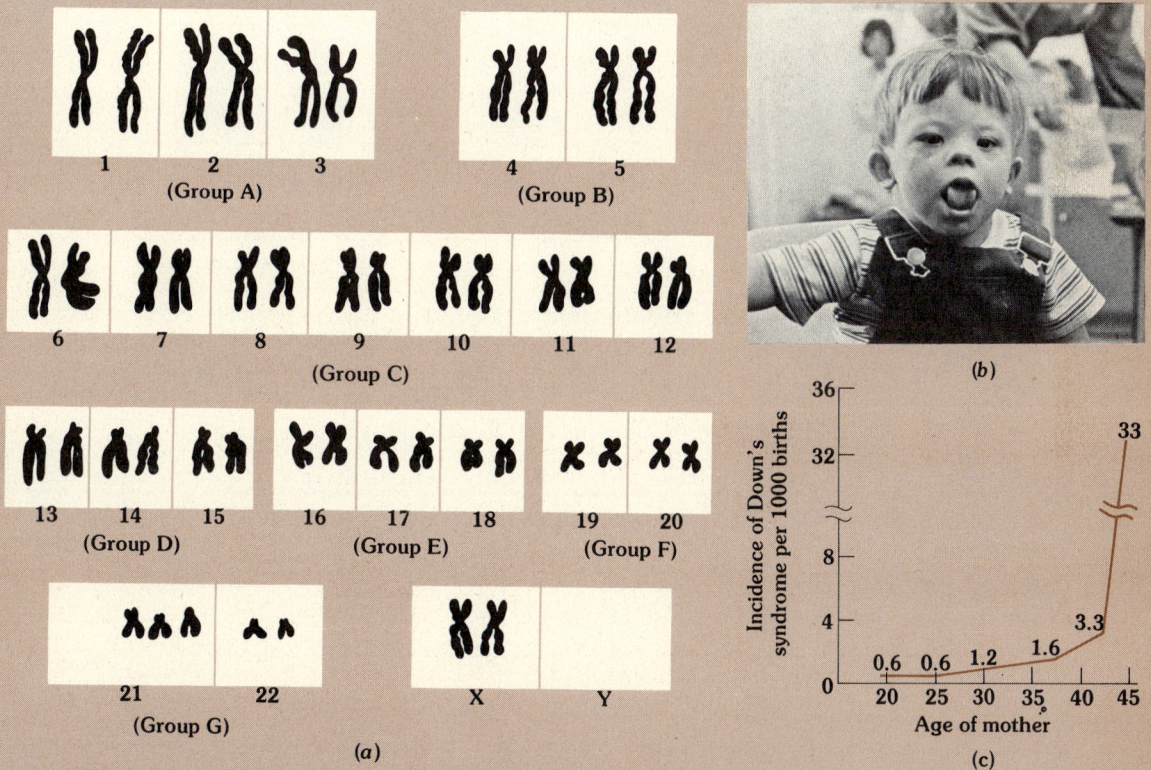

**Fig. 13.26**   (a) Chromosome pattern for Down's syndrome; (b) child with Down's syndrome; (Lebo, Jeroboam) (c) frequency of Down's syndrome as a function of maternal age.

45/XO
Turner's
streak gonads.
Dwarfism anomalies

+ X          + Y

46/XX
Normal
female

46/XY
Normal
male

+ Y          + X

+ X          + Y

47/XXX
Slight ↓ IQ

47/XXY
Klinefelter's
infertility.
Slight ↓ IQ

47/XYY
Tall stature.
Variable
phenotype

+ X          + Y          + Y          + X

48/XXXX
Mental
deficiency

48/XXXY
Klinefelter testis.
Mental deficiency.

48/XXYY
Mental deficiency.
Klinefelter testis.

+ X          + Y          + X          + Y          + X

49/XXXXX
Mental
and
physical
retardation

49/XXXXY
Severe mental
deficiency.
Skeletal anomalies.
Greater
testicular defect.

49/XXXYY
Tall stature
severe mental
deficiency.
Severe testicular
deficiency

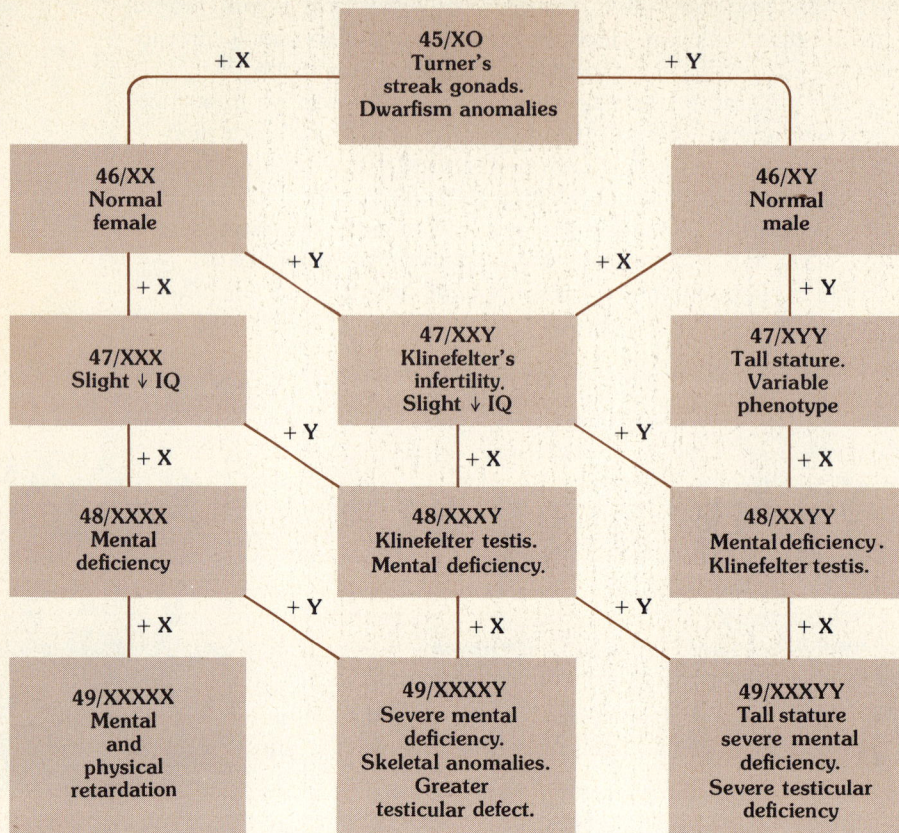

**Fig. 13.27  Anomalies based on presence or absence of various numbers of sex chromosomes.**

developmental irregularities. An extra X chromosome in males produces **Klinefelter's syndrome** (Fig. 13.29), typified by a narrow, elongate body and a slightly below-normal IQ. An extra Y chromosome has a somewhat similar effect, but there are indications at present that such individuals tend to be more violent than the average male, a situation that presents interesting legal problems—how different is genetically induced violence from insanity in terms of legal liability for crimes? The jury is still out on that question.

The most serious, yet survivable, chromosomal deletion is the **cri du chat syndrome** (Fig. 13.30), so named because of the peculiar mewing cry produced by the individual suffering from it. The syndrome also involves serious postural malformations and profound retardation.

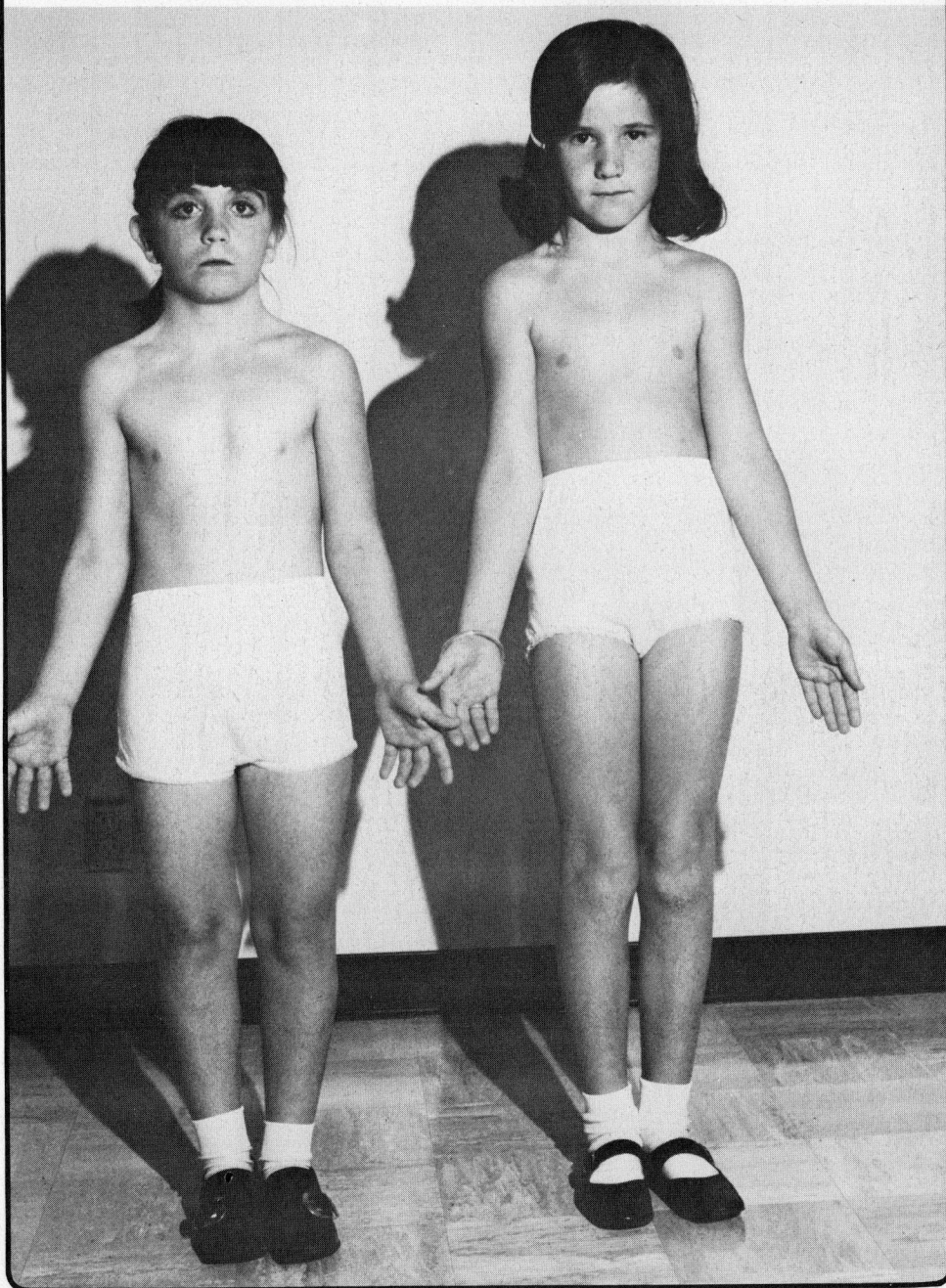

Fig. 13.28   Turner's syndrome in an eleven-year-old girl (*left*) shown next to her normal nine-year-old sister (*right*). (W. R. Centerwall)

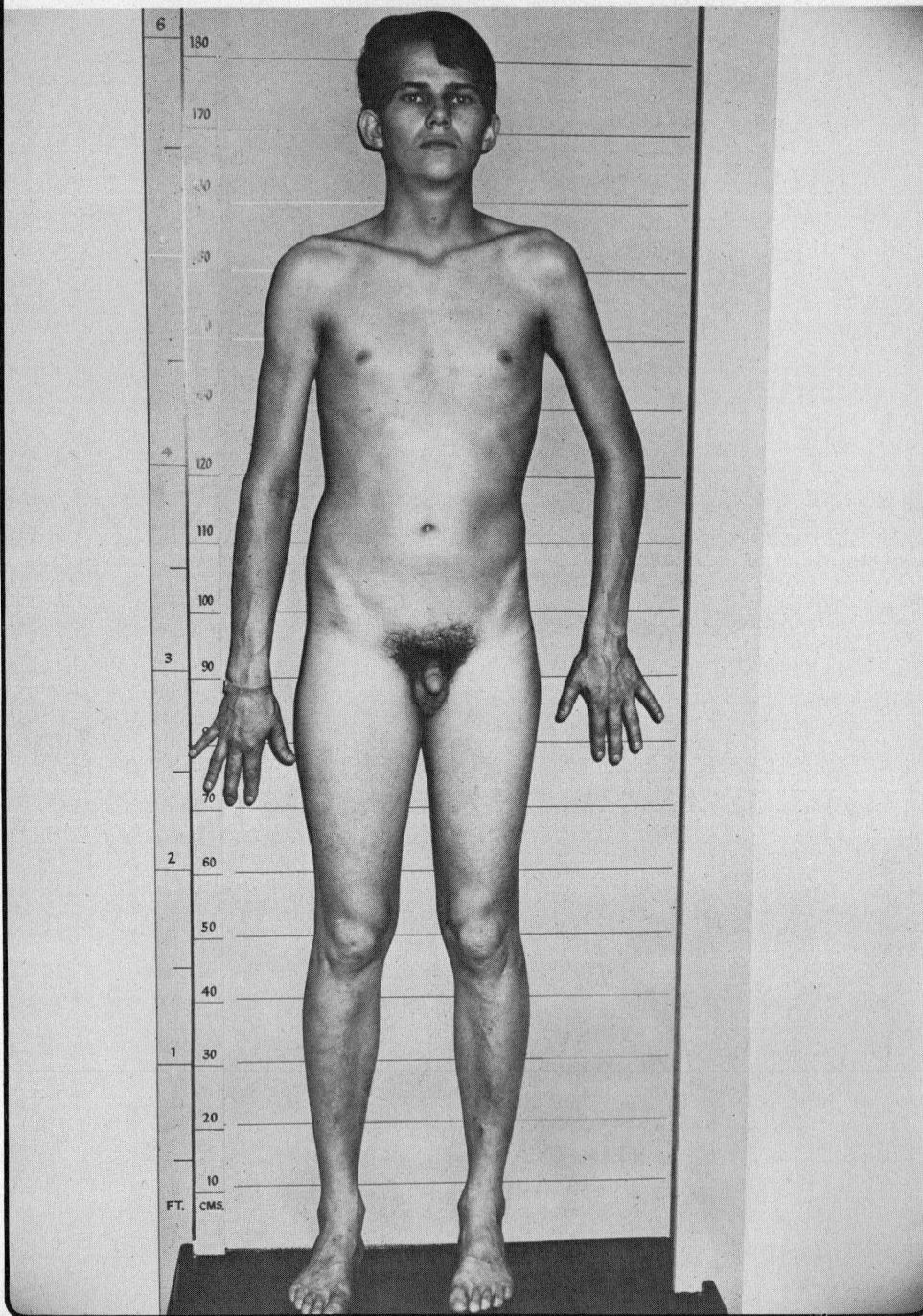

Fig. 13.29  Klinefelter's syndrome. Note long legs and sparse body hair. (W. R. Centerwall)

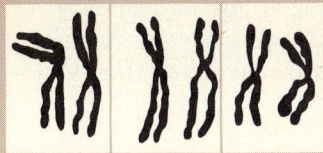

1    2    3       4    5

(Group A)       (Group B)

6    7    8    9    10    11    12

(Group C)

  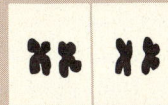

13    14    15       16    17    18       19    20

(Group D)       (Group E)       (Group F)

21    22       X      Y

(a)    (Group G)

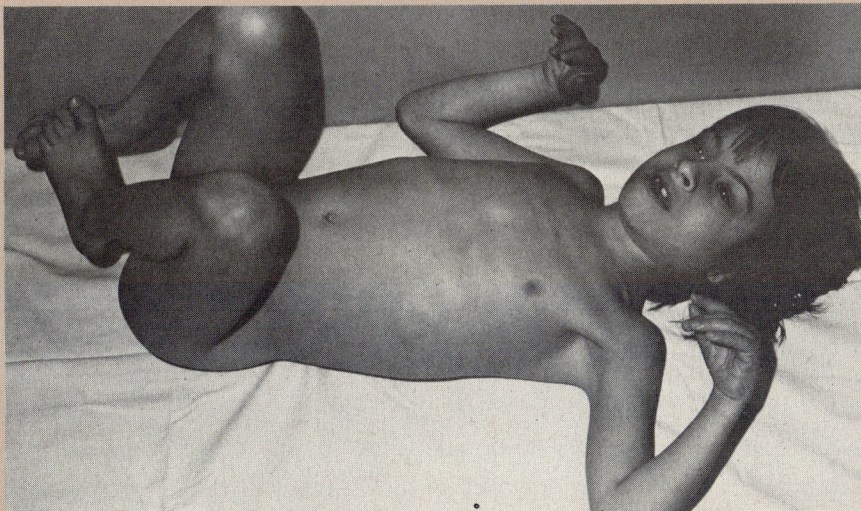

(b)

**Fig. 13.30**   (a) **Chromosomal arrangement for and** (b) **girl with Cri du chat syndrome. (W. R. Centerwall)**

**Environmentally induced defects**  Unlike the genetic conditions just mentioned, environmentally induced defects, even though they may be just as serious as genetic defects, do not, by definition, involve a change in the genetic constitution of the individual, but rather affect a given development pathway directly. As such, individuals who suffer from environmentally induced defects cannot transmit this defect to their children.

The four principle environmental factors responsible for inducing such defects are: (1) condition and health of the mother; (2) radiation; (3) viruses; (4) chemicals.

One important factor of the mother's health that affects the developing fetus is the presence or absence of maternal diabetes. Diabetic mothers tend to give birth to very large children (Fig. 13.31), sometimes with such developmental defects as improperly formed legs. Such children also have a higher death rate just before and just after birth than do children of nondiabetic mothers.

One of the blood factors which differs between individuals is the **Rh factor** (so called because it was first found in *rh*esus monkeys). When an Rh-negative woman (i.e., one lacking the factor) marries an Rh-positive man, and they produce an Rh-positive fetus, the mother may manufacture antibodies that react with the blood cells of the fetus, causing serious circulatory problems. Fortunately, this condition is now treatable, by vaccination of the mother or by a total blood transfusion of the infant at birth.

Dietary insufficiencies during pregnancy may cause a variety of conditions—or, more correctly, may allow the development of certain conditions in individuals genetically predisposed to them. Cleft palate and harelip, for example, are almost entirely preventable through adequate maternal nutrition (Table 13.2).

Radiation is extremely effective in inducing chromosomal breakage and, as such, is a powerful mutagenic agent. The effects of radiation vary with the strength of the exposure and the stage of embryonic development at the time of exposure, but can obviously be very profound, as has been amply demonstrated by the tre-

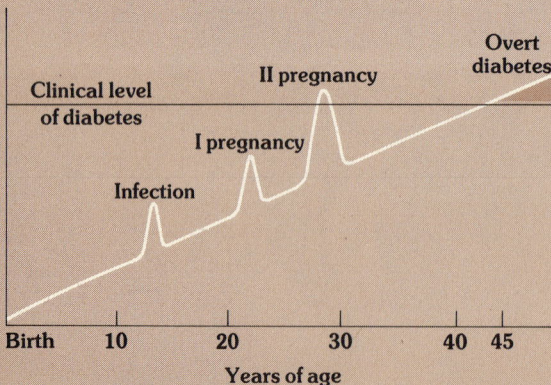

**Fig. 13.31
Development of adult onset diabetes as a function of repeated pregnancies.**

**Table 13.2**
**Risks of cleft palate in various family situations**

| | Risk of cleft palate (percent) |
| --- | --- |
| Frequency of cleft palate in the general population | 0.04 |
| Probability that the next child in the family will be affected if | |
| both parents are unaffected and | |
| they have an affected child | |
| but no affected relatives | 2.0 |
| they have an affected child | |
| and an affected relative | 7.0 |
| they have an affected child | |
| and are relatives | 2.0 |
| either of the parents has cleft palate and | |
| they have no affected children | 6.0 |
| they have an affected child | 15.0 |
| Risk of come other type of anomaly in any of the above combinations | Same as in the general population |

mendous rate of congenital defects in the children of the survivors of Hiroshima and Nagasaki.

Such viruses as chickenpox, mumps, hepatitis, measles, and German measles (rubella) are known to induce congenital defects. Rubella is particularly well studied, and it has been estimated that, as a result of the 1964 outbreak of rubella in this country, some 20,000 children were born with congenital defects. The consequences of rubella may be profound, depending on when the woman had the disease relative to her pregnancy (Figs. 13.32 and 13.33).

There are a large number of chemicals that induce congenital defects, the most notorious of which is the drug **thalidomide** (Figs. 13.34 and 13.35), which was originally recommended to pregnant women to control morning sickness. Ultimately, several thousand children were born deformed, principally in Germany and England. Fortunately for the United States, the Food and Drug Administration (FDA) banned its use here and we were spared this tragedy. Ever since, obstetricians have been chary of prescribing any drug to pregnant women. However, congenital defects continue to occur because the embryo is most sensitive to environmental insults in the first weeks of development, at a time when the woman may not know she is pregnant and therefore may be taking medications that are hazardous to the unborn child.

**GENETIC ENGINEERING** The evolution of human culture, which began in prehistoric times, has increasingly provided protection from the workings of natural selection. Culture may continue to evolve, but biologically we are gradually coming to a halt. For example, from prehistoric times on through the Renaissance, nearsightedness was a profound liability and was strongly selected against. Eyeglasses changed all that. Similarly, diabetes was a fatal disease 50 years ago, but

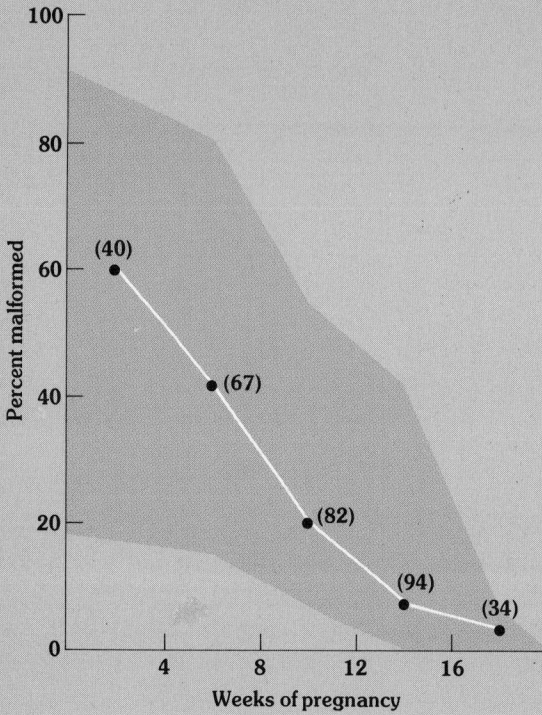

Fig. 13.32
Risk of fetal malformations following maternal rubella. Upper and lower limits represent the extremes of six independent estimates; the line marks the mean of the six studies.

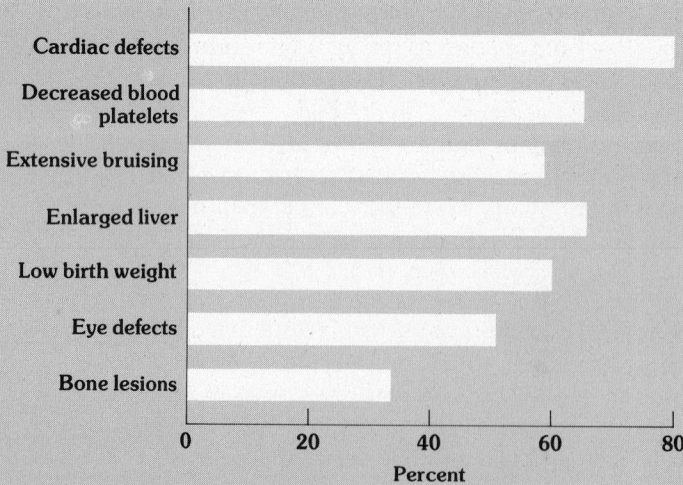

Fig. 13.33
Frequency of different birth defects in children born of mothers who had contracted rubella during pregnancy.

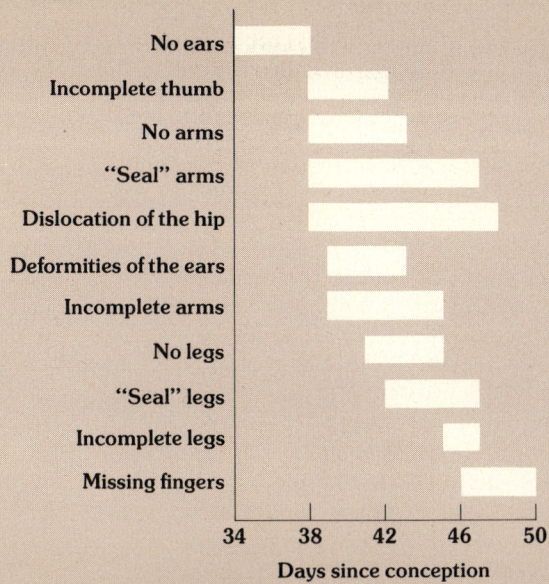

Fig. 13.34
Incidence of different malformations as a function of the stage of pregnancy at which the drug thalidomide was taken.

Days since conception

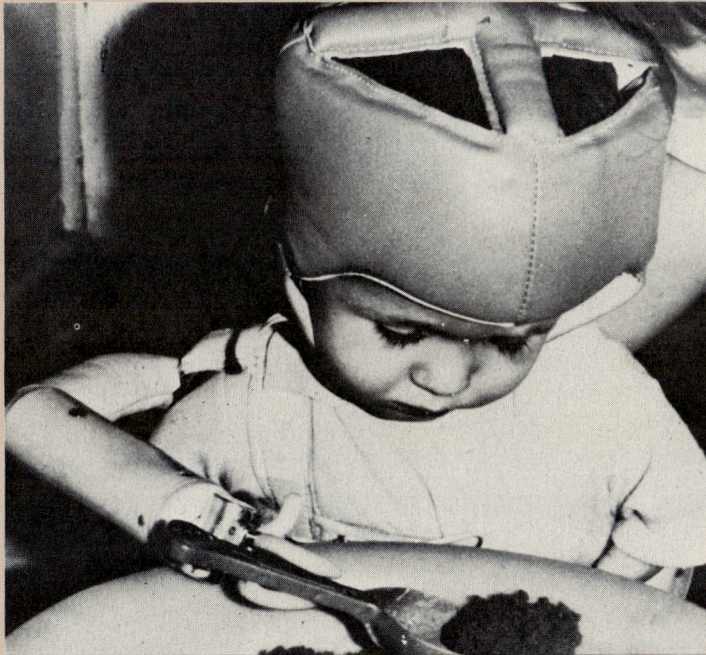

Fig. 13.35 Child born of mother who had taken the drug thalidomide. (Wide World)

With our increased knowledge of genes and enzymes (the products of genes), we are now able to detect the presence of certain genetically controlled congenital defects in a developing fetus *in utero*. This test is used with particular frequency in women over 40 who become pregnant. The chances of Down's syndrome are much higher in women over 40, and the extra chromosome is easily detected in embryonic cells floating in the amnionic fluid. Women who learn they are carrying a child with Down's syndrome may or may not elect to have an abortion, but in any case they know the facts, rather than having to worry for nine months, as previously.

With increasing frequency, married or engaged couples are seeking genetic counseling prior to becoming parents. Many genetic defects can be assayed for, and a prediction made as to the likelihood of having a child with some genetically controlled congenital defect. In deciding what syndromes and diseases to assay for, the geneticist will seek to compile a rather complete picture of the individual's ancestors. Many diseases are much more common in some populations than in others. People of Scottish descent would not be checked for sickle-cell anemia, for example, but people of Eastern European Jewish extraction would be checked for **Tay-Sachs disease** (a very severe condition of the nervous system which is generally fatal in early childhood), as this condition is limited to people of that ancestry. Other examples are given in Fig. 13.36.

For the most part, genetic counseling in the form of **amniocentesis** (Fig. 13.37) is ineffective in congenital defects caused by environmental insult for the simple reason that the problem in such cases is not genetic but developmental for which there is generally no biochemical abnormality which can be detected *in utero*.

As improved methods of detection continue to be developed, however, we will be faced with a series of difficult moral questions. Should genetic counseling be made mandatory, as are tests for syphilis? Does society have a right to prevent the birth—or the conception—of children who will be profoundly affected by a congenital defect? If there is a correlation between violent behavior and an extra Y chromosome, should all male babies be checked at birth? What do we do when we find a male baby with an extra Y chromosome—tell him he may become violent as he matures and thereby stigmatize him for life?

One of the frustrations of being a scientist is in recognizing that science is more effective in raising moral issues than in answering them.

**Genetic counseling**

now, millions of diabetics not only survive but reproduce. We tend to regard, with some justice, the saving of diabetics, hemophiliacs, and so on as a victory over nature—but, in the process, we are, as a population, becoming more and more reliant on pharmaceuticals to keep us alive. How much better things would be if we could but correct the genetic error at its source. A pipedream? Perhaps not. Scientists have actually been able to implant a gene into a mammalian cell by first tying it to a virus which invades the cell. The day may not be far off when point mutations may

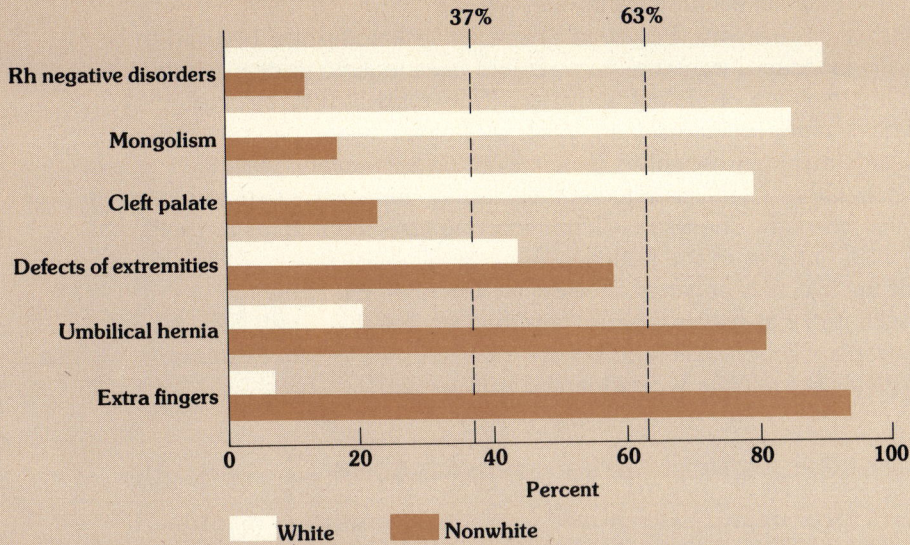

37%    63%

Rh negative disorders

Mongolism

Cleft palate

Defects of extremities

Umbilical hernia

Extra fingers

0    20    40    60    80    100

Percent

☐ White    ▨ Nonwhite

Fig. 13.36   Incidence of some selected congenital disorders by "race."

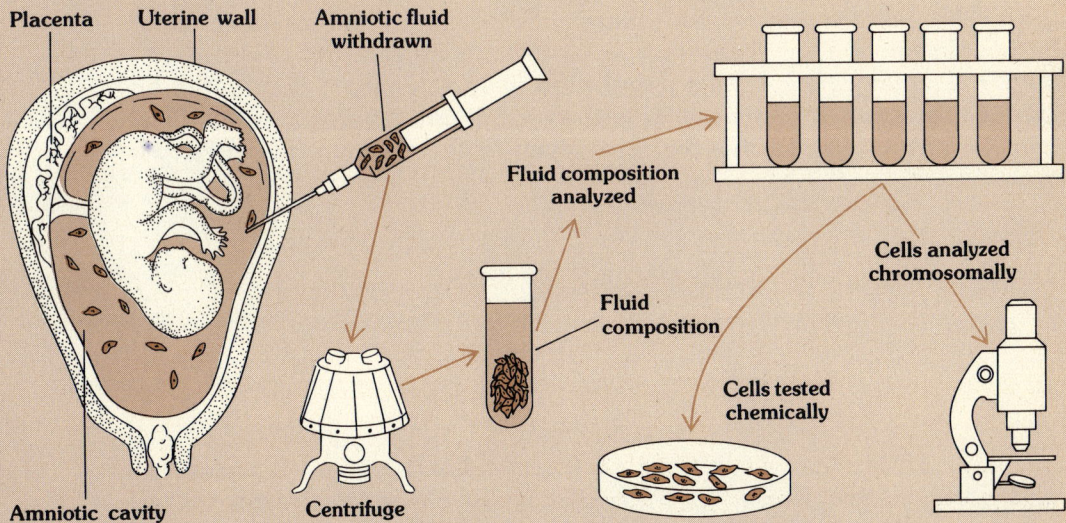

Placenta   Uterine wall   Amniotic fluid withdrawn

Fluid composition analyzed

Cells analyzed chromosomally

Fluid composition

Cells tested chemically

Amniotic cavity   Centrifuge

Fig. 13.37   In amniocentesis, a small amount of amniotic fluid containing cells is withdrawn and analyzed. This procedure is generally performed during the third or fourth month of pregnancy.

be substituted for, allowing the possessor (and his descendants!) complete and lasting freedom from the genetic disease in question.

Consider the other questions, however. The same techniques can be used to transform normally innocuous intestinal bacteria into carcinogenic (cancer-causing) killers, ushering in a new era of biological warfare. Superviruses may be created by the same techniques, albeit accidentally, to which we might have no defense. So severe are the hazards that many scientists have called for a voluntary abstention from such research. Other scientists argue equally loudly that such a demand interferes with scientific freedom. The issue will not soon be resolved and, as a result, "bioethics" will be a hot area in the future.

What about the injection of genes affecting intelligence? We know next to nothing about such genes at present, other than the fact that they exist. Still, the day will come when it will be possible to boost—or lower—intelligence by injection. Aldous Huxley's *Brave New World* may be with us yet.

## SUMMARY

During the past 125 years, we have moved from the assumption that inheritance involves the blending of characters, like mixing paints on a pallette, to discussions about the substitution of genes in human cells. We also now recognize that species are not immutable, but change over time. This change, which we call evolution, has a genetic basis. Organisms carrying different genes survive and reproduce at different rates, and the ones leaving the most offspring make the largest genetic contribution to the next generation, and thereby influence the direction in which the population evolves. Therefore, even though we think of evolution, like the species concept, as a population phenomenon, it is directed by genetic mutations in individuals, with the individual serving as the focal point for natural selection—the term given to the measure of genetic survival.

*What are the effects of burning all our fossil fuels? Is the world running down? Are there too many people in the world today? If not, at what point will there be? How do we decide what "too many" is? How are animal populations regulated? Does the answer fit the human problem?*

*These are the questions of today, but to answer them requires a basic knowledge of ecological principles, which is the topic of this chapter.*

chapter **XIV**

# chapter XIV
# Relationships with the
# environment–ecology

In the discussion of biological organization (Chapter 2), we briefly mentioned the upper categories of the series—**communities** (populations of different species in a common geographic area), **ecosystems** (communities plus the inorganic environment), and **biosphere** (the totality of the ecosystems of the world). **Ecology** is that branch of biology devoted to the study of the interrelationships within these three categories. It is an overview science, which requires extensive knowledge in a multiplicity of disciplines, and is still largely theoretical—the scope of ecological questions is so broad that experiments are often difficult to perform, both because of their size and because of the length of time required before definitive results occur.

At one time considered esoteric, ecology has recently become an area of intense interest, as public concern has increased regarding environmental problems. In fact, ecology has become synonymous with the environment, and consequently the term is in danger of losing its preciseness. Phrases like "ecologically compatible" on the sides of detergent boxes are nonsensical—"ecology" is not an entity, so how can anything be compatible with it? Ecology is the **study** of the environment, not the environment itself. Nevertheless, there is an obvious relationship between the two terms, and, for our purposes, it is desirable to examine the theoretical principles of ecology in order better to understand the nature of our practical problems with the environment.

**ECOLOGICAL HOMEOSTASIS AND PHYSICAL LAWS**  The pervasive themes of this book have been first, the limitations placed on biological systems by the laws of the physical sciences; and second, the mechanisms used by living organisms to ensure a stable internal environment (homeostasis). These same concepts are equally applicable to the study of communities, ecosystems, and the biosphere. Indeed, it is fundamental even to a cursory study of ecological principles to take note of two important facts:

1.  According to the **Second Law of Thermodynamics,** no conversion from one energy state to another is 100 percent efficient—some energy is always lost as

heat. (This is the law used to justify the conclusion that perpetual motion machines are impossible, for example.)

The Second Law of Thermodynamics applies equally well to individual organisms as to the biosphere (Fig. 14.1), and it accounts for the fact that all organisms, regardless of their level of activity, require some form of external energy source if they are to survive more than temporarily. But what of the biosphere? Because ecological processes involve the conversion of one form of energy to another, how does the biosphere continue to function without an external energy source? The answer, of course, is that it doesn't.

2. The biosphere is not a completely closed system, but rather, in order to maintain homeostasis, it requires a source of external energy—sunlight. However, as the biosphere has no more capacity than does a single organism to convert energy directly to matter, it must mean that energy is being used only to aid in the conversion of one set of molecules to another (Fig. 14.2). As such, because sunlight is not being used to increase the abundance of elements in the biosphere, the elements themselves must recirculate within the system. The homeostatic mechanisms at work, then, are designed to ensure that no required elements move in a dead-ended fashion—they *must* recirculate; otherwise the extinction of the biosphere is only a matter of time, in precisely the same way that the failure of homeostatic controls in an individual organism preordains its death.

No doubt these two principles seem high-blown and abstruse at present, but bear them in mind as we move through the chapter, because virtually *every* environmental problem we shall encounter has its basis in one or both of these two principles.

**BIOTIC RELATIONSHIPS**   Ecosystems and the biosphere include both **biotic** (living) and **abiotic** (physical) components. This is a rather basic distinction, and a useful one to choose in beginning an analysis of ecological principles.

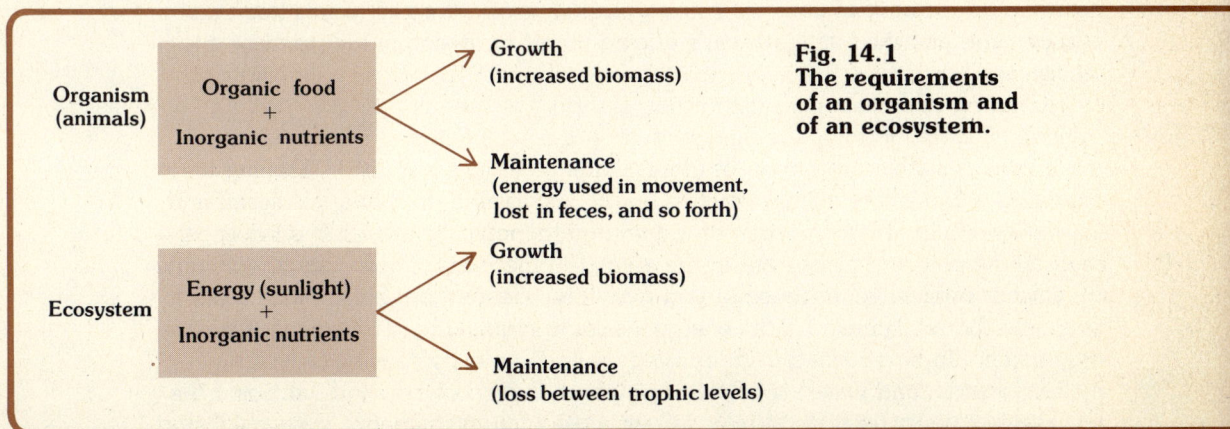

Organism (animals) → Organic food + Inorganic nutrients →
Growth (increased biomass)
Maintenance (energy used in movement, lost in feces, and so forth)

Ecosystem → Energy (sunlight) + Inorganic nutrients →
Growth (increased biomass)
Maintenance (loss between trophic levels)

**Fig. 14.1 The requirements of an organism and of an ecosystem.**

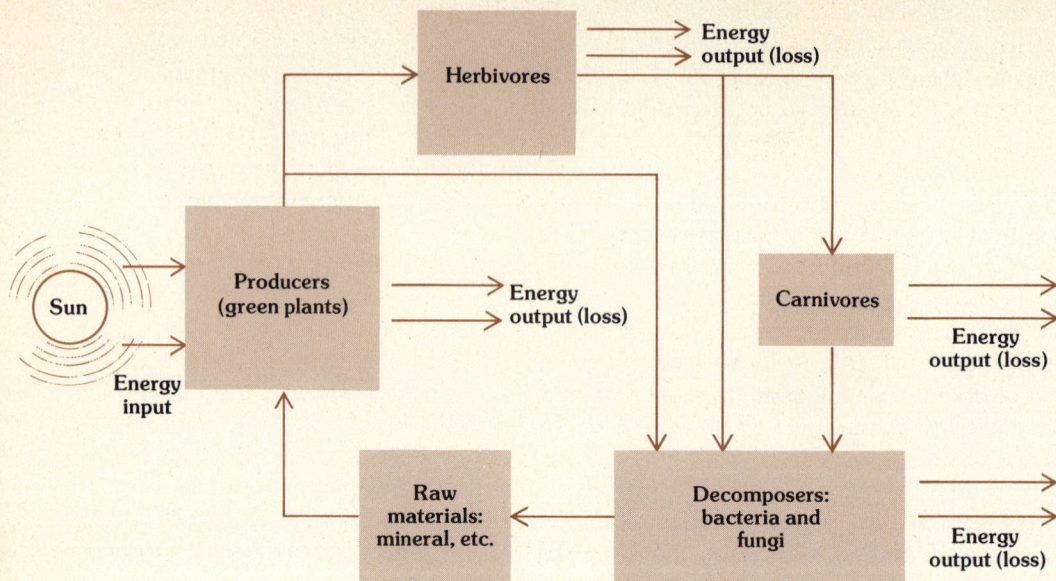

**Fig. 14.2  In an ecosystem, matter is recycled, but energy is not.**

**Energy flow**   As we have just seen, no conversion of energy can be accomplished without some waste (typically in the form of heat). By way of example, as you know, your car's engine becomes hot as it runs, owing to friction—the efficiency of the conversion of the energy stored in the molecules of gasoline to the energy of movement as your car glides gracefully along the highway is less than 100 percent, and the heat your car's engine builds up is simply the proof of this inefficiency.

Carrying the reasoning a bit further, it should also be apparent that not all of the energy stored in the grass being eaten by a cow is going to be turned into more cow. A sizable proportion of the available energy is lost just in keeping the cow alive—that is, the animal requires energy to breathe, to move, to digest, and so on. Moreover, a large portion of the energy theoretically available is never utilized by the cow and passes out of the animal in the form of cow pats, the volume of which is impressive, as those of you who have cleaned out barns well know. In fact, only about 10 percent of the energy available in grass (i.e., 10 percent of the energy that could be obtained from burning grass completely, leaving only ash, carbon dioxide, and water) is actually converted into more cow (growth); and the rest is either never used (fecal material) or is used for maintenance (Figs. 14.3 and

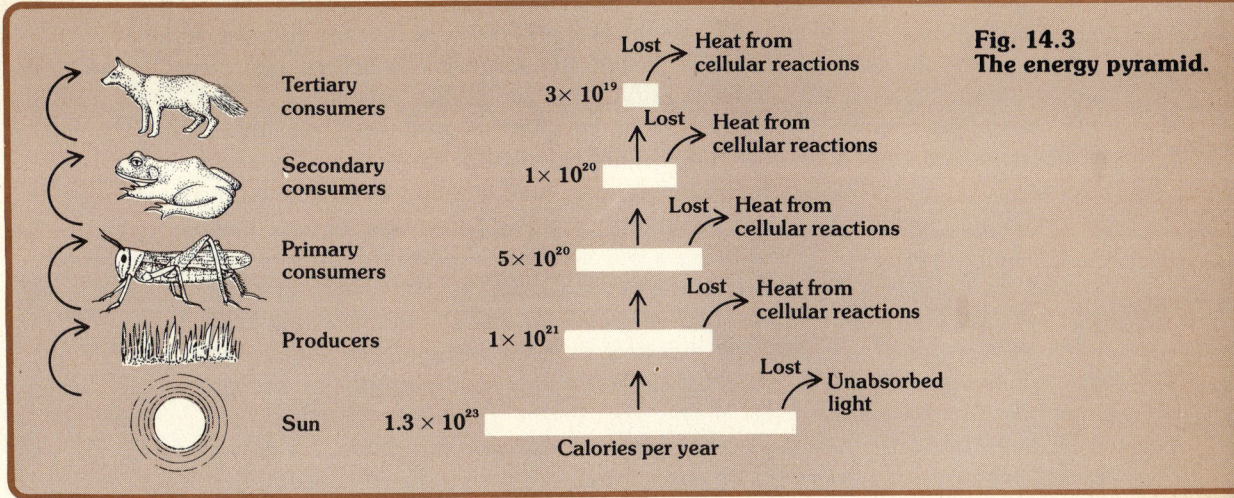

**Fig. 14.3
The energy pyramid.**

Tertiary consumers $3 \times 10^{19}$ — Lost → Heat from cellular reactions

Secondary consumers $1 \times 10^{20}$ — Lost → Heat from cellular reactions

Primary consumers $5 \times 10^{20}$ — Lost → Heat from cellular reactions

Producers $1 \times 10^{21}$ — Lost → Heat from cellular reactions

Sun $1.3 \times 10^{23}$ — Lost → Unabsorbed light

Calories per year

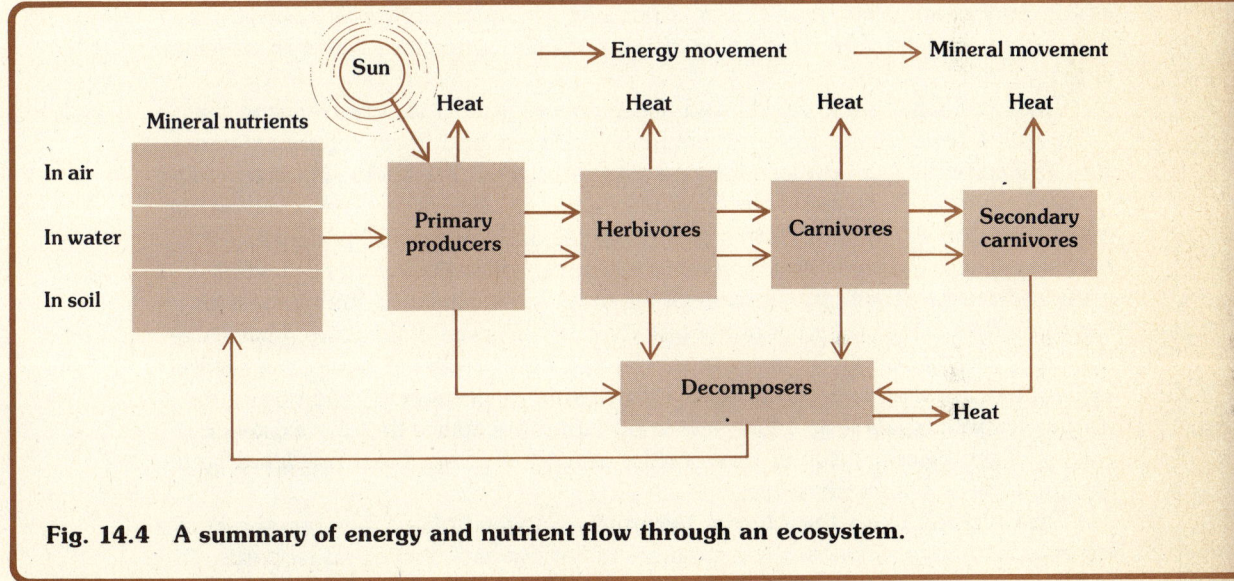

→ Energy movement        → Mineral movement

Fig. 14.4   A summary of energy and nutrient flow through an ecosystem.

14.4). Recall from the discussion of endothermy (Chapter 6) that the amount of energy required by endotherms for maintenance is many times that required by exotherms.

A similar picture could be painted for the wolf, which kills and eats the cow. Not all of the cow will be eaten—some of the blood may sink into the ground, the brains may remain inaccessible within the skull, most of the bones will be left uneaten—and the wolf must use some energy for maintenance as did the cow.

Again, only about 10 percent of the theoretically available energy will be utilized—and because this is only 10 percent of the 10 percent that the cow was able to extract from the grass to use for its growth, the wolf actually utilizes only 1 percent of the energy originally available in the grass for its own growth.

**Energy pyramids**   Organisms can be classified as **producers, consumers,** or **reducers,** depending on whether they synthesize organic materials out of inorganic materials (e.g., green plants), or obtain their organic requirements preformed by devouring other creatures (e.g., most animals), or break down dead organic material into its component inorganic compounds and elements (e.g., mushrooms, many bacteria, etc.). The reducers play a critical role in recycling putative waste material, such as feces, bone, dead leaves, and so on, and in making the inorganic components contained therein once again available to the producers. The nature of this cycle will be expanded upon shortly, but let us focus on the producers and consumers for the present.

Consumers eat either producers or other consumers. Thus, we can distinguish among **primary consumers** (those eating producers), **secondary consumers** (those eating primary consumers), and so on. The fact that many consumers, such as ourselves, may at any given moment be a primary, secondary, or tertiary consumer does not really detract from the value of our simplified scheme of things as just elucidated.

Based on what we have just said about energy flow, it stands to reason that there will be fewer primary consumers than producers and fewer secondary consumers than primary consumers. Raw numbers, of course, are not too intuitively satisfying—of course, there are fewer cows than there are blades of grass—but we can make the same statement in terms of dry weight, or, in ecological parlance, **biomass.** Thus, the biomass of the cows which can be supported by a field of grass is about one-tenth the biomass of the grass itself. Similarly, the biomass of the consumers of the cow totals about one-tenth the biomass of the cows. These relationships, whether of energy or of numbers, can be arranged in layer-cake fashion to form the so-called **energy pyramid** (or **pyramid of numbers**), made up of the various **trophic levels** (Fig. 14.5). It is important to recognize that the existence of this set of producer–consumer relationships in a pyramid style was predicted by the Second Law of Thermodynamics.

**Predictions from the energy pyramid**   Knowing about the energy pyramid explains why large secondary consumers, such as cougars, lions, and leopards, are relatively rare—they can only be one-tenth as numerous as the prey they live on. It also explains why it is that there are no really large tertiary consumers on land—that is, there are no animals whose primary food is lion meat, for example. Lions are simply too rare to be usable as a reliable food source. If the density of lions in an ideal location in Africa were $1.6/m^2$ $(1/mi^2)$, even a lion predator no larger than a lion itself could not have a density exceeding 1 per 10 $mi^2$, a density so low as to make the probability of finding a mate very unlikely (to say nothing of the energy demands involved in traversing such a large area in search of lions for food).

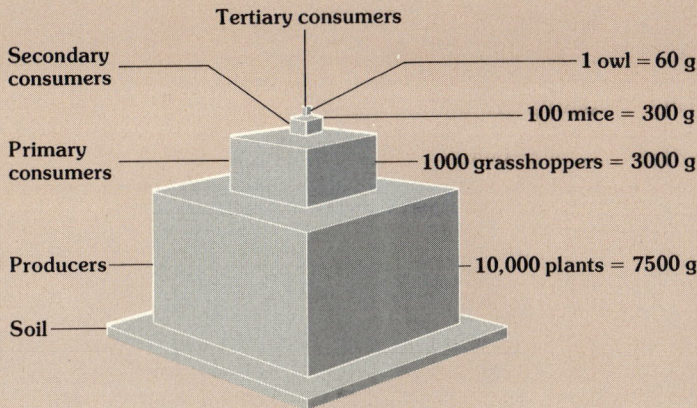

Fig. 14.5
Biomass pyramid.

Tertiary consumers

Secondary consumers

1 owl = 60 g

100 mice = 300 g

Primary consumers

1000 grasshoppers = 3000 g

Producers

10,000 plants = 7500 g

Soil

Let us now consider our own situation. As omnivores, we slide readily between various trophic levels. We can eat bread as easily as we can beef—or tuna, for that matter. (Tuna are tertiary consumers, for the most part, meaning that we become quaternary consumers when we use tuna as a food.) The crunch comes when we consider our obligation to eat more of the food that is low on the pyramid in order to allow sufficient food to people who at present are malnourished (Fig. 14.6). The question is not quite that simple, to be sure. Cows are actually a bit more than 10 percent efficient, in terms of conversion of grass, and pigs are even more efficient than cows. Moreover, as it is often pointed out, many cattle graze on lands too dry to support a cereal crop or other plant crop which would be used by us for food, so it just is not true that we could feed all that many more people by not eating beef—or so we are assured by the cattle people. The fact remains, however, that huge amounts of grains, notably corn, are fed to cattle in feed lots, and this is food that could just as easily be eaten by people. Even if we are conservative and allow just half the saving—that is, that only five people could feed on grain for every one person feeding on beef—do we have an obligation to eliminate beef from our diets to allow starving people in other countries to share our grain? That is a moral question, of course, not a biological one—but it is ecological theory that underlies our knowledge of the choice we face.

Another quite different problem may also be explained by an understanding of the energy pyramid. There are many elements and compounds which are not normally a part of biological systems and which organisms have a difficult time handling, once introduced into the system. These include such materials as the heavy metals, pesticides, and so on, which we shall discuss more fully later. The fact is, however, that such things are passed on in virtually full dose as they move up the pyramid, a process called **biomagnification.** Because there are relatively few animals at the top of the pyramid, they may have concentrations so great as to threaten life.

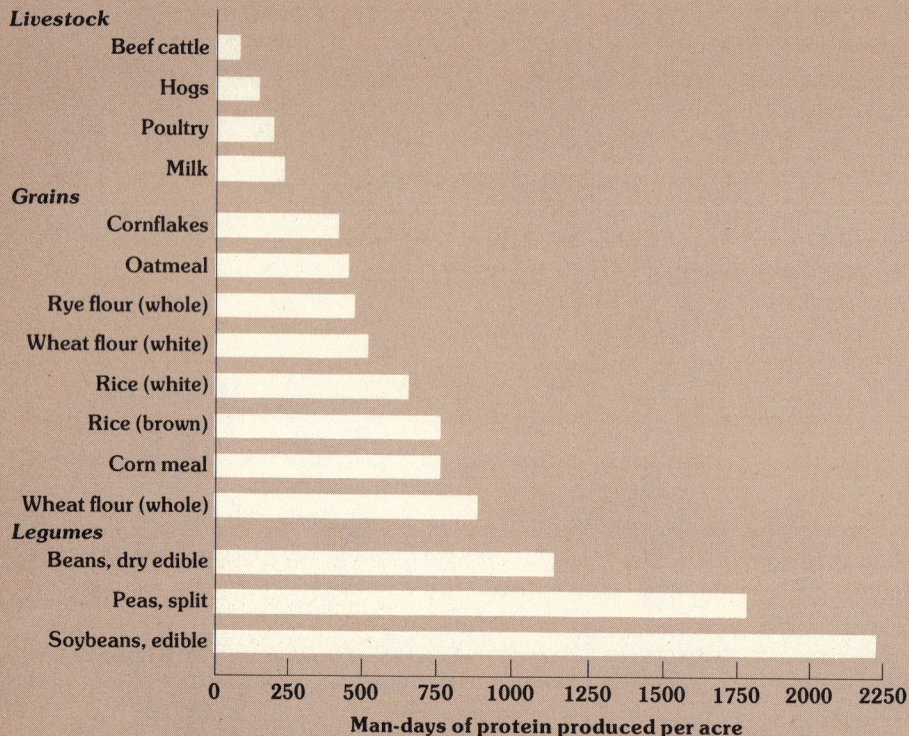

**Fig. 14.6** The efficiency of protein production of various foods, expressed as man-days (the amount required per adult per day) of protein produced per acre.

For example, the state bird of Louisiana, the brown pelican, has virtually disappeared from the coastal areas of that state. The residues of DDT sprayed on interior cropland are washed out to sea via the Mississippi, and are picked up by microorganisms in the Gulf of Mexico. These microorganisms are the food of small fish, which, in turn, are eaten by larger fish. Pelicans, which are tertiary consumers for the most part, feed on these fish, which, by this point, are heavily laden with DDT. The level of DDT in the pelicans is sufficient to interfere with the formation of the outer layers of the eggshell, causing the eggs to be very brittle, and leading to limited hatching success. Obviously, the population level of the pelican will therefore decline very quickly—and similar stories can be told for several other bird species, including eagles and fish hawks. This situation represents a breakdown in homeostatic controls with respect not only to the organisms but to the ecosystem as a whole, because there is no effective way of shunting these biologically destructive molecules out of the pyramid, and they therefore become progressively more concentrated in the upper tiers of the pyramid.

**Food webs**  Energy pyramids group species into different trophic levels, thereby serving to illustrate the efficiency of energy transfer from one category of species to another. A more fine-grained analysis can be obtained by considering each species of a given community individually, rather than as collective members of various trophic levels, and by analyzing their interrelationships in terms of which species eat or are eaten by which other species. In complex communities, where there are large numbers of species, these **food webs,** as the interrelationships are called, may be very complex (Fig. 14.7). Complexity, however, ensures stability, a desirable consequence in any community. To illustrate this point, consider the difference between a polar and a temperate community.

It is not altogether clear why the number of plant and animal species on land de-

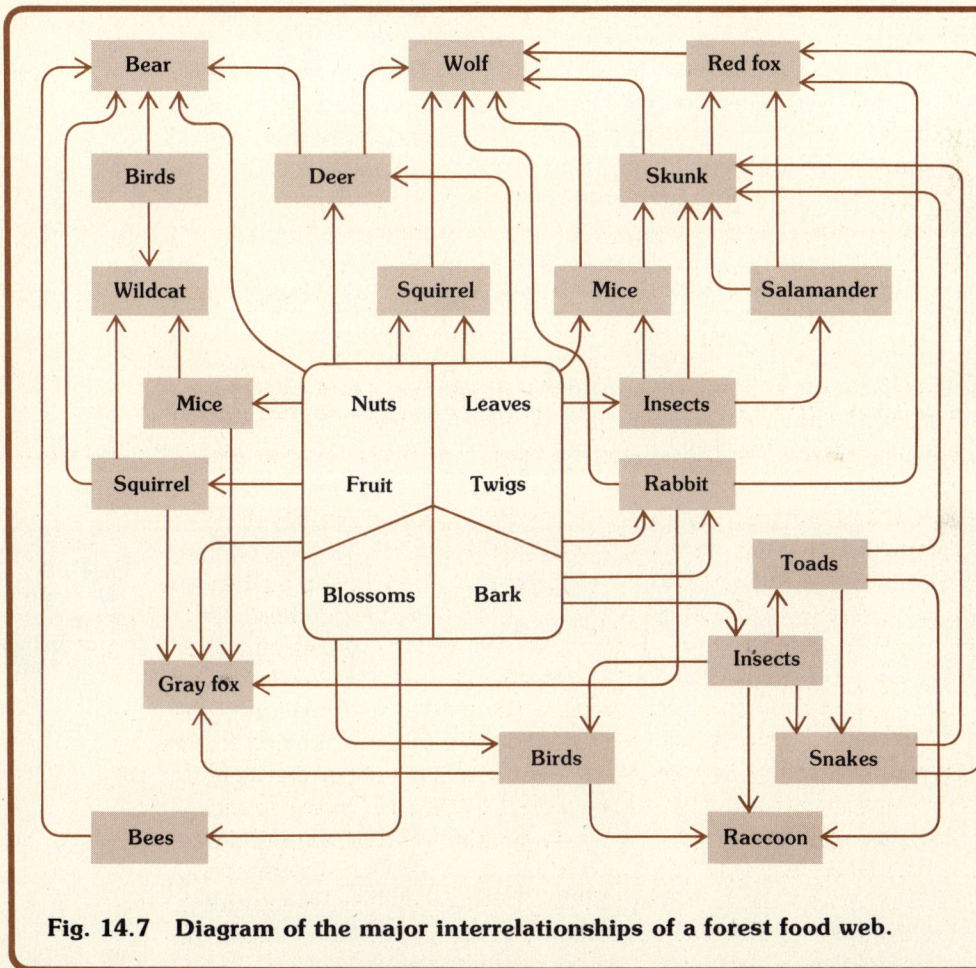

**Fig. 14.7  Diagram of the major interrelationships of a forest food web.**

clines as one moves away from the equator and towards the poles, although the fact that this does occur is not in dispute. In the subarctic regions, one of the major mammalian predators is the lynx, a kind of large bobcat. Although the lynx eats mice, lemmings, and the occasional bird, its principal prey is the arctic hare. Reciprocally, hares may also be eaten by foxes and wolves, but their primary predator is clearly the lynx. At first glance, it might seem that this relationship, which involves essentially only two species, would be extremely stable, but that is simply not the case. As relatively small mammals, hares mature quickly, and can thus reproduce much more rapidly than can the lynx. However, if there are a great many hares hopping around the landscape, fewer young lynx starve to death, the total number of lynx surviving increases, the hares are hard-pressed to survive long enough to reproduce, and their numbers fall. Without sufficient food, the lynx population also drops. At some point during the lynx die-off, the predation pressure on the remaining hares diminishes and their numbers begin to increase again. The result is a cycle involving overlapping population explosions and crashes (Fig. 14.8).

By contrast, consider the food web illustrated in Fig. 14.7. It is easy to see that, with multiple prey species, a given predator species will not suffer any substantial population decline should one of the prey species become scarce. Similarly, there is less reason to expect a population die-off in the prey species for the simple reason that predator numbers tend to remain stable.

What does all this mean to ourselves? Consider, for a moment, that it might well be possible to walk from Texas to Manitoba without ever being more than 30 m (100 ft) from a wheat field. Similarly, you could probably walk from Iowa to Indiana and always be within spitting distance of a corn field. This is an agricultural practice known as **monoculture.** In terms of community complexity, these wheat fields resemble a polar community more than a temperate community, a fact well illustrated when one considers the insect pests of wheat. There is a group of very primitive wasps called **sawflies,** which have a long fossil record, but which now are evolu-

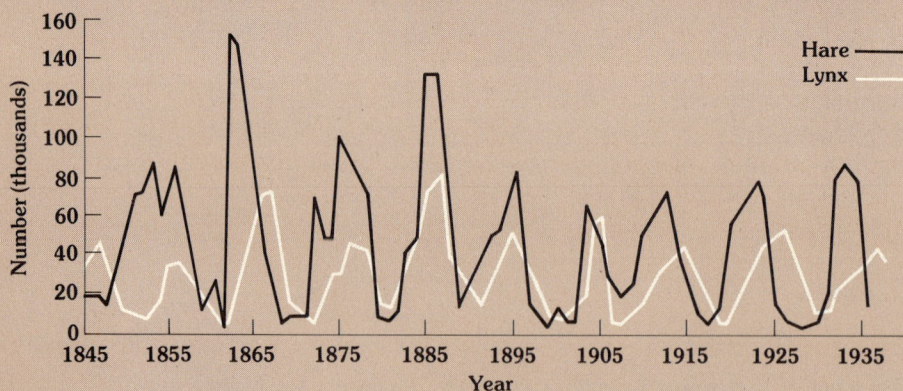

Fig. 14.8
Lynx and hare fluctuations based on pelts sold by trappers.

tionarily on their last legs, so to speak. One group of these sawflies spends its preadult life in the stems of grasses, but for the most part they are quite rare. A very prominent exception is the wheat-stem sawfly, a species that was given an evolutionary reprieve, once we decided to monoculture wheat. However, in simplifying the community, we have successfully eliminated most of the predator species of the sawfly simply because the predators have few other prey species on which to survive during those times when the sawfly is not available as food (i.e., while in the egg or pupa stage). As a consequence, we must share our wheat crop with the sawfly which, of course, ensures less wheat for ourselves.

In some areas of the country, there has been active reduction of monoculture, either by increasing the size of the fringe areas around the crop fields (such areas support a great diversity of animal species and provide a kind of reservoir for the predators), or by planting adjacent fields in different crops, to cut back on the near-epidemic outbreaks of single species of insect pests. Such techniques reduce efficiency somewhat, as slightly less acreage is planted and harvesting is somewhat more difficult, but they are ecologically more sound in that they ensure greater stability in the numbers of pest (and predator) species.

The same "diversity equals stability" argument can also be used to justify attempts at saving endangered species from extinction. Many people are troubled, perhaps out of a sense of collective guilt, over the incredible success of our own species at eliminating or endangering other species. Many other people are unconcerned, feeling that the earth is ours to do with as we please, plants and animals being here only to serve us. (One of these groups consists of enlightened idealists, whereas the other is comprised of insensitive clods, but I don't want to influence your thinking by saying which is which.) The fact is, however, that it is difficult to convince a doubter that *your* feeling of ill ease should change the doubter's thinking about *his* (or *her*) right to harpoon whales. What is needed is a pragmatic, nuts-and-bolts-style argument—and there is such an argument, based on the premise of diversity ensuring stability. The elimination of any species cuts a series of lines in the food web of the community to which the species belonged. Thus, just as a piece of cloth is weakened when several threads are cut, so, too, is the stability of the community weakened by the extinction of a species. Because we are ourselves part of the food web, in a very real sense we are endangering our own existence by bringing about the extinction of other species. This is not to say that there will necessarily be a sudden, dramatic change when a species becomes extinct, or that we will not be able to survive such extinctions. However, because we are subject to the same ecological principles as are other species, we are certainly jeopardizing ourselves by encouraging the extinction of other species—and this is hardly the best point in our history to jeopardize our future existence, an existence already seriously threatened by other problems.

**ABIOTIC RELATIONSHIPS**  As was mentioned at the outset of this chapter, it is imperative that those elements essential for life be recycled within the ecosystem, rather than coming to occupy some ecological cul-de-sac, where they may be-

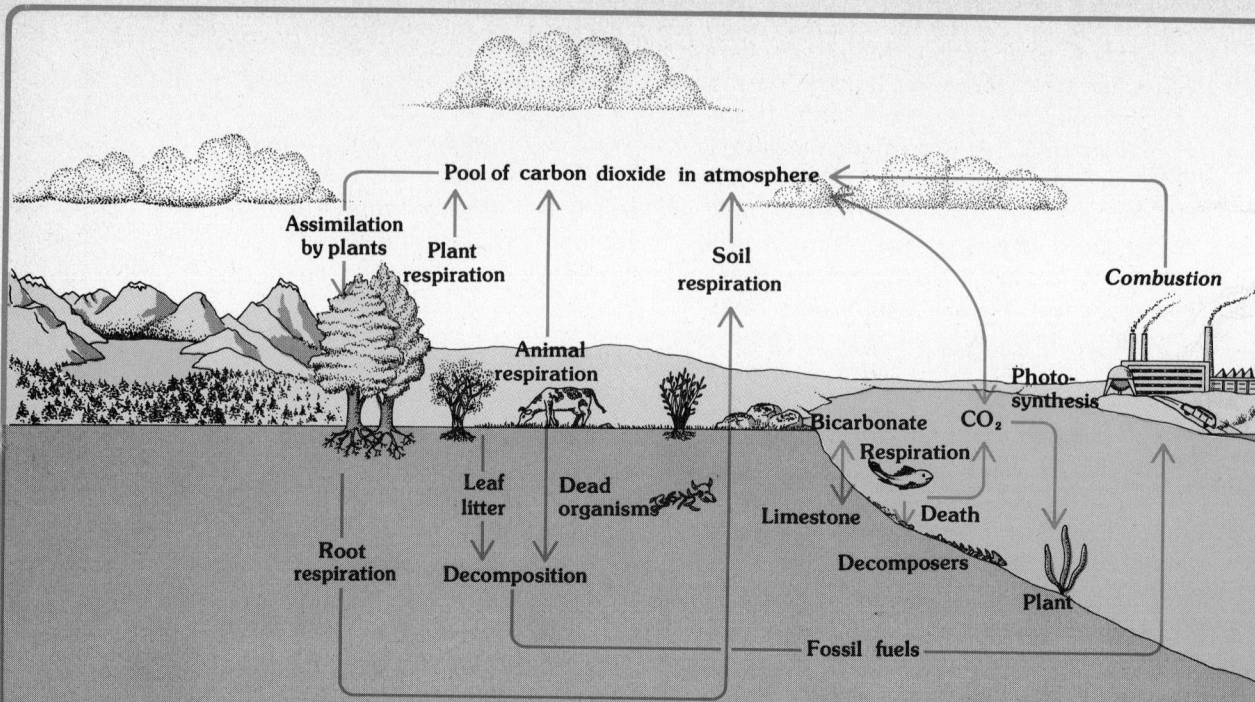

**Fig. 14.9   The carbon cycle.**

Pool of carbon dioxide in atmosphere

Assimilation by plants

Plant respiration

Soil respiration

*Combustion*

Animal respiration

Photo-synthesis

Bicarbonate    $CO_2$

Respiration

Root respiration

Leaf litter

Dead organisms

Limestone    Death

Decomposition

Decomposers

Plant

Fossil fuels

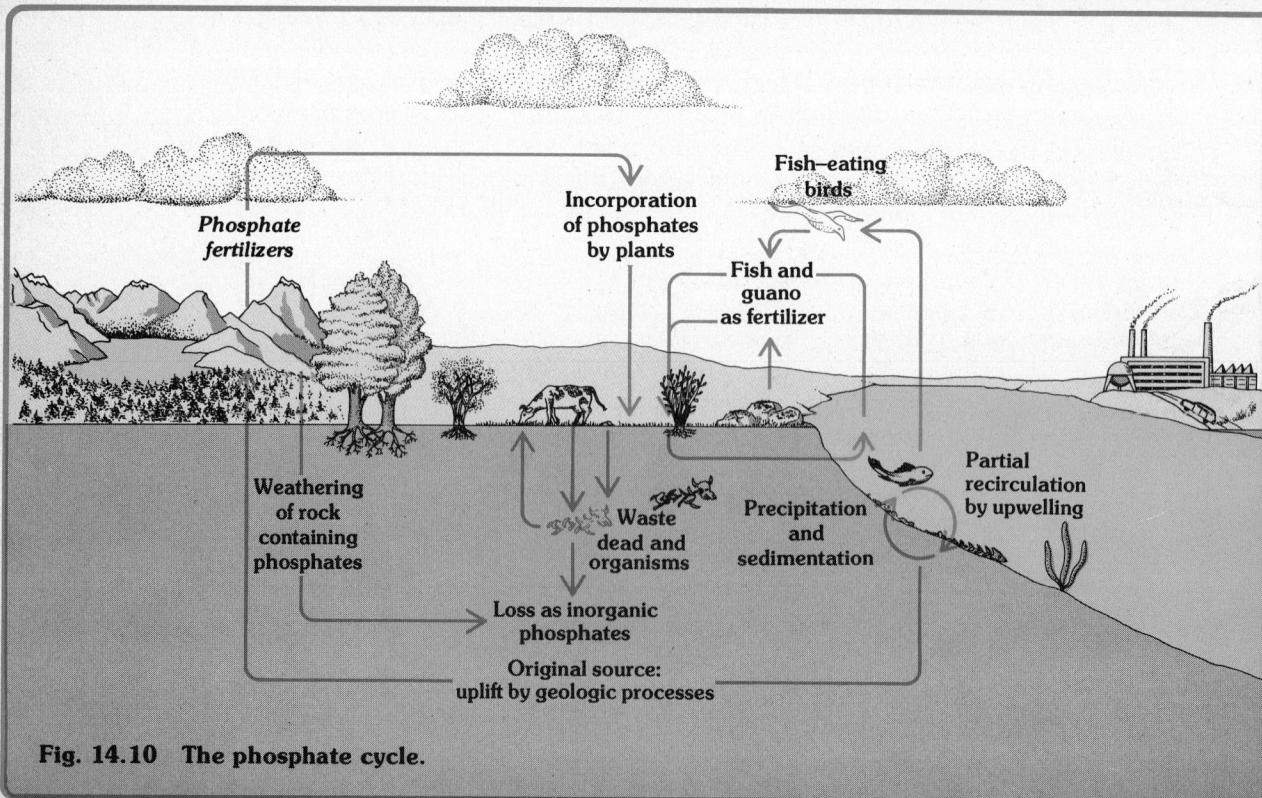

**Fig. 14.10   The phosphate cycle.**

*Phosphate fertilizers*

Incorporation of phosphates by plants

Fish–eating birds

Fish and guano as fertilizer

Weathering of rock containing phosphates

Waste dead and organisms

Precipitation and sedimentation

Partial recirculation by upwelling

Loss as inorganic phosphates

Original source: uplift by geologic processes

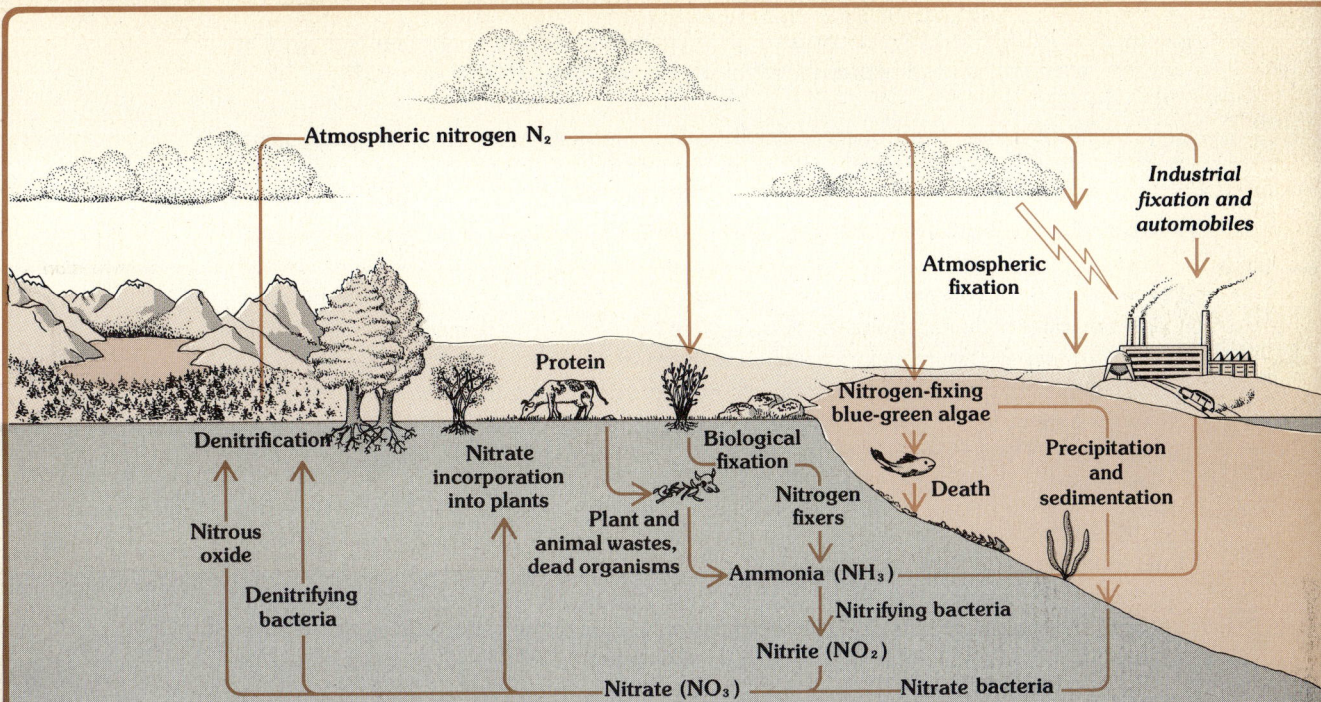

**Fig. 14.11   The nitrogen cycle.**

Atmospheric nitrogen $N_2$

*Industrial fixation and automobiles*

Atmospheric fixation

Denitrification

Protein

Biological fixation

Nitrogen-fixing blue-green algae

Death

Precipitation and sedimentation

Nitrate incorporation into plants

Nitrogen fixers

Nitrous oxide

Denitrifying bacteria

Plant and animal wastes, dead organisms

Ammonia ($NH_3$)

Nitrifying bacteria

Nitrite ($NO_2$)

Nitrate ($NO_3$)

Nitrate bacteria

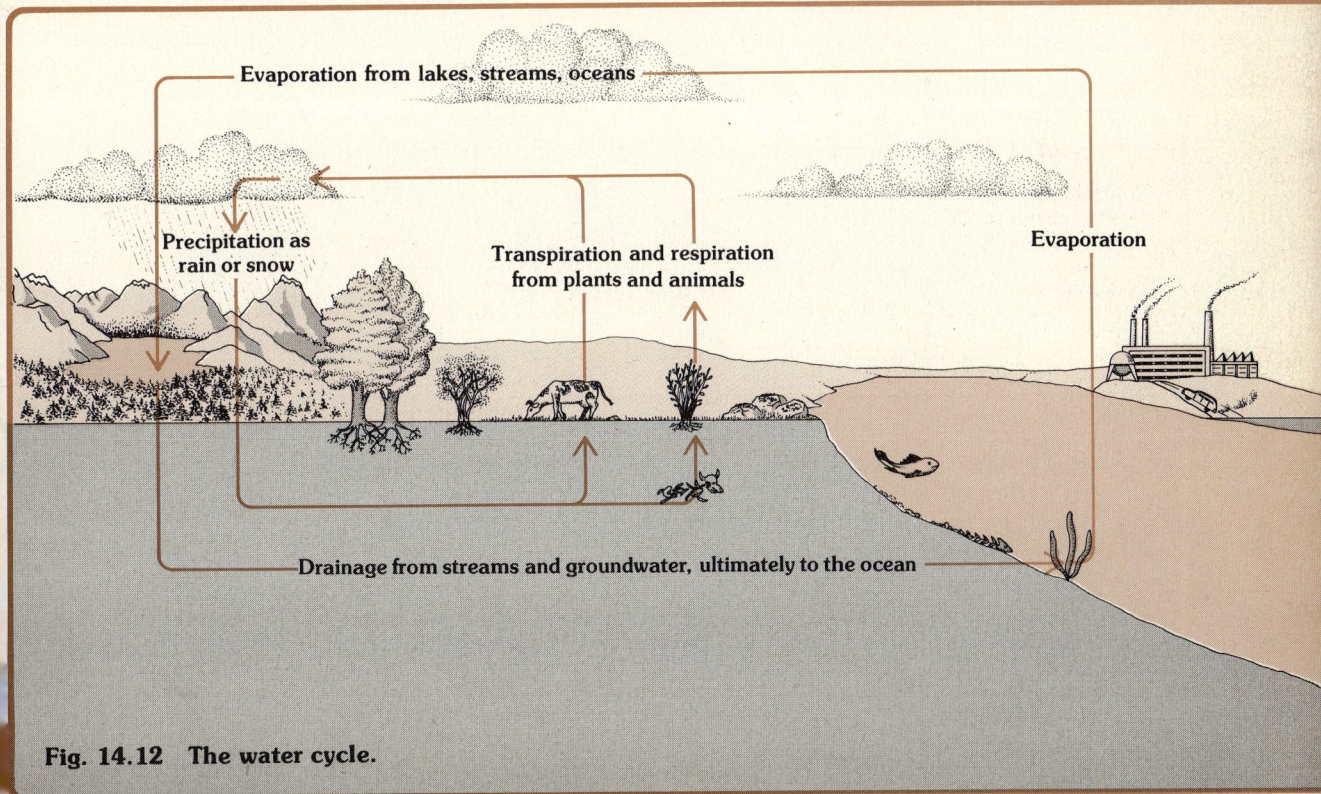

**Fig. 14.12   The water cycle.**

Evaporation from lakes, streams, oceans

Precipitation as rain or snow

Transpiration and respiration from plants and animals

Evaporation

Drainage from streams and groundwater, ultimately to the ocean

come inaccessible to living organisms. The cyclical nature of such vital elements as carbon, nitrogen, and phosphorus within the ecosystem is illustrated in Figs. 14.9–14.12.

Problems arise, from our standpoint, when (largely because of our own activities) the recycling is interfered with. Interference can take one of three forms:

1 *Pollution*—An excess production of any required material to the degree that the cycle responsible for recirculating the material is overwhelmed at some point, and the material piles up.

2 *Poison*—The introduction into the ecosystem of novel materials which are beyond the adaptive capacities of the organisms in the ecosystem to tolerate. This situation is similar to pollution, but here we are restricting pollution to excesses of materials that are normally handled by the ecosystem, as contrasted with materials novel (and therefore threatening) to the ecosystem.

3 *Depletion*—The sudden or gradual withdrawal from a cycle of some required material, resulting in deficiencies of the material at various points in the cycle.

All of these no doubt seem very abstract and hazy at this point. Let us therefore examine some specific situations.

**Heat pollution**   As we have seen from our earlier discussions of the Second Law of Thermodynamics, no conversion from one form of energy to another is accomplished without waste, and this waste typically takes the form of heat. However, the amount of heat produced by the energy transformations of living organisms, or

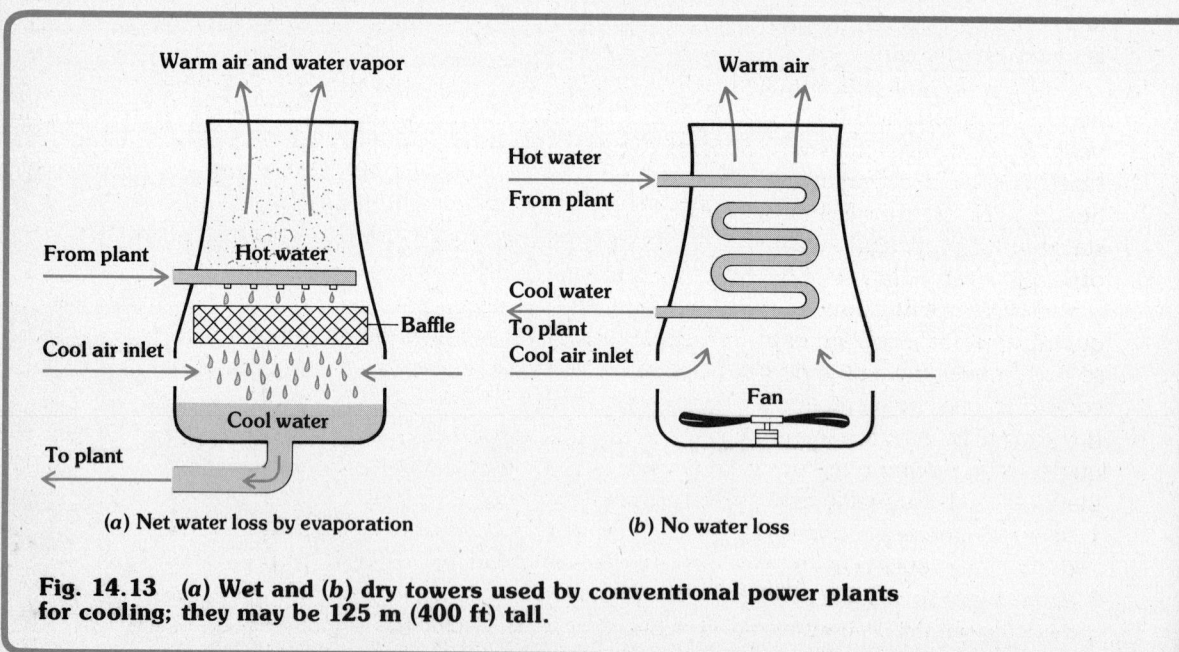

(a) Net water loss by evaporation          (b) No water loss

**Fig. 14.13**   (*a*) **Wet and** (*b*) **dry towers used by conventional power plants for cooling; they may be 125 m (400 ft) tall.**

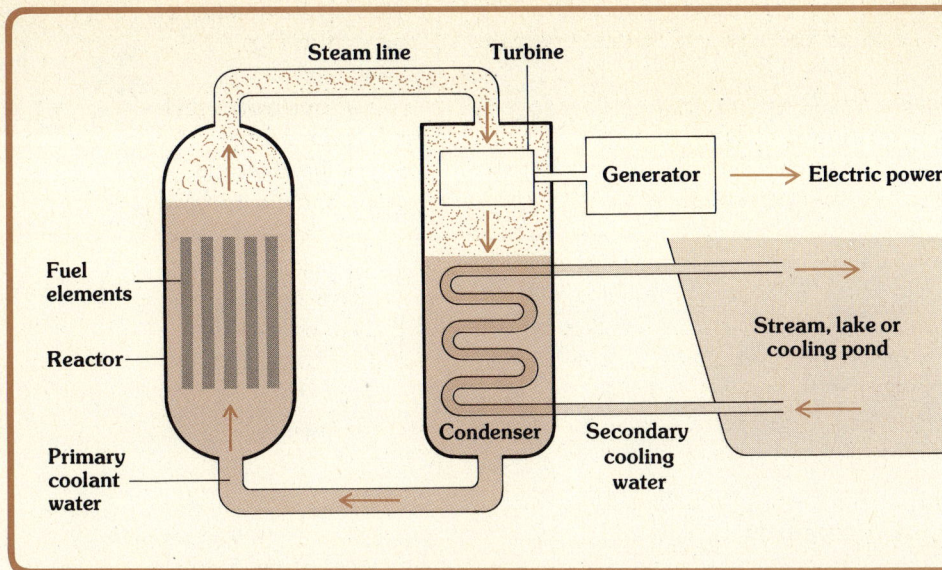

**Fig. 14.14 Diagram of cooling in a nuclear power plant.**

Labels in figure: Steam line · Turbine · Generator · Electric power · Fuel elements · Reactor · Primary coolant water · Condenser · Secondary cooling water · Stream, lake or cooling pond

even by geological events of the earth itself, are miniscule compared to the magnitude of the heat provided the earth by the sun. Put another way, the temperature of whatever environment an organism happens to live in is very rarely affected to any significant degree by anything other than climatic factors, the basis of which is heat from the sun.

The validity of this argument is severely strained, however, when we consider human activities since the beginnings of the Industrial Revolution, about 250 years ago. Increasingly, we have burned fossil fuels for heat and power of various types, and we are gradually reaching the point where our activities in terms of heat generation are beginning to affect the proper functioning of ecosystems. This statement is particularly well evidenced in the effects of heat pollution on bodies of water (heat pollution of the air occurs, but generally is dissipated quickly).

Because of its abundance and its capacity to absorb heat, water is very frequently used as a coolant in manufacturing plants, electrical generating facilities, and so on. Typically, water is pumped in from a lake or river, passed through a set of cooling coils in the plant, and is then pumped back into the river (Figs. 14.13–14.15). It may well be that the river water is never exposed to any plant impurities, and the quality of the water may remain the same in all respects save temperature, but that is enough to qualify as pollution. The effects may include:

1 An inability of certain cold-water organisms, such as trout, to survive, either because of death of the adults or, as is more frequently the case, failure of the eggs to hatch;

2 A lowering of the amount of oxygen dissolved in the water (cold water can hold considerably more oxygen than can warm water), with the result that those organisms that require abundant oxygen may die;

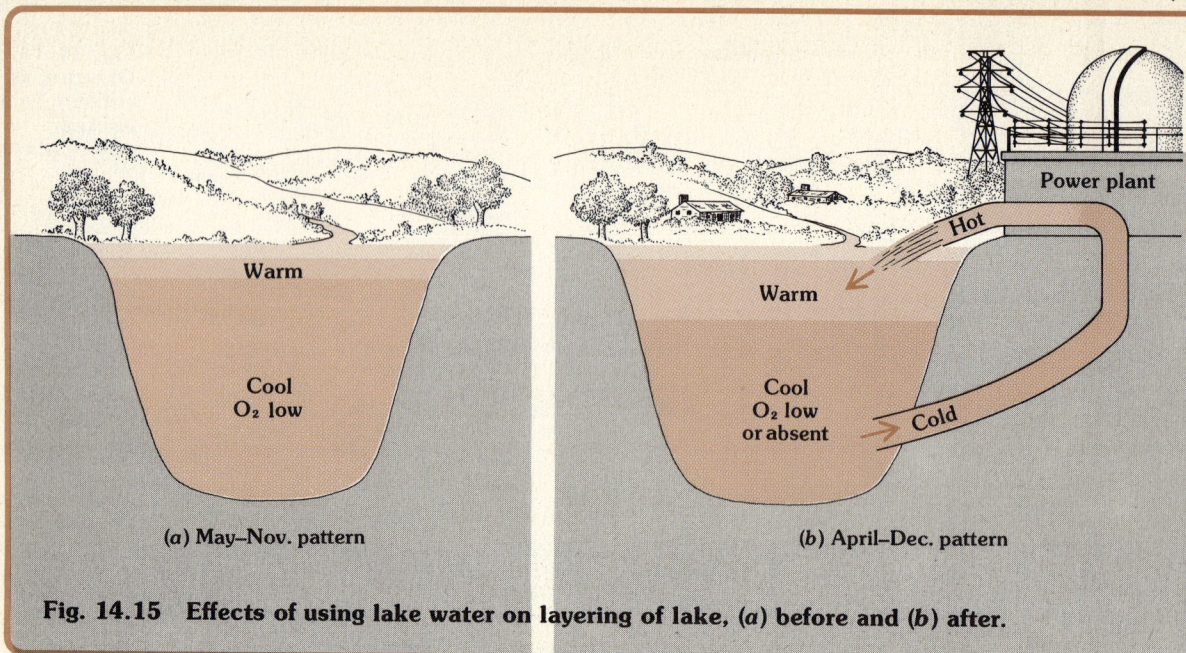

Fig. 14.15   Effects of using lake water on layering of lake, (a) before and (b) after.

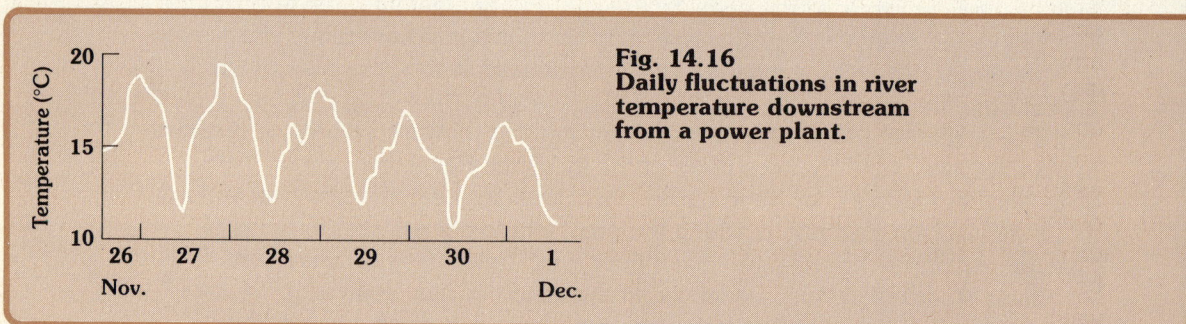

Fig. 14.16
Daily fluctuations in river
temperature downstream
from a power plant.

3 A periodic fluctuation in downstream water temperature, if the discharge of cooling water is not continuous; such fluctuations are typically disastrous for many aquatic organisms which frequently are able only to withstand the gradual temperature fluctuations seen in un-polluted water.

The situation in a small lake may be even more serious, as seasonal turnovers may be affected by the discharge of warm water into the lake, as can be seen in Fig. 14.16. Therefore, by altering the abiotic factors (e.g., temperature) we are elimi-nating organisms and are thereby affecting the ecosystem just as surely as if we were to cause extinction of species through any other route.

It must be borne in mind that what is happening in such situations is the introduc-tion into the ecosystem of a level of heat beyond the capacity of the ecosystem to accommodate. Consequently, the temperature within some or all of the ecosystem rises and profoundly affects the homeostatic balance of the ecosystem by dramatically changing the number and kind of species able to survive and function in that ecosystem.

**Essential element pollution**   In still other ways, we have managed to bring about pollution by essential elements (or compounds in some instances). For example, both the nitrogen (N) contained in synthetic fertilizers and the phosphate ($PO_4$) contained in detergents are capable of causing very rapid growth of bacteria, and especially algae, once these materials enter pond or lake water. As it happens, under natural conditions, it is usually the shortage of nitrogen and phosphates in water which keeps algal growth in check. Once these materials become available in quantity, however, the algae multiply very rapidly, choking waterways and preventing the penetration of sunlight to organisms living deeper in the water or at the bottom of the lake or pond (algae are typically found at or near the surface of the water). With lessened sunlight, these other organisms die, and in decaying they promote bacterial growth, which uses up most of the dissolved oxygen. The drop in oxygen in turn initiates the extinction of many species of active animals, including most game and commercial fish. This is essentially the situation which, in recent years, has affected even such large lakes as Lake Erie. The consequences of this type of pollution are sufficiently profound that the lakes have been described as being dead, in the sense that both the numbers and kinds of living organisms have been markedly reduced.

   A second example of essential substance pollution involves $CO_2$. Green plants and animals live in an interesting interrelationship whereby $CO_2$ is utilized by green plants during photosynthesis, and $O_2$ is released as a byproduct. (Green plants also respire, and in so doing utilize $O_2$ and release $CO_2$, but in terms of gas volumes this is a minor process.) In contrast, animals depend on green plants for their $O_2$ and release $CO_2$ as the principal respiratory waste product.

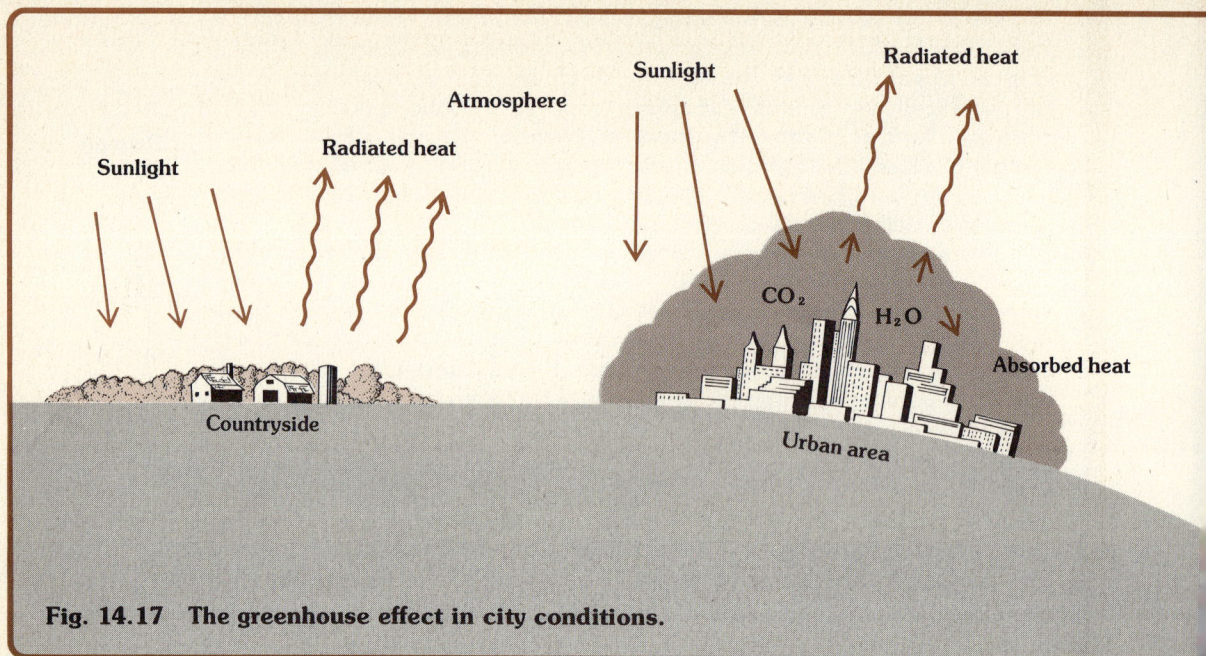

**Fig. 14.17   The greenhouse effect in city conditions.**

Within broad limits, this interrelationship is in homestatic balance. However, the combustion of fossil fuels for power releases vast amounts of $CO_2$ into the atmosphere as a byproduct of the combustion process. For hundreds of years, we have burned large amounts of wood and coal (and, more recently, oil and natural gas) for home and industrial use, but during the last 75 years, the amount of such combustion has been enormously increased by automobiles and the generation of huge amounts of electricity. The net result has been a steady rise in the amount of $CO_2$ present in the atmosphere, indicating a production of this compound in amounts beyond the capacity for use by green plants.

Thus far, no species has become extinct as a result of $CO_2$ poisoning—the change has been minimal, albeit steadily upward. However, there is intense speculation regarding the effects of increased $CO_2$ production on the earth's temperature. There are two schools of thought on this question. One group believes that an increase in atmospheric $CO_2$ will cause less sunlight to penetrate to the earth's surface, resulting in a drop in the earth's temperature. The other group, which probably has somewhat broader support, believes just the opposite—that $CO_2$ will act to capture heat radiated away from the earth's surface, much as glass allows sunlight to pass through but will trap heat, the basis of the operation of a greenhouse. Indeed, this phenomenon, with respect to atmospheric $CO_2$, has been termed the **greenhouse effect** (Fig. 14.17). As an increase in the temperature of the earth of only 1° or 2°C (2° or 3°F) would be sufficient to cause melting of the polar ice-

Most pollutants entering the atmosphere are quickly and widely dispersed. As a result, air quality is diminished perceptibly over a broad area, a fact which limits the effectiveness of purely local efforts to control air pollution. However, except for extremely noxious agents in areas of high industrial density, air pollutants are generally not concentrated enough to be a serious short-term threat to most people's health.

Freedom from air pollution hazards is a function of atmospheric dynamics. Normally the temperature of air drops slowly but regularly with an increase in altitude (*a*). However, periodically (the actual frequency

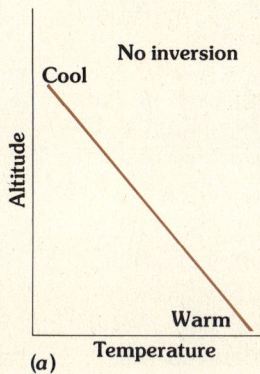

*Wide World*

(*a*)

varies according to local geography) **temperature inversions** may occur, in which a layer of warm air lies *above* a layer of cold air, rather like an inverted saucer (*b*). Such inversions prevent dissemination of pollutants and

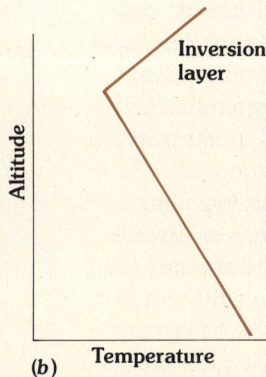

*Wide World*

(*b*)

**Temperature inversions**

provide the local citizenry with graphic—and frequently dangerous—proof of the amount of pollutants they are regularly introducing into the atmosphere.

caps, and a substantial increase in sea level. To those of you living below an elevation of 250 m (800 ft) (with respect to the present sea level) the greenhouse effect is of particular importance—unless you tread water well.

**Trace elements as poisons**   During the course of evolutionary history, organisms developed the capacity to incorporate many of the smaller relatively common elements into their metabolic machinery. Similarly, they also devised mechanisms to eliminate the other common elements which were not required when these happened to be accidentally ingested. For example, salts of the metal **lithium,** the smallest of the metals in terms of atomic size, are frequently used in treating certain types of severe depression. Clearly, there is a metabolic effect on the part of lithium, but, in small doses, it is excreted from the body without producing lasting ill effects. Many other elements, especially most of the heavy metals (lead, mercury, cadmium, and others) are either sufficiently rare in the environment, or are found in a chemically inert form, and have posed no historical threat to living organisms. As such, organisms typically did not evolve the means of dealing with these elements in terms of being able to excrete them even in moderate amounts, should they be exposed to them.

The activities of various industries have operated to bring about an increase in our exposure to these heavy metals and to a variety of organic compounds which contain them, as such compounds frequent show a deleterious biological activity. Coupled with **biomagnification,** discussed earlier, we are increasingly threatened with potentially dangerous doses of these elements. This latter situation was graphically illustrated by the Minamata Bay disaster in Japan, which began in 1953. Fishermen and their families living in the area began to show a variety of strange symptoms, beginning with loss of coordination, which led to paralysis, mental retardation, and death in more than 40 cases. The cause was mercury poisoning in the form of organic mercury compounds, which have a strong and destructive affinity for proteins in the cell membrane. Inorganic mercury was being dumped into the bay by a variety of industrial firms as a waste product; bacteria then converted this mercury into organic compounds which were picked up in minute amounts by microorganisms, and then in increasing quantities in the secondary and tertiary consumers. The fisherman ate the fish and received toxic levels of the mercury compounds. Human consumption of the fish was ultimately banned (as it has been in a number of areas, including portions of the Great Lakes—and, for the same reason, swordfish is no longer available in your food market), but the damage to the families of the fisherman was already done. Moreover, their means of making a living was also taken away. The problem is not quickly resolved, even if the dumping of mercury were to stop immediately, because the compounds tend to be recycled, in classic ecological fashion.

A similar picture could be painted for lead. Lead interferes with a variety of metabolic processes. For example, the presence of lead in the blood in concentrations of less than 1 ppm (part per million) is sufficient to inhibit the activity of the enzyme responsible for the synthesis of hemoglobin. In addition, lead causes damage to many organs, including the liver, the kidneys, and the brain, with extensive (and frequently irreversible) damage occurring before the appearance of obvious symptoms. Where does the lead come from (Fig. 14.18)? At one time, lead was extensively used in paint, and many children became seriously ill from eating peeling paint. The amount released into the air from the burning of gasoline containing lead (lead was originally added to gasoline to stop engine "knock") was sufficiently impressive that all new cars are now required to use lead-free gasoline. Nevertheless, substantial amounts of lead continue to be introduced into the atmosphere, especially from lead smelting operations—and, what goes up the chimney of the smelter with the smoke ultimately rains back down on the surrounding countryside, with the result that areas lying near lead smelters contain very high concentrations of lead in the soil and water. The long-term effects of these high concentrations has yet to be determined.

Perhaps the most dangerous of all of the heavy metals is cadmium, an element sufficiently obscure that most people have never even heard of it. Cadmium affects virtually every organ in the body, and we have essentially no way of excreting it. Thus, the amount of cadmium in our bodies slowly increases over the years, with clinical symptoms often not appearing until the age of 50, by which time the poisoning may be acute. We are exposed to cadmium in a variety of ways. Cadmium is found in the plastic used in plastic water pipe, and tends to be leached out of the pipe in minute, but measurable, amounts. Consequently, some environmentalists have suggested letting the water run for a few seconds in the morning before drinking it to allow the water that has been sitting in the pipe all night, and which therefore contains relatively large amounts of cadmium, to go down the drain. Cadmium is also found in cigarette smoke and as an adjunct to lead mining operations. Industrially, it is used in the manufacture of batteries and, in smaller amounts in a variety of

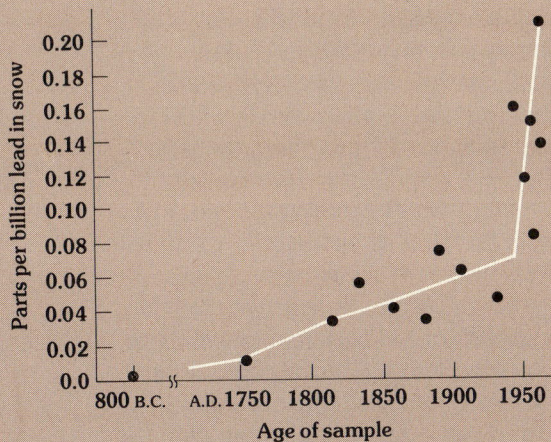

Fig. 14.18
History of atmospheric lead pollution as shown through analysis of snow from the glaciers of Greenland.

other industries. It is certain to become an element more commonly recognized as the hazards it poses become more broadly known.

**Radioactive poisons**   We are constantly exposed to radiation, primarily from the decay of naturally occurring radioactive elements and from the relatively small amount of ultraviolet light that manages to pass through our atmosphere. The biological significance of this amount of radiation is difficult to measure for the simple reason that every organism is constantly being exposed to it. Because higher levels of radiation are known to induce mutations, and still higher levels cause cell damage to the point that mitosis is impeded, it is assumed that this naturally occurring or **background** radiation may cause some deleterious effects. Nevertheless, these harmful effects must be minimal at best, a conclusion which is substantiated by the very fact that life evolved and thrived while being constantly exposed to background radiation.

Everything changed when X-rays were discovered around the turn of the century, and their importance to medical diagnosis became apparent. For many years, there was essentially no concern about the high-energy particles making up X-rays, ultraviolet rays, and radioactive materials. For example, radium-impregnated paint was used to coat the numerals on watch faces so they would glow in the dark. The amount of radium is so low that the wearer is not injured—but, at one time, the workers who painted the numerals used to moisten their brushes in their mouths. Repeated instances of loosened teeth, falling hair, and other symptoms of radiation poisoning put a stop to that practice. Similarly, 25 years ago, shoe stores routinely used a device employing X-rays to examine children's feet as they were trying on new shoes. A child was told to put his newly shod feet into two portals at the bottom of a drinking fountain-sized machine, and the child could then watch through a viewer, along with his mother and the clerk, as he wiggled his toes inside his new shoes to be sure the shoes fit properly. Although no child ever had his feet fall off as he walked out of the store as a consequence of using the viewing machine, the cumulative effects of repeated high doses of X-rays are certain to be deleterious. As a final example, consider that even today sun worshippers routinely expose themselves to the ultraviolet rays of the sun in order to achieve that most desirable of beauty marks—a **tan**. (Incidentally, it is no coincidence that this same word is used to describe the toughening of hides to make leather, for that is precisely the effect of repeated exposures of the human skin to strong sunlight.) All too often, the initial consequences are a severe burn which, in later years, may lead to skin cancer. Perhaps 50 years from now, there will be less fascination with becoming tanned, just as workers no longer lick their radium-coated paintbrushes or shoe stores no longer use X-ray machines, but at present the appeal of the sun is still strong.

All of this pales beside the threat posed by the release of radioactive materials from nuclear explosions. A great many different types of radioactive materials, most of which emit radiation for many years, were released into the atmosphere during the halcyon days of the 1950s when the United States and the USSR competed with each other for the largest bomb. Faced with an enraged world community which objected to the fallout, neither country now practices aboveground testing. (France, China, and, more recently, India have attempted to fill the void, however.)

The primary concern regarding fallout was the production of **strontium 90** (90 being the atomic weight of this isotope of strontium). The atomic structure of strontium is similar to that of calcium, although the atom itself is much larger, and, as a consequence, strontium 90 tends to be incorporated into the growing bones of young children in place of calcium. The amounts actually incorporated were very small of course, but because strontium 90 emits radiation for many years, and because the bone marrow is the site for the manufacture of blood cells, it was feared that a lifetime of exposure to this adjacent radioactivity might induce leukemia.

Returning to the ecological principle involved, we can see that radioactive materials have no place in the scheme of things in the sense that there is no ecological mechanism for distinguishing between normal and radioactive isotopes of a given element. Therefore, any required element (or a mimic of a required element, as strontium is of calcium) is retained in the cycle, even if radioactive, despite the fact that living cells are very intolerant of the high-energy particles emitted by radioactive substances.

**Organic compounds as poisons**  We have come a long way since a German chemist of the last century first synthesized urea in a test tube and declared organic chemistry open to human experimentation. In the process, we have repeatedly stumbled onto newer and more exotic organic compounds, which represent, as trace elements and radioactive materials, novel exposures to which organisms frequently have no tolerance.

The magnitude of the threat posed by these new organics is demonstrable in numerous ways. For instance, most plastics are notoriously resistant to biological breakdown. Long after our monuments have crumbled, it is reassuring to know that traces of our civilization will remain in the form of relics of plastic. However, simply because of their biological inertness, plastics pose little direct threat to ecosystems. In addition, we might consider the **fluorocarbons** (various combinations of the gas **fluorine** with carbon), which we use extensively as propellants in aerosol cans and which have been implicated in the destruction of the **ozone layer** at the outer edge of the atmosphere. This would be a formidable effect, if true, because the ozone layer (ozone, $O_3$, is a molecule comprised of three atoms of oxygen) is responsible for filtering the great majority of the ultraviolet light from the sun, preventing all but a small fraction from reaching the earth's surface. We could also consider the various types of organic compounds produced from the interaction of sunlight, the atmosphere, and automobile exhausts, the full extent of which is only now being realized. However, perhaps the best single example is provided by pesticides.

**Pesticides**  Pesticides, which include materials used primarily to kill weeds, fungi, and insects, embody a wide range of materials. One of the major types of weed-killers **(herbicides)** is 2,4-D, a mimic of a natural plant hormone which kills plants by causing a disruption in the plant's metabolism. A related compound, 2,4,5-T, was widely used in Vietnam as a **defoliant** (causing the leaves to fall off; in larger doses, it causes the death of the trees). More than one-tenth of the whole country was defoliated and large sections of forests destroyed. To state that destruction of the forests affects the ecosystem deleteriously is to state the obvious.

The **insecticides** are of particular interest, because of their extensive use. Most of the modern insecticides fall into one of two major groups of chemicals: the **chlori-**

nated **hydrocarbons,** including **DDT** and **chlordane** among many others, and the **organophosphates,** including **dieldrin, aldrin, parathion,** and many others. A number of the newest insecticides are from still different groups, the historical problem with insecticides being that once a population of insects develops resistance to one insecticide within a given group, it generally is also resistant to most of the rest of the group. Most of the organophosphates break down much more quickly than do the chlorinated hydrocarbons, and in that sense do not pose a persisting threat to the ecosystem; however, they partially detract from that saving grace by being a great deal more toxic to mammals, including ourselves. In addition, at least one of their number, **endrin,** was found to induce cancer in experimental animals and has consequently been banned.

It is difficult to read with a straight face the pronouncements of the chemists who developed these compounds during and after World War II to the effect that the days of the housefly were numbered—DDT would render them extinct. As you know, the housefly managed to adapt (see Chapter 13) rather well to this threat. Unfortunately, the totality of the effects of these insecticides within the ecosystem was not recognized for many years. The fact that these were novel organic compounds with which no living organism had ever come in contact should have suggested that many organisms might have a problem in dealing with them, but such a suggestion does not ever seem to have been made. We now know that biomagnification can pose a threat to many species in the top trophic levels, but this has been a recent discovery. Perhaps more importantly, we also now know that such things as DDT interfere with the proper functioning of the cellular organelles responsible for photosynthesis in green plants. Levels of only 1–2 ppm, which are routinely found in nature (human breast milk has been found to contain up to 5 ppm) reduce photosynthetic activity by 20 percent. When you consider that with a half-life of 30 years DDT is very persistent in the environment, that more than 10,000,000 kg (22,000,000 lb) of DDT are still exported annually from the United States for use in malaria control in tropical areas, and that photosynthetic activity is required to produce the oxygen we breathe and to use up the carbon dioxide we excrete, the effects of DDT are sobering indeed.

Perhaps the most significant ecological effect of widespread insecticide use (and certainly the most obvious effect) has been that insecticides do not discriminate between "good" insects and "bad" insects. It is estimated that fewer than 3000 species of insects (out of more than 1,000,000 described species) are pests, yet DDT sprayed over an area will kill most of the insects with which it comes in contact. This is particularly unfortunate as regards predator and parasite species, as these species tend to be in lower numbers, have longer generation times, and are slower to return to the sprayed areas than are the pest species (recall the asynchrony of the lynx and hare population curves, and you will see why this should be so). Therefore, the decision to spray is frequently like the first shot of heroin—things are never the same again, and repeated doses may be necessary just to maintain the status quo. Ironically, as we have already seen, monoculture is frequently responsible for "creating" a pest species (in that it provides such a suitable habitat that the population of the insect reaches pest levels), and in that sense the eco-system has already been thrown out of balance. Insecticides represent an attempt to compensate for this error, but in reality frequently compound the problem.

Fortunately, as we have learned more about how insects live, other more specific methods of control have been discovered. These include the use of selected parasites and predators, the application of insect viruses, and the utilization of insect hormones, all of which are far less threatening to the ecosystem because each is a "natural" product, meaning one to which the ecosystem has long been exposed. The use of chemical pesticides continues at a very high rate, however, primarily because it is easier and less expensive than are alternative methods at present.

**Depletion of resources**  The whole concept of a stable ecosystem hinges on the necessity of recycling the essential elements. The rate of recycling may vary, to be sure, and frequently certain elements are in short supply, but the cycle con-tinues nonetheless. Earlier we discussed how, in our "wisdom," we have come to aug-ment certain parts of different cycles, often resulting in a buildup at some point in the cycle. Clearly, any material introduced must come from somewhere, and we must consider the consequences of running out of our supply of these materials at some point in the future. For the economists out there, you can think of the cycles in an ecosystem as being analogous with living off the interest in a savings account, in the sense that, despite having spent the interest, the principal remains and gener-ates interest anew. Our introduction of more material at different points in the cycle is analogous to spending the principal, for ultimately we will have none left.

The most obvious example is our use of fossil fuels as an energy source. We have already seen how the carbon dioxide produced by their combustion and how the nitrogen leached from the fertilizers made from natural gas have deleteriously af-fected the ecosystems; let us now consider the consequences of running out of fossil fuels.

Natural gas is a very clean fuel, in that the only major pollutant produced is $CO_2$. Oil contains somewhat greater amounts of impurities which are not consumed when the oil is burned, and the efficiency of the internal combustion engine is such that much of the gasoline is incompletely consumed, giving rise to the extensive air pollution associated with automobile exhausts. These fuels are in increasingly shorter supply. Where do we go in the future?

We have vast reserves of coal in this country, almost all of which are of rather poor quality and difficult to reach, except by strip mining. Few things could be consid-ered more disruptive to an ecosystem than strip mining, the ramifications of which are rather well known by every educated person. An equal problem may be pre-sented by greater combustion of the coal, an incredibly dirty fuel in terms of the re-lease of poisons and pollutants. Coal contains large amounts of sulfur, which is oxi-dized during combustion of the coal to produce **sulfur dioxide** ($SO_2$), one of the most noxious of gases. Much of this gas can be trapped by filters or the sulfur may be removed before burning, although both are expensive processes and are much re-sisted by power companies. But what of the huge amounts of trace elements and heavy metals released up the stack, most of which will pass through any of the filters now available? Transforming all our present stationary power sources (electrical generation and industrial plants) to allow them to burn coal would double the amount of these toxic materials presently being released into the atmosphere. Nuclear power, at least in the form of the fission reactors now in operation, is, as we have seen, very inefficient and very demanding of our increasingly scarce supply of water as a coolant, and, in any event, we only have about a 30-year reserve of uranium. The magnitude of the threat from leakage of radioactive wastes is a matter of dispute, but the possibility of leakage poses, at the very least, a very real concern.

Increasingly, environmentalists are looking at new and better ways of tapping wind and especially solar power. The utilization of such energy sources is still sometime off, and the mechanics of harvesting them are beyond the scope of this book, but they are less ecologically obtrusive than are present methods of generating energy, and have the added benefit of being never-ending, for all intents and purposes.

Many other materials and foods are in short supply as well (Figs. 14.19–14.21). Not all of these are of direct importance to the functioning of the ecosystem, which

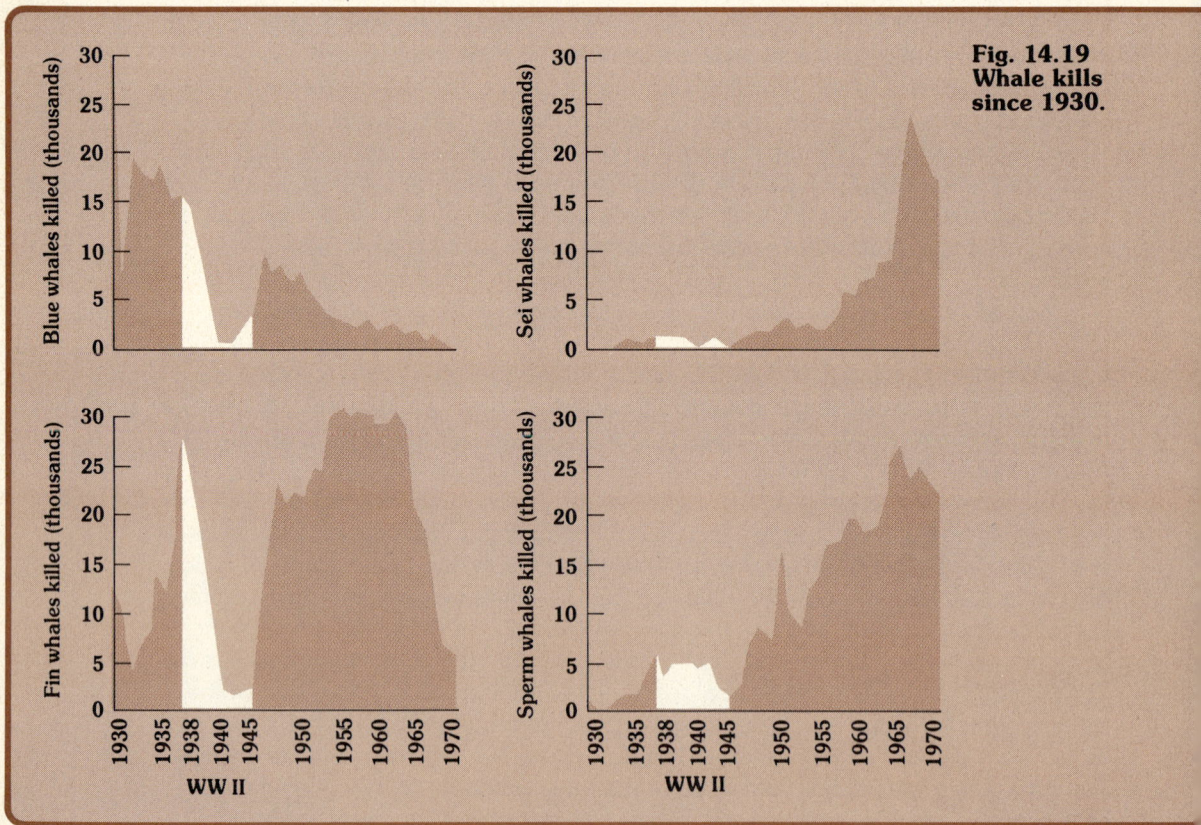

Fig. 14.19
Whale kills
since 1930.

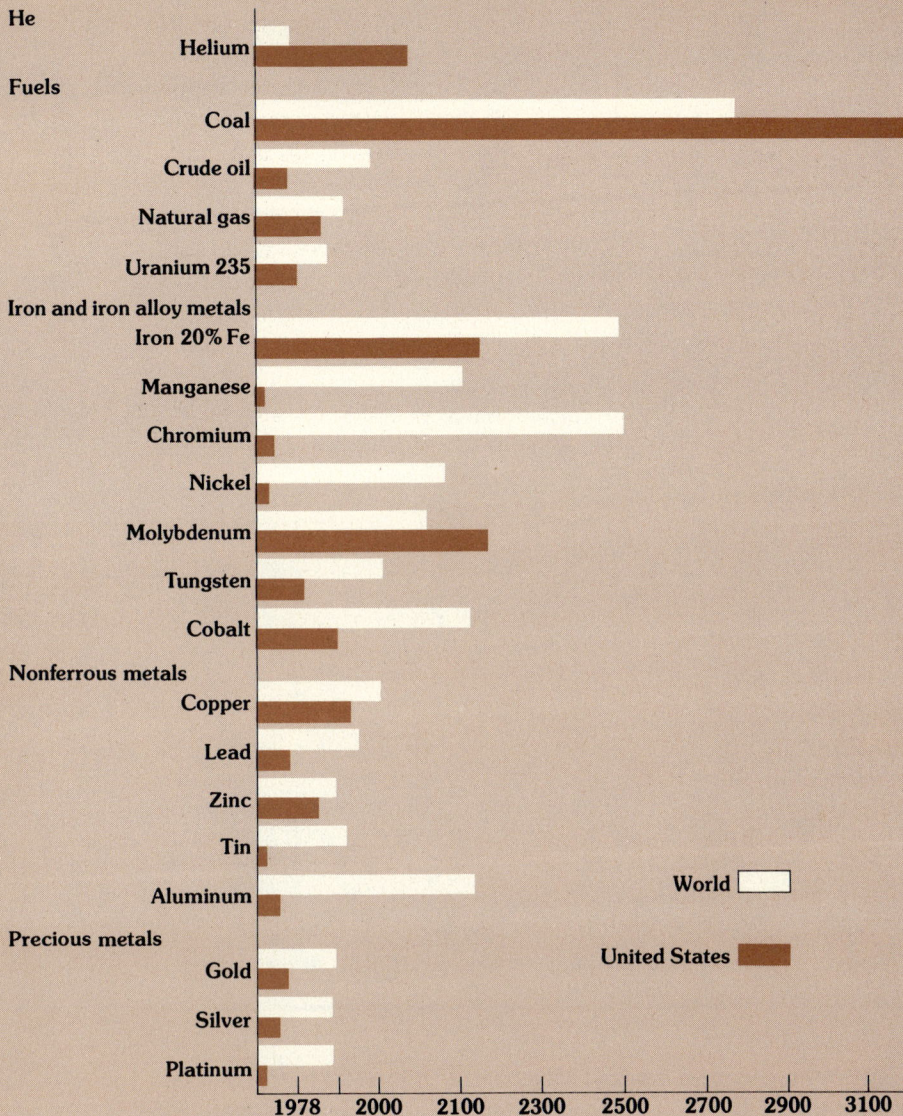

Fig. 14.20  Estimated natural resource reserves for the world and for the United States.

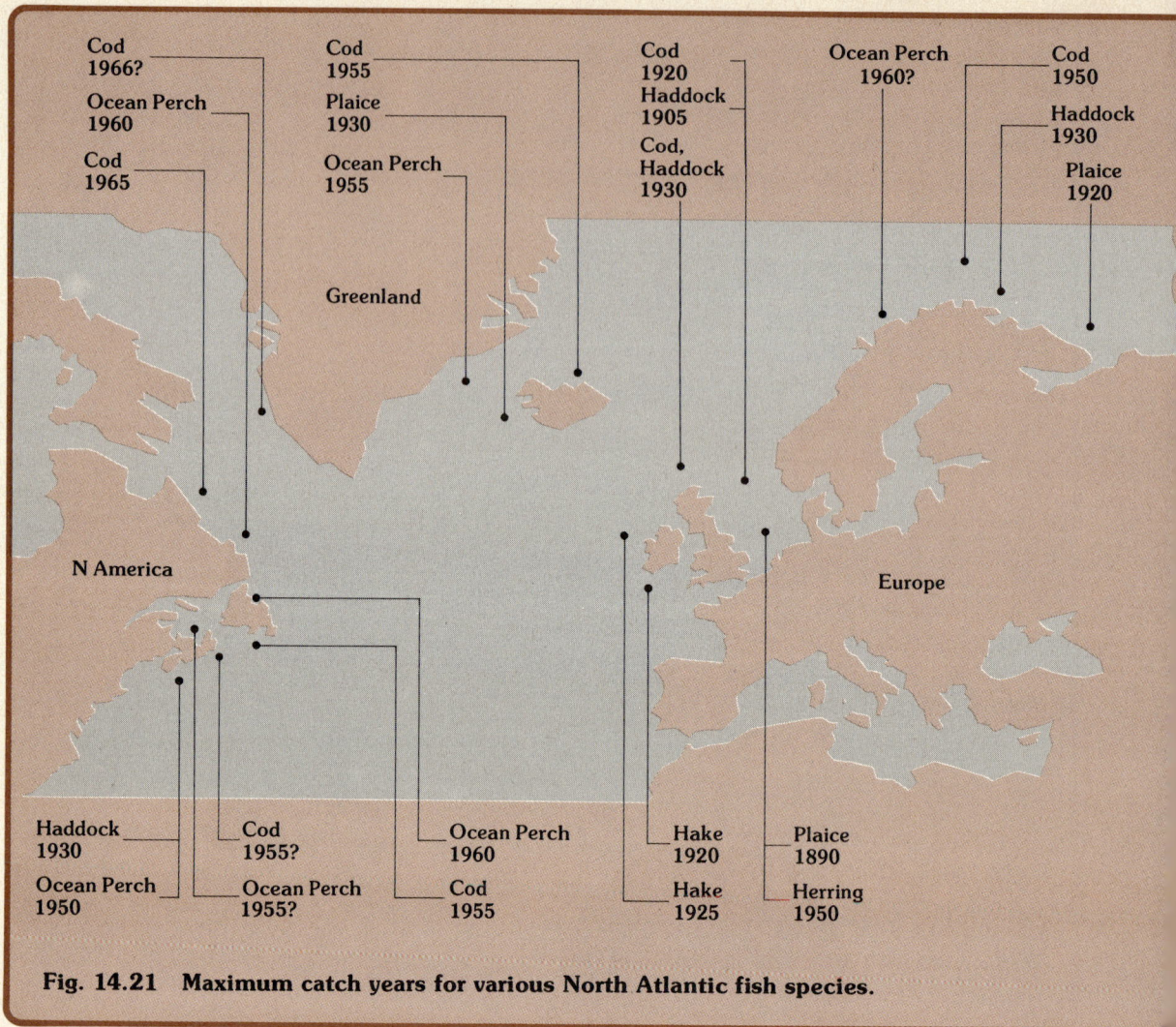

**Fig. 14.21 Maximum catch years for various North Atlantic fish species.**

Map labels:

Cod 1966?
Ocean Perch 1960
Cod 1965
Cod 1955
Plaice 1930
Ocean Perch 1955
Greenland
Cod 1920
Haddock 1905
Cod, Haddock 1930
Ocean Perch 1960?
Cod 1950
Haddock 1930
Plaice 1920
N America
Europe
Haddock 1930
Ocean Perch 1950
Cod 1955?
Ocean Perch 1955?
Ocean Perch 1960
Cod 1955
Hake 1920
Hake 1925
Plaice 1890
Herring 1950

---

is our primary concern here, but all are of at least indirect interest to the degree that our present level of functioning in the ecosystem will have to be altered once these materials are gone.

**REPRODUCTIVE STRATEGIES**   A totally different way of analyzing the impact of a species on its environment (and vice versa) is to analyze population size and turnover rates.

**Reproductive potential**   All species, if they are to survive, must possess some reproductive mechanism that will guarantee the survival of at least the same number of reproductives every generation. The manner by which this is achieved differs markedly among different species. Oysters produce huge numbers of eggs, because the mortality rate of juvenile oysters is enormous; most large mammals have very few offspring, by comparison, but a much higher percentage of these survive to adulthood than do oysters. The result is the same in all cases, namely the perpetuation of about the same number of reproductives as in the previous generation.

Suppose *all* the offspring were to survive. If all the offspring of a single fertilized female housefly were to survive and reproduce, in seven generations there would be over 5 trillion houseflies, even if no fly survived for more than one generation. Such a growth rate is said to be **exponential,** or **logarithmic,** as it is an accelerating process. Of course, all the houseflies do not survive. Some are eaten by birds and spiders, others are swatted by housewives, still others never reach adulthood because the substrate they are living in while in the larval stage dries out. Put another way, there is some **limiting factor** (or set of factors) which prevents the flies from ever achieving their **reproductive potential.**

**Carrying capacity**   As we have just seen, the environment in which a population lives is less than perfect, and at some point, limiting factors prevent the species from reaching its reproductive potential. This point of stability, in terms of population size, is called the **carrying capacity** of the environment (Figs. 14.22–14.23).

Carrying capacity and limiting factors are not constants, but rather are subject to considerable variation over time. Obviously, the housefly population is larger in the summer than in the winter, because the carrying capacity of the environment is greater in the summer (there are fewer limiting factors). Food is available in quan-

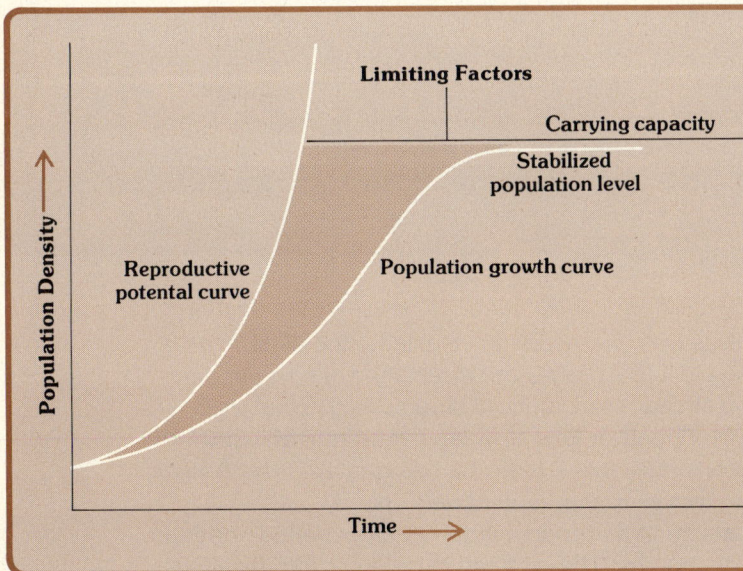

**Limiting Factors**

Carrying capacity

Stabilized
population level

Reproductive
potential curve

Population growth curve

Population Density →

Time →

**Fig. 14.22
Relationship
between repro-
ductive poten-
tial and carrying
capacity.**

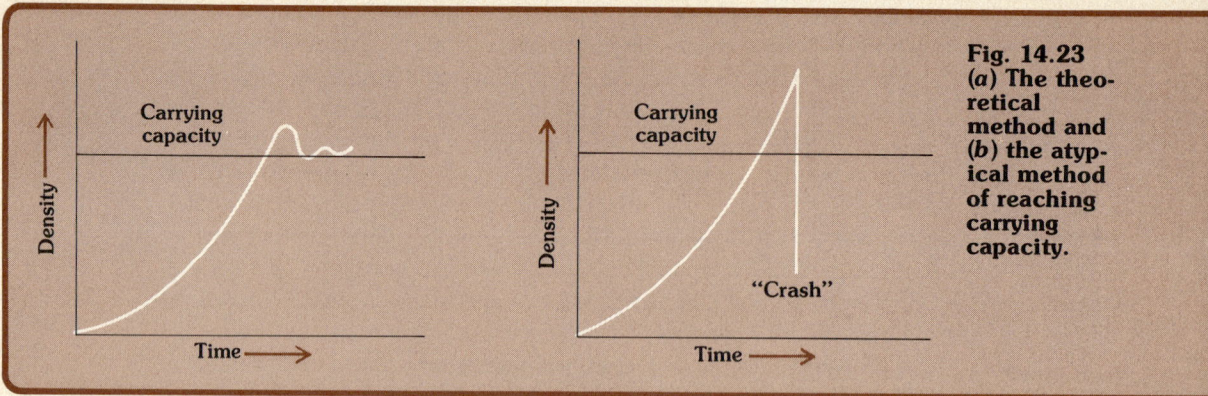

Fig. 14.23 (a) The theoretical method and (b) the atypical method of reaching carrying capacity.

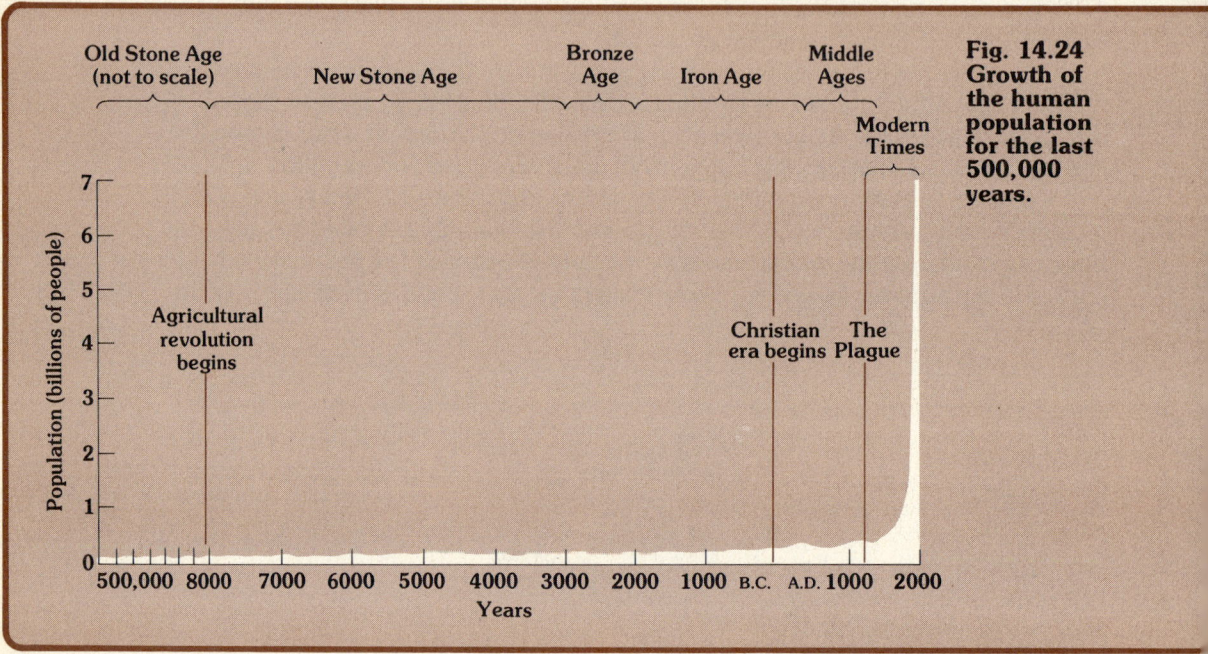

Fig. 14.24 Growth of the human population for the last 500,000 years.

tity, and shelter from the elements is less of a necessity. Similarly, the algal "blooms" that appear every spring are a reflection of the elimination of cold temperature as a limiting factor. As we have already seen, artificial blooms may occur when nitrogen or phosphates are added to a pond or lake as wastes from human activities. This increase in algal growth demonstrates that it must be nitrogen and phosphorus availability which were the limiting factors in the "unfertilized" pond.

Other organisms exhibit variability in their population size over periods of time longer than merely a few weeks or months. The carrying capacity of the environment in which the hares and lynx live, to use an earlier example, is a reflection of

**Table 14.1**
**Population growth rate, doubling time, and population size for selected countries**

| Region or country | Annual growth rate (%) | No. of years to double population | 1971 population (millions) | 1985 population estimate (millions) |
|---|---|---|---|---|
| Costa Rica | 3.8 | 19 | 1.9 | 3.2 |
| Colombia | 3.5 | 20 | 22.1 | 35.6 |
| Mexico | 3.4 | 21 | 52.2 | 84.4 |
| Syria | 3.3 | 21 | 6.4 | 10.5 |
| Pakistan | 3.3 | 21 | 141.6 | 224.2 |
| Algeria | 3.3 | 21 | 14.5 | 23.9 |
| Iran | 3.0 | 23 | 29.2 | 45.0 |
| Lebanon | 3.0 | 23 | 2.9 | 4.3 |
| Guatemala | 2.9 | 24 | 5.3 | 7.9 |
| Brazil | 2.8 | 25 | 95.7 | 142.6 |
| Egypt | 2.8 | 25 | 34.9 | 52.3 |
| India | 2.6 | 27 | 569.5 | 807.6 |
| Haiti | 2.5 | 28 | 5.4 | 7.9 |
| World | 2.0 | 35 | 3706.0 | 4993.0 |
| Australia | 1.9 | 37 | 12.8 | 17.0 |
| China | 1.8 | 39 | 772.9 | 964.0 |
| Canada | 1.7 | 41 | 21.8 | 27.0 |
| Argentina | 1.5 | 47 | 24.7 | 29.6 |
| Puerto Rico | 1.4 | 50 | 2.9 | 3.4 |
| Japan | 1.1 | 63 | 104.7 | 121.3 |
| United States | 1.1 | 63 | 207.0 | 241.0 |
| Spain | 1.0 | 70 | 33.6 | 38.1 |
| Russia | 1.0 | 70 | 245.0 | 286.9 |
| France | 0.7 | 100 | 51.5 | 57.6 |
| Denmark | 0.5 | 140 | 5.0 | 5.5 |
| Sweden | 0.5 | 140 | 8.1 | 8.8 |
| Britain | 0.5 | 140 | 56.3 | 61.8 |
| West Germany | 0.4 | 175 | 58.9 | 62.3 |

From Population Reference Bureau. *1969 World Population Data Sheet,* Population Reference Bureau, Washington, D.C., 1969.

the limitations placed by each on the other—that is, an increase in the number of hares allows an increase in the carrying capacity of the environment with respect to the lynx.

It is nonetheless true that, over a sufficient period of time, the average population size of most species is relatively constant. There may be periodic changes in population size (the cycle is about 11 years in the rabbit–lynx example), but over large periods of time the populations remain about the same size.

All of these concepts are equally applicable to human populations. Although it is difficult to determine, it would appear that human populations were essentially stable until the introduction of agriculture about 10,000 years ago (Fig. 14.24). From that point, there was a gradual increase until about the beginning of the Industrial Revolution, at the turn of the eighteenth century. Growth since then has been increasingly of the exponential type (Table 14.1).

What brought about this huge population increase in so short a period of time? It was almost certainly a combination of factors—the expansion of a crowded Europe into their colonies starting from about the beginnings of the sixteenth century; industrialization, especially in the harvesting and transportation of food and other commodities; and perhaps most important of all, the great advances in medicine during the past 100 years.

**What is the carrying capacity for humans?**   One of the more important debates of modern times has an ecological basis. In everyday terms, it involves the question of whether or not the world (or the United States) is overpopulated—or, in ecological terms: What is the carrying capacity for humans? An admission that there is, in fact, a carrying capacity short of the point at which we are physically piled on top of each other is an important first step, but even this concession is not made willingly by some adherents of unlimited growth. Let us examine the evidence.

First of all, we need to define the conditions of our existence. Do we define the needs of our population in terms of the life-style of an affluent American (or, even more dramatically, an Arab oil shiek), or are they defined in terms of a Bombay beggar? Rather clearly, we are already overpopulated if our criteria are based on the American. Just in terms of fossil fuels, the developed countries, which total less than 25 percent of the earth's population, use over 80 percent of the

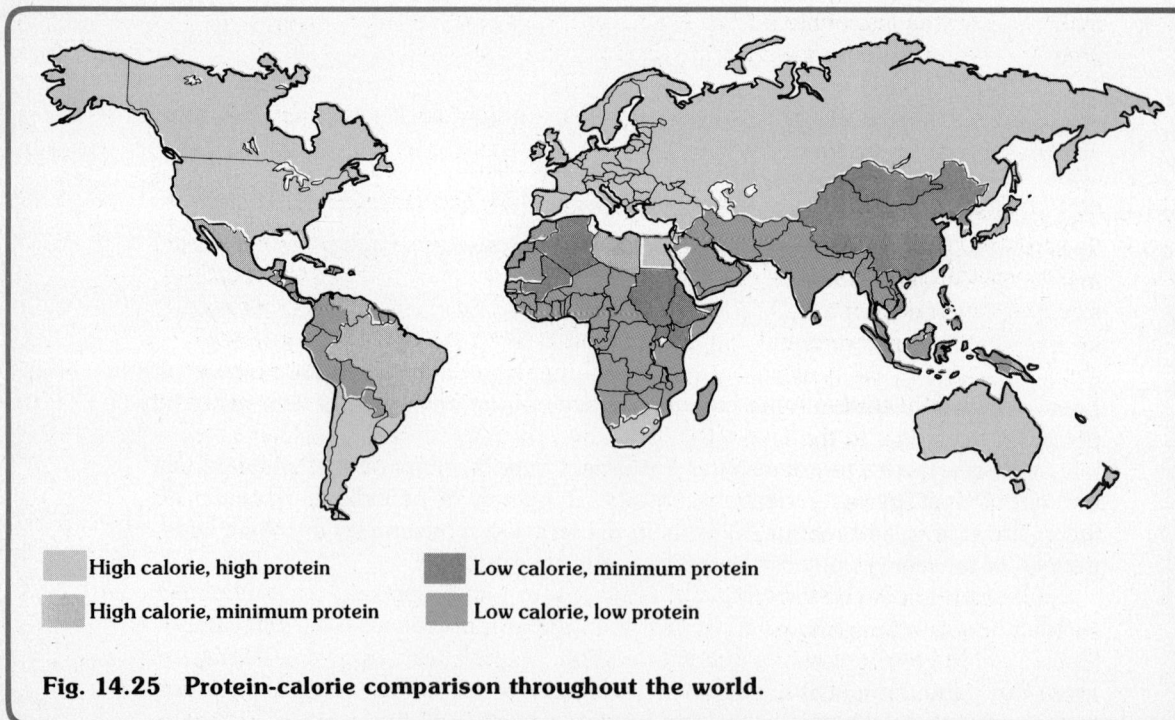

| | |
|---|---|
| High calorie, high protein | Low calorie, minimum protein |
| High calorie, minimum protein | Low calorie, low protein |

**Fig. 14.25   Protein-calorie comparison throughout the world.**

coal, oil, and natural gas. To bring the rest of the world up to our level would require an immediate tripling of the use of these materials, and they simply do not exist. Overpopulation is even more dramatic in terms of food—there is just not enough animal protein available to allow food consumption worldwide in the manner in which it occurs in the United States (Fig. 14.25). Indeed, much of the reason for the environmental problems discussed earlier comes from our (i.e., the Western World's) insistence on stretching limiting factors artificially.

If we accept that the carrying capacity of the earth is something less than the present population (assuming we want everyone to have our present standard of living), what are the ramifications of the continuing rapid growth?

What we can extrapolate from animal populations does not bode well. Typically, in those relatively few species that show rapid population explosions, there is an equally dramatic crash—and not just to the carrying capacity, but considerably below it (Fig. 14.23b). Because of population imbalances in many countries (Figs. 14.26–14.27), many authorities already consider us to be overripe for a plague of some type—and the possibilities of war caused by overcrowding and food shortages is by no means beyond the realm of possibility. Surely such drastic means of population reduction are unnecessary, given more humane alternatives. But are the alternatives viable? Is there time to save the overpopulated countries, or will it become necessary, as some have suggested, simply to turn our backs? What amount of sacrifice (e.g., taking up a wholly vegetarian diet) will the affluent societies be willing to make to save the less affluent? These are the tough kinds of policy questions that your generation is going to have to help decide.

**BIOLOGY AND LAW**   One of the ironies of modern life is that at a time when more and more individuals are desirous of increased personal freedom, laws are becoming more pervasive and restrictive. Yet there is a certain logic to this fact. The growth of laws is a function of an increasingly mobile and crowded society, for as the number of contacts we have with other individuals increases, the chances of inimical encounters also increase, making it necessary for society to provide standards to govern such encounters. Thus, we are finding ourselves suddenly surrounded by laws regulating environmental concerns, wheeas a decade ago, such laws were almost unknown. The National Environmental Policy Act, which established the Environmental Protection Agency (EPA), dates only to 1969; effective regulation of air pollution dates only to the Clean Air Amendments of 1970; water quality has a longer history, but it was not until the 1972 Amendments to the Water Pollution Control Act that the EPA was given a broad charge with respect to water quality (specifically, it was "to restore and maintain the chemical, physical, and biological integrity of the Nation's water").

Laws tend not to be formed in a vacuum, but rather are typically created by reasonable people attempting to find reasonable answers to widely recognized problems. The last phrase should be emphasized for, particularly as regards such esoteric entities as "environmental quality," an informed (and irate) citizenry is the primary force behind the establishment of laws and regulations designed to effect remedies.

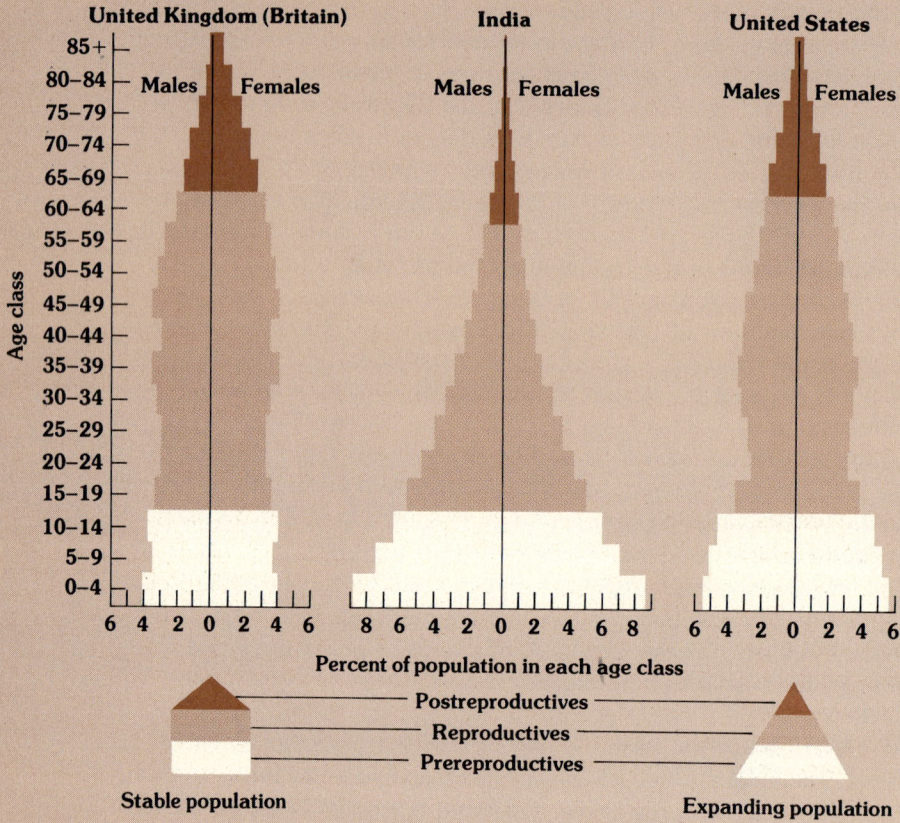

Fig. 14.26 Age distribution pyramids from Great Britain, India, and the United States as of 1960.

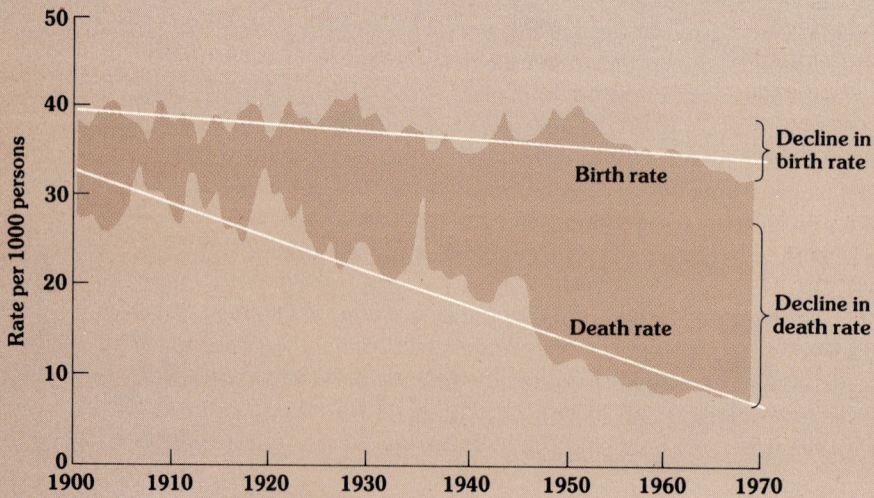

Fig. 14.27 Changes in the birth and death rates in Ceylon since 1900.

However, to state the obvious, there is a world of difference between having laws, or having the power to set regulations, and actually enforcing or setting them. Some cynical observers have claimed, with more than a little validity, that regulatory agencies very often become tools of those they were erected to regulate. The EPA has had mixed reviews. For example, it was the courts, not the EPA, which interpreted and set standards for the phrase "no significant deterioration" in the Clean Air Act. The EPA has taken a go-slow posture with respect to the trace elements and heavy metals as well. At this writing, the discharges of fewer than ten are regulated, and regulation is woefully incomplete in some instances. For example, the discharge of mercury is regulated in plants that work directly with mercury and mercury compounds, but much less than half the mercury that enters the environment does so through these manufacturing plants. The huge amount of mercury that enters the atmosphere through the stacks of power plants burning coal (which contains trace amounts of mercury) is totally unregulated. Equal reluctance was shown with respect to the proposed ban on fluorocarbons, a ban widely recommended because of preliminary reports indicating their deleterious effects on the ozone layer. The EPA argued that it was unwilling to act until the deleterious effects were proved. Would it be unreasonable to suggest that perhaps those who propose to release novel substances into the atmosphere should have the onus placed on them to prove, at least to a reasonable degree, that the substances are *not* going to cause environmental problems? After all, there are other propellants that can be used in spray cans, but there is only one ozone layer. Many environmentalists therefore suggest that fluorocarbons be banned until such time as they are proved *not* to have a negative effect on the ozone layer.

Similarly, the EPA action with respect to DDT can best be described as cautious. Most of the 5,000,000-kg (11,000,000 lb) domestic consumption has been eliminated, but 10,000,000 kg (22,000,000 lb) is still exported. Admittedly, there is a tradeoff here. DDT is a very effective and inexpensive insecticide, and malaria, which is transmitted by mosquitoes, remains a serious and widespread disease. Nevertheless, it is interesting to note that when the developed countries took the opportunity to aid the underdeveloped countries, they chose the cheapest insecticide—and one that had been banned in the United States. Does that signify our intentions with respect to the larger problems facing the underdeveloped nations in the future? Moreover, there are strong pressures now operating to allow use of DDT to control tussock moths in the western coniferous forests so important to the major lumber companies.

In the final analysis, laws generally will be structured to reflect the will of the majority (so long as the rights of the minority are not infringed). As such, it is necessary to have citizens who are informed as to the tradeoffs of a given policy and who therefore can make their wishes known based on knowledge and not ignorance. The importance of biological information in making these decisions in the future can hardly be overemphasized. It is hoped that the topics discussed in this book have helped in some way to make you more aware of the forces underlying the functioning of biological systems.

A principal tenet of biology is that levels of biological organization are subject to the same rules of the physical sciences. Clearly, the complex organization at the ecosystem level is just as influenced by the Second Law of Thermodynamics as is any individual organism. Similarly, the response of the ecosystem to the physical laws is the same as that of the individual organism—the development of homeostasis. All levels of biological organization require stability, or at least predictability, and as a measure of this requirement, homeostatic control mechanisms have evolved. Indeed, most of our present environmental problems have arisen as a consequence of our inadvertent (although successful) efforts to alter the network of homeostatic cycles of elements upon which the ecosystems depend for their continued existence.

All of these problems are exacerbated by a burgeoning human population possessing rising expectations. Fewer people and lessened demands may suffice to save the organization of ecosystems as we now know them, but we are running out of time.

# Index

Condoms (or "rubbers"), 197
Cones, retina, 265
Conflict situations, 279
Congenital defects, 362–371
    chromosomal abnormalities, 365–366
    enviromentally induced, 370–371
    single gene, 362
Congestive heart failure, 124
Constipation, 107
Contact inhibition, 308
Control animal, 14
Controlled conditions, 13, 14
Convoluted tubule, 161
Cornea of the eye, 263
Coronary arteries, 125
Coronary thrombosis, 128
Corpus luteum, 186
Correlation, scientific, 14
Cortex, 256
Cortical hormones, ACTH and, 221–223
Cortisol, 220
Cortisone, 26, 220
Countercurrent principle, 169–171, 321
Cranial nerves, 249
Cretinism, 217
Cri du chat syndrome, 366
Cro-Magnon man, 300, 348–349, 352, 359
Crossing over, 314
Culture, 302
Curare (poison), 244, 245
Cushing's disease, 223
Cuteness response, ritualized behavior and, 298
Cybernetics (Wiener), 35
Cystic fibrosis, 146
Cytoplasm, 28

Darwin, Charles, 332–336, 348
Deadly nightshade, 245
Death, 4, 308
    leading causes of, 109
Deductive reasoning, 14
Defecation, 105–107
Deficiency hepatitis, 104
Deletions, chromosomal, 338
Demerol (meperidine), 261
Denatured enzyme (protein), 25

Dendrites, 231
Deoxyribonucleic acid (DNA), 26, 149, 209, 336, 337
Deoxyribose, 26
Depolarized cell membrane, 235
Depressants, 260
Descent of Man (Darwin), 348
Development, 315–326
    cloning, 324
    events at birth, 325–326
    extraembryonic membranes, 319–321
    recapitulation theory, 321–325
    totipotency, 324
Diabetes insipidus, 210, 224–225
Diabetes mellitus, 210, 223, 224–225
Diaphragm, 138–139
Diaphragm (contraceptive), 197
Diarrhea, 107
Dichlorodiphenyltrichloroethane (DDT), 384, 400, 411
    resistance to, 346
Dieldrin, 400
Dietary necessities, 77–79
    essential components of, 78
Diethylstilbesterol (DES), 198
    cancer and, 200
Diffusion, 38, 94
    and cell size, 39
    facilitated, 94–95
Diffusion gradient, 94, 232
Digestive system, 42, 74–108
    accessory organs, 98–104
    defecation, 105–108
    dietary necessities, 77–79
    esophagus, 82–83
    in infants, 84
    introduction, 74–77
    large intestine, 104–105
    mouth, 79–82
    small intestine, 90–98
    stomach, 83–90
Dihybrid cross, 330
Disaccharide, 23
Dislocation, 58
Displacement activity, 279
Dissociation, 20
Division of labor, 300
Dominant-recessive relationships, 340

78 79 80 9 8 7 6 5 4 3 2 1